# Chaotic Dynamics
## Theory and Practice

# NATO ASI Series

## Advanced Science Institutes Series

*A series presenting the results of activities sponsored by the NATO Science Committee, which aims at the dissemination of advanced scientific and technological knowledge, with a view to strengthening links between scientific communities.*

The series is published by an international board of publishers in conjunction with the NATO Scientific Affairs Division

| | | |
|---|---|---|
| A | **Life Sciences** | Plenum Publishing Corporation |
| B | **Physics** | New York and London |
| | | |
| C | **Mathematical and Physical Sciences** | Kluwer Academic Publishers |
| D | **Behavioral and Social Sciences** | Dordrecht, Boston, and London |
| E | **Applied Sciences** | |
| | | |
| F | **Computer and Systems Sciences** | Springer-Verlag |
| G | **Ecological Sciences** | Berlin, Heidelberg, New York, London, |
| H | **Cell Biology** | Paris, Tokyo, Hong Kong, and Barcelona |
| I | **Global Environmental Change** | |

### Recent Volumes in this Series

*Series B: Physics*

# Chaotic Dynamics
## Theory and Practice

Edited by

## T. Bountis

University of Patras
Patras, Greece

Plenum Press
New York and London
Published in cooperation with NATO Scientific Affairs Division

Proceedings of a NATO Advanced Study Institute on
Chaotic Dynamics: Theory and Practice,
held July 11–20, 1991,
in Patras, Greece

**NATO-PCO-DATA BASE**

The electronic index to the NATO ASI Series provides full bibliographical references (with keywords and/or abstracts) to more than 30,000 contributions from international scientists published in all sections of the NATO ASI Series. Access to the NATO-PCO-DATA BASE is possible in two ways:

—via online FILE 128 (NATO-PCO-DATA BASE) hosted by ESRIN, Via Galileo Galilei, I-00044 Frascati, Italy.

—via CD-ROM "NATO-PCO-DATA BASE" with user-friendly retrieval software in English, French, and German (© WTV GmbH and DATAWARE Technologies, Inc. 1989)

The CD-ROM can be ordered through any member of the Board of Publishers or through NATO-PCO, Overijse, Belgium.

Library of Congress Cataloging-in-Publication Data

```
Chaotic dynamics : theory and practice = Chaotikē dynamikē : theōria
  kai praktikē / T. Bountis [editor].
       p.   cm. -- (NATO ASI series. Series B, Physics ; v. 298)
     "... Workshop held within the program of activities of the NATO
  Special Program on Chaos, Order, and Patterns"--P. [3].
     "Proceedings of a NATO Advanced Study Institute on Chaotic
  Dynamics, Theory and Practice, held July 11-20, 1991, in Patras
  Greece"--T.p. verso.
     "Published in cooperation with NATO Scientific Affairs Division."
     Includes bibliographical references and index.
     ISBN 0-306-44247-7
     1. Chaotic behavior in systems--Congresses.   I. Bountis, Tassos.
  II. North Atlantic Treaty Organization.  Scientific Affairs
  Division.   III. Nato Advanced Study Institute on Chaotic Dynamics,
  Theory and Practice (1991 : Patrai, Greece)  IV. Title: Chaotikē
  dynamikē.  V. Series.
  Q172.5.C45C438  1992
  003'.7--dc20                                    92-21986
                                                       CIP
```

ISBN 0-306-44247-7

© 1992 Plenum Press, New York
A Division of Plenum Publishing Corporation
233 Spring Street, New York, N.Y. 10013

# SPECIAL PROGRAM ON CHAOS, ORDER, AND PATTERNS

This book contains the proceedings of a NATO Advanced Research Workshop held within the program of activities of the NATO Special Program on Chaos, Order, and Patterns.

# SPECIAL PROGRAM ON CHAOS, ORDER, AND PATTERNS

# PREFACE

Many conferences, meetings, workshops, summer schools and symposia on nonlinear dynamical systems are being organized these days, dealing with a great variety of topics and themes - classical and quantum, theoretical and experimental. Some focus on integrability, or discuss the mathematical foundations of chaos. Others explore the beauty of fractals, or examine endless  possibilities of applications to problems of physics, chemistry, biology and other sciences.

A new scientific discipline has thus emerged, with its own distinct philosophical  viewpoint and an impressive arsenal of new methods and techniques, which  may be called Chaotic Dynamics. Perhaps its most outstanding achievement so far has been to shed new light on many long-standing issues involving complicated, irregular or "chaotic" nonlinear phenomena. The concepts of randomness, complexity and unpredictability have been critically re-examined and the fundamental importance of scaling, self-similarity and sensitive dependence on parameters and initial conditions has been firmly established.

In this NATO ASI,  held at the seaside Greek city of Patras, between July 11-20, 1991,  a serious effort was made to bring together scientists representing many of the different  aspects of Chaotic Dynamics. Our main aim was to review recent advances, evaluate the current state of the art and identify some of the more promising directions for  research in Chaotic Dynamics.

Another important goal of the meeting was to offer the opportunity to as many graduate students and young researchers as possible, working in different areas of Chaotic Dynamics, to present their recent results and discuss them with  more "senior" colleagues. Judging from the number of those who attended (there were 80 students from 12 countries including the U.S., India and Australia) and the high quality of their contributions to this volume, one may safely conclude that Chaotic Dynamics is alive and well, and with a bright future indeed.

In fact, as was pointed out at the beginning of the meeting, it seemed a good idea to make this volume more of a "young researchers' proceedings" than a collection of reviews,  presented already by some of the main speakers, recently, at other conferences. The "new generation" of chaotic dynamicists at the meeting were, therefore, strongly encouraged to submit their work to the proceedings, with the understanding, of course, that it would be subject to refereeing.

The result is a volume that - at least in the view of this editor - is a well-balanced account of the areas of research represented at the meeting. The division of the book into chapters and the ordering of different subjects has no special significance and is only meant as a guide for the reader.

If I may be allowed to make a philosophical remark (and for a meeting held in Greece this does not seem entirely inappropriate), the study of dynamical systems is characterized by a number of interesting dialectical relationships: order vs. chaos, determinism vs. randomness, models vs. physical reality, analysis vs. computation. It was in this spirit that the words "theory" and "practice" were chosen for the title of the meeting and the heading of this book. That they finally turned out to divide its contents into two nearly equal parts can only be attributed to an intervention by the two protecting deities of our field: Apollo and Dionysos.

I am sure I speak for all of us when I express our sincere *"ευχαριστίες στους Αρχαίους Ελληνες για την πληθώρα των φιλοσοφικών ιδεών του κόσμου και του χάους, του μέτρου και της υπερβολής, της θεωρίας και της πρακτικής για την ανάλυση δυναμικών συστημάτων και χαοτικών φυσικών φαινομένων".(*)*

Finally, I wish to thank Mrs. L. Kordopati and her team for their excellent secretarial services and express my gratitude to all colleagues and graduate students of the Universities of Patras and Athens  who took part in the organization and helped make this meeting a pleasant and rewarding experience for all.

Tassos Bountis

March, 1992


(*) Translation  (though not so necessary):

"thanks to the Ancient Greeks for a plethora of philosophical ideas on order and chaos, meter and hyperbole, theory and practice for the analysis of dynamical systems and chaotic physical phenomena".

# CONTENTS

## STATISTICAL PHYSICS, CELESTIAL MECHANICS AND COSMOLOGY



## PART II

## CHAOTIC DYNAMICS : PRACTICE

## CONTROLLING DYNAMICAL SYSTEMS



## SEMICONDUCTORS, SUPERCONDUCTORS, LASERS AND ELECTRONIC CIRCUITS

## BIOLOGY, CHEMISTRY, ATMOSPHERIC AND MAGNETOSPHERIC DYNAMICS



## HAMILTONIAN DYNAMICS, DISSIPATIVE DYNAMICS AND NORMAL FORMS

# COMPLEXITY AND UNPREDICTABLE SCALING OF HIERARCHICAL STRUCTURES


R. Badii

Paul Scherrer Institute, LUS
CH-5232 Villigen PSI, Switzerland


## ABSTRACT


The problem of characterizing complexity is formulated in the framework of a hierarchical modelling of physical systems. Complexity is related to the task of predicting the asymptotic scaling behaviour of system's observables from the available finite-resolution measurements. The analysis is performed on one-dimensional stationary symbolic patterns such as those encountered in nonlinear dynamics, cellular automata, spin systems, making use of suitable coding methods in a statistical-mechanical environment. The asymptotic limit of thermodynamic sums is estimated within a grand-canonical formalism, with the help of transfer-matrix and renormalization techniques: the convergence of the associated scaling functions is directly related to the previously defined complexity measure.


## 1.  Introduction: what is a complex system?

The concept of complexity arises spontaneously in the study of a large number of systems which are the object of disciplines as diverse as physics, mathematics, chemistry, biology, computer science. The term "complex" is often associated with chaotic dynamics, cellular automata, hydrodynamic flows, neural networks, spin glasses and, of course, with living organisms [1]. Apparently, these systems have very little in common. However, with a more careful inspection, it is possible to recognize in all of them the simultaneous presence of both "irregular" and "ordered" structures. Complexity is hence related to some form of "organization", which is neither too strict, as in a crystal, nor too loose, as in a gas of weakly interacting particles [1]. Indeed, when confronted with a "complex" system, the observer is unable to discriminate clearly its "fundamental constituents" and to characterize their mutual relations in a concise way: that is, the behaviour is so "intricated" that any specifically designed, finite model description eventually departs from it, either when time proceeds or when the demand for accuracy (resolution) is sharpened [2,3,4].

Such elusiveness has constituted the main hindrance to finding a quantitative definition of complexity which meets universal agreement. The fields in which most efforts have been originally concentrated are automata [5] and information [6] theory and computer science [7,8]. More recently, physical research in this topic has received a considerable impulse, especially after the understanding of phase transitions and of chaotic dynamics [9]. Further interest has been raised by the discovery of "glassy" behaviour and by the construction of the first mathematical models in evolutionary biology and neuroscience [1,10].

The list of the presently available definitions of complexity is so long and variegated [11] that one is quickly diverted from the original meaning of the term. Therefore, it is essential to recall it, in order to base the discussion on a solid ground [12]:

- *a complex system is composed of* **parts, interrelated** *in a way which is* **hard to understand.**

Three points stand out: first of all, complexity intrinsically implies the presence of a **subject** whose task it to "understand" the **object**; secondly, the object must be conveniently divided into **parts** (by the subject) and, third, the **interactions** among them must be measured and incorporated into a model (the accuracy of which is a measure of the understanding). Clearly, the "disassembling" of the object can be indefinitely iterated by further splitting each element of the partition into subelements, thus obtaining a **hierarchy** of subsystems. This is best visualized as a **tree**, the levels of which represent increasingly refined coarse-grained pictures of the system: in fact, the values of the **interactions** among the subsystems, measured at all achievable resolution scales, may be assigned to the corresponding branches. Within this formulation of the problem, complexity is immediately related to the difficulty of **predicting** the asymptotic (infinite-resolution) **scaling** behaviour of the system with a model $\mathcal{M}$, using the (obviously non-asymptotic) data stored on the hierarchical tree [13], thus establishing a full correspondence with the "literal" definition of complexity given above, as summarized in the following scheme:

$$
\begin{array}{ll}
(\text{parts} \longrightarrow \text{subparts} \longrightarrow \text{subsubparts} \longrightarrow \quad ... \quad) & \longrightarrow \textbf{hierarchy} \\
(\text{interactions at different resolution levels}) & \longrightarrow \textbf{scaling} \\
(\text{subject vs. object}) & \longrightarrow \textbf{model predictions}.
\end{array}
$$

This approach to the problem leads to considering complexity as a **relative** quantity, which depends on properties of both system (object) and observer (subject), in agreement with the common view of general systems research [4,14] (notice, however, that the best known complexity measures [11] do not satisfy this primary requirement).

Before continuing the general discussion, it is useful to consider two concrete cases. The periodic "message" $aabaabaab\ldots$ could be decomposed, for example, into a sequence of two words (i.e., $aa$ and $b$ or $ba$ and $a$) which may "interact" only in one way: namely, $aa$ ($ba$) must be followed by $b$ ($a$) and not again by $aa$ ($ba$). If, instead, the same signal is split into words of length 3, the composition rule is even stronger: there is simply no alternative, even just in principle, to writing always $aab$ (or $aba$ or $baa$, depending on the initial phase), since the signal is the repetition of that single word. The resulting hierarchical trees have no side-branches. A crystalline structure exhibits a rigorous order, corresponding to very strong interactions. Viceversa, $\delta$-correlated random sequences, such as those produced by a coin toss, are completely structureless: with any choice of the parts (strings of symbols), all combinations of them are possible. No rule forces or prohibits any particular concatenation of (a finite number of) words. There are no interactions (formal definitions can be found in section 2) and the associated tree contains all possible ramifications, without exceptions. In the following paragraph it will be clarified whether one should consider these two extreme cases as simple or not.

While the relative character of complexity has not been incorporated into previous definitions, nearly all approaches stress the importance of some kind of **unpredictability** which is associated with the concept of complexity. Several measures, referring to the analysis of spatio-temporal patterns (portions of text, time-series, fluid surfaces, spin configurations, DNA chains), deal with predictions of *local* behaviour, such as the conformation of a (space- or time-) neighbourhood of the currently scanned region in the sample. These are entropy-like quantities [6,7,8] which attribute large complexity values to a random uncorrelated process [11]. In fact, no matter how much information is collected about such a system, the outcome of the next (either in space or time) observation is totally unpredictable. However, if an exact reproduction of a given pattern is, in general, impossible, the description of the class of all sub-patterns compatible with it may be rather easy. In fact, the set of all possible realizations of a (discrete) $\delta$-correlated process can be described very concisely [11]. This simple consideration is naturally taken into account in the hierarchical approach outlined above, in which the predictions refer to the *asymptotic scaling* of system's observables: i.e., given (measured) the interactions among parts with characteristic sizes $\varepsilon_i$, the model $\mathcal{M}$ is asked to predict those among subsystems with sizes $\varepsilon'_i$, scaled by some (generally position-dependent) factor $\xi_i = \varepsilon'_i/\varepsilon_i$. This view is *orthogonal* to the previous one and classifies both ordered (periodic and quasiperiodic) and disordered (random) systems as simple, since they can be finitely specified. As an illustration, consider again the periodic signal *aabaabaab*.... Not only its future is exactly predictable after a rapid investigation, but also the set of all admissible words (parts) of any size: $a$ and $b$, of length 1, $aa$, $ab$ and $ba$ of length 2, and so on. Hence, periodic structures are trivially classified as simple with both predictability criteria. A random uncorrelated signal, instead, can be recognized to be simple (as it should, since it has absolutely no structure) only within the hierarchical approach. In fact, the admissible outcomes of a coin-tossing experiment can be ordered on a complete binary tree (see sections 2 and 3 for more details):

$$(a, b) \rightarrow ((aa, ab), (ba, bb)) \rightarrow (((aaa, aab), (aba, abb)), ((baa, bab), (bba, bbb))) \rightarrow \dots ,$$

where $(aa, ab)$ are the descendents of $a$, $(ba, bb)$ those of $b$ and so on. It is hence a trivial task to predict that all combinations of heads ($a$) and tails ($b$) of any length $n > n_0$ can occur, since a standard statistical analysis performed for all $n \leq n_0$ does not reveal any exception to this rule.

The above qualitative observations, present in almost all studies of complexity [1,11] but not put into practice very often, lead to the following important (although simple) remark: complexity is "far" from *both* pure order and complete disorder. It is possible to reinforce this concept further, by saying that complexity is far from any *known* model. This is the distinctive feature of what has been called "organized complexity" in Ref. ([3]), situated between phenomena which can be described by low-dimensional deterministic dynamical equations (e.g., Newtonian mechanics) and many-particle systems which are studied with statistical methods. The most striking aspect of this intermediate "world" is the ability to exhibit, at a macroscopic scale, properties which do not exist at a microscopic one (or viceversa): they may concern the symmetries of the objects (parts) involved or of the forces (interactions) acting on them [10], or even the development of coherent patterns from a microscopic dynamics with short-ranged interactions. These "emergent" properties are the hallmark of the so-called "self-generated" complexity [9,10,11] and are common to a large number of systems, all characterized by the appearance of different structure at different resolution levels. Such behaviour is currently believed to stem from the infinite iteration of some (generally unknown) basic rule which, although probably simple, is very hard to discover, since it is related to the macroscopic observables only through the above mentioned infinitely long "history": there is no "short cut" connecting them. This implies, in turn, the existence of memory effects, with long-ranged (space-time) correlations, and of anomalous scaling behaviour. Notice, however, that similar phenomena can be observed also in other fields of investigation, which do not share the peculiar properties of self-generated complexity mentioned above: for example, optimization methods in $N \gg 1$ dimensions (neural networks), computationally "hard" problems (travelling salesman), localization and disorder [1,10]. Complexity probably always implies the existence of anomalous scaling laws, but the opposite need not be true. In the next sections, these ideas will be formalized in a precise way.

## 2. Symbolic encoding and shift dynamics: parts and interactions

It is possible to treat very different systems within a unique framework with the introduction of suitable encoding methods. The analysis is considerably simplified if the resulting representations yield one-dimensional symbolic signals (patterns) [11,13], typical examples of which are provided by DNA chains, spin and cellular automata configurations. Discrete-time dynamical systems (mappings) can also be encoded in such a form with the following procedure. Let the time evolution be specified by [15]

$$\mathbf{x}_{n+1} = \mathbf{F}(\mathbf{x}_n) \ , \quad n \in Z \ , \tag{1}$$

where $\mathbf{x}_n$ is a point in a $d$-dimensional phase-space $X$ and $\mathbf{F}$ a nonlinear function. We introduce a *partition* $\mathcal{D}$ of $X$ consisting of a finite number $r$ of disjoint subsets $\Delta_i$ ($i = 0, 1, \ldots, r - 1$): i.e., $\Delta_i \cap \Delta_j = \emptyset$, for $i \neq j$, and $X = \bigcup_{i=0}^{r-1} \Delta_i$. We further assume that the transformation $\mathbf{F}$

admits a *natural* invariant measure $m$ [16,17], so that $m[\mathbf{F}^{-1}(\Delta_i)] = m(\Delta_i)$, for all $\Delta_i \in \mathcal{D}$, where $\mathbf{F}^{-1}$ denotes the inverse of $\mathbf{F}$. A generic orbit $\omega = \{\mathbf{x}_0, \mathbf{x}_1, \ldots, \mathbf{x}_n\}$ visits various elements $\Delta \in \mathcal{D}$. Denoting with the symbol $s_n \in A = \{0, \ldots, r-1\}$ (where $A$ is the *alphabet*) the index of the domain $\Delta$ visited at time $n$, the trajectory $\omega$ is mapped to the symbolic sequence $S = s_0 s_1 \ldots s_n$. It is important to notice that sequence $S$ can be produced (in $n$ iterates) only by the points $\mathbf{x}_0$ which belong to the intersection $\Delta_S \equiv \Delta_{s_0} \cap \mathbf{F}^{-1}(\Delta_{s_1}) \cap \ldots \cap \mathbf{F}^{-n}(\Delta_{s_n})$ ($\mathbf{F}^n$ being the $n$-th iterate of $\mathbf{F}$). Since the map admits an invariant measure $m$, the signal $S = \ldots s_0 s_1 s_2 \ldots$, produced by infinitely iterating $\mathbf{F}$, is stationary [17,18,19]. The probability $P(S)$ of each finite subsequence $S$ can then be evaluated as the frequency of occurrence of $S$ in $\mathcal{S}$. The normalization is, as usual, $\sum_{|S|=n} P(S) = 1$, $\forall n$, where $|S|$ denotes the length (number of symbols) of $S$. Obviously, $P(S) = m(\Delta_S)$: i.e., the probability of a sequence equals the mass contained in the phase-space region with the same label. Therefore, symbolic strings $S$ with increasing length $|S| = n$ identify smaller and smaller sets in $X$ and their probability decreases accordingly. The collection of all admissible two-symbol strings $s_i s_j$ indexes the elements $\Delta_{s_i} \cap \mathbf{F}^{-1}(\Delta_{s_j})$ of the first *refinement* [15] $\mathcal{D}_1$ of the partition $\mathcal{D}$ under $\mathbf{F}$; three-symbol strings label the second refinement $\mathcal{D}_2$, and so on. If every infinitely long symbol-sequence corresponds to a single point, the partition $\mathcal{D}$ is called *generating* [15] and the study of the symbolic signal $\mathcal{S}$ is "equivalent" to that of the real trajectories of the system. Approximations to such partitions can be obtained in various ways [20,21,22].

It is well known that systems of ordinary differential equations can be transformed into mappings with the help of Poincaré sections [15]; moreover, several techniques (inertial manifold, amplitude equations, mode expansions) have been developed to reduce the study of partial differential equations to that of ordinary ones [23,24], so that the range of problems which can be encoded symbolically is further extended. It is important to notice that interesting behaviour does not necessarily follow from the infinite dimensionality of phase-space: the "reduced" systems are required to conserve the essential features of the original ones. Fluid and optical instabilities, propagation of "defects", chemical reactions belong to the class of systems which can be analyzed with these techniques. Growth processes under restrictive conditions and some turbulent phenomena are also amenable to symbolic encodings, possibly through *ad hoc* procedures [25,26].

One is therefore led to the study of one-dimensional stationary symbolic patterns (or "signals" if they are parametrized by a time-coordinate) as a very convenient mathematical representation of the behaviour observed in many different physical situations. Of course, systems on a lattice like cellular automata or Montecarlo spin processes produce patterns in several dimensions: $d \geq 1$ in space and 1 in time (although the interest is often in the spatial configuration after a long time or just in the time evolution at a fixed spatial position). *A fortiori* do all continuous spatio-temporal systems the dynamics of which cannot be reduced with any of the methods mentioned above. Although the general discussion of section 1 applies also to these objects, intrinsic difficulties appear in their analysis which do not have a counterpart in the one-dimensional case. First of all, the choice of the parts is more arbitrary, since not only their size must be determined but also their shape. Secondly, the "tiles" which one obtaines cannot be interpreted as the labels of regions of some associated phase-space, as for one-dimensional signals. Finally, any attempt to scan a high-dimensional pattern along a one-dimensional path in general produces objects whose properties are different from those of the original one, unless the system has trivial symmetries which can be easily exploited: it is possible to generate, for example, non-stationary signals (i.e., non-invariant under translations) even though the whole pattern is stationary [11]. In order to avoid these ambiguities, we concentrate on one-dimensional symbolic signals with translationally invariant statistics. So far, no general complexity analysis of symbolic patterns of arbitrary dimensionality has been undertaken.

Whatever the origin of the signal $\mathcal{S} = \ldots s_0 . s_1 s_2 \ldots$ is (the dot denoting an arbitrary origin), one can attribute it to a dynamical process, the so-called *shift map* $\hat{\sigma}$, whose action takes place in the space $A^Z$ of all bi-infinite sequences over the alphabet $A$ and is defined by $\hat{\sigma}(\ldots s_0 . s_1 s_2 \ldots) = \ldots s_1 . s_2 s_3 \ldots$. It is also useful to define the set $\Sigma$ of all finite sequences over $A$ which, for the binary alphabet $\{0, 1\}$, reads $\Sigma = \{0, 1, 00, 01, 10, 11, 000, 001, \ldots\}$. Of course, a generic signal $\mathcal{S}$ need not contain, as subsequences, all elements of $\Sigma$. For example, in most natural languages one never encounters more than three consecutive consonants; the letter "q" is usually followed by the "u"; in a musical score abrupt changes from high to low notes (or viceversa) or among different tonalities are avoided, as well as long repetitions of a single note. The set $\mathcal{L}(\mathcal{S})$ of all admissible subsequences of $\mathcal{S}$ is called the *language* and is usually properly contained in $\Sigma$. The set $\mathcal{L}$ is invariant under $\hat{\sigma}$: i.e., $\mathcal{L}(\mathcal{S}) = \mathcal{L}(\hat{\sigma}(\mathcal{S}))$ [19]. The pair $(\mathcal{L}, \hat{\sigma})$ constitutes a symbolic dynamical system [27]: when $\mathcal{L} = A^{\mathbb{Z}}$ one has the so-called *full shift*; a *subshift* is defined as the restriction of the shift to a closed shift-invariant subspace.

Symbolic dynamical systems are extensively studied in ergodic theory [18,27], where the interest is on the long-time behaviour of the associated nonlinear transformation $\mathbf{F}$. The signals are distinguished, according to the properties of the stationary measure $m$, into *ergodic*, *weakly mixing*, *strongly mixing* and other types [17]. It is then spontaneous to ask which of these kinds of dynamics is the most "complex" one. Another classification of the various possible signals can be made with reference to the shape of the corresponding power spectra [28]: at the leftmost end (see the table below) one finds periodic signals (non-ergodic, with discrete spectrum), "followed" at a higher level of "structure" by quasiperiodic ones (ergodic, non-mixing, with singular-continuous spectrum) with increasing number of incommensurate frequencies. At the rightmost end of this ordering are random ($\delta$-correlated) processes and, to their left, Markov processes with increasing memory $n$ (defined later and indicated as M$n$ in the table, with absolutely continuous spectrum).

| Nonergodic | | ... w-Mixing ... | s-Mixing | *Ergodic Theory* |
|---|---|---|---|---|
| Periodic | Quasiper. ... | ? | ... M$n$ ... M1 M0 | *Power Spectra* |
| $K(1) = 0$ | ... | ? | ... $K(1)$ growing $\nearrow$ | *Information Theory (KS-entropy)* |
| $\infty$ (crystal) | ... | ? | ... $n$ ... 1    0 | *Thermodynamics (interaction range)* |

As already discussed in the introduction, a complex system can be neither "too ordered" nor "too random". This consideration excludes therefore the extreme regions of the two upper diagrams in the table. Also quasiperiodic signals and finite-memory Markov processes should not be regarded as complex, since well-known finite models exist for them: the former ones are still too regular and the latter ones not enough structured. Therefore, ergodicity appears as a necessary (but certainly not sufficient) condition for a system to be complex, since quasiperiodic structures are ergodic and yet simple. On the other hand, strongly mixing systems should not be ultimately classified as complex, since they smooth away any "interesting" feature (uneveness) of a non-equilibrium measure $m$ very quickly [17]. As discussed in the introduction, they are complex if considered from the point of view of criteria based on the time-predictability. Our aim, however, is to obtain accurate models for their *scaling* behaviour. Intuitively, the asymptotic factorization of the probabilities [17,18], as $\lim_{n\to\infty} m(\Delta_i \cap \mathbf{F}^{-n}(\Delta_j)) = m(\Delta_i)m(\Delta_j))$, for all $i,j \in A$, is expected to facilitate this task. In fact, the two events $i$ and $j$, occurring $n$ time steps apart, can be treated as independent for large $n$ (for higher degrees of mixing, see [29,30]).

The degree of independence of *subsequent* events (symbolic sequences $S$) in time is measured by the Kolmogorov-Sinai (KS or *metric*) entropy [18]

$$K(1) = \lim_{n \to \infty} H(s_1 s_2 \ldots s_n)/n = \lim_{n \to \infty} -\frac{1}{n} \sum_{|S|=n} P(S)\ln P(S) \, , \qquad (2)$$

where $S = s_1 s_2 \ldots s_n$ and $H_n = H(s_1 s_2 \ldots s_n)$ is the "ordinary" entropy of the *product* of $n$ events [19]. The quantity $K(1)$ is maximal (equal to $\ln r$) for random uncorrelated signals and vanishes for systems without exponential spreading of phase-space regions (if the size of $\Delta_S$ shrinks exponentially with $|S|$, the probability $P(S)$ scales as $P(S) \sim \exp[-n\kappa(S)]$, with $0 < \kappa(S) < +\infty$, and $K(1) = \langle \kappa \rangle$ is the expectation value of the "local" entropy $\kappa(S)$). Let us consider the probability $P(ab)$ of a product event $ab$ (joint observation of symbol $a$ at time $n$ and of symbol $b$ at time $n + 1$), which can be expressed as $P(ab) = P(a)P(b|a)$, in terms of the conditional probability $P(b|a)$ of event $b$, given $a$ [19,28]. If $a$ and $b$ are mutually dependent, $P(b|a) \neq P(b)$ and $P(ab)$ does not factorize into the product $P(a)P(b)$ of single-event probabilities. It can be shown [19] that $H(ab) \leq H(a) + H(b)$ (the equality holding only for independent events): that is, the knowledge of one past event ($a$) reduces the uncertainty on the future outcomes. When the *memory* extends over $0 < k \leq n$ steps in the past, the conditional probability to observe event $s_{n+1}$ simplifies as $P(s_{n+1}|s_1 \ldots s_n) = P(s_{n+1}|s_{n-k+1} \ldots s_n)$: this defines a Markov process of order $k$. *Memory increase always leads to entropy decrease*[19] (see again the table). Once more, we see that the emergence of "organization", i.e. of an increasing number of "rules", corresponds to a departure from extreme cases such as perfect order or stochastic behaviour. The most interesting (and puzzling) behaviour lies in the middle. Finally, in order to complete this brief review of the necessary mathematical tools for the complexity analysis of symbolic signals, we define the interactions and show their relation with memory and entropy. This identification is straightforward once it is realized that there is a striking analogy between the mathematical structure of discrete stochastic processes and dynamical systems on a side and lattice spin systems on the other. The statistical mechanical (or thermodynamic) formalism for dynamical systems has been introduced by Sinai, Bowen and Ruelle[31] and is currently applied to a wide range of problems in nonlinear dynamics[30,32,33]. Let us consider a generalized Bernoulli shift[15] (an unfair coin-tossing) in which the symbols 0 and 1 appear in a random uncorrelated fashion with probabilities $P(0)$ and $P(1)$. The probability of a string $S = s_1 s_2 \ldots s_n$ containing $k$ 0's and $n - k$ 1's is simply $P(S) = P(0)^k P(1)^{n-k}$, because of the independence of the events. By formally defining a "Hamiltonian"[31,34] $\mathcal{H}(S) = -\ln P(S)$, and noting that $k = \sum_i (1 - s_i)$ and $n - k = \sum_i s_i$, one gets

$$\mathcal{H}(S) = nJ_0 + J_1 \sum_{i=1}^{n} s_i \, , \tag{3}$$

where $J_0 = -\ln P(0)$ and $J_1 = -\ln[P(1)/P(0)]$. Expression (3) is easily recognized as an Ising Hamiltonian (by setting $\sigma_i = 2s_i - 1$) in an external field ($J_1$), *without exchange interaction*. If, instead, the symbol (spin) at site $i + 1$ depends on its predecessor at site $i$ (as in a Markov process), the probability $P(S)$ reads $P(s_1 s_2 \ldots s_n) = P(s_1)P(s_2|s_1) \cdot \ldots \cdot P(s_n|s_{n-1})$, so that

$$\mathcal{H}(S) = -\ln P(s_1) - \sum_{i=1}^{n-1} \ln P(s_{i+1}|s_i) \, .$$

Only four different terms appear in the sum, corresponding to the possible pairs $(s_i, s_{i+1}) = (0,0)$, $(0,1)$, $(1,0)$ and $(1,1)$: they occur a number of times given by $n_{00} = \sum (1 - s_i)(1 - s_{i+1})$, $n_{01} = \sum (1 - s_i)s_{i+1}$, $n_{10} = \sum s_i(1 - s_{i+1})$ and $n_{11} = \sum s_i s_{i+1}$, respectively. By reordering the sums one immediately realizes that the Hamiltonian describes an Ising model with nearest-neighbour interactions, represented by the term $J_2 \sum s_i s_{i+1}$, where $J_2 = \ln[P(1|0)P(0|1)] - \ln[P(0|0)P(1|1)]$ is determined by the conditional probabilities. Hence, interactions are synonymous with memory in this approach. To complete the parallel with classical spin systems on a one-dimensional lattice, one can define a canonical partition function

$$Z_n(\beta) \equiv \sum_{|S|=n} P^\beta(S) = \langle P^{\beta-1} \rangle_n \sim e^{-n(\beta-1)K(\beta)} \, , \tag{4}$$

where $\beta - 1$ plays the role of an inverse temperature and the generalized metric entropy $K(\beta)$[33,35] that of the free energy. Taking the limit $\beta \to 1$, expression (2) for the KS entropy is recovered; for $\beta \to 0$ the topological entropy[36]

$$K(0) = \lim_{n \to \infty} \ln N(n)/n \tag{5}$$

is obtained, where $N(n)$ is the number of allowed sequences of length $n$. Different thermodynamic representations can be defined in terms of other observables than the probabilities $P(S)$: for example, the Hamiltonian $\mathcal{H}(S)$ may be taken as the logarithm of the size $\delta(S)$ of the phase-space domain $\Delta_S$[33,37], or as the largest Lyapunov exponent $\lambda_1(S)$ of a map, provided that these quantities are available, either analytically (as functions of $s_1$, $s_2$, ..., $s_n$) or numerically, for every $S$. If the data consist of the symbolic signal $S$ only, the most natural choice is, however, $\mathcal{H}(S) = -\ln P(S)$. When the interactions have infinite range and decay sufficiently slowly with the inter-spin distance[38], the (symbolic) dynamical system may undergo various types of phase-transitions, associated with non-analytic behaviour of the thermodynamic functions at critical $\beta$-values $\beta_c$[39,40,41]. The estimates of the asymptotic behaviour in the "thermodynamic limit" $n \to \infty$

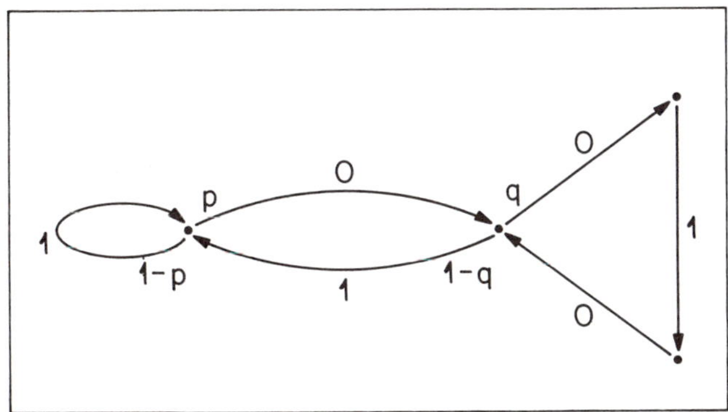

Figure 1. Directed graph for the regular language specified by the set $\mathcal{F} = \{000, 0011\}$ of irreducible forbidden words. The parameters $0 < q < 1$ and $0 < p < 1$ represent the probabilities to follow the corresponding arcs.

are then particularly difficult, since the system exhibits a characteristic critical slowing-down [40]. These phenomena, which have been observed in several nonlinear systems [33,35,37,42], suggest a possible connection between long-range order (strictly related to slowly decaying interactions for one-dimensional lattices [39,40]) and complexity, as it will be discussed in section 6.

## 3. Hierarchical Modelling of Languages

In the previous section we have briefly discussed how to identify the "parts" of a physical system, once encoding into a symbolic form has been performed, and how to define the interactions. In the present section, we show how this information can be incorporated into a hierarchical model which can be used to infer asymptotic properties of the system by extrapolation to the infinite-resolution (thermodynamic) limit.

In order to understand what is to be meant by the word "model" in the context of the characterization of complexity, it is necessary to recall a few notions from the theory of *formal languages* [43]: in fact, every such language has a model counterpart in a discrete automaton. A classification based on the size of the memory required by the automata can be made [43]; a further distinction is introduced by the existence or not of some explicit stochastic mechanism. The lowest-order class consists of the *subshifts of finite type* (SFT) which are specified by a finite list $\mathcal{F}$ of finite forbidden words: i.e., they are subsets of symbolic sequences of a full shift which do not contain any of the forbidden blocks in $\mathcal{F}$. For example, $\mathcal{F} = \{00\}$ defines the Fibonacci shift: it is easy to verify that any (random) concatenation of the two words $w_1 = 1$ and $w_2 = 01$ is allowed.

Generalizations of subshifts of finite type are the *sofic systems* (SS) [44,45] (the typical example being the language in which $\mathcal{F} = \{w : w = 0(11)^n 0, n \in \mathbb{N}\}$, where $v^n$ denotes the $n$-th consecutive repetition of word $v$) and *renewal systems* (RS) [46], consisting of all infinite concatenations of a finite set of words. Some sofic systems are conjugate neither to SFTs nor to RSs: they are called strictly sofic [45]. In computer science, both SFTs and SSs have the same model representation (a finite automaton [43]), although in the second case the list $\mathcal{F}$ of irreducible forbidden words is infinite (irreducible meaning not containing shorter ones). These languages are called *regular* and constitute the lowest member of the Chomsky hierarchy [43]. The signal is produced by means of a random walk on a finite directed graph (i.e., the mechanism is sequential, stochastic: see Fig. 1). A signal $S$ is accepted by the automaton (which, then, operates as a language-recognizing *machine*) only if it corresponds to a possible path on the graph [43]. Formal languages are defined independently of the arc-probabilites $p_i$: the same language (set of allowed words) is produced by two graphs with the same shape and different $p_i$'s. The output of a nonlinear dynamical system, such as the logistic map $x_{n+1} = 1 - a x_n^2$ (with $s_n = [1 + \mathrm{sgn}(x_n)]/2$) in the chaotic regime is a regular language only at special parameter values (for which a *Markov partition* exists [15]): in general, it is not.

Higher-order generation schemes are defined by means of parallel mechanisms, called *grammars*, which transform symbols in a string $S$ into words chosen from a list $\mathcal{W} = \{w_1, w_2, \ldots\}$, to yield a new string $S'$. If all symbols in $S$ are substituted simultaneously, one has a so-called D0L language [43b]. For example, starting with $S = ab$ with the rule $(a \to \psi(a) = w_1 = ab; b \to \psi(b) = w_2 = bba)$, one obtains $S' = \psi(ab) = abbba$, $S'' = \psi^2(ab) = abbbabbabbaab$, and so on. The two substitutions $\psi_{PD} : (0 \to 01; 1 \to 10)$ and $\psi_{QP} : (0 \to 1; 1 \to 01)$ are particularly relevant in nonlinear dynamics since they correspond to the period-doubling (PD) accumulation point dynamics[47,48] and to the golden-mean quasiperiodic (QP) transition to chaos[50], respectively.

The second class in the Chomsky hierarchy is constituted by the *context-free* grammars[43], in which the symbol-substitutions are non-simultaneous and the rules are stochastic (any symbol $a$ may have several images $\psi_1(a)$, $\psi_2(a)$, ...): the corresponding model is a pushdown automaton. In Ref. [51] it has been shown that D0L schemes can be seen as an extension of context-free Chomsky grammars. All these languages (of parallel origin, because of the substitutions) cannot be modelled in terms of regular ones since the associated graphs would be unbounded, thus requiring an infinite amount of memory[11,52,53].

By extending the "grammatical rules" to depend on a number of nearest-neighbours of the symbol to be rewritten, one obtains a so-called *context-sensitive*[43] grammar. The machine representation is a linear-bounded automaton, which has still higher memory-requirements than the previously described languages. Finally, the highest class in the Chomsky hierarchy is constituted by the *unrestricted* (or recursively-enumerable) languages[43], produced by *universal computers* (Turing machines) using an indefinitely large memory.

In computer science, complexity is usually expressed in terms of the size of the fastest program (usually, a universal Turing machine) which is able to reproduce *exactly* the first $N$ bits of an input string[6,7,8], in the limit $N \to \infty$. Alternatively, one may consider the time needed by the shortest program which performs the same task[54]. These measures of complexity, apart from not being effectively computable because of the need for an optimal (either shortest or fastest) program, are essentially indicators of randomness[11]: the most random sequence is the most unpredictable upon increasing its length $N$. The inadequacy of this approach towards a characterization of complexity stems from the misunderstanding of the term "model". In fact, the study of the complexity of a signal $S$ does not require trying to construct a machine which reproduces (models) a *single* specimen of a language (the input sequence), even though arbitrarily long ($|S| = N \to \infty$). Rather, one should split $S$ into parts (all the subwords thus obtained and their concatenations constituting the language $\mathcal{L}$) and predict the form and intensity of the interrelations among them with a model. In this way, the stress is on the whole ensemble (members of $\mathcal{L}$) of possible realizations of the process which underlies the formation of $S$. In other words, discovering the "microscopic" rules which govern the dynamics, although very valuable if successful, is not necessarily the primary commitment to be undertaken when dealing with a presumably complex system. Indeed, modelling thermal convection with the Navier-Stokes equations certainly yields very accurate results; however, the much simpler Lorenz equations[55] also provide an excellent description of the dynamics; in turn, asymptotic properties of the system are better understood in terms of a third model (the so-called geometrical model) which can be further simplified into a one-dimensional map of the form (1), by taking a Poincaré section of the flow[15]. After this chain of reductions, ergodic and spectral properties of the system can be understood analytically or computed numerically with much more ease than by using the first description. Another example is provided by cellular automata (CA)[5]: they may produce spatial patterns with a macroscopic structure (e.g., distribution of blocks of $n$ symbols with $n \gg 1$) which has no apparent relationship with the microscopic dynamical rules. Although the latter ones are, in general, short-ranged (involving, for example, only nearest-neighbours), the limit set may exhibit algebraically decaying correlations[11]. Similar considerations are valid for Ising systems[56] or chains of oscillators[57] with competing interactions and for statistical mechanical systems exhibiting phase-transitions[39,40]. In all these cases, the knowledge of the details of the dynamics allows finding short-cuts neither for the generation of the limit set $S$ (spatial configuration reached after an infinite iteration of the dynamical rules), nor for the evaluation of ergodic, spectral or thermodynamical properties of it. Such kinds of spatio-temporal (i.e., parallel) dynamics are hence good candidates for the attribute of "complex", both in a computer-scientific view and in the sense exposed in the present work.

The most general way to construct a hierarchical model which is required to perform predic-

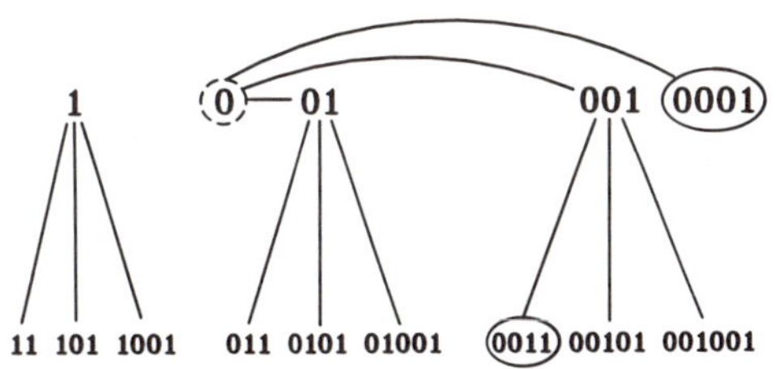

Figure 2. First two levels of the "primitive"-tree for the regular language of Fig. 1.

tions directly for macroscopic observables is to split the system into subsystems and to organize them on a tree. The simplest implementation of this technique when dealing with a symbolic signal $S = \ldots s_0 s_1 s_2 \ldots$ is achieved by considering the symbols of the alphabet $A = \{0, \ldots, r-1\}$ as the elementary parts and concatenations $S$ of them as compound objects which approach "the complex" for $|S| \to \infty$. In this way, an $r$-nary tree is obtained, where the symbols $\{0, \ldots, r-1\}$ are assigned to the vertices of the first level and strings $S$ composed of $|S| = l$ symbols to those of level $l$. All branches (at most $r$) leaving vertex $S$ point to the allowed extensions $S0$, $S1$, ... of sequence $S$. If the signal is aperiodic, there are branching vertices, corresponding to strings with more than one possible continuation. Of course, it may happen that some branches are missing altogether because the corresponding words are forbidden. The tree then presents an irregular structure. However, it is often possible to reduce this asymmetry with a more careful choice of the parts. Let us consider again the Fibonacci shift: sequence 00 is forbidden and the only possible continuation after symbol 0 is 1. Therefore the symbolic dynamics yields concatenations of the "words" $w_1 = 1$ and $w_2 = 01$: in terms of them, a complete binary tree is recovered. If, on the one hand, such an ideal situation is rare, on the other it is clear that a "clever" choice of words $w_i$ may lead to a considerable decrease of defects on the tree. In the case illustrated in Fig. 1, the choice of the three "primitive" words $w_1 = 1$, $w_2 = 01$ and $w_3 = 001$ leads to a tree with only one defect (the irreducible forbidden concatenation $w_3 w_1 = 0011$), as displayed in Fig. 2.

Therefore, it is convenient to split the signal $S$ into *variable-length* "primitive" words and construct trees in which concatenations of $l$ primitives appear at level $l$[13]. The set (not necessarily finite) of primitives constitutes a "code"[58]: that is, a conversion table (such as $w_1 = 1$, $w_2 = 01$, $w_3 = 0011101$, ...) between symbols from two different alphabets (the $w$'s on a side and $\{0, 1\}$ on the other, for example). Variable-length codes are commonly employed in information processing for their superior efficiency with respect to *block-codes* (where all words have the same length)[58]. Among the various advantages they provide, there is an increased compactness in the representation of the *topology* of the problem (as it is easily verified for the case of Fig. 2)[13] with respect to ordinary block-code trees[59]. Of course, when the language is "uniform", i.e. it admits all the $r^n$ possible sequences of length $n$, there is no advantage in using variable-length code, unless the probabilities of the sequences differ substantially from one another (see Huffman codes[58]). In nonlinear dynamics the languages are "irregular" for generic parameter values. The symbolic signal produced by the logistic map in a finite region around $a = 1.85$, can be rewritten completely in terms of the three words $w_1 = 1$, $w_2 = 01$ and $w_3 = 001$. At $a = 1.85$ the (infinite) list $\mathcal{F}$ of irreducible forbidden words begins with $w_3 w_1$, $w_3 w_2 w_3$, $w_3 w_2 w_2 w_1 w_3 w_2$. The whole language $\mathcal{L}$ can be generated from the tree based on these primitives, although not all subsequences (such as 10, 100, 010, ...) are allocated on its vertices: the increased conciseness implies no loss of information about the composition of the language since these strings still appear in the body of longer sequences, across the "junction" between primitives (as, e.g., in $\ldots w_2 w_2 w_1 w_2 w_3 w_2 w_2 \ldots = \ldots 0\overline{1}0\overline{11}0\overline{1}0\overline{01}0\overline{1}01 \ldots$).

Before presenting criteria for the choice of suitable primitive words, we complete the discussion of the hierarchical model by mentioning the basic metric (probabilistic) features associated with the tree. The probabilities $P(w_i)$ of the primitives $w_i$ (or of the symbols $s_i$, for a block-code) must satisfy the normalization condition $\sum_i P(w_i) = 1$. Furthermore, owing to the Kolmogorov consistency condition $P(S) = \sum_i P(Sw_i)$, every complete level of the tree has total mass 1: it represents a full covering of the phase-space (in fact, $1 = \sum_i P(w_i) = \sum_{i,j} P(w_i w_j)$, and so on). Trees constructed in this way are equivalent to "generalized" Markov models, which describe the dynamics as a sequence of events $w_i$, occurring with measurable transition probabilities, according to unknown rules (determined by a dynamical map $\mathbf{F}$ or by iteration of a CA or by a Monte-carlo simulation, for example). Since the sequences have variable-length at every level, in general, the memory extent depends on the probabilities of the orbits. The order of such models can be estimated by the "average Markov time" per level [13]

$$\bar{\theta} = \lim_{l \to \infty} \theta_l = \lim_{l \to \infty} \frac{1}{l} \sum_{\text{level } l} |S| P(S) \tag{6}$$

which equals 1 for ordinary block-code trees. In section 6 it will be shown that expression (2) for the metric entropy becomes, in the new formulation,

$$K(1) = \lim_{l \to \infty} h_l = \lim_{l \to \infty} - \sum_{\text{level } l} P(S) \ln P(S)/(l\theta_l) \,. \tag{7}$$

For a periodic signal, this immediately yields $h_l \equiv 0$, $\forall\, l$, since only one primitive (the basic block) exists. A description based on such Markov trees provides a tool for the understanding of asymptotic scaling properties of the system. An application to the analysis of power spectra can be found in Ref. ([60]). In the next sections, the topological and metric features of the trees will be related to the convergence of thermodynamic averages of the same type as those introduced in section 2 in connection with ergodic theory. Complexity will be quantified by the discrepancy between predicted and actual asymptotic behaviour.

Finally, let us briefly discuss a criterion for the choice of primitive words. It is known that a particularly relevant role in determining the recurrence (long-term) properties of nonlinear dynamical systems is played by the set $\Omega$ of "non-wandering" points [15]: a point $\mathbf{x}$ is non-wandering for the map $\mathbf{F}$ if, for any neighbourhood $U$ of $\mathbf{x}$, there exists a number $n > 0$ such that $\mathbf{F}^n(U) \cap U \neq \emptyset$ (an analogous definition can be given for a generic subshift $\hat{\sigma}$). Correspondingly, a symbolic sequence $S$ which labels a domain $\Delta_S$ in $\Omega$ will be observed in the signal $\mathcal{S}$ with a well-defined frequency of occurrence. The set $\Omega$ consists of points with a weak recurrence property: in particular, all periodic points of $\mathbf{F}$ belong to $\Omega$. This property suggests a useful criterion to distinguish primitive words from "transient" (i.e., non-recurrent) ones [13]: we define a primitive as a string $w$ which can be periodically extended up to the maximum investigated block-length $n_{max}$ and which does not contain a prefix with the same property (e.g., 001 is a primitive if 001001... is allowed and 000... is not; see Refs. ([13,60]) for details). This is a "strong" condition: namely, one which does not depend on tuning parameters or on statistical weights. Weaker conditions (depending, e.g., on the probabilities $P(S)$) are discussed in Refs. ([13,61]). The first few periodic primitives of the Hénon map [62] $(x_{n+1}, y_{n+1}) = (a - x_n^2 + by_n, x_n)$ at $(a = 1.4, b = 0.3)$ are [13,60] $w_1 = 1$, $w_2 = 01$, $w_3 = 0011101$, $w_4 = 0011111$, $w_5 = 00111101$, $w_6 = 00011101$, $w_7 = 00011111$, $w_8 = 000111101$, $w_9 = 0011110011101$. The list is infinite, as well as that of the irreducible forbidden words (beginning with 0000, 0010, 0110, 0101000, 0111000 [21]).

## 4. Complexity and Predictions

A hierarchical tree of the type introduced in the previous section is a purely topological frame on which a full metric model can be constructed. The evaluation of the probabilities $P(S)$ of each string $S$ on the tree leads to the construction of a hierarchy of Markov models as a description of the actual dynamics. Each branch, in fact, corresponds to an allowed transition: a vertex labelled $Sw$ can be associated with the conditional probability $P(w|S) = m[\Delta_S \cap \mathbf{F}^{-|S|}(\Delta_w)]$ of observing event $w$ after event $S$, computed as $P(w|S) = P(Sw)/P(S)$. The language $\mathcal{L}$ can be generated with the help of a transfer matrix $\mathbf{T}$[63] which, at level $l+1$, reads

$$T_{w_0 w_1' \ldots w_l'; w_1 \ldots w_{l+1}} \equiv \frac{P(w_1 \ldots w_{l+1})}{P(w_1 \ldots w_l)} \delta_{w_1 w_1'} \cdot \ldots \cdot \delta_{w_l w_l'} \, . \tag{8}$$

Matrix $\mathbf{T}$ records the conditional probabilities for transitions between the two strings $w_0 w_1' \ldots w_l'$ and $w_1 w_2 \ldots w_{l+1}$ upon (left) shifting of the signal by one primitive at a time [13,60]: given any "start" string, its "follower" in the signal $S$ can be chosen at random from all admissible ones, according to the values of the conditional probabilities. The procedure can be iterated to yield a signal (or an ensemble of signals) with statistical properties close to those of the original one. The accuracy depends on how well direct products of low-order matrices $\mathbf{T}$ approximate higher-order ones. Rather than focusing on this time-sequential[60] capability of the model, a complexity analysis requires employing the information stored on the tree to "predict" the scaling behaviour of the $P(S)$'s in the asymptotic limit $|S| \to \infty$, for all sequences $S$. For this purpose it is essential to make the best use of the probability values $P(S)$, known only for $|S| \leq n_{max}$, to approximate as accurately as possible the coefficients of $\mathbf{T}$, for $|S| > n_{max}$. The conditional probabilities, in fact, express the interactions among the parts of the system.

The simplest kind of predictor evaluates the expected probability $P_0(Sw)$ of a level-$l$ sequence $Sw$ as the product $P(S)P(w)$: i.e., the events $S$ and $w$ are treated as independent. This assumption is equivalent to approximating the original map $\mathbf{F}$ with a piecewise linear map $\mathbf{G}_l$ having the following characteristics: it is topologically described by a tree identical to that of $\mathbf{F}$ up to level $l$ and it generates a natural measure $m_0$ which assigns precisely the probability $P_0(Sw)$ to each string $Sw$, although the associated phase-space domains $\Delta_S$ for $\mathbf{F}$ and $\Delta_S^{(0)}$ for $\mathbf{G}_l$ may have different size[64]. A more accurate predictor consists of a finite-order Markov model, defined by

$$P_0(w_1 \ldots w_n) = P(w_1 \ldots w_k) \cdot P(w_{k+1}|w_2 \ldots w_k) \cdot \ldots \cdot P(w_n|w_{n-k+1} \ldots w_{n-1}) \, , \tag{9}$$

where probabilities are assumed to be known for sequences consisting of no more than $k$ primitives (and are used to infer those of longer strings). The most efficient scheme of this type is obtained by letting the memory $k$ increase with the level $n$, as $k = n - 1$[60]. This predictor can also be related to a sequence $\mathbf{G}_l$ of maps[64]. The accuracy of a predictor in estimating the actual probability values $P(S)$ of the system can be quantified by introducing a measure of complexity for this specific task. Accordingly, the *metric complexity* of the system (either a nonlinear map $\mathbf{F}$ or a subshift $\hat{\sigma}$), relative to the model (set of predictions $P_0$), is defined as the information gain [13]

$$C_1 = \lim_{l \to \infty} C_1(l) \equiv \lim_{l \to \infty} \sum_{\text{level } l} P(S) \ln \frac{P(S)}{P_0(S)} \, , \tag{10}$$

where the sum extends over all sequences at level $l$. The quantity $C_1(l)$ can be interpreted as an information-theoretic "distance" between maps $\mathbf{F}$ and $\mathbf{G}_l$ (or between the true subshift $\hat{\sigma}$ and an approximate $\hat{\sigma}_l$): it is non-negative and vanishes only if the predictions coincide with the actual probabilities $P(S)$, for all $S$. At variance with previous studies [65,66], where the accuracy provided by *periodic* maps $\mathbf{G}_l$ has been investigated, here the unknown dynamics is approached from "above", through a sequence of Markov models which include increasingly longer memory effects. The metric entropy $K(1)$ of map $\mathbf{G}_l$ may not increase with $l$ and is larger than (or equal to) that of $\mathbf{F}$. If the model agrees with the actual system closely or even exactly, for large $l$, the transformation $\mathbf{F}$ is "simple". In this sense, complex behaviour "lies between order and disorder" and is, therefore, "hard" to model (both with periodic and Markov approximations: see again the diagram in section 2).

The metric complexity characterizes the difficulty of obtaining an asymptotically accurate model. This can be easily illustrated from the point of view of ergodic theory. Consider for simplicity a block-code tree. The quantity $C_1(l)$ can be related to the convergence properties of the block-entropy (eqs. (2) and (7))

$$h_l = H_l/l \sim K(1) + f(l) \tag{11}$$

to the metric entropy $K(1)$, for $l \to \infty$: the function $f(l) \geq 0$ (tending to 0 for $l \to \infty$) represents the speed of convergence of $h_l$ to $K(1)$[64]. The first predictor considered above yields (by substituting the expressions for $P_0$ and $h_l$ into eq. (10) and summing over dummy symbols)

$$C_1(l) = -lh_l + (l-1)h_{l-1} + h_1 \to h_1 - K(1) \geq 0 , \tag{12}$$

for $l \to \infty$[64]: i.e., the complexity is the difference between the coarse-grained and the fine-grained metric entropy[67]. Therefore, this simple predictor yields zero complexity just for ordered (using a periodic-primitive code) and delta-correlated processes, for which $h_1 = K(1)$, and a positive value in all other cases. A more interesting example is provided by predictor (9) which yields

$$C_1(l) = -lh_l + (l - k + 1)kh_k - (l - k)(k - 1)h_{k-1} . \tag{13}$$

Here the $l$-symbol (or $l$-primitive) sequence is split into fixed-length ($k$ steps) blocks, at variance with the previous $(l-1)$-to-1 decomposition. Assuming a convergence speed $f(l) \sim e^{-\gamma l}$, one finds $C_1(l) \sim l\left[ke^{-\gamma k} - (k-1)e^{-\gamma(k-1)}\right]$. Hence, the metric complexity diverges for $k < k^*$ and vanishes for $k > k^*$, where $k^* = e^\gamma/(e^\gamma - 1)$ can be interpreted as the memory-range of the signal, from the point of view of entropy evaluations. In case $k^*$ is an integer, $C_1$ may be finite at $k = k^*$. This behaviour has been experimentally observed in[68]. Notice that a "model size" can be defined as a quantity proportional to the amount of information $kh_k$ necessary to "reproduce" the system with accuracy $C_1$. The complexity of a system is large (infinite) if an inadequate model is used and is zero if the information is redundant [64]. The optimal description is given by the minimal model which is able to yield $C_1 = 0$. A different type of sharp transition is observed for the speed $f(l) \sim l^{-\alpha}$ (i.e., $\gamma \to 0$, $k^* \to \infty$), which is believed to be generic [64]: $C_1$ vanishes for $\alpha > \alpha_l^* = 1 + O(1/\ln l)$, diverges for $0 < \alpha < 1 - O(1/\ln l)$, and may be finite otherwise. No dependence on $1 \leq k \leq l-1$ exists in this case. Different predictors can be designed to educe and analyize particular asymptotic behaviour of the entropies (e.g., at intermittency [15]).

When predictor (9) is used with $k = l - 1$, the terms $P/P_0$ appearing in eq. (10) are ratios between *conditional* probabilities: in fact, by setting $S = uV$ (with $u$ being a primitive and $V$ a sequence of level $l - 1$), one has $P_0(Sw) = P(S)P(w|V)$ and $P(Sw)/P_0(Sw) = \sigma^{(l+1)}(uVw)/\sigma^{(l)}(Vw)$ where the expressions $\sigma^{(l+1)}(uVw) = P(uVw)/P(uV)$ and $\sigma^{(l)}(Vw) = P(Vw)/P(V)$ are terms of a generalized scaling function $\sigma(t)$ [13] for the probabilities, evaluated at two consecutive levels of resolution (see section 6 for a precise definition). Therefore, $C_1$ has the meaning of a global measure of the convergence of the scaling function, and systems with predictable scaling properties are simple [13]. The factorization assumption, for example, yields positive values of $C_1$ for most non-hyperbolic [15] attractors, since it disregards memory effects, whereas piecewise-linear systems are metrically simple also within this approximation (the natural invariant measure is non-singular). In this sense, $C_1$ is also a measure of nonlinearity.

Finally, it is worth mentioning that equation (10) for $C_1$ can be generalized to a function $C_q$ of a parameter $q$ [13]. For $q = 0$, one obtains the topological complexity

$$C_0 = \lim_{l \to \infty} \lim_{n \to \infty} \ln \frac{N_0(l,n)}{N(l,n)} , \tag{14}$$

where $N_0(l,n)$ is the number of sequences predicted at level $l$, given the knowledge of all words of length $|S| < n$, and $N(l,n)$ the number of those with length $|S| \leq n$ allocated at the same level. Notice that $C_1$ reduces to $C_0$ if all existing and predicted sequences are separately equiprobable (i.e., $P(S) = 1/N$ and $P_0(S) = 1/N_0$, $\forall\, S$). The topological complexity is identically zero if all predicted orbits exist. Of course, one never achieves an optimal description of the language (unless a finite one exists): the understanding of the object can always be improved, in general. Virtually, this process may only end when a perfect match between model and system is found, i.e., when $C_q = 0$.

## 5. Recoding and renormalization

The hierarchical procedure described above implies a coarse-graining of the signal (or, equivalently, of phase-space). However, the arbitrariness of the choice of the primitives may prevent the whole analysis from approaching the asymptotic properties of the system in the optimal way. A considerable improvement is obtained by resorting to a higher-level modelling procedure, endowed with a parallel unfolding mechanism. This consists of renaming the primitive words $w_1, w_2 \ldots$, identified in the analysis of the original signal $S_0$ with alphabet $A_0$, with symbols from a new alphabet $A_1 = \{0, 1, \ldots\}$. The whole analysis is repeated on the transformed signal $S_1$ thus obtained. The iteration of this procedure yields a progressive coarse-graining of the signal (corresponding to an increase of resolution per symbol in phase-space). The description of the image-signal $S_k$, obtained after $k$ recoding steps, consists of the derived tree and of the code which keeps track of the previous block-renaming cascade (i.e., of the relations between each symbol in the alphabet $A_k$ and its pre-image string in $S_0$). The recoding procedure is equivalent to a renormalization-group transformation on the nonlinear map $\mathbf{F}$ associated with the subshift[49]: in fact, renaming a sequence $S$ of length $n$ with a single symbol is equivalent to considering the $n$-th iterate of $\mathbf{F}$ in the phase-space element $\Delta_S$. Accordingly, recoding $S_0 \rightarrow S_1$ and recomputing the probabilities $P'$ in $S_1$ is a completely analogous operation to obtaining a renormalized block-Hamiltonian $\mathcal{H}' = -\ln P'(w_1 \ldots w_m)$. Notice that no recoding is possible with an ordinary block-code tree: renormalization in that case is tantamount to considering levels 2, 4, ..., $2l$ (or multiples of any other integer $k > 2$) on the tree, but *does not change* the values of the attached probabilities. Also, no iteration of the process is possible.

For perfectly self-similar languages, such as those of PD and QP, the trees obtained at each recoding step are identical: i.e., an exact renormalization is readily achieved[13]. In many cases the dynamics itself is explicitly of parallel type, as in Montecarlo updates of spin chains or in cellular automata. The identification of the long-ranged "coherent" structures appearing in such systems is particularly difficult and requires consideration of increasingly long blocks of symbols, as in part achieved by means of the renormalization (recoding) technique. For signals of purely sequential nature, only a few recoding steps are usually possible and useful[60]. Finally, it is worth mentioning that weakly-mixing subshifts can be obtained by means of compositions of D0L-substitutions[46,27]: an example is provided by a single application of the rule $\psi(1) = 11$ and $\psi(0) = 0$ to the language of the period-doubling accumulation point.

## 6. Thermodynamic formalism and complexity

A system is complex if it reveals different laws (interactions) at different resolution (coarse-graining) levels. This behaviour, in fact, renders the predictions in the *scaling direction*, i.e. from level to level on the tree, particularly difficult (inaccurate). At the opposite extremum are self-similar processes which present exactly the same structure at each level: they are the *fixed points* of the scaling dynamics[69]. The algorithms employed to estimate the asymptotic scaling behaviour of the system's observables (like the sequence probabilities $P(S)$ or the nearest-neighbour distances[47] $\delta(S)$) determine the convergence properties of thermodynamic averages[31,33,59] such as $Z_n(\beta)$ in eq. (4). The evaluation of the free energy $K(\beta)$ in the thermodynamic limit $n \rightarrow \infty$ is particularly difficult when phase-transitions[33,37] are present. These are caused either by anomalous exponential scaling of the observables for some sequences $S$, or by mixed exponential-algebraic scaling: the former mechanism is ubiquitous in non-hyperbolic dissipative dynamical systems, whereas the latter one is typical of intermittent phenomena and of conservative chaos. Accordingly, various kinds of transitions may appear[70]. The approach based on variable-length codes, combined with the recoding technique, yields a much more compact representation of the scaling behaviour, since it reconstructs the hierarchy of phase-space in terms of elements (words) which satisfy a certain distinguishing condition (for example, periodic extendibility) and allows to work directly with the renormalized hamiltonian (evaluated numerically for each sequence $S$ in the recoded signal $S_k$). The variability of the length of the sequences at each level leads immediately to a grand-canonical formulation of the problem. Since there is no more a unique word-length $n$ to be singled out as in eq. (4), we introduce the following level-$(l+2)$ grand-partition sum

$$\Omega_{l+2}(z;\beta) \equiv \sum_{w_0 w_1 \ldots w_{l+1}} P^\beta(w_0 w_1 \ldots w_{l+1}) z^{|w_0 w_1 \ldots w_{l+1}|} \,. \tag{15}$$

The term $z^{|S|}$ has been introduced to provide a detailed compensation for the (generally exponential) decrease of $P(S)$ with $|S|$ [13]: if $P(S) \sim \exp[-|S|\kappa(S)]$, taking $z \approx \exp[\beta\langle\kappa\rangle]$ nearly renders stationary the sum in eq. (15). More precisely, we define $K(\beta)$ in terms of the special value $z(\beta)$ of $z$ which keeps $\Omega_l(z(\beta);\beta) \approx O(1)$ in the limit $l \to \infty$. The free energy is then given by $K(\beta) = \ln z(\beta)/(\beta-1)$. Indeed, if $|S| = n$, for all $S$, we recover relation (4). Relation (7) can be easily obtained by inserting $z(\beta) = \exp[(\beta-1)K(\beta)]$ into expression (15), taking the limit $\beta \to 1$ and equating the result to 1.

A more accurate estimate of $K(\beta)$ is obtained by comparing the partition functions at two consecutive levels:

$$\sum_{w_0 w_1 \ldots w_{l+1}} P^\beta(w_0 w_1 \ldots w_{l+1}) z^{|w_0 w_1 \ldots w_{l+1}|} = \lambda(z;\beta) \sum_{w_0 w_1' \ldots w_l'} P^\beta(w_0 w_1' \ldots w_l') z^{|w_0 w_1' \ldots w_l'|} \,. \tag{16}$$

In the limit $l \to \infty$, $z(\beta)$ is now determined from the relation $\lambda[z(\beta);\beta] = 1$, which states the stability of the sum. For $z > z(\beta)$, $\Omega_l(z;\beta)$ diverges (for $z < z(\beta)$ it vanishes). In fact, eq. (16) can be rewritten as an eigenvalue equation in the following way [63,13]: the argument of the l.h.s. sum is multiplied and divided by $P^\beta(w_0 w_1' \ldots w_l') z^{|w_0 w_1' \ldots w_l'|}$, and then a second sum over dummy indices $w_1' \ldots w_l'$ is taken, after inserting the Kronecker deltas $\delta_{w_1' w_1} \cdot \ldots \cdot \delta_{w_l' w_l}$. Finally, one obtains the eigenvalue equation

$$\sum_{w_0 w_1' \ldots w_l'} \sum_{w_1' w_2' \ldots w_{l+1}} \sigma^\beta(w_0 w_1 \ldots w_{l+1}) z^{|w_{l+1}|} \delta_{w_1' w_1} \cdot \ldots \cdot \delta_{w_l' w_l} P^\beta(w_0 w_1' \ldots w_l') z^{|w_0 w_1' \ldots w_l'|} =$$

$$\tag{17}$$

$$= \lambda(z;\beta) \sum_{w_0 w_1' \ldots w_l'} P^\beta(w_0 w_1' \ldots w_l') z^{|w_0 w_1' \ldots w_l'|}$$

for a generalized transfer matrix $\mathbf{T}(z;\beta)$ [13] defined by

$$T_{w_0 w_1' \ldots w_l'; w_1 \ldots w_{l+1}} \equiv \sigma^\beta(w_0 w_1 \ldots w_{l+1}) z^{|w_{l+1}|} \delta_{w_1' w_1} \cdot \ldots \cdot \delta_{w_l' w_l} \tag{18}$$

where $\sigma(w_0 w_1 \ldots w_{l+1}) \equiv P(w_0 w_1 \ldots w_{l+1})/P(w_0 w_1 \ldots w_l)$ is a term of the level-$l$ probability scaling function already mentioned in section 4. With this formulation, the condition for the evaluation of $K(\beta)$ is expressed by saying that the largest eigenvalue $\lambda_1(z;\beta)$ of $\mathbf{T}$ must equal 1 for $z = z(\beta)$. For the language of Figs. 1 and 2, the topological entropy $K(0)$ is obtained by solving the following characteristic equation (for $l = 1$)

$$\mathrm{Det}(\mathbf{T} - \mathbf{I}) = \begin{vmatrix} z - 1 & z^2 & z^3 \\ z & z^2 - 1 & z^3 \\ 0 & z^2 & z^3 - 1 \end{vmatrix} = 0 \,, \tag{19}$$

thus determining $z(0)$ as the smallest zero of the polynomial in eq. (19). Consideration of higher levels does not change the value of $z(0)$, since no other prohibitions exist in this example. More in general, for a Markov process of (finite) order $k$, matrices corresponding to levels $l > k$ can be obtained as products of order-$k$ matrices. The simplest case is a purely self-similar system (memory $k = 0$).

A detailed description of the asymptotic behaviour of the conditional probabilities can be graphically represented by means of a scaling function $\sigma(t)$ [69] which, for our purposes, is defined as follows. The level-$l$ sequences $S = w_1 w_2 \ldots w_l$ are mapped onto the unit interval by associating a value $t = t(S)$ to each of them. For every $t = t(S)$ the level-$l$ approximation $\sigma_l(t)$ of the scaling function $\sigma(t)$ is given by the conditional probability $P(S = w_1 \ldots w_l)/P(w_1 \ldots w_{l-1})$. The definition of the ordering parameter $t = t(S) \in [0,1]$ can be easily understood by referring to the

14

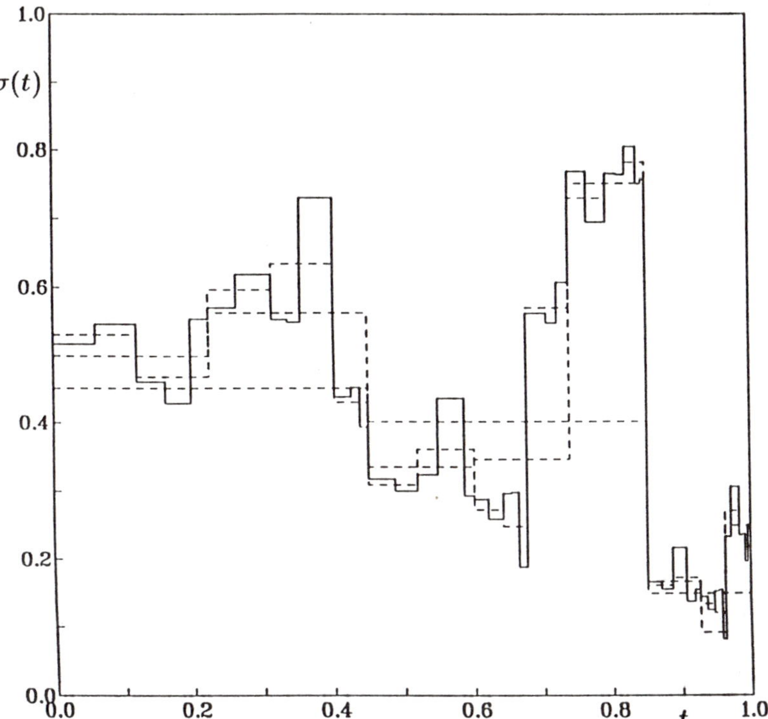

Figure 3. First four levels of approximations to the scaling function $\sigma(t)$ of the logistic map at $a = 1.85$. Dashed lines: levels 1-3, solid line: level 4.

logic tree of Fig. 2. At level $l = 1$, the primitives $w_k$ are considered in the same order as on the tree (e.g., from left to right) and $t(w_k) = t(w_{k-1}) + P(w_k)$, $k = 1, \ldots, N(1, n_{max})$ ($w_0$ being the empty string, with $t(w_0) = 0$). Hence, $\sigma_1(t)$ is piecewise constant over $N(1, n_{max})$ intervals. The generic $k$-th interval is split, at level 2, into subintervals labelled by all sequences $w_j w_k$ ending with $w_k$ and ordered from left to right according to the order of the $w_j$'s on the first level of the tree. In this way, we have $t(w_j w_k) = t(w_{j-1} w_k) + P(w_j w_k)$ and the index $k$ is increased by one after that all $j$'s have been scanned. At level 3, all subintervals of $w_j w_k$ are labelled as $w_i w_j w_k$, with the $w_i$'s ordered as usual, and so on. For example, with a binary tree, the values of $t$ correspond, from left to right, to the sequences 0 and 1 at level 1, to 00, 10, 01 and 11, at level 2, to 000, 100, 010, 110, 001, 101, 011 and 111, at level 3. The widths of the intervals are just the corresponding sequence-probabilities, so that forbidden strings do not appear at all [13]. As an illustration, we display in Figure 3 a schematic plot of the first four approximations to $\sigma(t)$ for the logistic map at $a = 1.85$. The topology of the tree is the same as in Fig. 2, for the first two levels. Notice that some intervals are split only into two parts and not three, due to the prohibitions. A two-scale Cantor set would appear as a two-plateau curve. With this definition of $\sigma(t)$, the metric complexity can be rewritten as $C_1 = \lim_{l \to \infty} \langle \ln(\sigma_{l+1}/\sigma_l) \rangle$, where $\langle f(t) \rangle = \int_0^1 f(t) dt$. Once more, a complex system is associated to irregular (i.e., non-converging) scaling behaviour. It becomes therefore clear that self-similarity is just a fixed point of the *scaling dynamics*. In Fig. 4, convergence to a limit curve already appears at level 4, although there are still some large deviations (associated however with low-probability sequences). The convergence is confirmed by the analysis of the higher levels, not displayed for clarity reasons. Similar measures have been performed on the Lorenz[55] system, the Hénon map[13,60] and the filtered baker map[37].

The occurrence of a phase-transition in the thermodynamical formalism described above is only possible if the scaling function assumes a continuous range of values, although this condition alone is not sufficient [37]. In order for a system to be complex, even using predictor (9) with $k = n - 1$, a stronger condition is necessary: namely, the non-convergence of the scaling function. Examples of such a behaviour are difficult to find (there are no "simple" models for complex systems, of course); however, non-convergence of $\sigma_l(t)$ with $l$, caused by inappropriate orderings of the $t$ axis has been reported in Ref. ([71]).

## 7. Conclusions

From the above discussion, it emerges that complex systems present, at increasingly finer levels of resolution, properties that cannot be accurately predicted from those measured at coarser scales. In other words, different structure is observed at different depth in the hierarchical description. Complexity is neither order, nor disorder. Markov processes are also simple. Any system for which a good (hierarchical) model can be constructed is simple. It is then natural to ask oneself whether "true" complexity exists at all, since we are unable to produce workable examples of it and, when we succeed in doing it, complexity vanishes. It is not surprising, therefore, that there is a large number of open questions arising from the multiform manifestations of this phenomenon. Its properties suggest a possible connection with chaotic (unpredictable) renormalization-group schemes [72], on the thermodynamical side, and with weakly-mixing transformations (examples of which are indeed rare[27,29,30]), for what concerns ergodic theory. Another major question is whether "ultimate" complexity (i.e., refractory to any modelling attempt) is compatible with stationarity or thermodynamical equilibrium. In this respect, very interesting examples of "complex behaviour" are provided by Ising models [56] and lattice systems [73,57] with competing interactions. Finally, life itself, the paradigm for complexity, is certainly a non-equilibrium, non-stationary process in time, but its "instantaneous spatial encoding", a DNA chain $S = s_1 s_2 \ldots s_N$, need not be non-stationary (apart from some initial and final portion), in general, when studied as a (space-) signal; the same remark applies to cellular automata and to other (possibly stochastic) lattice systems. The increasing interest in these topics which we are presently witnessing in many areas of scientific research promises a considerable progress towards a better understanding of complexity, in many of its aspects, in a near future.

## Acknowledgements

Very useful discussions with A. Politi are acknowledged. The topics reviewed in the present contribution will be treated more extensively in a future work [74], together with other approaches to the characterization of complexity.

## REFERENCES

[1] *"Lectures in the Sciences of Complexity"*, edited by D.L. Stein, Santa Fe Institute, Addison-Wesley, Reading, Mass. (1989).

[2] D.L. Stein, in [1] (preface).

[3] W. Weaver, Am. Sci. **36**, 536 (1968).

[4] G.J. Klir, Systems Res. **2**, 131 (1985).

[5] J. von Neumann, *"Theory of Self-Reproducing Automata"*, edited by A. Burks, University of Illinois Press, Urbana, Ill. (1966).

[6] A.N. Kolmogorov, Probl. Inf. Transm. **1**, 1 (1965).

[7] R.J. Solomonoff, Inf. Control **7**, 1 (1964).

[8] G. Chaitin, J. Assoc. Comp. Math. **13**, 547 (1966).

[9] L.P. Kadanoff, Physics Today, p. 9, March 1991.

[10] P.W. Anderson, Physics Today, p. 9, July 1991.

[11] P. Grassberger, Int. J. Theor. Phys. **25**, 907 (1986) and Wuppertal preprint B 89-26 (1989).

[12] *"Webster's New International Dictionary of the English Language"*, Merriam-Webster, Springfield, Mass. (1986);
*"The Oxford English Dictionary"*, Clarendon Press, Oxford (1989).

[13] R. Badii, Europhys. Lett. **13**, 599 (1990); Weizmann Inst. Preprint (1988); in *"Measures of Complexity and Chaos"*, p. 312 edited by N.B. Abraham et al., Plenum Press, New York (1990); R. Badii, M. Finardi and G. Broggi, in *"Information Dynamics"*, p. 35, edited by H. Atmanspacher et al., Plenum, New York (1991); R. Badii, M. Finardi and G. Broggi, in *"Chaos, Order and Patterns"*, edited by P. Cvitanović et al., Plenum Press, New York (1991).

[14] L. Löfgren, Int. J. General Systems **3**, 197 (1977) and in *"Systems and Control Encyclopedia"*, M. Singh Ed., Pergamon Press, Oxford (1987).

[15] J. Guckenheimer and P. Holmes, *"Nonlinear Oscillations, Dynamical Systems and Bifurcations of Vector Fields"*, Springer, New York (1986).

[16] J.-P. Eckmann and D. Ruelle, Rev. Mod. Phys. **57**, 617 (1985).

[17] A. Lasota and M.C. Mackey, *"Probabilistic Properties of Deterministic Systems"*, Cambridge University Press, Cambridge (1985).

[18] I.P. Cornfeld, S.V. Fomin and Ya.G. Sinai, *"Ergodic Theory"*, Springer, New York (1982).

[19] A.I. Khinchin, *"Information Theory"*, Dover, New York (1957).

[20] P. Grassberger and H. Kantz, Phys. Lett. **113A**, 235 (1985).

[21] P. Grassberger, H. Kantz and U. Moenig, J.Phys. **22A**, 5217 (1990).

[22] L. Flepp, R. Holzner, E. Brun, M. Finardi and R. Badii, Phys. Rev. Lett. (1991).

[23] D.K. Campbell, in [1].

[24] A.C. Newell, in [1].

[25] I. Procaccia and R. Zeitak, Phys. Rev. Lett. **60**, 2511 (1988).

[26] M.H. Jensen, in *"Information Dynamics*, edited by H. Atmanspacher et al., p. 103, Plenum, New York (1991).

[27] S. Kakutani, *"Selected Papers"*, Vol. 2, edited by R.R. Kallman, Birkhäuser, Boston (1986).

[28] A. Papoulis, *"Probability, Random Variables and Stochastic Processes"*, McGraw-Hill, Singapore (1984).

[29] K. Petersen, *"Ergodic Theory"*, Cambridge University Press, Cambridge (1983).

[30] *"Ergodic Theory, Symbolic Dynamics and Hyperbolic Spaces"*, edited by T. Bedford, M. Keane and C. Series, Oxford University Press, New York (1991).

[31] Ya.G. Sinai, Russ. Math. Surv. **27**, 21 (1972);
R. Bowen, Lecture Notes in Math. **470**, Springer, Berlin (1975);
D. Ruelle, *"Thermodynamic Formalism"*, Vol. 5 of *Encyclopedia of Mathematics and its Applications*, Addison-Wesley, Reading, Mass. (1978).

[32] T.C. Halsey, M.H. Jensen, L.P. Kadanoff, I. Procaccia and B. Shraiman, Phys. Rev. **A33**, 1141 (1986).

[33] R. Badii, Riv. Nuovo Cim. **12**, No 3, 1 (1989).

[34] Ya.G. Sinai, *"Theory of Phase Transitions: Rigorous Results"*, Pergamon Press, Oxford (1982).

[35] P. Grassberger, R. Badii and A. Politi, J. Stat. Phys. **51**, 135 (1988).

[36] R.L. Adler, A.G. Konheim and M.H. McAndrew, Trans. Amer. Math. Soc. **114**, 309 (1965).

[37] P. Paoli, A. Politi and R. Badii, Physica **D36**, 263 (1989).

[38] F.J. Dyson, Commun. Math. Phys. **12**, 91 (1969); **21**, 269 (1971).

[39] K. Huang, *"Statistical Mechanics"*, Wiley, New York (1987).

[40] S.K. Ma, *"Modern Theory of Critical Phenomena"*, Benjamin, New York (1975).

[41] A.S. Wightman, Introduction to *"Convexity in the Theory of Lattice Gases"* by R.B. Israel, Princeton University Press, Princeton (1979).

[42] T. Bohr and T. Tél, in *"Directions in Chaos"*, Hao Bai Lin Ed., World Scientific, Singapore (1988).

[43] (a) J.E. Hopcroft and J.D. Ullman, *"Introduction to Automata Theory, Languages and Computation"*, Addison-Wesley, Reading, Mass. (1979); (b) G. Rozenberg and A. Salomaa, *"The Mathematical Theory of L Systems"*, Academic Press, London (1980).

[44] B. Weiss, Monatshefte für Mathematik **77**, 462 (1973).

[45] R.L. Adler in [30].

[46] M. Keane in [30].

[47] M.J. Feigenbaum, J. Stat. Phys **19**, 25 (1978).

[48] The actual symbolic dynamics of the logistic equation at PD is described by a different transformation which is, however, completely equivalent to the more symmetric one described here (called Morse-Thue substitution and first considered in Ref. [49] in connection with PD).

[49] I. Procaccia, S. Thomae and C. Tresser, Phys. Rev. **A35**, 1884 (1987).

[50] M.J. Feigenbaum, L.P. Kadanoff and S.J. Shenker, Physica **5D**, 370 (1982).

[51] J.P. Crutchfield and K. Young in *"Complexity, Entropy and Physics of Information"*, W. Zurek Ed., Addison-Wesley, Reading, Mass. (1989).

[52] J.P. Crutchfield and K. Young, Phys.Rev.Lett. **63**, 105 (1989).

[53] D. Auerbach and I. Procaccia, Phys.Rev. **A41**, 6602 (1990).

[54] C.H. Bennett in [1].

[55] E.N. Lorenz, J. Atmos. Sci. **20**, 130 (1963).

[56] M. Droz, Z. Rácz and P. Tartaglia, Phys. Rev. **A41**, 6621 (1990).

[57] P. Häner and R. Schilling, Europhys. Lett. **8**, 129 (1989).

[58] R. Hamming, *"Coding and Information Theory"*, Prentice-Hall, Englewood Cliffs, NJ (1986).

[59] R. Artuso, E. Aurell and P. Cvitanović, Nonlinearity 3 325-359 and 361-386 (1990).

[60] M.A. Sepúlveda and R. Badii, in *"Measures of Complexity and Chaos"*, p. 257, edited by N.B. Abraham et al., Plenum Press, New York (1990); R. Badii, M. Finardi, G. Broggi and M.A. Sepúlveda, "Hierarchical resolution of power spectra", Physica **D**, to appear (1992).

[61] Z. Kovacs, this issue.

[62] M. Hénon, Comm.Math.Phys. **50**, 69 (1976).

[63] M.J. Feigenbaum, M.H. Jensen and I. Procaccia, Phys. Rev. Lett. **57**, 1503 (1986).

[64] R. Badii, "Quantitative characterization of complexity and predictability", Physics Letters **A**, to appear (1991).

[65] P.R. Halmos, Trans. Amer. Math. Soc. **55**, 1 (1944).

[66] A.B. Katok and A.M. Stepin, Soviet Math. Dokl. **6**, 1638 (1966) and Russ. Math. Surv. **22**, 77 (1967).

[67] S. Lloyd and H. Pagels, Ann. Phys. (N.Y.) **188**, 186 (1988).

[68] D. Hennequin and P. Glorieux, Europhys. Lett. **14**, 237 (1991).

[69] M.J. Feigenbaum, J. Stat. Phys. **52**, 527 (1988).

[70] P. Szépfalusy, T. Tél and G. Vattay, Phys. Rev. **A43**, 681 (1991); Z. Kovács and T. Tél, Eötvös preprint, Budapest (1991).

[71] M.H. Jensen, L.P. Kadanoff and I. Procaccia, Phys.Rev. **A36**, 1409 (1987).

[72] S.R. McKay, A.N. Berker and S. Kirkpatrick, Phys. Rev. Lett. **48**, 767 (1982); B. Derrida, J.-P. Eckmann and A. Erzan, J. Phys. **A16**, 893 (1983).

[73] N.M. Švrakić, J. Kertész and W. Selke, J. Phys. **A15**, L427 (1982).

[74] R. Badii and A. Politi *"Complexity, Hierarchical Structures and Scaling in Physics"*, to be published by Cambridge University Press, Cambridge (1993).

# ASYMPTOTIC STATE ESTIMATION USING OBSERVERS
# IN DYNAMICAL AND CONTROL SYSTEMS


I. Kupka

Department of Mathematics

University of Toronto

Toronto, Ontario M5S 1A1


## Introduction

In this paper I want to report on work done in collaboration with J.P. Gauthier and A. Hammouri in several publications. (see [GHK], [GK]).

## PART I

## DYNAMICAL SYSTEMS

**0) Position of the problem**: we are given a physical system $S$ whose evolution is governed by a smooth (i.e. infinitely differentiable or real analytic) dynamical system whose state space is a smooth manifold $X$ and whose dynamic $\frac{dx}{dt}(t) = F(x(t))$ is given by a smooth vector field $F$ on $X$. We perform measurements on this system whose result at time $t$ is represented by the value $h(x(t))$ at the state $x(t)$ of the system at time $t$, of a smooth function $h : X \to Y$ called the output function of the system. $Y$ is called the output space. The function $t \to h(x(t))$ is called the output of the trajectory $x(t)$.

The evolution of the system $S$ is represented by a path $\{x(t)|t \geq 0\}$ in the state space $X$. At any time $t$ we know the structure $F$ governing the evolution of the system and the information $\{h(x(s))|t_0 \leq s \leq t\}$ about $S$ collected in the time interval $(t_0, t)$. Our problem is to reconstruct the state $x(t)$ of the system $S$ at time $t$ from these data.

Such a problem splits into two:

**Problem 1**: do the data determine $x(t)$ uniquely?

**Problem 2**: construct an algorithm allowing us to compute $x(t)$ from the information $\{h(x(s))|t_0 \leq s \leq t\}$.

These are the questions we are going to study here.


*Chaotic Dynamics: Theory and Practice*
Edited by T. Bountis, Plenum Press, New York, 1992

## 1) Necessary and sufficient conditions for problem 1

An obvious and simple necessary and sufficient condition for the answer to problem 1 to be positive is that the couple $(F, h)$ be observable:

**Definition 1:** A couple $(F, h)$ is called observable if the following condition is satisfied: given any two trajectories $x_i : [0, e_i[ \to X, i = 1, 2, \quad e_1, e_2 > 0$, such that their initial states $x_1(0)$ and $x_2(0)$ are distinct, then for some time $t$, $0 \le t < min(e_1, e_2)$

$$h(x_1(t)) \neq h(x_2(t)).$$

In other terms: what we ask is that if two trajectories are distinct, the available informations about them are different.

## 2) Some heuristics about the second problem

Let us assume now that the pair $(F, h)$ is observable and that $Y = \mathbb{R}^c$. How would we go about computing $x(t)$ knowing $y(s) = h(x(s))$ for $0 \le s \le t$? An easy way is to consider the successive derivates $y(t)$, $\frac{dy(t)}{dt}, \ldots, \frac{d^n y}{dt^n}$ of the output $y$. Denoting the Lie derivative in the direction of $F$ by $\theta(F)$ and introducing the functions $h^n : X \to \mathbb{R}^c$, $h^0 = h$, $h^{n+1} = \theta(F)h^n$, i.e. $h = (\theta(F))^n h^0$ we have the relations:

$$(OB) \quad \begin{cases} y(t) & = & h^0(x(t)) \\ \frac{dy(t)}{dt} & = & h^1(x(t)) \\ \vdots & & \\ \frac{d^n y(t)}{dt^n} & = & h^n(x(t)) \\ \vdots & & \end{cases}$$

Under some generic assumptions, one can solve the system $(OB)$ for $x(t)$ and determine $x(t)$ as a function $\Phi\left(y(t), \frac{dy(t)}{dt}, \ldots, \frac{d^N y(t)}{dt^N}\right)$ of $y$ and its derivatives up to order $N$.

The problem with this kind of procedure is that $y(t)$ is not known exactly since it is colored by noise. Hence the real output $\tilde{y}(t)$ is not differentiable. Even if it were, its derivatives would be very different from those of $y$ and the estimation of $x(t)$ thus obtained would not be very good. Hence we need a different kind of procedure. This is supplied by the concept of observers.

## 3) Observers and asymptotic estimations

The procedure we shall describe does not give an exact estimate of $x(t)$ but rather an asymptotic one in the sense that the longer the time span the better the approximation. We define an observer for the pair $(F, h)$:

**Definition 2:** An observer for the pair $(F, h)$ is:
  (i) a control system driven by the output of $S$ defined on a state space $Z$ (smooth manifold) and whose dynamics is represented by

$$\frac{dz}{dt} = G(z, y) \qquad z \in Z \quad y \in Y,$$

$G$ being a smooth vector field on $Z$ parametrized by $Y$, the dependence on both $y$ and $z$ being smooth.

(ii) an imbedding $\tau : \ X \to Z$ of the state space of the system $S$ into the state space of the observer.

(iii) a Riemannian metric $d_Z$ on $Z$ having the following property: given *any* trajectory $\hat{x} : \ [0, +\infty[ \ \to \ X$ of $S$, *any* trajectory $z : \ [0, +\infty[ \ \to \ Z$ of the time-dependent system

$$\frac{dz}{dt}(t) = G(z(t), \hat{y}(t)) \text{ where } \hat{y}(t) = h(\hat{x}(t))$$

satisfies the condition

$$d_Z(z(t), \ \tau(\hat{x}(t))) \to 0 \quad \text{as} \quad t \to +\infty.$$

Usually one asks for more namely that there exist constants $\alpha > 0$, $C > 0$ such that

$$d_Z(z(t), \ \tau(\hat{x}(t))) \leq Ce^{-\alpha t}.$$

A very nice feature of this concept is that generically a pair $(F, h)$ is observable and has an observer.

## 4) The case of a linear system

To illustrate the concepts introduced above, let us discuss presently the linear case which has been known for a long time. Here $X = \mathbb{R}^d$, $F(x) = Ax$, $h(x) = Cx$ where $A : \ \mathbb{R}^d \to \mathbb{R}^d$ and $C : \mathbb{R}^d \to \mathbb{R}^c$ are linear mappings. The observability of the pair $(F, h)$ is equivalent to

$$\bigcap_{n=0}^{d-1} Ker C \circ A^n = 0.$$

This also implies the pole placement property: for any number $\alpha > 0$ there exists a linear mapping $K_\alpha : \ \mathbb{R}^c \to \mathbb{R}^d$ such that all the eigenvalues of the endomorphism $A - K_\alpha C : \mathbb{R}^d \to \mathbb{R}^d$ have their real parts $< -\alpha$.

This in turn implies by the Theorem of Lyapounov that there exists a symmetric positive-definite linear mapping: $S_\alpha : \ \mathbb{R}^d \to \mathbb{R}^d$ such that

$$S_\alpha(A - K_\alpha C) + (A - K_\alpha C)^* S_\alpha = -2\alpha S_\alpha - Id.$$

Now an observer for our system can be obtained as follows: $Z = X = \mathbb{R}^d$, $\tau = Id,$

$$G(z, y) = (A - K_\alpha C)z + K_\alpha y$$

$$d_Z(z_1, z_2) = \sqrt{< S_\alpha(z_1 - z_2), \ z_1 - z_2 >}.$$

It is easy to see that we get the following estimate in this case:

$$d_Z(z(t), \ \hat{x}(t)) \leq e^{-\alpha t} d_Z(z(0), x(0)).$$

## 5) The General Case

We are going to state a general result which solves both problems 1 and 2 for "generic" systems. We shall make the following assumptions

**AS 1**: the state space $X$ is compact.

**AS 2**: $Y = \mathbb{R}$.

On the function spaces $C(X)$ (the smooth functions on $X$) and $VF(X)$ (the smooth vector fields on $X$) we put the usual Frechet space topologies. Then we can state the theorem:

**Theorem 1.** *The set of all couples $(F, h)$ in $VF(X) \times C(X)$ which are observable and admit an observer, is open and dense.*

We are not going to prove this theorem but rather describe the observer. As its state space $Z$ we take $\mathbb{R}^{2d+1}$ where $d = dimX$ and the imbedding $\tau : X \to Z$ will be the mapping $x \in X \to (h_0(x), h_1(x), \ldots, h_{2d+1}(x))$. The proof shows that for an open dense set of couples $(F, h)$, this mapping is really an imbedding. Then it can be shown that the image $\tau_*(F)$ of $F$ by $\tau$ is the restriction to $\tau(X)$ of a field $\widetilde{F}$ on $Z$ of the form:

$$\widetilde{F} = \sum_{i=0}^{2d-1} z^{i+1} \frac{\partial}{\partial z^i} + f(z) \frac{\partial}{\partial z^{2d}}$$

where $z^0, \ldots, z^{2d} : Z \to \mathbb{R}$ are the canonical coordinates on $Z = \mathbb{R}^{1+2d}$ and $f : Z \to \mathbb{R}$ is a smooth function with compact support. The restriction of $f$ to $\tau(X)$ is just $h_{2d+1} \circ \tau^{-1}$ and $h \circ \tau^{-1}$ is the restriction of the coordinate $z^0$ to $\tau(X)$. In other words, the system $(F, h)$ considered on $\tau(X)$, is the restriction to $\tau(X)$ of the following system on $Z$:

$$(\text{EX}) \quad \frac{dz}{dt} = Az + f(z)e_{2d}.$$

with the output function $k = z^0$.

$e_{2d}$ is the constant vector field on $Z$ with value $(0, 0, \ldots, 1)$ and $A : Z \to Z$ the linear mapping $A^*(z^i) = z^{i+1}$, $0 \leq i \leq 2d - 1$, $A^*(z^{2d}) = 0$. $(A^*(z^i) = z^i \circ A)$.

The system $EX$ above can be considered as a linear system $\frac{dz}{dt} = Az$ with a nonlinear perturbation $f(z)e_{2d}$. Since $k_i = z^i$, $0 \leq i \leq 2d - 1$, in this case, the linear system $Az$ with the output function $k = z^0$ is observable and we can apply what we have discussed in §4: for any $\alpha > 0$, there exists a vector $K_\alpha \in Z$ (here $Y = \mathbb{R}$, so a linear mapping $\lambda : \mathbb{R} \to Z$ is characterized by the vector $\lambda(1)$ image of 1) such that the eigenvalues of the endomorphism $A - K_\alpha \otimes z^0$ of $Z$ have their real parts $< -\alpha$.

Let $\theta$ be a positive parameter. Define $K_\alpha(\theta)$ as the vector in $Z$ having as components $K_\alpha^0, \theta K_{\alpha^1}, (\theta)^2 K_\alpha^2, \ldots, (\theta)^{2d} K_\alpha^{2d}$. Then our observer for $(F, h)$ has $Z$ as state space, $\tau : X \to Z$ as imbedding and the following equation as dynamics:

$$\frac{dz}{dt} = Az - z^0 K_\alpha(\theta) + y K_\alpha(\theta) + f(z)e_{2d}.$$

The metric $d_Z$ will be as in §4:

$$d_Z(z_1, z_2) \;=\; \sqrt{< S_\alpha(z_1 - z_2),\; z_1 - z_2 >}.$$

We can show that taking $\theta$ sufficiently large to kill the effect of the nonlinearity of the original system, the system just defined is indeed an observer for $(F, h)$.

In the second part of this paper we turn our attention to control systems.

## PART II

## CONTROL SYSTEMS

**0) Introduction**: Now we assume that the evolution of our system $S$ can be controlled. The controls are represented by functions $u$ of a time interval $[a, b]$ into a subset $U$ of a euclidean space $\mathbb{R}^m$, called control space, which are measurable and bounded. The evolution $x(t)$ of $S$ corresponding to a control $u(t)$ is governed by the smooth dynamics

$$\frac{dx(t)}{dt} \;=\; F(x(t),\; u(t))$$

where $F$ is a smooth vector field on $X$, parametrized by $U$. By a trajectory of the system we shall denote a couple of curves $(x, u) : [a, b] \to X \times U$ such that $u$ is measurable and bounded, $x$ is absolutely continuous and:

$$\frac{dx(t)}{dt} \;=\; F(x(t),\; u(t)) \quad \text{for almost every } t \in [a, b].$$

As for dynamical systems, the measurements we perform on the system $S$ are represented at time $t$, by the value $h(x(t), u(t))$ on the state $x(t)$ and control $u(t)$ of $S$ at time $t$, of a smooth function $h : X \times U \to Y$. $Y$ is a manifold and is called the output space. $h$ is called the output function. The function $t \to h(x(t), u(t))$ is called the output of the trajectory $(x, u)$.

The statement of the estimation problem is similar: given a trajectory $(x, u) : [a, b] \to X \times U$ of the system, from the knowledge of the input $u(s)$ and of the output $y(s) = h(x(s),\; u(s))$ for a time interval: $t_0 \leq s \leq t$, reconstruct the trajectory $(x(t),\; u(t))$.

As in the case of dynamical systems, we assume that we know the law $F$ governing the evolution of $S$.

Again the problem stated above splits into two:

**Problem 1:** does the input-output couple $\{(u(s), y(s)) | \; t_0 \leq s \leq t\}$ determine $x(t)$ uniquely?

**Problem 2:** construct an algorithm computing $x(t)$ from $\{(u(s),\; y(s)) | \; t_0 \leq s \leq t\}$.

A necessary and sufficient condition for observability states:
given any $L^\infty$ control function, $u : [0, T[ \to U$ and any two trajectories $(x_i, u) : [0, e_i[ \to X \times U$, $i = 1, 2$ $(e_i \leq T \; i = 1, 2)$ of the system, the set of all $t \in [0, \; \min(e_1, e_2)[$ such that $h(x_1(t), u(t)) \neq h(x_2(t),\; u(t))$, has positive Lebesgue measure.

Now the new fact is that this necessary and sufficient condition is not generic contrary to what happens in the dynamical system case. In other terms a control system is not observable in general. This raises the question of characterizing those systems which are observable.

## 1) Infinitesimal Observability

In the linear case, i.e. $X = \mathbb{R}^d$, $Y = \mathbb{R}^c$ $U = \mathbb{R}^m$ $F(x, u) = Ax + Bu$ $A : \mathbb{R}^d \to \mathbb{R}^d$, $B : \mathbb{R}^m \to \mathbb{R}^d$ both linear and $h(x, u) = Cx + Du$ $C : \mathbb{R}^d \to \mathbb{R}^c$ $D : \mathbb{R}^m \to \mathbb{R}^c$ everything is the same as if there were no controls i.e. the system is observable if and only if $\bigcap_{n=0}^{d-1} Ker\, CA^n = 0$ and the observer is similar. Observability has also been characterized for other classes of systems. (see [W])

In the general case the concept of observability is unstable under perturbations of the system. This has led us to introduce a seemingly weaker condition of observability, which enabled us to obtain fairly complete results about problem 1 and 2. This new concept is infinitesimal observability.

Before giving the definition we have to introduce the lifting of a system $S$. The control space of the lifting $LS$ is the same $U$ as for $S$, its state space the tangent space $TX$ of $X$, its dynamics: $\frac{d\xi}{dt} = T_X F(\xi, u)$ where $T_X F : TX \times U \to TTX$ (tangent space of $TX$) is the tangent mapping of $F : X \times U \to TX$ and its output mapping is $T_X h : TX \times U \to TY$ the tangent mapping of $h : X \times U \to Y$. ($u$ is considered as a parameter).

**Definition 3:** A system $S$ is called infinitesimally observable if any trajectory $(\xi, u) : [0, T[ \to TX \times U$ of the lifting $LS$ of $S$ whose output $T_X h(\xi, u)$ is zero for almost every $t \in [0, T[$ is itself the zero trajectory, i.e.

$$T_X h(\xi(t),\ u(t)) = O_{y(t)} \text{ in } T_{y(t)}Y \text{ for almost every } t \in [0, T[$$

$(y(t) = \pi_Y T_X h(\xi(t), u(t)),\ \pi_Y : TY \to Y$ canonical projection) implies that:

$$\xi(t) = O_{x(t)} \text{ in } T_{x(t)}X \text{ for every } t \in [0, T]$$

$(x(t) = \pi_X(\xi(t)),\ \ \pi_X : TX \to X$ canonical projection).

**Remark 1:** This condition is not generic for the system $S$, i.e. very few systems satisfy it.

**Remark 2:** There are very intimate relations between observability and infinitesimal observability. But we shall not go into this topic here.

## 2) Characterization of Infinitesimal Observability

From now on, we shall assume that the output space $Y$ is the real number line $\mathbb{R}$. Before stating the main theorem, we need a few notations. For each integer $n \geq 0$, let $h^n : X \times U \to \mathbb{R}$ denote the $n^{\text{th}}$ iterated Lie derivative:

$$h_u^n = \theta(F_u)^n h_u \quad h^0 = h.$$

These functions define a flag of distributions $\mathcal{D}_n(u)$, $n \geq 0$, for each $u \in U$:

$$\mathcal{D}_n(u) = \bigcap_{k=0}^{n} Ker \, d_X h_u^k \subset TX$$

(i.e. at each $x \in X$  $\mathcal{D}_n(u)_x = \bigcap_{k=0}^{n} Ker \, d_X h^k(x, u) \subset T_x X$). Clearly $\mathcal{D}_0(u) \supset \mathcal{D}_1(u) \supset \mathcal{D}_2(u) \supset \cdots$.

Finally we introduce the singular sets $\widetilde{M}$ and $M$:

$$\widetilde{M} \subset X \times U, \quad M \subset X, \quad d = dim X$$
$$\widetilde{M} = \{(x, u) | d_X h^0(x, u) \wedge \ldots \wedge d_X h^{d-1}(x, u) = 0\}$$
$$M = \text{projection of } \widetilde{M} \text{ into X.}$$

## Main Theorem: Geometric Form

Assume that:

(i) $Y = \mathbb{R}$

(ii) $U$ is a compact connected subanalytic set (resp. $U$ is an affine subvariety of some $\mathbb{R}^N$)

(iii) $X, F, h$ are real analytic (resp. $X, F, h$ are real analytic and for any $x \in X$, both mappings: $u \in U \to F(x, u) \in T_x X$, $u \in U \to h(x, u) \in \mathbb{R}$ are polynomial).

(iv) the system $(F, h)$ is infinitesimally observable, then:

1) $M$ is a closed subanalytic subset of $X$ (resp. a semi-analytic subset of $X$) of codimension *at least 1*. Call $\overline{M}$ the closure of $M$ (in the first alternative $\overline{M} = M$).

2) On $X - \overline{M}$, the distribution $\mathcal{D}_n(u)$ is independent of $u \in U$ and has constant rank equal to $d - n - 1$, $0 \leq n \leq d - 1$.

## Main Theorem: Algebraic Form

Under the same assumptions as in the geometric form we have:

1) same statement as in the geometric version

2) for any $a \in X - \overline{M}$, any $v \in U$ there exists an open neighborhood $V_a \subset X - \overline{M}$ of $a$ such that the restrictions $x^0 = h_v^0 | V_a, \ldots, x^{d-1} = h_v^{d-1} | V_a$ of $h^0, h^1, \ldots, h^{d-1}$ to $V_a$ form a coordinate system on $V_a$ and on $V_a \times U$, $h^n$ is a function of $u, x^0, \ldots, x^n$ only for $0 \leq n \leq d - 1$:

$$h^n = H^n(u, x^0, \ldots, x^n) \quad 0 \leq n \leq d - 1$$

3) everywhere on $V_a \times U$

$$\frac{\partial H^n}{\partial x^n} \neq 0 \text{ for } 0 \leq n \leq d - 1.$$

The preceding theorem shows that in $V_a \times U$ the dynamics of the system have the following form:

$$\frac{dx^0}{dt} = F^0(u, x^0, x^1), \ldots, \frac{dx^n}{dt} = F^n(u, x^0, x^1, \ldots, x^{n+1})$$

$$\frac{dx^{d-1}}{dt} = F^{d-1}(u, x^0 \ldots x^{d-1})$$

and the output is: $y = H^0(u, x^0)$.

Moreover: $\frac{\partial H^0}{\partial x^0}$, $\frac{\partial F^n}{\partial x^{n+1}}$, $\quad 0 \leq n \leq d-2$, never take the value 0.

**Remark 3:** If we drop the compactness assumption about $U$ in the first assumption (iii) or the polynomiality assumption about $F$ and $h$ in the second assumption (iii), the theorem is false.

The preceding theorem goes a long way toward solving the problem of the characterization of infinitesimally observable systems. What remains to be done is to study the behaviour of the system at the points of the bad set $M$.

Now we turn our attention to the second problem that of constructing an observer.

### 3) Observers for observable systems

Due to our ignorance of the behaviour of the system at the points of the bad set $M$ we can only construct observers on open relatively compact subsets of $X$ contained in $X - \overline{M}$.

**Theorem 2: Existence of Observers**

Assume that $X$ carries a global coordinate system $x^0, \ldots, x^{d-1} : X \to \mathbb{R}$ such that the dynamics of the system can be expressed as follows:

$$\frac{dx^n}{dt} = F^i(u, x^0, \ldots, x^{n+1}) \quad 0 \leq n \leq d-2$$

$$\frac{dx^{d-1}}{dt} = F^{d-1}(u, x^0, \ldots, x^{d-1})$$

and the output function $h$ is a function $H(u, x^0)$ of $u$ and $x^0$ only, where $H$ and the $F^i$ satisfy the following conditions.

(i) They are globally Lipschitz in $x$ on $U \times X$.

(ii) There exist constants $A, B, \quad B > A > 0$ such that on $X \times U :$ $A \leq \left| \frac{\partial H}{\partial x^0} \right| \leq B$ $A \leq \left| \frac{\partial F^n}{\partial x^{n+1}} \right| \leq B \quad 0 \leq n \leq d-2.$

Then for each $\alpha > 0$, the system admits an exponential asymptotic observer having $X$ as state space and $\alpha$ as time constant.

Now we shall exhibit the observer. Its construction is based on the following lemma used by Dayawanza in stabilization theory.

**Lemma 1(Dayawanza).** *Let $A, B$ be two constants $B > A > 0$. Let $\Sigma(A, B, m+1)$ denote the class of all time dependent linear dynamical systems on $\mathbb{R}^m$ of the form:*

$$\begin{cases} \frac{dx^i}{dt} = a_i(t)x^{i+1}(t) \quad 0 \leq i \leq m-1 \\[2mm] \frac{dx^m}{dt} = 0 \end{cases}$$

with time dependent output function $h(t, x) = c_0(t)x^0(t)$. The $a_0, \ldots, a_{m-1}, c$ are measurable, essentially bounded real-valued functions defined on $[e, +\infty[$ satisfying the following boundedness conditions: for almost every $t \in [e, +\infty[$,

$$A \leq a_n(t) \leq B, \quad 0 \leq n \leq m - 1$$
$$A \leq c(t) \leq B.$$

Then for any $\alpha > 0$ there exist a vector $K_\alpha \in \mathbb{R}$ and a $m+1 \times m+1$ symmetric positive-definite matrix $S_\alpha$, $K_\alpha$ and $S_\alpha$ depending *only on $A$ and $B$* such that for any system $(a_0, \ldots, a_{m-1}, c)$ in $\Sigma(A, B, m+1)$ and solution $x : [0, +\infty[ \to \mathbb{R}^{m+1}$ of the system:

$$\begin{cases} \frac{dx^n}{dt}(t) = & a_n x^{n+1}(t) - c(t) K_\alpha^n x^0(t) \quad 0 \leq n \leq m-1 \\[2mm] \frac{dx^m}{dt}(t) = & -c(t) K_\alpha^m x^0(t) \\ & (K_\alpha^0, \ldots, K_\alpha^m \text{ are the components of } K_\alpha) \end{cases}$$

decays exponentially to 0 with time constant $\alpha$ in the $S_\alpha$ - norm: let $\|X\|_\alpha = \sqrt{<S_\alpha X, X>}$ for $X \in \mathbb{R}^{m+1}$

$$\|x(t)\|_\alpha \leq e^{-\alpha t} \|x(0)\|_\alpha \quad \text{for all} \quad t \geq 0.$$

Now for any positive value of the parameter $\theta$ we define a system $OS(\theta)$ having $U \times \mathbb{R}$ as control space, $X$ as state space and the following dynamic:

$$\frac{dz}{dt}(t) = F(z(t), u(t)) - [h(z(t), u(t)) - y(t)]K_\alpha(\theta)$$

where $(u, y) \in U \times \mathbb{R}$ is the control, $K_\alpha(\theta) \in \mathbb{R}^d$ is the vector whose components are $(K_\alpha^0, \theta K_\alpha^1, (\theta)^2 K_\alpha^2, \ldots, (\theta)^{d-1} K_\alpha^{d-1})$, $K_\alpha = (K^0, \ldots, K_\alpha^{d-1}) \in \mathbb{R}^d$ being the vector associated to the class $\Sigma(A, B, d)$ in the Lemma 1 above and $A$ and $B$ having the values stated in Theorem 2.

One can show that there exists a function $\theta_0(\alpha, L)$ of the parameters $\alpha > 0$, $L > 0$, such that if $\theta \geq \theta_0(\alpha, L)$, where $L$ is a Lipshitz constant for $H$ and the $F^n$, $0 \leq n \leq d - 1$, the system $OS(\theta)$ is an exponentially asymptotic observer for the system $S$ with time constant $\alpha$. The metric $d_Z$ associated to this observer is: $d_Z(z_1, z_2) = \sqrt{<S_\alpha(z_1 - z_2), z_1 - z_2>}$, $S_\alpha$ being the matrix that Lemma 1 associates to the values of $A, B$ given in Theorem 2.

## CONCLUSIONS

**1) What remains to be done?** The main limitation of our study is that it deals only with single output systems. It is very important to extend these results to multi-output systems. In that direction very interesting results have been obtained by Krener and Respondek. (see [K-R]) But they are restricted to linearizable systems and this limits their scope.

**2) Robustness:** The observers constructed above are robust with respect to noises with bounded variance. Then the variance of the observer is uniformly bounded in time.

These observers are also robust with respect to variations of the structural constants in the dynamic $F$ or the output function $h$.

For the proof of these fact we refer the reader to the publication [DG].

**3) Applications:** We have applied our observers to the stabilization of control
systems in the publication [GK].

On the practical side these observers have been applied to biological reactors
(see [GHO]) and to distillation columns (see [DG]). In this last case the results
presented here do not apply as they stand since the systems considered there have
several outputs. Nonetheless the methods introduced in this paper can be applied
to construct a very efficient observer in this last case too.

## REFERENCES


[BZ]     D. Bestle, M. Zeitz, Canonical form design for nonlinear observers with
linearizable error dynamics; Int. J. Control, 23, 1981, 419-413.

[D]     F. Deza, PhD. Thesis. INSA de Rouen, France, 1991.

[DA]     W. Dayawanza, personal communication.

[DG]     F. Deza, J.P. Gauthier, Observers for nonlinear systems and applications
to distillation columns; to appear in Chemical Engineering Science.

[GB]     J.P. Gauthier, G. Bornard, Observability for any $u(t)$ of a class of nonlinear
systems; IEEE Trans. Aut. Control, 26, 1981, 922-926.

[GHK]     J.P. Gauthier, H. Hammouri, I. Kupka, Observers for nonlinear systems;
to appear at IEEE CDC Conference, December, 1991, Brighton England.

[GHO]     J.P. Gauthier, H. Hammouri, S. Othman, A simple observer for nonlinear
systems, application to bioreactors; to appear IEEE Trans. Aut.
Control, 1991.

[GK]     J.P. Gauthier, I. Kupka "Separation principle for bilinear systems
with dissipative drift" to appear in I.E.E.E. Trans. Aut. Control.

[H]     H. Hironaka, Subanalytic Sets, Number Theory, Algebraic Geometry and
Commutative Algebra, in honor of Y. Akizuki; Kinokuniya, Tokyo, 1973,
453-493.

[HG1]     H. Hammouri, J.P. Gauthier, Bilinearization up to output injection;
Syst. and Control letters, 11, 1988, 139-149.

[HG2]     H. Hammouri, J.P. Gauthier, Global time varying linearization up to output
injection; to appear, SIAM Journal of Control.

[KI]     A. Krener, A. Isidori, Linearization by output injection and nonlinear
observers; Syst. and Control letters, 3, 1983, 47-52.

[KR]     A. Krener, W. Respondek, Nonlinear observers with linearizable error
dynamics; SIAM J. on Control and Optimization, 23, 1985, 197-216.

[L]     S. Lojasievicz, Triangulation of semi analytic sets; Annal. Sc. Nor. Sup.
PISA, 1965, 449-474.

[LU]     D.G. Luenberger, Observers for multivariable systems; IEEE Trans. Aut.
Control 11, 1966, 190-197.

[NA]     R. Narasimhan, Introduction to the theory of analytic spaces, Springer Verlag,
Lecture Notes in Mathematics 25, 1966.

[NI]     H. Nijmeijer, Observability of a class of nonlinear systems, a geometric
approach; Report of University of Twente, 1982.

[NTT]   S. Nicosia, P. Tomei, A. Tornambe, A nonlinear observer for elastic robots; IEEE J. of Robotics Automat, Vol. RA-4, 45-52, 1988.

[T]     A. Tornambe, Use of asymptotic observers having high gains in the state and parameter estimation; 28th IEEE CDC Conference, December, 13-15, 1989, Tampa, Florida, USA.

[W]     D. Williamson, Observability of bilinear systems, with applications to biological control; Automatica, 13, 1977, 243-254.

# LEARNING OPTIMAL REPRESENTATIONS

Joseph L. Breeden and Norman H. Packard

Center for Complex Systems Research - Beckman Institute
and the Physics Department, University of Illinois
405 North Mathews Avenue, Urbana, IL. 61801, USA

## 1. INTRODUCTION

In the analysis of most experimental time series, the question of how to best represent the data has always had an obvious answer – as a time series. However, with the discovery of chaotic attractors, we have learned that the data from many experimental systems can be better represented in a reconstructed state space [1]. This reveals features of the dynamics that were not apparent in the original time series. These state space reconstructions may simply be considered a generalization of the phase space plots used in classical mechanics.

With the realization that state space reconstructions can be beneficial to data analysis, the question of which representation is best becomes preeminent. Work by Takens [2] has shown that if the dynamics are on an $m$-dimensional manifold, an embedding of the system can be obtained with a $2m + 1$-dimensional reconstructed state space using derivatives or delay coordinates. These theorems have been extended [3] to account for noise and fractals, but they state only that embeddings may be obtained rather than which coordinates are best. In fact, different reconstructions of experimental data will not be equivalent for most goals. This early work is in part responsible for the current emphasis upon obtaining embeddings of the dynamics of experimental systems.

We take a more general view of what an *optimal representation* should be. Prior to choosing a representation for the data, one should determine the ultimate goal. It is unlikely that any single representation will be optimal for all possible objectives. For instance, minimizing mutual information [4] selects representations that are well spread in the state space and thus advantageous for calculating certain invariants of the dynamics: dimension, entropy, Lyapunov exponents, etc. Mutual information gives a good representation of the current state of the system; however, this representation will not be optimal for doing forecasting [5] or noise reduction [6] where optimization is done with respect to the future state of the system. Minimal average noise reduction may itself be used as a criterion, as suggested by Farmer and Sidorowich [7]. Another possibility might employ an Occam's razor criterion to select the representation from which the simplest model of the dynamics can be constructed.

Model-based control of one variable of a dynamical system [8] is another interesting possibility, because the stability of the control is determined within the reconstructed

---

*Email address: [breeden,n]@complex.ccsr.uiuc.edu

state space. This means that even with ideal data, not all representations are acceptable. Thus Takens' assurances of topological equivalence are insufficient to guarantee a useful state space reconstruction.

Another important aspect of this technique is the concept of generalized coordinates. State space reconstructions are almost always made using derivatives $\{x(t), \dot{x}(t), \ldots\}$ or delay coordinates $\{x(t), x(t - \tau_1), \ldots\}$ [1]. These are by no means the only possibilities. In fact, any functions on the original time series that incorporate information about the state of the system at different times should be considered as candidates for reconstruction coordinates, such as moving integration windows, moving averages, etc. In some situations, the optimal reconstruction may employ a mixture of different types of coordinates.

We propose a technique for searching through the space of possible coordinates to find the representation that best suits the goals of the experimenter. Possible coordinates for reconstruction, goals to reconstruction, and appropriate quality functions to suit these are discussed. We have developed a learning algorithm to find these optimal representations. Examples of the application of these ideas are given for the Hénon and Ikeda maps, and the Rössler and Mackey-Glass systems.

## 2. METHODOLOGY

We consider the situation of an experimenter who has made a sequence of measurements of a physical system providing a single time series of fixed length. We have developed a general procedure for searching the space of possible dimensions and coordinates with which to reconstruct the state space. This search attempts to optimize a quality function, relating to the experimenter's objectives, which is designed to place a numerical value upon the usefulness of a given trial reconstruction. We have developed a learning algorithm based upon the genetic algorithm [9,10,11] to search for optimal representations because of the potential complexity of the fitness landscape being searched. Since we are searching through dimensions, the dimensionality of the quality landscape is not even known. The natural manner in which this problem can be cast in the context of the genetic algorithm is also an advantage. A search is necessary, because situations arise in which the best D-dimensional representation is not a subspace of the best (D+1)-dimensional representation [12]. In such cases, building a representation by sequentially adding new coordinates will be ineffective.

Once we have decided upon the space of coordinates from which we wish to choose, we need a functional criterion for defining an optimal representation. This criterion may be either globally or locally defined. A globally defined quality function arises for "equations of motion" modelling [13,14,15,16,17], $\dot{x} = f(x; c)$. In this case the quality of a representation can be defined as a function of both $N_c$, the number of free coefficients $c$, and $\eta$, the variance of the model with respect to the data. Thus, the best representation would be one that gives both a simple and accurate model.

Locally defined quality functions are applied to small regions of state space and then combine for the overall quality. For example, local linear maps can be fit to the state space locally and the predictive variances summed. To accomplish this, we select neighborhoods from the state space for evaluation. In some cases it is advantageous to use k-D trees [18]. A k-D tree is a tree-like data structure which adapts to the distribution of the data points in the state space. We use a recursive algorithm that scans along each axis to find the best location for a partition such that the quality over the two resulting bins is greater than that of the original bin. This gives enormously improved results over a simple square grid, but at a computational cost. The quality for the representation is obtained by combining the qualities of the final bins.

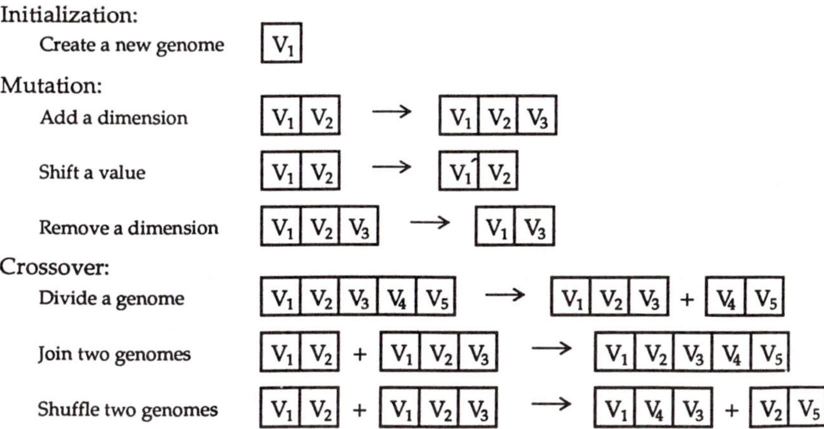



Figure 1. The operators used for our genetic algorithm-based search are shown. A particular representation is encoded by a "genome", where each "gene" specifies a choice for one coordinate. The *mutation* operators move a genome a small distance in the quality landscape, while the *crossover* operators can generate large moves in the quality landscape. The divide and join operators help in searching quality landscapes with high symmetry. The shuffle operator is the equivalent of the crossover operator of standard genetic algorithm techniques.

Having decided upon a quality function with which to compare representations, we choose the space of coordinates over which to search and implement our search algorithm. For this, a representation will be described by a genome with each gene identifying a coordinate. For example, $\{V_1, V_2, V_3\}$ is a 4-dimensional representation (including $x(t)$) where $V_i$ indicates how the $i^{th}$ coordinate is generated from $\{x(t)\}$: for a derivative, $V_i$ is the order of the derivative; for a delay, $V_i$ is the time delay; for a smoothed coordinate, $V_i$ is the size of the averaging window; etc.

The search begins by constructing an initial random population of genomes. These representations are evaluated according to the quality function and ranked. The highest quality members of the population are mutated using the operators defined in figure 1. These new members are then evaluated, ranked, and so forth.

This iterative procedure of mutation, evaluation, and ranking continues until the population converges. Finally, the highest quality member of the population is used to obtain the necessary coordinates for an optimal representation of the data. Note that at no point have we given a minimum number of data points or maximum noise level permitted for this procedure to succeed. That is because this technique finds the optimum representation for the data at hand. If the dynamics are on an attractor but less than a few orbits of the attractor are observed, then the resulting reconstruction may be uninteresting regardless of the number of observations made. This is characteristic of all nonlinear analysis techniques.

## 3. EXAMPLES

We demonstrate our search through representations (both coordinates and dimensions) via numerical studies of the Hénon and Ikeda maps and the Rössler and Mackey-Glass systems. These examples are purely illustrative and in no way an exhaustive demonstration of the possibilities.

Our first example demonstrates a search for the *inferable* dimension, $d_i$, of a data set [19]; the minimal embedding dimension which can be inferred from the available time series. We consider a neighborhood of points in a $D$-dimensional reconstructed state space at time $n$ and compute the volume of the smallest ellipsoid containing these points at times $n$ and $n+1$, $V^n$ and $V^{n+1}$ respectively. Performing this calculation on $N$ neighborhoods, we compute the average divergence rate over the attractor as

$$\zeta = \prod_{i=1}^{N} \left( \frac{V^{n+1}}{V^n} \right)^{-N \cdot D} \tag{3.1}$$

By plotting $\zeta$ as a function of $D$, we observe a power law behavior for large $D$. The largest dimension for which $\zeta$ is above this power law determines $d_i$. Note that this procedure is different from most dimension calculations because we examine the dynamics indicated by the data and not just the static distribution. This helps us distinguish between noise and dynamics.

We discuss this example because, in general, a search over representations for a fixed $D$ is necessary to compute the minimal value of $\zeta$. For the Hénon map (figure 2a),

$$\begin{aligned} x_{n+1} &= y_n - 1.4x_n^2 + 1 \\ y_{n+1} &= 0.3x_n \end{aligned}$$

and the Ikeda map (figure 2b),

$$\begin{aligned} s &= 0.4 - 6/(1 + x_n^2 + y_n^2) \\ x_{n+1} &= 1 + 0.9 \left( x_n \cos\left(s\right) - y_n \sin\left(s\right) \right) \\ y_{n+1} &= 0.9 \left( x_n \sin\left(s\right) + y_n \cos\left(s\right) \right) \end{aligned}$$

as with all maps $\mathbf{x}^{n+1} = f\left(\mathbf{x}^n\right)$, the best $D$-dimensional representation is

$$\zeta = \left\{ x^n, x^{n-\tau_1}, x^{n-\tau_2}, \ldots, x^{n-\tau_{D-1}} \right\} \tag{3.2}$$

where $\tau_j = j$. For the Rössler system [20] (figure 2c),

$$\begin{aligned} \dot{x} &= -y - z \\ \dot{y} &= x + ay \\ \dot{z} &= b + (x - c)z \end{aligned}$$

and continuous systems generally, $\tau_j = j\tau$ is not adequate to minimize $\zeta$.

After minimizing $\zeta$ for each $D$, the results were collected and $d_i$ determined, figure 3. Since these are noiseless examples, $d_i = m$ for the Hénon map and Rössler system. However, for the Ikeda map, $d_i = 4$ whereas $m = 2$. This is the expected result for delay coordinates because of the problems discussed by Takens. The inferable dimension is discussed in detail by Breeden and Packard [19] with further examples.

In our last example we consider the Mackey-Glass system,

$$\dot{x} = -\gamma x + \beta x_\tau \frac{\theta^n}{\theta^n + x_\tau^n} \tag{3.3}$$

which is a model for the white blood cell population in humans [21]. For this study, we have used $\gamma = 0.1$, $\beta = 0.2$, $\theta = 1$, $n = 10$, and $\tau = 30$. Here we compare the optimal two dimensional reconstructions. We restrict our attention to two dimensional reconstructions simply for purposes of illustration. In figure 4 we show the quality as a function of delay for three different optimization criteria: 1. Mutual information [4,22],

$$M\left(\tau\right) = \sum_{partitions} p\left(x\left(t\right), x\left(t+\tau\right)\right) \log \frac{p\left(x\left(t\right), x\left(t+\tau\right)\right)}{p\left(x\left(t\right)\right) p\left(x\left(t+\tau\right)\right)} \tag{3.4}$$

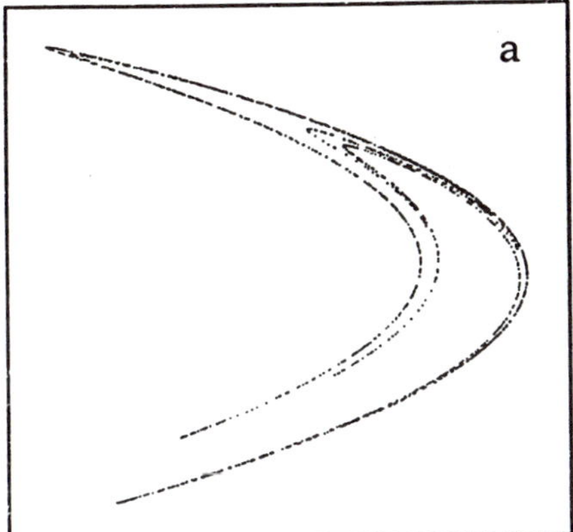
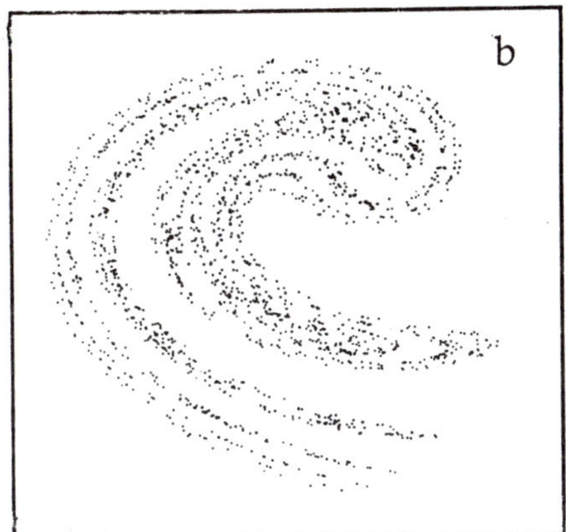
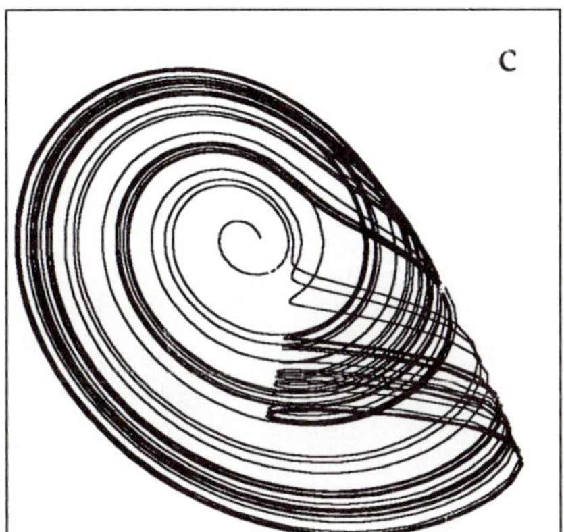

Figure 2. The Hénon map, Ikeda map, and Rössler system are shown in (a), (b), and (c) respectively, plotted as $y_n$ vs. $x_n$.

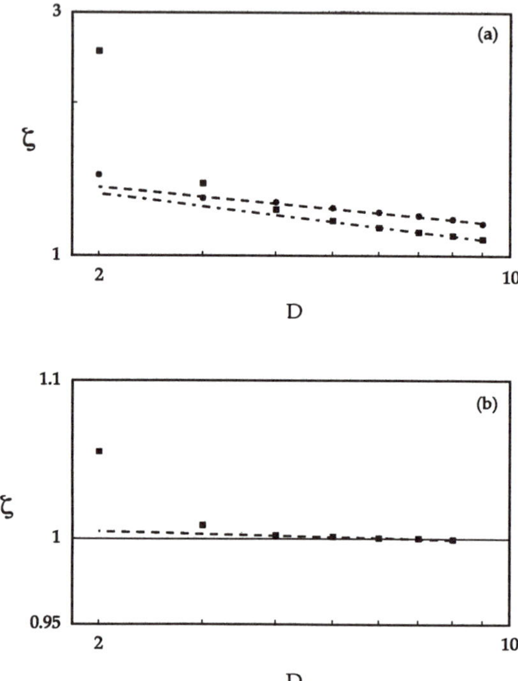

Figure 3. A graph of $\zeta$ vs. $D$ is shown for (a) the Hénon map (circles) and Ikeda map (squares), and (b) the Rössler system. The dashed lines indicate the power law for behavior observed for dimensions greater than the minimal embedding dimension. The inferable dimension is largest $D$ for which $\zeta$ is above this line.

where $Q_{MI} \equiv 1/M\,(\tau)$. 2. Modelling with local linear maps [23],

$$x_i\,(t_{n+1}) = \mathbf{m_i} \cdot \mathbf{x}\,(t_n) + b_i, \quad \forall i \in [1, D] \tag{3.5}$$

with $\mathbf{m_i}$ and $b_i$ fit to the data in each partition of a D-dimensional reconstruction. The quality function is $Q \equiv 1/\eta$ where

$$\eta \equiv \frac{1}{N \cdot D} \sum_{partitions} \sum_{i,n} \left(x_i\,(t_{n+1}) - \mathbf{m_i} \cdot \mathbf{x}\,(t_n) - b_i\right)^2 \tag{3.6}$$

and $N$ is the total number of data points. 3. Forecasting only $x(t)$ with local linear maps for which

$$\eta \equiv \frac{1}{N} \sum_{partitions} \sum_{n} \left(x_0\,(t_{n+1}) - \mathbf{m_0} \cdot \mathbf{x}\,(t_n) - b_0\right)^2. \tag{3.7}$$

We find different preferred delays for these criteria. The optimal reconstruction according to mutual information is shown in figure 5a, which is optimally spread in the state space, and thus useful for calculating invariants of the dynamics. Figure 5b is attempting to minimize the estimation error from local linear forecasting which would prefer a deterministic state space had we not restricted it to two dimensions. If we restrict our goal to forecasting only $x(t)$, we find figure 5c is optimal. Other goals would lead to different optimal representations.

## 4. CONCLUSIONS

We have presented an outline for a method of generating optimal global representations of experimental data. The brief examples above did not demonstrate the use of generalized coordinates, but the method developed is well suited to these. The learning algorithm is based upon the genetic algorithm and is sufficiently general to accommodate a variety of optimality criteria.

This work considers the analysis of a single time series, but the methods described can be easily modified to handle multivariate or spatiotemporal [24,12] systems. This could have several interesting applications. For instance, time and space derivatives can be used for coordinates to find an approximately deterministic representation of spatiotemporal data. From this representation, the most important terms for a model or even equations of motion in the form of partial differential equations can be provided for theoretical considerations. Another possibility would be to take one variable of a multivariate data set and search to find which other variables make the original one most predictable, or controllable. These are just a few of the possibilities.

The range of conceivable objectives to representation and coordinates for reconstruction is just beginning to be developed, and the learning algorithm presented here is an effective technique for investigating a wide variety of these.

## 5. ACKNOWLEDGEMENTS

This work was supported in part by National Science Foundation grant number NSF PHY86-58062.

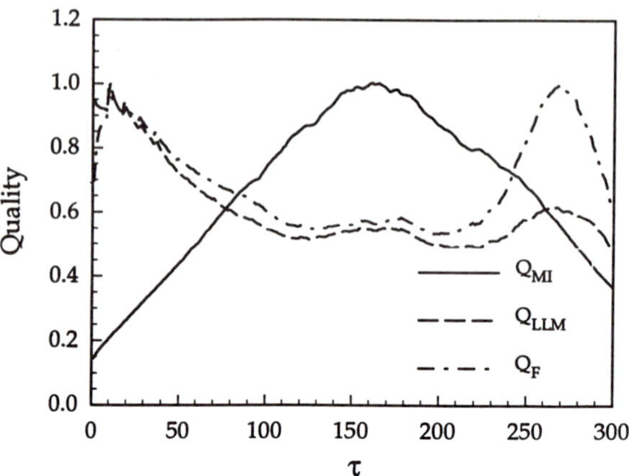

Figure 4. The quality as a function of delay is shown for data from the Mackey-Glass system for three different goals. The scale is arbitrary, so each has been normalized to the maximum quality observed. $Q_{MI}$ is the quality corresponding to a mutual information criterion, eq. 3.4; $Q_{LLM}$ is the quality for making predictions from local linear maps, eq. 3.5 & 3.6; and $Q_F$ is the quality for trying to forecast just $x(t)$ with local linear maps, eq. 3.7. The maxima indicate the preferred delay coordinates, which are clearly different for the various goals. For purposes of visualization, the search was restricted to two dimensional reconstructions.

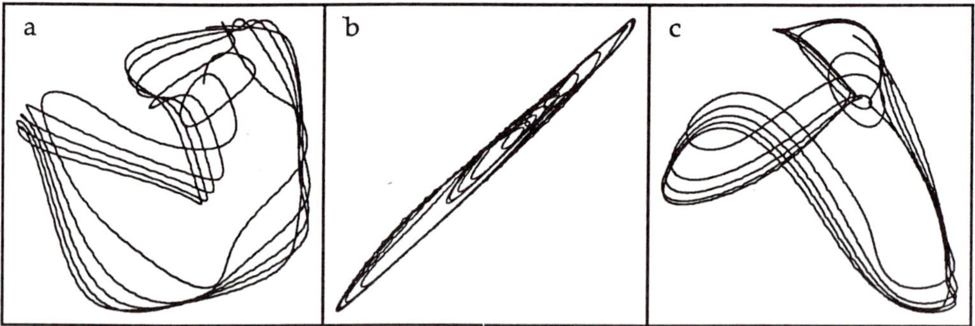

Figure 5. The delay coordinate reconstructions shown are those selected as optimal by the criteria in fig. 4. These optimal reconstructions correspond to: (a) mutual information, (b) local linear maps, and (c) forecasting only x(t).

# REFERENCES

[1] N. H. Packard, J. P. Crutchfield, J. D. Farmer, and R. S. Shaw, "Geometry from a time series", *Phys. Rev. Lett.*, **45**, 712–715, 1980.

[2] F. Takens, "Detecting strange attractors in turbulence", in: D. Rand and L.-S. Young, eds, *Dynamical Systems and Turbulence, Warwick, 1980, Lecture notes in mathematics vol. 898*, pages 366–381, Berlin, 1981. Springer-Verlag.

[3] T. Sauer, J. A. Yorke, and M. Casdagli, "Embedology", Preprint, 1991.

[4] A. Fraser and H. Swinney, "Independent coordinates for strange attractors from mutual information", *Phys. Rev. A*, **33A**, 1134–1140, 1986.

[5] M. Casdagli, S. Eubank, J. D. Farmer, and J. Gibson, "State space reconstruction in the presence of noise", Technical Report LA-UR-91-1010, Los Alamos National Laboratory, 1991.

[6] E. J. Kostelich and J. A. Yorke, "Noise reduction: Finding the simplest dynamical system consistent with the data", *Physica D*, **41**, 183–196, 1990.

[7] J. D. Farmer and J. J. Sidorowich, "Optimal shadowing and noise reduction", Technical Report LA-UR-90-653, Los Alamos National Laboratory, 1990.

[8] J. L. Breeden and N. H. Packard, "Model-based control of nonlinear systems", preprint, 1991.

[9] J. Holland, *Adaptation in Natural and Artificial Systems*. University of Michigan Press, 1975.

[10] D. E. Goldberg, *Genetic Algorithms in Search, Optimization, and Machine Learning*. Addison-Wesley, 1989.

[11] N. H. Packard, "A genetic learning algorithm for the analysis of complex data", *Complex Systems*, **4**, 543, 1990.

[12] F. C. Richards, T. P. Meyer, and N. H. Packard, "Extracting cellular automaton rules directly from experimental data", *Physica*, **45D**, 189–202, 1990.

[13] J. Cremers and A. Hübler, "Construction of differential equations from experimental data", *Zeit. Naturforsch.*, **42a**, 797–802, 1986.

[14] J. P. Crutchfield and B. S. McNamara, "Equations of motion from a data serie", *J. Complex Sys.*, **3**, 417–452., 1987.

[15] T. Eisenhammer, A. Hübler, N. Packard, and J. A. S.Kelso, "Modeling experimental time series with ordinary differential equations", Technical Report CCSR-89-7, Center for Complex Systems Research, 1989.

[16] J. L. Breeden, F. Dinkelacker, and A. Hübler, "Using noise in the modelling and control of dynamical systems", *Phys. Rev. A*, **42**, 5827, 1990.

[17] J. L. Breeden and A. Hübler, "Reconstructing equations of motion using unobserved variables", *Phys. Rev. A*, **42**, 5817–5826, 1990.

[18] S. M. Omohundro, "Efficient algorithms with neural network behavior", *J. Complex Sys.*, **1**, 273–347, April 1987.

[19] J. L. Breeden and N. H. Packard, "Computing the observable number of degrees of freedom from experimental data", preprint, 1991.

[20] O. Rössler, *Phys. Lett.*, **57A**, 397, 1976.

[21] L. Glass and M. C. Mackey, *From Clocks to Chaos*. Princeton University Press, 1988.

[22] A. M. Fraser, "Information and entropy in strange attractors", *IEEE Trans. Information Theory*, **35**(2), 245–262, March 1989.

[23] J.-P. Eckmann and D. Ruelle, *Rev. Mod. Phys.*, **57**, 617, 1985.

[24] T. P. Meyer, F. C. Richards, and N. H. Packard, "Learning algorithm for modeling complex spatial dynamics", *Phys. Rev. Lett.*, **63**, 1735–1738, 1989.

# MULTIFRACTAL CODING MEASURES IN DYNAMICS

Giorgio Mantica

Service de Physique Théorique, CEN–SACLAY
F-91191 Gif-sur-Yvette, Cedex, France

Symbolic dynamics is a powerful tool to prove analytical results in dynamical systems. However, this tool is – in most instances – non-constructive, in the sense that no simple prescription links the symbol sequence with the real-space trajectory of an evolution.

In this paper, I review some work of mine on this theme, that I first began at the summer '89 Les Houches session, thanks to the lectures of Martin Gutzwiller. As such, the perspective presented here is highly personal, and does not claim any generality.

The initial coordinates of a deterministic system can be quite obviously expressed as a sequence of symbols. Since it contains all the future evolution (granted by suitable existence – uniqueness theorems), this sequence can be regarded as a particular coding of the trajectory. A partition of phase space (not necessarily Markovian) defines on the other hand a different symbolic sequence associated to the same trajectory, where the evolution in time is more evident.

These two symbolic sequences are obviously related, but their relation is fairly non – trivial. Gutzwiller and Mandelbrot [1, 2] have shown that they give rise to multifractal measures. By analyzing these measures, I have shown that the multifractal properties follow from a set of exact renormalization equations, which can be equivalently used to define such measures.

A first example of this procedure is given by the geodesic motion in a singular triangle $\mathcal{T}$ in the hyperbolic upper half plane $(ds^2 = (dx^2 + dy^2)/y^2)$ [3]. $\mathcal{T}$ is enclosed by the circle $\mathcal{R}$ of radius $1/2$, centered at $x = \frac{1}{2}$, and by the lines $x = 0$ $(\mathcal{O})$, and $x = 1$ $(\mathcal{L})$. Geodesics in hyperbolic space are represented by the upper half of circles centered on the real axis. Their intersections with this latter are denoted in Fig. 1 by $\xi$ (the infinite past point), and $\eta$ (the infinite future).

The motion in this triangle is obtained by mapping the intersections of the trajectory with $\mathcal{R}$ on $\mathcal{O}$ via $z \rightarrow z/(1 - z)$ and $\mathcal{L}$ on $\mathcal{O}$ via $z \rightarrow z - 1$. A coding of this motion can therefore be obtained by the sequence of symbols $r$ and $l$, denoting the temporal sequence of such remappings. The symbolic code (s.c) of a trajectory is therefore the doubly-infinite (past and future) sequence of symbols $\{s_i\}$, $s_i = r, l$, $i = -\infty, \infty$. Following Series [3], there is a one-to-one relation (modulo sets of zero measure) between s.c. and geodesics.

A multifractal function can be associated to this coding. To do this, we follow [1] and take $\eta \leq 1$. We then consider only the motion following an arbitrary initial time: $(s_0, s_1, \ldots)$, with $s_0 = r$. This sequence depends *only* upon the infinite future point $\eta$. A *coding function* $F$ can be defined so that $F = F(\text{code}) = F(\eta))$ and that it be possible to uniquely recover the s.c. from the numerical value of the function: to do

*Chaotic Dynamics: Theory and Practice*
Edited by T. Bountis, Plenum Press, New York, 1992

this, write

$$F(\eta) = \sum_{i=1}^{\infty} \varepsilon(s_i)\, 2^{-i}, \tag{1}$$

where $\varepsilon(r) = 0$, $\varepsilon(l) = 1$. $F$ is best seen as the *distribution function* of a measure, $\mu$ : $F(\eta) = \int_0^\eta d\mu(x)$. This measure (pictured in Fig. 2) has interesting multifractal properties. We now show that they follow from a semigroup of exact symmetries, which offer an equivalent definition of $F$.

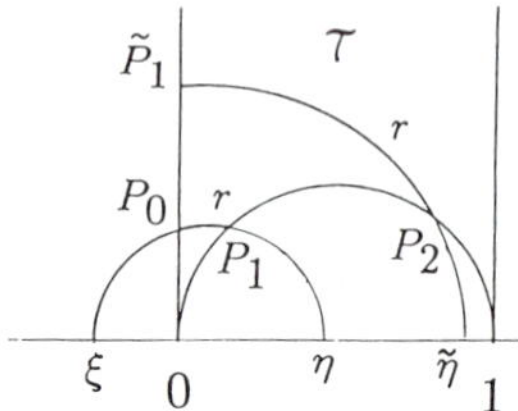

Fig. 1. The singular triangle $\mathcal{T}$.

Fig. 2. Grey tone representation of the measure $d\mu$ of hyperbolic motion in the singular triangle $\mathcal{T}$.

The main tool to unveal the similarity structure of this measure is given by the Möbius mappings of the unit interval into itself

$$\begin{aligned}
M_0(x) &= 1 - x, & P_0(y) &= 1 - y, \\
M_1(x) &= \frac{x}{1+x}, & P_1(y) &= \frac{y}{2}, \\
M_2(x) &= \frac{1}{2-x}, & P_2(y) &= \frac{y+1}{2}.
\end{aligned} \tag{2}$$

Infact, it can be proven [4] that the following functional relations hold:

$$F(M_i(\eta)) = P_i(F(\eta)), \quad i = 0, 1, 2, \tag{3}$$

These relations can be extended to the full semigroups generated by the $M_i$, and $P_i$. These similarity properties can be used to provide an alternative definition of $F$, via the formalism of Iterated Functions Systems [5]. To do this, we consider the random process over $[0, 1]$ given by

$$x_n \longrightarrow x_{n+1} = M_{\sigma n}(x_n), \quad \sigma = 1, 2, \tag{4}$$

where $\sigma_n$ are random variables taking the values 1, and 2, with equal probability. This is the definition of an I.F.S. generated by the two maps $M_1, M_2$. In [4] I have proven that its equilibrium measure is $\mu$; as a consequence, a fast way of computing the density of $F$ is obtained by sampling a realization of the random process (4).

The function $F$ itself can be obtained as an attractor of a suitable I.F.S. To do this, we use the mappings $w_i$ from $[0, 1] \times [0, 1]$ into itself given by

$$w_i(x, y) = (M_i(x), P_i(y)), \tag{5}$$

for $i = 1, 2.$ These mappings possess a *unique* invariant set, the attractor $G$, which verifies

$$G = \cup_i w_i(G). \tag{6}$$

I have also proven that eqs. (3) imply that $G$ coincides with the *graph* of the function $F$, i.e. the set of points $(x, F(x))$. A random process analogous to (4)

$$(x_{n+1}, y_{n+1}) \rightarrow w_{\sigma n}(x_n, y_n) \tag{7}$$

can be used to draw the graph of $F$.

An interesting consequence of the above formalism regards the Hölder exponent $\alpha(\eta)$ of $F(\eta)$: $F(\eta + h) - F(\eta) \sim h^{\alpha(\eta)}$, as $h > 0$ tends to zero [6]. Due to eqs. (3), and the fact that $M_i$ are smooth maps, one finds the crucial relation $\alpha(\eta) = \alpha(M_i(\eta))$: $\alpha(\eta)$ is therefore invariant under the semigroup of Möbius maps generated by $M_i$, $i = 0, 1, 2$.

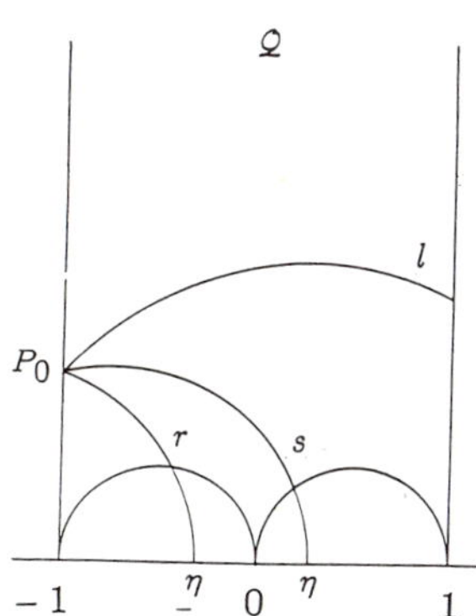


*Fig. 3. The hyperbolic torus Q.*

A generalization of this technique pertains to to more complex hyperbolic billiards like the one pictured in Fig. 3. A Möbius analysis can be similarly performed, obtaining the coding measure of Fig. 4 as attractor of an I.F.S. [4]. This exact analysis can be extended to all hyperbolic billiards with angles at infinity.

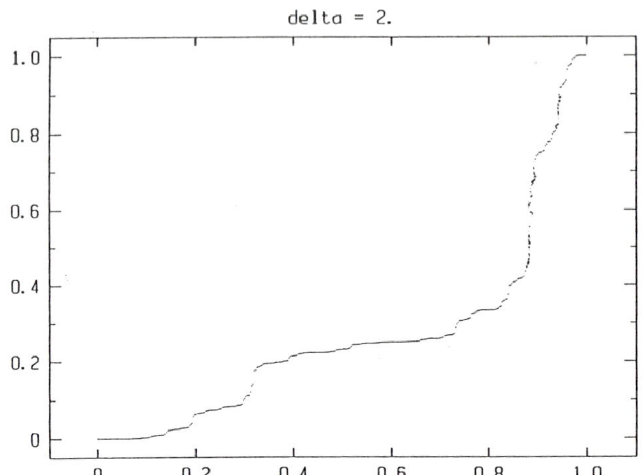


*Fig. 4. Point on the attractor for the coding function of the motion in the torus of Fig. 4.*

The same technique can also approximate the coding measures of arbitrary chaotic systems. The well known anisotropic Kepler problem [2], with Hamiltonian

$$H = \frac{p_x^2}{2\mu} + \frac{p_y^2}{2\nu} - \frac{1}{\sqrt{x^2 + y^2}}, \tag{8}$$

is a significant example.

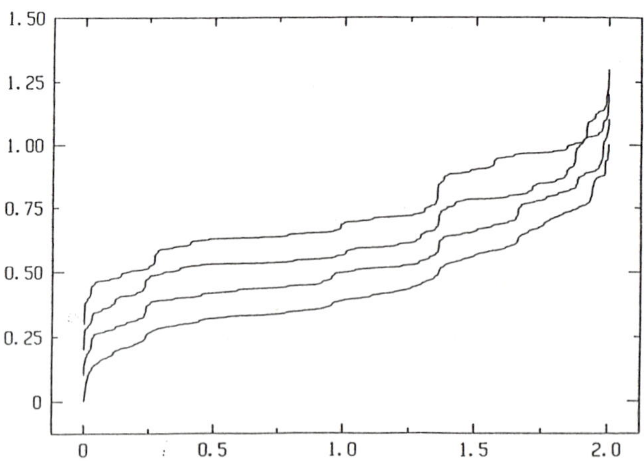

*Fig. 5. Exact (bottom) and approximate (shifted) coding functions for the anisotropic Kepler motion, constructed after [4].*

46

In the A.K.P., the effective masses $\mu$, and $\nu$, are different. This system is chaotic when the mass ratio $\frac{\mu}{\nu}$ is sufficiently high ($\geq 5$). A symbolic coding can be obtained following Gutzwiller [7]: a set of trajectories starting on the $x$ positive axis, with zero initial $p_x$ momentum is considered. A constant energy surface, e.g. $H = -1/2$ implies that $x$ labels an unique trajectory, for any $0 \leq x \leq 2$. The time evolution determines the symbolic sequence $b_j$: $b_i = 0$ if the $i$-th intersection with the $x$-axis occurs for $x \leq 0$, and $b_i = 1$ otherwise. A coding function $F$ is then defined via $F(x) = \sum b_i 2^{-i}$. This function is non-decreasing, and shows multifractal features.

Suitable Möbius maps $M_i$ and linear maps $P_i$ can be found so that eqs. (3) (and the associated I.F.S.) define a good approximation of $F$ [8, 9, 10]. In Fig. 5 the coding functions corresponding to the exact Kepler problem and its I.F.S. approximations are shown.

One can expect that employing such approximate similarity properties good abstract models of realistic systems can be costructed. This should enable us a more "constructive" use of symbolic dynamics.

## References


[1] M.C. Gutzwiller, and B.B. Mandelbrot, *Phys. Rev. Lett.* **60** (1988) 673-676.

[2] M.C. Gutzwiller, *Physica* **38 D** (1989) 160-171.

[3] C. Series, *Ergodic Theory of Dynamical Systems* **6**, (1986), 601, *J. London Math. Soc.* (2), **31**, (1985) 69.

[4] D. Bessis and G. Mantica, "Construction of Multifractal Measures in Dynamical Systems from their Invariance Properties", *Phys. Rev. Lett.* **66**, 2939 (1991).

[5] M.F. Barnsley and S.G. Demko, *Proc. R. Soc. London* **A 399** (1985) 243-275.

[6] B.B. Mandelbrot, *The Fractal Geometry of Nature*, (Freeman, New York, 1982).

[7] M.C. Gutzwiller, *J. Math. Phys.* **18**, (1977), 806.

[8] M.F. Barnsley, *Fractals Everywhere*, (Academic Press, New York, 1988)

[9] C.R. Handy, and G. Mantica, *Physica* **D 43**, 17-36 (1990).

[10] G. Mantica and Allan Sloan, *Complex Systems* **3**, (1989) 37-62.

# DETERMINATION OF FRACTAL DIMENSIONS AND

# GENERALIZED ENTROPIES FOR STRANGE ATTRACTORS

Ágnes Fülöp

Institute for Solid State Phys., Eötvös University
Budapest, P.O. Box 327, 1445 Hungary

## ABSTRACT

Two independent quantities of chaotic behaviour, the generalized fractal dimensions and the generalized entropies are detemined for the strange attractor of the Lozi map. We fine that both exhibit multifractal structures.

## INTRODUCTION

Strange attractors play central role in studying chaotic motion. A huge number of articles discuss the fractal properties and generalized entropies as main characteristics of chaotic systems[1..8]. In this paper we concern the questions whether the strange attractor of the Lozi map shows geometrical multifractal behaviour or not, and if yes, we calculate the generalized entropies as dimensions in the history space. We will study the Lozi map

$$\left. \begin{array}{rcl} x_{n+1} & = & 1 - a * |x_n| + b * y_n \\ y_{n+1} & = & x_n \end{array} \right\} \tag{1}$$

by numerical methods.

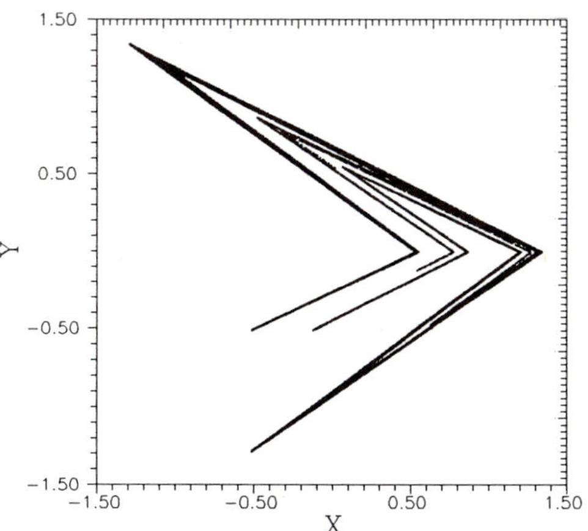



Fig.1.Lozi map with a=1.7, b=0.5.The coor−
dinates of initial point are x=0.3, y=0.3.

The fractal dimension of a deterministic construction is calculated from the well-known expression $M(R) \sim R^D$ , where $M(R)$ is the number of particles within a circle of radius $R$ centered at an arbitrary point of the strange attractor and $D$ is the fractal dimension.

Let us define geometrical multifractality of structures according to Ref[9]. Put a grid on the object. Its linear size is $L$, its mass is $M_0$, and l is the box size of grid. We can determine $M_i$ as the mass (the number of particles) of the $i$ -th box ($i = 1, 2...$). Let us define the $\alpha$ mass index

$$M_i \sim M_0(l/L)^{\alpha_i} \tag{2}$$

where $l/L \to 0$. The number of those boxes, wich suit to the mass index by expression (2) is $N(\alpha)$. The centers of the boxes form a fractal structure with dimension $f(\alpha)$

$$N(\alpha) \sim (l/L)^{-f(\alpha)} \tag{3}$$

If there exists a set of various mass indices, the object is a geometrical multifractal.

The generalized fractal dimension $D_q$ [10] is defined for sud a multifractal as

$$\sum_i M_i^q \sim M_0^q(l/L)^{(q-1)D_q} \tag{4}$$

in the limit $l/L \to 0$.

Then we use the generalized sandbox method [11] to determine the generalized fractal dimension $D_q$ . This procedure is particularly suited in the case of geometrical multifractaly. This technique was used for another systems as well, e.g. for the Assymetric Cantor Set[11], DLA[12,13,14,15], or the repellor of chaotic scattering[16]. This method is based on observing the average of $M(R)$ (and its q-th moments). $M(R)$ is the mass in boxes of size $R$. The centers of the boxes are randomly situated on the fractal and have to choose them with a uniform distribution. So we determine the $D_q$ over this fractal by

$$\langle (M(R)/M_0)^{(q-1)} \rangle \sim (R/L)^{(q-1)D_q} \tag{5}$$

Let us note that the centers of boxes are situated on the fractal and that $R \ll L$ .

I have calculated the $D_q$ spectrum of the Lozi attractor on a 1000 by 1000 lattice. After 2.5 million iterations of an initial point $M_0$ equals to 23800 boxes. The averaging was made over 5000 centers of the boxes which were distributed randomly on the fractal. Then I have determined $\langle (M(R)/M_0)^{(q-1)} \rangle$ and plotted its logarithm divided by $(q-1)$ versus $\ln(R/L)$ for various $q$ values.

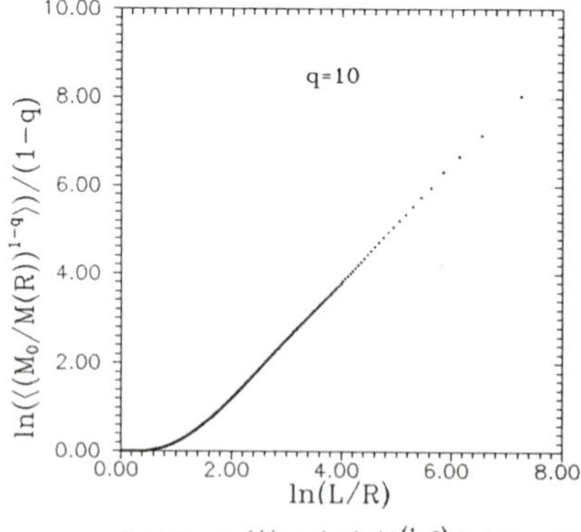

Fig.2. $\ln(\langle (M_0/M(R))^{(1-q)} \rangle)/(1-q)$ vs $\ln(L/R)$ for q=10.

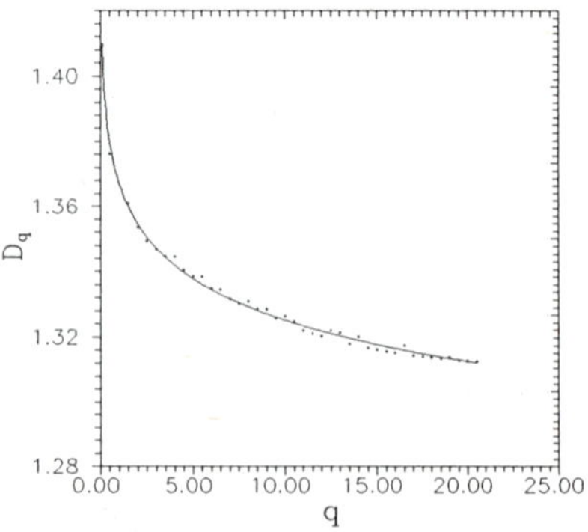

Fig.3. This figure shows the function $D_q$ vs q for q)0.

In Fig.2. the slope of the graph yields $D_q$ for $q = 10$ in the range $R \ll L$. Changing $q$ we obtain the $D_q$ spectrum of generalized fractal dimension, which monotonously decreases as a function of $q$.


## ENTROPY

The generalized entropies can be best treated by introducing first the so-called history space[17]. The construction is the following.

We gain information from series of measurements through observation. Make a partition of the phase space. Next take record which piece is visited at time $k$. The pieces are labelled from 0 to $m - 1$. The sequence $O_n = (i_1...i_n)$ , gives coarse grained characterization of the trajectory, where $i_k$  $k = 1...n$ can take any number between $0...m - 1$. A given sequence occurs with probability $P(O_n)$. Each sequence of length $n$ can be coded by a number $z_n^{(j)}$ one to one in the following manner.

$$z_n^{(j)} = \sum_{k=1}^{n} i_k m^{-k} \qquad j = 0...m^{-n} - 1 \tag{6}$$

$z_n^{(j)}$ are in the intervall $[0,1)$. Let us define the probability density function:

$$P_n(z) = P(i_1...i_n)m^n \quad if \quad (z_n^{(j)} \le z \le z_n^{(j)} + m^{-n}) \tag{7}$$

which is normalized on $(0,1]$. In the limit $n \to \infty$ we get $P(z) = \lim_{n \to \infty} P_n(z)$. In the history space in which z is a point $P(z)$ defines a measure.

The order q dynamical information of the history space for the fixed length $n$ [18] is given by

$$I_n(q) = \frac{1}{(q-1)} \ln \sum_{O_n} P(O_n)^q \quad where \quad q \ne 1 \tag{8}$$

Generally it diverges like (const) $*$ n. In that case the finite quantity

$$K_q = \lim_{n \to \infty} I_n(q)/n \tag{9}$$

is called Renyi entropy of order $q$ [19]. The order $q$ dimension of the history space $D^*$ is directly connected to $K_q$ via

$$K_q = D^*(q) \ln(m) \tag{10}$$

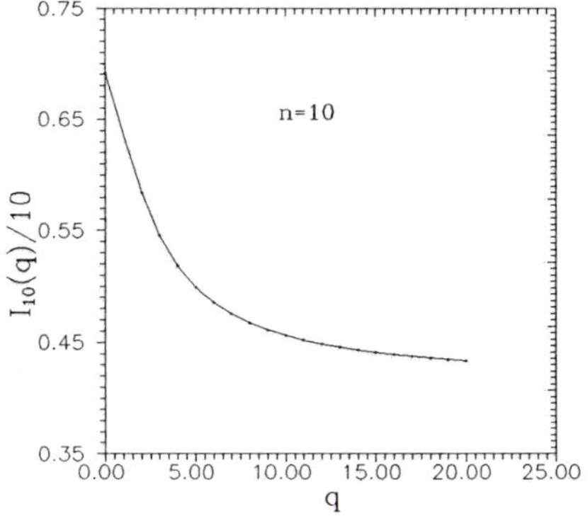



Fig.4.The graph displays
$\ln(\Sigma P(O_n)^q)/(1-q)/n$ vs q for n=10.

I have calculated the Renyi entropy of order $q$ for the Lozi map by numerical methods. I used again the 1000 by 1000 partition as earlier to determine the fractal dimension. Sequences of length n were studied between 1 and 15. After    2.5 million iterations I determined the ratio $I_n(q)/n$ . This is shown on Fig 4. as a function of $q$ for $n = 10$. In the limit $n \to \infty$ we gain $K_q$ (9).

Comparing the spectra of $D_q$ and $K_q$ one notices the similar behaviour for the Lozi map. In summary, we calculated two independent characteristics of the Lozi map.

REFERENCES


1. B. B. Mandelbrot, The Fractal Geometry of Nature
   (Freeman, San Francisco, CA) 1982.
2. Hiroki Hata, Prog. Theor. Phys. 78(1987)721.
3. P. Grassberger, Phys. Lett. A 97(1983)227.
4. G. Paladin and A. Vulpiani, Phys. Rep. 156(1987)147.
5. P. Grassberger, J. Phys. A 22(1989)585.
6. D. Auerbach, B. Oshughnessy and I. Procaccia, Phys Rev. A 37(1988)2234.
7. T. C. Halsey, M. H. Jensen, L. P. Kadanoff, I. Procaccia and B. Shtaiman,
   Phys. Rev. A 33(1086)1141.
8. P. Grassberger and I. Procaccia, Phys. Rev. A 28(1983)2591.
9. T. Tél and T. Vicsek, J. Phys. A 20(1987)L835.
10. H. G. E. Hentschel, I. Procaccia, Phys. D 8(1983)435.
11. T. Tél, Á. Fülöp, T. Vicsek, Phys. A 159(1989)155.
12. P. Meakin, Prog. Solid. 20(1990)135.
13. T. Vicsek et.al., Europh Lett. 12(1990)217.
14. T. Vicsek, Phys. A 168(1990)490.
15. J. F. Fernandez, J. M. Albarrán, Phys. Rev. Lett. 64(1991)2133.
16. C. Jung and T. Tél, J. Phys. A 24(1991)2793.
17. J. D. Farmer, Z. Naturforsch. 37a(1982)1304.
18. A. Csordás and P. Szépfalusy, Phys. Rev. A39(1989)4767.
19. A. Renyi, Probability Theory (North-Holland, Amsterdam 1970).

# ON THE QUATERNIONIC JULIA SETS

Peter Petek

Dept. of Mathematics, University of Ljubljana
Jadranska 19, 61000 Ljubljana, Slovenia

## 1.  Introduction: The field of quaternions

Mandelbrot in his book [5] already offers computer generated images of quaternionic Julia sets
by A. Norton. This author contributed more images with explanations in [6, 7], and Holbrook
did similarly in [4].

The family of all quadratic functions in quaternions is broad and can due to noncommutativity be reduced to one parameter as is the case in the complex field. We thus take the essential
restriction to the family $f(X) = X^2 + Q$. Further restrictions are imposed on the parameter
$Q$ by the *one equator property*, which makes it possible to divide $QJ$ in two hemispheres and
admits symbolic dynamics.

Periodic points appear not only on the complex Julia set, but the question of their density
in $QJ$ is not yet resolved.

In fact there are more questions opened than answered. Maybe one should even look for a
counterexample to the Poincaré conjecture in the iteration of quadratic functions in quaternions.

The field of quaternions can be represented as a direct sum

$$\mathcal{H} \simeq \mathcal{R} \oplus \mathcal{R}^3 \tag{1}$$

and we shall write for a quaternion

$$X = (\xi, \vec{x}); \ \xi \in \mathcal{R}, \ \vec{x} \in \mathcal{R}^3 \tag{2}$$

Multiplication is defined by

$$AB = (\alpha\beta - \vec{a}\vec{b}, \ \alpha\vec{b} + \beta\vec{a} + \vec{a} \times \vec{b}) \tag{3}$$

and is associative, distributive, yet not commutative. The norm $\|A\| = \sqrt{\alpha^2 + \vec{x}^2}$ is related to
multiplication $\|AB\| = \|A\|\|B\|$.

*Chaotic Dynamics: Theory and Practice*
Edited by T. Bountis, Plenum Press, New York, 1992

Squaring in quaternions thus reduces to

$$X^2 = (\xi^2 - \vec{x}^2,\ 2\xi\vec{x})$$

the square root has exactly two values except in the case of zero or purely real negative radicand:

$$\sqrt{(-a^2,\ \vec{0})} = (0,\ \vec{x});\ \vec{x}^2 = a^2$$

Two quaternions commute iff their vector parts are colinear, therefore the center for multiplication is isomorphic to reals, imbedded as

$$r : \mathcal{R} \to \mathcal{H},\ \ r(\xi) = (\xi,\ \vec{0})$$

Complex numbers can be imbedded depending on a unit vector $\vec{x} \in \mathcal{R}^3$:

$$i_{\vec{x}}(a + bi) = (a,\ b\vec{x})\ .$$

Later we shall be using the canonical basis $\{\vec{i}, \vec{j}, \vec{k}\}$ and canonical imbedding $i_{\vec{i}} : \mathcal{C} \to \mathcal{H}$. Also we shall consider $\mathcal{R}$ and $\mathcal{C}$ as subspaces of $\mathcal{H}$, while the space spanned by $(1, \vec{i}, \vec{j})$ shall be denoted by $\mathcal{V}$.

Contrary to the complex case, where the iteration of quadratic functions can be reduced to a one-parameter family, this can not be done in quaternions, due to noncommutativity. Compare [3].

Here we shall consider only the family $f_Q(X) = X^2 + Q$ and its iteration

$$X_{n+1} = X_n^2 + Q \tag{4}$$

As has been shown already in [6, 7, 4], it is sufficient to take $Q = c = \gamma + iq \in \mathcal{C}$ and $X_0 \in \mathcal{V}$ since the orbit then stays in $\mathcal{V}$. Henceforth we apply this (inessential) restriction.

For the squaring $q : \mathcal{H} \to \mathcal{H}$

$$q(X) = X^2$$

There is the obvious candidate for the quaternionic Julia set , the unit sphere

$$S^3 = \{X;\ X^2 = \xi^2 + \vec{x}^2 = 1\}.$$

If $\|X_0\| > 1$, the sequence of iterates quickly diverges to infinity, for $\|X_0\| < 1$ it converges to zero, and $S^3$ is $q$-invariant.

It can easily be proved that for any $Q$ the sequence (4) for large enough $X_0$ diverges to infinity. We set, to comply with the squaring case as well as with the complex analogue

*Definition.* The quaternionic Julia set $QJ$ of the quadratic function $f(X) = X^2 + Q$ is the boundary of the basin of attraction of infinity.

First suppose the parameter real $Q = \gamma$. Then the sequence stays in the complex plane $\mathcal{C}$ and the quaternionic Julia set is obviously obtained by rotating the complex Julia set

$$QJ = \bigcup_{\|\vec{x}\|=1} i_{\vec{x}}(J_\gamma) \tag{5}$$

We shall call each of the copies $i_{\vec{x}}(J_\gamma)$ a *meridian.*

# 2. Restrictions on the parameter

From now on we work with the additional restriction that $c \in Int(M_1)$, inside the big cardioide of the Mandelbrot set. The Julia set in this case is a fractal circle, intersecting the imaginary axis at least twice.(Compare [4],[7]). We choose one of the intersection points $ai \in J$ . The whole sphere

$$S^2 = \{(0, \vec{x}); \ \vec{x}^2 = a^2\} \tag{6}$$

belongs to the $QJ$ and we shall call this sphere *equator of $QJ$* in $\mathcal{H}$. Equator of course depends on the possible choice of $a$. There are exactly two points on the equator in the complex Julia set $\pm ai$. To eliminate the choice of the equators we set

*Definition.* The complex number $c \in M_1$ has the *one equator property* (OEP), if the Julia set of $z \to z^2 + c$ intersects the imaginary axis exactly twice.

The property holds trivially for real $c$, for values off the real line $c = \gamma + qi, q \neq 0$ it is however not easy to check. Figure 1 shows in black the points inside the big cardioide that do not possess OEP - the result of a computer experiment of course. We see lots of "thickened curves" black, among them the parabola $\gamma = q^2$ (connecting the origin with the radix of the bubble of four-cycles), though we do not have an analytic proof of it. Later we shall see the important role of this parabola. Also we see a white region of small parameters of negative real part, likewise no analytic result is known. The structure seems interesting but not yet understood.

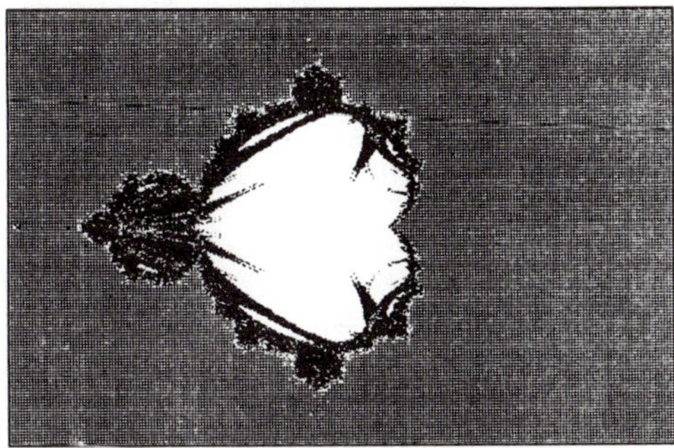

Figure 1. Points inside the big cardioide, not having the OEP, are marked black. (computer generated by prof. J.Virant)

Since $c \in Int(M_1)$, there is a homeomorphism between the corresponding Julia set $J_c$ and the unit circle $S^1 = \{z = e^{2\pi it}\}$:

$$\phi : J_c \to S^1 \ ,$$

satisfying

$$\phi(z^2 + c) = (\phi(z))^2$$

and the mapping

$$t : J_c \to [0,1); \ t(z) = \frac{ln(\phi(z))}{2\pi i}$$

attributes to points in the Julia set a number – the external angle – in binary notation

$$t = 0.t_1 t_2 t_3 \dots \quad t_i = 0,\ 1$$

with

$$t(z^2 + c) = 2t(z) \mod 1 \tag{7}$$

What we are interested in is the angle $\tau = t(-a^2 + c)$.

*Definition.* The complex number $c$ with OEP has the *nonperiodic angle property* (NAP), if the binary expansion of $\tau = t(-a^2 + c)$ is not pure periodic.

The definition only excludes angles of the form $\tau = \frac{m}{2^k - 1}$. And if $c$ has NAP, no Bernoulli shift $2^k \tau \mod 1$ can coincide with $\tau$.

The NAP ensures that the inverse images of the equator are all disjoint. The property is likewise difficult ot check, but since the complementary set has measure zero, in computer experiments we can assume it.Suppose now the corresponding $c$ has OEP and NAP. We shall give "geographical" names to some points and sets of $QJ \cap \mathcal{V}$.

*North pole $N$:* the repreprepelling fixed point
$\quad N = \frac{1 + \sqrt{1 - 4c}}{2}$ (positive real part of the square root)

*South pole $S = -N$*

*Magnetic pole $M = -a^2 + c$*

*Equator $\mathcal{E} = f^{-1}(M) \cap \mathcal{V}$*

*Northern (southern) hemisphere $QJ^\pm$:*
$\quad QJ^\pm \cap \mathcal{V} = \{X = (\xi, \vec{x}) \in QJ \cap \mathcal{V};\ \xi > 0\ (\xi < 0)\}$

The inverse image of the equator consists of two smooth circles (not geometric circles!), one in each $QJ^\pm$. We call them first parallels $\mathcal{P}^+$ and $\mathcal{P}^-$. Likewise the inverse image of each first parallel consists of two spheres, the second parallels, and so on.

$$f^{-1}(\mathcal{P}^+) = \mathcal{P}^{++} \bigcup \mathcal{P}^{-+}$$

Each finite sequence of plus and minus signs defines a parallel of the corresponding order.

The name *parallel* however is slightly misleading. These sets do not in general lie on parallel hyperplanes or even approximately so. In the sequel this shall be discussed in a section.

## 3.  Dissolution of meridians

In the case the parameter is real every copy $i_{\vec{x}}(J_\gamma)$ of the corresponding Julia set was called a meridian $\mathcal{M}_{\vec{x}}$. The meridian inersects the equator in two points $\pm a\vec{x}$ and can be represented due to the well known properties of Julia sets, as the closure of the union of inverse images:

$$\mathcal{M}_{\vec{x}} = Cl(\bigcup_{n \geq 0} f^{-n}(a\vec{x})).$$

The above formula can be used in a more general situation for any $Q$.

*Definition.* Let the parameter $Q = \gamma + iq$ have OEP, $a$ being the radius of the unique equator. The *meridian* of a unit vector $\vec{x} \in V$ is defined

$$\mathcal{M}_{\vec{x}} = Cl(\bigcup_{n \geq 0} f^{-n}(a\vec{x})) \tag{8}$$

We expected the meridian to be likewise a fractal circle, connecting the north and south poles. The experiment however showed that it was not so. The projection showed as if the whole interior of the Julia set were filled with inverse images. Since $2^n$ inverse images on the $n$-th level would represent storage problems, we applied a random algorithm. $X_0$ was chosen on the equator, but not on the complex plane and then only one inverse image was chosen randomly at each step:

$$X_{n+1} = \varepsilon\sqrt{X_n - Q} \ , \quad \varepsilon = \pm 1$$

Several thousand iterations vere computed and it seems that the meridian covers the part of $QJ \cap V$

*Conjecture.* Let $Q = \gamma + iq$ have OEP, $a$ being the equator's radius and $\vec{x}$ a unit vector with $X_0 = (0, \vec{x}) \in V$. If $X_0 \notin C$

$$\mathcal{M}_{\vec{x}} = QJ \cap V$$

## 4. Periodic points

For a given Q, periodic points of order $n$ are solutions of the equation

$$f_Q^n(P) = P \ ,$$

i.e. they are functions of the parameter $P = P(Q)$ and as such commute $PQ = QP$. But since they commute, they lie in the plane of Q, the complex plane of the Julia set. Thus the only periodic points are those on the Julia set. False!

The point is that $P = P(Q)$ is not a uniquely defined function and the chain of reasoning breaks there. In fact we do find for each $n > 1$ a whole geometric circle of periodic points off the Julia set.

Fixed point. Since from $P^2 + Q = P$ follows that $P$ and $Q$ commute, the fixed points lie in the plane defined by $Q$ and therefore in the original Julia set. Only if $Q = (\gamma, \vec{0})$ is real and $\gamma > \frac{1}{4}$ – thus otside the Mandelbrot set – we have a whole sphere of repelling fixed points $P = (\frac{1}{2}, \vec{p}); \ \|p\| = \sqrt{\gamma - \frac{1}{4}}$.

Periodic points of order two and higher are solutions of the equation $f_Q^n(X) = X$. We are interested in the nontrivial case $\vec{q} \neq 0$ and solutions $XQ \neq QX$, i.e. $\vec{x}$, $\vec{q}$ not colinear. Writing

$$f_Q^i(X) = (\xi_i, \alpha_i\vec{x} + \beta_i\vec{q}); \ i \geq 0$$

recurrence relations are obtained

$$\xi_{i+1} = \xi_i^2 - (\alpha_i\vec{x} + \beta_i\vec{q})^2 + \gamma$$

$$\alpha_{i+1} = 2\xi_i \alpha_i$$

$$\beta_{i+1} = 2\xi_i \beta_i + 1$$

with initial condition

$$\alpha_0 = 1, \ \beta_0 = 0$$

and periodicity condition

$$\xi_n = \xi, \ \alpha_n = 1, \ \beta_n = 0.$$

The above conditions reduce to a polynomial system for $\xi$, $\vec{x}^2$, $\vec{x}\vec{q}$. In the case $n = 2$ solutions are

$$\xi = -\frac{1}{2}, \ \vec{x}^2 = (\vec{x} - \vec{q})^2 = \frac{3}{4} + \gamma,$$

and thus form a circle for the values of the parameter in the big cardioide. In $\mathcal{V}$ we have the two periodic points

$$-\frac{1}{2} + \vec{i}\frac{q}{2} + \vec{j}\sqrt{\frac{3}{4} + \gamma - \frac{q^2}{4}} \tag{9}$$

the last square root being real because $c$ lies inside the cardioide.

# References


[1] M. Barnsley, Fractals Everywhere. *Academic Press 1988*

[2] P. Blanchard, Complex analytic dinamics on the Riemann sphere. *Bull. Amer. Math. Soc.* **11**, 1 (1984) 85-141

[3] J.E. Fornaess and N. Sibony, Complex Hénon mappings in $C^2$ and Fatou Bieberbach domains. *Prepublications Université de Paris-Sud, Mathématiques* 31 (1990) 1-41

[4] J.A.R. Holbrook, Quaternionic Fatou-Julia sets. *Ann. sc. math. Québec* **11**, 1 (1987) 79-94

[5] B.B. Mandelbrot, The Fractal Geometry of Nature. *Freeman and company, New York 1983*

[6] A. Norton, Generation and Display of Geometric Fractals in 3-D. *Computer Graphics* **16** (1982) 61-67

[7] A. Norton, Julia sets in the Quaternions. *Comput. and Graphics* **13**, 2 (1989) 267-278

# PERTURBATION THEORY AND ANALYTICITY OF NORMALIZING TRANSFORMATIONS FOR AREA PRESERVING MAPS

Armando Bazzani and Giorgio Turchetti

Department of Physics, University of Bologna
Via Irnerio 46, 40126 Bologna Italy

## 1. INTRODUCTION

The normal forms theory developed by Poincaré , Birkhoff [1] and Bruno [2] is a basic tool for the study of flows or maps in the neighborhood of critical points. The properties of the transformation which conjugates a given system with its normal form are related to the integrability properties. Limiting the analysis to area preserving maps, the case of elliptic fixed points is highly non-trivial becouse of the presence of small divisors and relevant in many problems of celelstial mechanics, plasma physics [3] and beam dynamics [4] . The stability problem was solved by the KAM [5] theory, but the dynamics and geometry of non-linear resonances and the related integrability conditions are not yet completly understood.
The singularities of the normalizing transformations can be investigated in the complexified phase space where they are an obstruction to the integrability of the system. Holomorphic dynamics was developed by Poincaré but general results in the hamiltonian case are still missing; perturbation theory only discriminates between analyticity on discs, as in the hyperbolic case, and its absence, as in the elliptic case, where the series have an asymptotic character [6]. The leading singularities are poles in the radial coordinate and their contribution is significant on a range given by the width of the corresponding islands since its  residue  is  just  its square [7].
The perturbation theory allows to construct a functional equation for the normalizing tranformation, which provides in principle the analyticity properties of the normalizing transformation as it was recently shown for the volume preserving maps [8]. The functional equation, first proposed by Siegel and Moser [9], cannot be solved but a sequence of approximations can be obtained using an iterative scheme.
We propose here an alternative approach based on the use of perturbation theory at a finite order to provide a quasi integrable map and the use of the KAM theory to make it integrable. The normalizing tranformations are analytic in the plane of the complex radial coordinate apart for cones containing the real and imaginary axis (if the quadratic term in the nonlinear frequency does not vanish). If the linear frequency satisfies a diophantine condition the envelope of cones is a domain whose boundary has an exponential tangency at the origin.
The analyticity domain is extended to reach any point of the real axis which corresponds to a KAM curve in the real phase space. From the domain are excluded only the neighborhhoods of points of the real axis corresponding to resonant frequencies. The measure of these neighborhoods decreases exponentially fast by approacing the origin, just as the measure of the chain of islands, in agreement with a theorem of Neishtadt [10].
The key point for proving analyticity is the boundedness of the divisors off the real radial axis. This can be more easily understood in the Siegel problem [11] where the small divisors are present, if we consider the complexification of the frequency. In this case the normal form is linear and the conjugation analytic in a disc whenever the frequency is diophantine. An addition of an imaginary part to the frequency preserves the analyticity in a disc even when

the real part of the frequency becomes resonant. This is a simplified model for understanding the analyticity properties of the conjugation of complex KAM curves and therefore the local properties of the model discussed above.

The plan of the work is the following: in section 2 we introduce the normal form for a polynomial map and bound the conjugation function In section 3 we describe the Siegel-Moser functional equations [12] for the Fourier components and analyze the singularities of its iterative solution, discussing the simplified functional equation for the majorant function. In section 4 we consider the Siegel problem for complex frequencies quoting the estimates on the radius obtained with majorant series and a KAM method. In section 5 we discuss the results for the complexified area preserving maps by quoting the basic theorem on the analyticity of the conjugation function.

## 2. NORMAL FORMS AND PERTURBATIVE BOUNDS

We consider an area preserving analytic map with an elliptic fixed point at the origin

$$\mathbf{M}: \qquad \begin{pmatrix} X' \\ Y' \end{pmatrix} = \mathcal{R}(\omega) \begin{pmatrix} X + F(X,Y) \\ Y + G(X,Y) \end{pmatrix} \tag{2.1}$$

where $\mathcal{R}(\omega)$ is the rotation matrix of an angle $\omega$, $F$ and $G$ are of order $\geq 2$ in $xy$ and satisfy the condition

$$\frac{\partial F}{\partial X} + \frac{\partial G}{\partial Y} + \frac{\partial F}{\partial X}\frac{\partial G}{\partial Y} - \frac{\partial F}{\partial Y}\frac{\partial G}{\partial X} = 1 \tag{2.2}$$

so that $M$ is area preserving. We assume $F$ and $G$ analytic in the unit polydisc. For $F = 0$, $G = X^2$ one obtains the well known Hénon map which corresponds to a kicked cubic oscillator.

An integrable map $U$ is given by

$$\mathbf{U}: \qquad \begin{pmatrix} x' \\ y' \end{pmatrix} = \mathcal{R}\left[\Omega(x^2 + y^2)\right] \begin{pmatrix} x \\ y \end{pmatrix} \tag{2.3}$$

where

$$\Omega = \omega + (x^2 + y^2)\Omega_2 + \dots \tag{2.4}$$

is an amplitude dependent frequency. The invariant curves of $U$ are circles. The map $M$ is said to be *integrable* if there is a change of coordinates $\boldsymbol{\Phi}$ tangent to the identity

$$\boldsymbol{\Phi}: \qquad \begin{cases} X = \Phi(x,y) = x + \phi(x,y) \\ Y = \Psi(x,y) = y + \psi(x,y) \end{cases} \tag{2.5}$$

which conjugates $\mathbf{M}$ with $\mathbf{U}$ according to

$$\mathbf{M} \circ \boldsymbol{\Phi} = \boldsymbol{\Phi} \circ \mathbf{U} \tag{2.6}$$

For a generic map $\mathbf{M}$ with $F$ and $G$ analytic in a polydisc, there is no analytic solution of (2.6) in an open neighborhood of the origin as a consequence the perturbative solutions are given by *divergent series*.

In order to understand this divergence we investigate the complexified version of the map, namely we consider $X$ and $Y$ complex and $\mathbf{M}$ as map of $\mathbf{C}^2$. The final goal is to discover the singularities of $\Phi$ and to relate them to the non-integrability of the initial map and to the presence of orbits topologically different from circles, whose existence is predicted by the Poincaré-Birkhoff theorem.

We first perform a linear change of coordinates (whose jacobian is constant even though not unit)

$$\begin{cases} Z = X + iY \\ W = X - iY \end{cases} \qquad\qquad \begin{cases} z = x + iy \\ w = x - iy \end{cases} \tag{2.7}$$

so that the map $\mathbf{M}$ now reads

$$\mathbf{M}: \begin{cases} Z' = e^{i\omega}[Z + f(Z,W)] \\ W' = e^{-i\omega}[W + g(Z,W)] \end{cases} \tag{2.8}$$

where $f = F + iG$, $g = F - iG$. When $X$ and $Y$ are real the two equations in (2.8) are complex conjugate. A generic map $\mathbf{M}$ can be conjugated with a map $\mathbf{U} = (U,V)$ in normal form up to any order $N$ if $\omega/2\pi$ is irrational, or up to the order $N = q - 2$ if $\omega/2\pi = p/q$ is rational, by a map $\boldsymbol{\Phi} = (\overline{\Phi},\overline{\Psi})$ where $\overline{\Phi} = \Phi + i\Psi = z + \overline{\phi}$ and $\overline{\Psi} = \Phi - i\Psi = w + \overline{\psi}$. If $\boldsymbol{\Phi}$ is a polynomial of order $N$ the map $\mathbf{U}$ will have remainder terms starting with order $N + 1$ namely, dropping hereafter the bar on the complex components of $\boldsymbol{\Phi}$

$$\mathbf{U} = \boldsymbol{\Phi}^{-1} \circ \mathbf{M} \circ \boldsymbol{\Phi}: \begin{cases} z' = e^{i\Omega(zw)}[z + f_N(z,w)] \\ w' = e^{-i\Omega(zw)}[w + g_N(z,w)] \end{cases} \tag{2.9}$$

Here $\Omega$ is a polynomial of order $[(N-1)/2]$ in $zw$.

*Recurrence.* Let $\Delta_\pm$ denote the operator

$$\Delta_\pm H \equiv H(e^{i\omega}z, e^{-i\omega}w) - e^{\pm i\omega}H(z,w)$$

where $H(z,w)$ denotes a function defined in $\mathbf{C}^2$ and $\Pi_\pm$ the projector on the subspace of functions such that $\Delta_\pm H = 0$.
We say that a map $\mathbf{U} = (U,V)$ of $\mathbf{C}^2$ is *in normal form* if $\Delta_+ U = 0$, $\Delta_- V = 0$ namely if $U = \Pi_+ U$, $V = \Pi_- V$. The recursive equations for the conjugation then read

$$\begin{cases} (1 - \Pi_+)[\Phi]_n = \Delta_+^{-1}(1 - \Pi_+)[Q]_n & [U]_n = \Pi_+[Q]_n \\[2ex] (1 - \Pi_-)[\Psi]_n = \Delta_-^{-1}(1 - \Pi_-)[Q']_n & [V]_n = \Pi_-[Q']_n \end{cases} \tag{2.10}$$

for $n \leq N$ and $[\ ]_n$ denotes a homogeneous polynomial of order $n$. In (2.10) $Q$ and $Q'$ are remainders defined by

$$\begin{cases} Q = e^{i\omega}f(\Phi,\Psi) - \phi(U,V) + \phi(e^{i\omega}z, e^{-i\omega}w) \\[2ex] Q' = e^{-i\omega}g(\Phi,\Psi) - \psi(U,V) + \psi(e^{i\omega}z, e^{-i\omega}w) \end{cases} \tag{2.11}$$

We notice that the remainder $[Q]_n, [Q']_n$ depend only on lower order terms of $\Phi, \Psi, U, V$: for instance $[Q]_2 = [f]_2, [Q']_2 = [g]_2$.

*Small Divisors:* the operator $\Delta_\pm$ generates, for an area preserving map, the so called small divisors which cause the divergence of the conjugation process. Indeed

$$\Delta_+^{-1} z^k w^j = \frac{z^k w^j}{e^{i\omega(k-j)} - e^{i\omega}} \qquad k - j \neq 1 \tag{2.12}$$

*Norms.* In order to estimate the conjugation function and the normal form map we introduce the following norms: if $\mathbf{z} = (z,w) \in \mathbf{C}^2$ then $||\mathbf{z}|| = \max\{|z|, |w|\}$ so that $||\mathbf{z}|| \leq R$ is a polydisc. Writing the expansion of $f(z,w)$ as

$$f = \sum_n [f]_n \qquad [f]_n = \sum_{k=0}^{n} f_{n,k} z^k w^{n-k} \tag{2.13}$$

we define

$$||[f]_n||_1 = \sum_{k=0}^{n} |f_{n,k}|, \qquad ||f||_R = \sum_n ||[f]_n||_1 R^n \tag{2.14}$$

and we assume $||[f]_n||_1 \leq 1$, $||[g]_n||_1 \leq 1$. Letting

$$D(n) = \min_{m \in \mathbb{Z}} |n\frac{\omega}{2\pi} - m|, \qquad S(n) = \frac{1}{\min_{1 \leq k \leq n} D(k)} \tag{2.15}$$

and observing that $|e^{ik\omega} - 1| \geq 4D(k)$ we can write

$$\begin{cases} ||(1 - \Pi_+)[\Phi]_n||_1 \leq \frac{1}{4}S(n+1)||[Q]_n||_1 \\ ||[U]_n||_1 \leq ||[Q]_n|_1 \end{cases} \tag{2.16}$$

and a similar equation where $\Phi$, $U$, $Q$ are changed into $\Psi$, $V$, $Q'$. Taking into account that $||[Q]_2||_1 \equiv ||[f]_2||_1 \leq 1$ and that $[U]_2 = 0$, we make the inductive hypothesis $||[Q]_n|| \leq \delta_n \mu_n c^{n-2}$ (the same for $Q'$), where

$$\mu_n = \sum_{s=2}^{n-1} \mu_s \sum_{m_1+m_2+\ldots+m_s=n,\ m_i \geq 1} \mu_{m_1}\mu_{m_2}\ldots\mu_{m_s}, \qquad n \geq 3, \quad \mu_1 = \mu_2 = 1 \tag{2.17}$$

and

$$\delta_n = S(3)\,S(4)\ldots S(n), \qquad n \geq 3, \quad \delta_1 = \delta_2 = 1 \tag{2.18}$$

and the following result holds

**Theorem** If $f$ and $g$ are analytic in the unit polydisc, then

$$\begin{cases} ||[\Phi]_n||_1 \leq S(n+1)\delta_n \mu_n c^{n-2} \\ ||[U]_n||_1 \leq \delta_n \mu_n c^{n-2} \end{cases} \tag{2.19}$$

for $n \geq 2$ and $||[\Phi]_1||_1 = 1$, with

$$c = \max\left\{3,\ \frac{3}{4} + 2S(3)\right\} \tag{2.20}$$

*Proof.* We first observe that (2.19) is trivially satisfied for $n = 2$ and assuming it true for $n - 1$ then we obtain $||[Q]_n||_1 \leq c^{n-2}c^{-1}[1 + 2\delta_n \mu_n]$, provided that

$$\delta_n = \max\begin{cases} \max_{2 \leq s \leq n-1} S(s+1)\delta_s \max_{m_1+m_2+\ldots+m_s=n,\ m_i \geq 1} \delta_{m_1}\ldots\delta_{m_s}, \\ \\ \max_{2 \leq s \leq n-1} \max_{m_1+m_2+\ldots+m_s=n,\ m_i \geq 1} S(m_1+1)\ldots S(m_s+1)\delta_{m_1}\ldots\delta_{m_s}, \end{cases} \tag{2.21}$$

having for convenience defined $S(2) = 1$. It then not hard to prove, taking into account the monotone character of the sequence $S(k)$, that (2.21) is satisfied by (2.18). As a consequence $||[Q]_N||_1 \leq c^{n-2}\delta_n \mu_n (3/c)$ and the induction holds for $c \geq 3$. To complete the proof we have to estimate $\Pi_1 \Phi$ which is not determined by the functional equation. To this end we impose that $\Phi$ is also area preserving (even though not strictly necessary) and obtain

$$||\Pi_+[\Phi]_n||_1 \leq S(n+1)\delta_n \mu_n A^{n-2}\frac{2S(3)}{c}\frac{(n-1)}{S(n+1)} \tag{2.22}$$

Finally recalling that

$$S(n)^{-1} = D(q_k) \leq \frac{1}{q_{k+1}} \leq \frac{1}{n}, \qquad \text{for } q_k \leq n < q_{k+1} \tag{2.23}$$

where $q_k$ are the denominators of the continued fraction truncation of $\omega/2\pi$ we can bound the fraction on the l.h.s. of (2.22) with $c^{-1}[3/4 + 2S(3)]$ and imposing this is less than 1 finally the induction is proved.

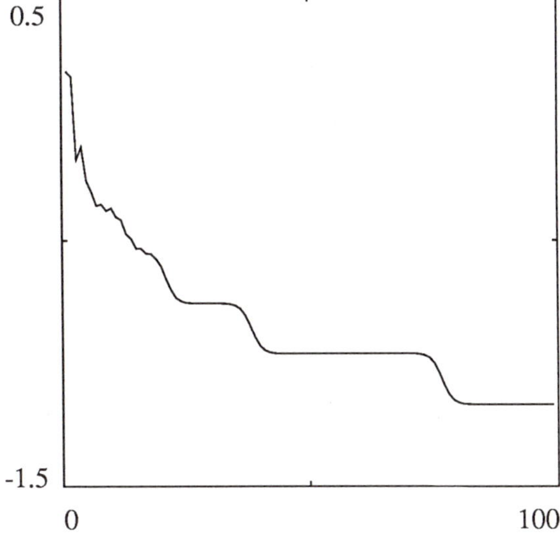

**Fig. 1** Ratios $||[\Phi]_{2n+1}||/||[\Phi]_{2n-1}||$ for the golden mean in the Hénon map

It was shown that $\mu_n$ is bounded by $n!$. Scaling the coordinates so that $||[F]_2||_1 \leq S(3)^{-1}$ we have $c = 3$ and $\delta_n = S(4)\ldots S(n)$. If we truncate the series at order $N$, using the diophantine estimate

$$D(n)^{-1} \leq \gamma n^\eta \quad \Longrightarrow \quad S(n) \leq \gamma n^\eta$$

and $\mu_n \leq c'n! \leq (N+1)^{n-2}$ (valid for $3 \leq n \leq N$ if $N+1 \geq 6c'$) then

$$||[\Phi]_n||_1 \leq \left[ c\gamma(N+1)^{1+\eta} \right]^{n-2}, \qquad ||[U]_n||_1 \leq \left[ c\gamma(N+1)^{1+\eta} \right]^{n-2} \tag{2.24}$$

The map in the new variables $z, w$ takes the form (2.9) and one can check that

$$||[\Omega]_n| \leq \left[ 2c\gamma(N+1)^{1+\eta} \right]^{n-1} \tag{2.25}$$

and prove the bound

$$||[f_N]_n|| \leq \frac{4^{-N}}{R_N^n}, \qquad ||[g_N]_n|| \leq \frac{4^{-N}}{R_N^n}, \qquad R_N^{-1} = 18\, c\gamma(N+1)^{1+\eta}$$

where $n \geq N+1$. As a consequence $f_N$ and $g_N$ are analytic in a polydisc of radius $R$.

*Nekhoroshev stability estimates.* [12] In the polydisc of radius $R$ we can estimate the remainders $f_N, g_N$ according to

$$||f_N||_R \leq 4^{-N} \left( \frac{R}{R_N} \right)^{N+1} \left( 1 - \frac{R}{R_N} \right)^{-1} \tag{2.26}$$

In a disc of radius $R \leq R_N/2$ the remainder is bounded by $8^{-N} \sim \exp(- [c\gamma R_N]^{-1/(1+\eta)})$ which is exponentially small with $1/R_N$. As a consequence one has stability for exponentially long times as it was proved in reference 5.

*Singularities and numerical results* The estimates obtained for the norms are close to the numerical results [13,14]: indeed using the estimates (2.24) the ratios $||[\Phi]_{n-1}||_1/||[\Phi]_n||_1$ are proportianal to

$$\frac{1}{n\,S(n+1)} = \frac{D(q_j)}{n}, \qquad \text{for } q_j \leq n < q_{j+1} \tag{2.27}$$

63

The numerical results exhibit constant ratios $\propto D(q_j)/q_j$ in intervals logarithmically shifted $[n_j, n_{j+1}[$ where $n_j \sim q_j \log q_j$.

A direct heuristic analysis of the series confirmed by the numerical results shows that any of these convergence discs is associated to a leading (continued fraction) resonance which produces a pole in $zw$. Letting

$$\begin{cases} z = re^{i\theta} & w = re^{-i\theta} \\ r = \sqrt{zw} & \theta = \dfrac{1}{2i} \log \dfrac{z}{w} \end{cases} \tag{2.28}$$

be polar coordinates, the map reads

$$U : \begin{cases} \theta' = \theta + \Omega(r^2) + a_N(r,\theta) \\ r' = r + b_N(r,\theta) \end{cases} \tag{2.29}$$

where $a_N$ and $b_N$ are analytic in

$$|\text{Im}\theta| \leq \Delta, \qquad |r| \leq \frac{R_N}{4} e^{-\Delta} \tag{2.30}$$

and both bounded by $4^{-N+2} \sim \exp\left[-(c'\gamma R_N)^{-1/(1+\eta)}\right]$.

We notice that the non-linear frequency $\Omega$ has the resonant value $2\pi p/q$ (where $p$ is the nearest integer to $q\omega/2\pi$) when

$$\Omega(r^2) \sim \omega + r^2 \Omega_2 \left[1 + O(N^{-2})\right] = 2\pi \frac{p}{q} \tag{2.31}$$

namely for $r = r_q$ where

$$r_q^2 \equiv \frac{2\pi}{-\Omega_2} \epsilon_q, \qquad \epsilon_q = \frac{\omega}{2\pi} - \frac{p}{q} \tag{2.32}$$

The resonance $p/q$ has the following relevant features

   i) it is *real* if $\epsilon_q \Omega_2 < 0$
  ii) it affects the *Fourier components* $k = 1 \pm q$ of $\Phi$
 iii) it corresponds to a *pole* of $\Phi$ whose residue is proportional to $r_q^{q-2}$ and $r_q^q$ for $k = 1 - q$ and $k = 1 + q$ respectively.

The residue of a real resonance ($r_q$ real) for $k = 1 - q$ is just the square of the width of the corresponding chain of islands. When $r_q$ is imaginary we say that the resonance is virtual since it does not appear in phase space but still corresponds to a pole and causes the divergence of the series.

The divergence mechanism of the perturbation series is clear: when a resonance occurs and the topology of orbits changes, according to the Poincaré -Birkhoff theorem, a pole arises in the conjugation function of the corresponding perturbation order. As long as the pole $r_q$ dominates the ratio of subsequent terms in the perturbation series is given by $|r_q| \propto |\epsilon_q| = D(q)/q$. In the next section we show the precise mechanism of pole generation.


## 3. THE SIEGEL-MOSER EQUATIONS


These are a system of functional equations written for the Fourier components of the map, which allow to describe, at least at an heuristic level, the singularities of the normalizing transformation. Let $H$ be a function from $\mathbf{C}^2$ to $\mathbf{C}$ and define $\{H\}_k$ according to

$$\{H(e^{i\omega} z, e^{-i\omega} w)\}_k = e^{ik\omega} \{H(z,w)\}_k \tag{3.1}$$

and notice that $\{H\}_k$ is a series not a polynomial. If $\omega$ is non resonant then $\{H\}_k$ is the $k$-th Fourier component of $H$. We shall then write the conjugation function as

$$\Phi = \sum_k \{\Phi\}_k \tag{3.2}$$

We observe also that the normal forms have only the $k = \pm 1$ components $U = \Pi_+ U = \{U\}_1$ and $V = \Pi_- V = \{V\}_{-1}$ and by choosing $\{\Phi\}_1 = z$, $\{\Psi\}_{-1} = w$ the following functional equations must be satisfied

$$\begin{cases} (e^{ik\Omega} - e^{i\omega})z = e^{i\omega}\{f(\Phi, \Psi)\}_1 & \\[2mm] \{\Phi\}_k = \dfrac{e^{i\omega}\{f(\Phi, \Psi)\}_k}{e^{ik\Omega} - e^{i\omega}} & k \neq 1 \\[2mm] \{\Psi\}_k = \dfrac{e^{-i\omega}\{f(\Phi, \Psi)\}_k}{e^{ik\Omega} - e^{-i\omega}} & k \neq -1 \end{cases} \qquad (3.3)$$

These equations can be solved recursively starting with $\Phi = z$ and $\Psi = w$ and the following result can be proved.

**Theorem** At step $n$ the functions $\Phi, \Psi$ agree with the perturbative solution of order $n$ up to the same order.

As a consequence the order $n$ polynomial truncation of $\Omega$ obtained after $n$ iterations of (3.3) is real.
Starting at order 1 with $\Phi = z, \Psi = w$ and $\Omega^{(1)} = \omega$ at order 3 Siegel-Moser equations read

$$\{\Phi\}_k = \frac{e^{i\omega}\{f(z + \phi^{(2)}, w + \psi^{(2)})\}_k}{e^{ik\Omega} - e^{i\omega}} \simeq c\frac{r^{|k|}}{e^{ik[\omega + \Omega_2 r^2]} - e^{i\omega}}\left(1 + O(r)\right) \qquad (3.4)$$

The poles of $\Phi$ are located at $(k-1)\omega + k\Omega_2 r^2 + 2\pi p = 0$ and for $k = 1 \pm q$ we have $r = r_q$ where

$$(r_q^{\pm})^2 = \frac{2\pi}{-\Omega_2}\frac{q\,\epsilon_q}{q \pm 1} \qquad (3.5)$$

and the residues $\gamma_q^{\pm}$ for $k = 1 \pm q$

$$\gamma_q^- = c_q^- \frac{(r_q^-)^{q-2}}{2(1-q)\Omega_2}, \qquad \gamma_q^+ = c_q^+ \frac{(r_q^+)^q}{2(1+q)\Omega_2}, \qquad (3.6)$$

In the first case $k = 1 - q$ it is found that the residue is proportional to the distance of the separatrices of the pendulum like orbits associated with the $2\pi p/q$ resonance which one can compute using the resonant perturbation theory [16]. If we use the diophantine estimate we see that the the bound for the closest pole is obtained from $|r_{N+1}^-|^2 \geq (2\pi/|\Omega_2|)D(N+1)/N$ and is proportional to $(N+1)^{-(1+\eta)/2}$.
It is interesting to compute the norms of $\Phi_<, \Psi_<$ defined by the sum of a finite number of Fourier components $|k| \leq N$ and $\Omega$ which satisfy the functional equation

$$\mathbf{M} \circ \boldsymbol{\Phi}_< = \boldsymbol{\Phi}_< \circ \mathbf{U} + \mathcal{R}_N \qquad (3.7)$$

where $\mathbf{U} = (e^{i\Omega}z, e^{-i\Omega}w)$ and $\mathcal{R}_N$ is a remainder term. Indeed letting $\Omega = \omega + \hat{\Omega}$ and defining

$$W = \max\left\{\|\Phi\|_R, \|\Psi\|_R, \|\hat{\Omega}\|_R\right\}$$

one finds a functional equation for $W$ which suggests a possible mechanism to generate non-polar singularities in the limit process.
One crucial property of our norms, shared by the so called majorant series, is that $\|F \circ G\|_R = \|F\|_R \circ \|G\|_R$. Assuming that $f$ and $g$ start with third order terms as it is always the case after one perturbative step and that they are analytic in the unit polydisc, we can write

$$\|f\|_R \leq \frac{R^3}{1 - R} \qquad (3.8)$$

As a consequence we have

$$\max_{|z|\leq R,|w|\leq R}|\hat{\Omega}| \leq ||\hat{\Omega}||_R \leq ||\log\left[1+\frac{1}{z}\{f(\Phi,\Psi\}_1]||_R \leq$$
$$\frac{R}{1-R}\circ\frac{1}{R}||\{f(\Phi,\Psi\}_1||_R \leq \frac{R}{1-R}\circ\frac{1}{R}\frac{W^3}{1-W}\leq\frac{W^3}{R(1-W)-W^3} \tag{3.9}$$

We observe also that we can write

$$|e^{ik\Omega}-e^{i\omega}|\geq\left[r_k^2-||\hat{\Omega}||_R\right],\qquad r_k^2=\frac{4D(k-1)}{|k|},\qquad k\neq 0,\ r_0=4D(1) \tag{3.10}$$

provided that $||\hat{\Omega}(r)||_R < \rho_{\min}$ where $\rho_{\min}=\min r_k^2$. In the same way we obtain

$$||\Phi||_R \leq R+\frac{1}{\rho_{\min}-||\hat{\Omega}||_R}\sum_{k\neq 1}||\{f(\Phi,\Psi\}_k||_R \leq R+\frac{1}{\rho_{\min}-||\hat{\Omega}||_R}||\{f(\Phi,\Psi\}||_R =$$
$$R+\frac{1}{\rho_{\min}-||\hat{\Omega}||_R}\frac{W^3}{1-W} \tag{3.11}$$

Finally we have

$$W < r+\frac{W^2}{\rho_{\min}-W}\left[\frac{W}{(1-W)}+\frac{W(\rho_{\min}-W)}{R(1-W)-W^3}\right] \tag{3.12}$$

Making the a priori estimates $W\leq cR\leq\rho_{\min}/4$ with $c>1$ assuming $\rho_{\min}\leq c\rho_{\min}\leq 1/2$ the square bracket is bounded by 1 and we have

$$W\leq R+\frac{W^2}{\rho_{\min}-W}\leq R+\frac{c^2R^2}{\rho_{\min}-W} \tag{3.13}$$

The inequality is satisfied with

$$W\leq W_-(R)=\frac{R+\rho_{\min}}{2}\left[1-\sqrt{1-4R\frac{\rho_{\min}+c^2R}{(\rho_{\min}+R)^2}}\right]\leq cR\sqrt{2\frac{cR+\rho_{\min}/c}{\rho_{\min}+R}}<cR \tag{3.14}$$

provided that $4R(\rho_{\min}+c^2R)(\rho_{\min}+r)^{-2}\leq 1/2$ which holds for $c\geq 4$, as one can check using $\sqrt{1-x}\geq 1-x/\sqrt{2}$ for $x\leq 1/2$. Choosing $c=4$ our a priori estimates are satisfied; since $W_-(R)$ is defined by a series whose convergence radius is $\rho_0=\rho_{\min}/(1+2c)$, our estimates hold for $R\leq\rho_{\min}/1(4c)=\rho_{\min}/16$.
We notice that any approximation to majorant function $W(R)$ obtained with a finite number $n$ of iterations has only poles while in the limit $n\to\infty$ has algebraic singularities are obtained. This shows the difficulty of a correct description of the singularity structure of the functions satisfying the Siegel-Moser equations.
The estimates on $||[\Phi]_n||_1$ and $||[\Omega]_n||_1$ for $n\leq N$ are proportional to $\rho_{\min}^{-n}$ and agree with (2.24). However by refining the procedure these bounds become $\rho_{\min}^{-n/2}$ in agreement with the presence of the nearest pole at $r=r_{\min}$ if $\Omega_2\neq 0$; if $\Omega_{2k}=0$ for $k\leq m-1$ the bounds become $\rho_{\min}^{-n/(2m)}$ and they are obviously optimal.


## 4. COMPLEX FREQUENCIES


Before quoting the KAM results on the analyticity of the conjugation of the complex area preserving map with an integrable map we wish to discuss the conjugation of a complex map

$$z'=e^{i(\omega+i\alpha)}z+f(z),\qquad f(z)=O(z^2) \tag{4.1}$$

with its linear part. If $\omega$ is non-resonant and $\alpha = 0$ this is the well known Siegel problem and the conjugation function $z = \Phi(\varsigma)$ with the linear map $\varsigma' = e^{i\omega}\varsigma$ is analytic in the Siegel disc whose radius has lower bound $R$ given by the Bruno function

$$R = Ce^{-2B(\omega)}, \qquad B(\omega) = \sum_{k=0}^{\infty} \frac{\log q_{k+1}}{q_k} \tag{4.2}$$

where $q_k$ are the denominators of the continued fraction truncations. When $\omega = 0$ the majorant series estimates gives $R = C'e^{-\alpha}(1 - e^{-\alpha})$ so that as $\alpha \to 0$ the lower bound to the radius vanishes linearly with $\alpha$. When the frequency is complex as in (4.1) the estimate is slightly more tricky if we require that it should be continuous as $\alpha \to 0$. In this case letting $s$ be an integer such that

$$D(q_{s+1}) \le \frac{2\alpha}{e} < D(q_s) \tag{4.3}$$

by slightly modifying Bruno's proof [2] we find

$$R \ge Ce^{-2B(\omega,\alpha)}, \qquad B(\omega,\alpha) = \sum_{k=0}^{s} \frac{\log q_{k+1}}{q_k} + \log\left(\frac{\alpha}{e}\right) \sum_{k=s+1}^{\infty} \frac{1}{q_k} \tag{4.4}$$

If $\omega/2\pi$ approaches zero or a rational we see that the lower bound turns out to be proportional to $\alpha^{2m}$ where $1 < m = \sum_{k\ge 0} q_k^{-1} \le \sum_{k\ge 0} F_k^{-1} \le 7/2$ rather than $\alpha$ ($F_k$ Fibonacci numbers), so that it is not optimal in this respect. The Bruno estimate can be related to the diophantine parameters of $\omega$ taking into account the relation

$$q_k + q_{k+1} \le \gamma q_k^{\eta} \implies \gamma = \min_k \frac{q_k + q_{k+1}}{q_k^{\eta}} \tag{4.5}$$

which gives $R \ge C_1 \gamma^{-2m}$. The KAM estimate is sharper $R \ge C_1[\gamma^{-1} + \alpha/(2e)]$ and the limit for both $\alpha \to 0$ and $\gamma^{-1} \to 0$ is continuous. The crucial point is the solution of the homologic equation we sketch in the case $\eta = 1$ for simplicity; following the proof [16] which gives accurate estimates and letting $\lambda = e^{i(\omega+i\alpha)}$ we write

$$\phi(\lambda\varsigma) - \phi(\varsigma) = f(\varsigma) \tag{4.6}$$

which is solved by

$$\sum_{n\ge 2} \frac{f_n \varsigma^n}{\lambda^n - \lambda} \tag{4.7}$$

The norm of the perturbation is measured by

$$\epsilon = ||df/dz||_r = \sum_{n=2}^{\infty} n||f_n||r^{n-1} \tag{4.8}$$

The homologic equation requires an estimate of the divisors. We observe that

$$D(k) = |e^{ik(\omega+i\alpha)} - 1| = \left[(1 - e^{-k\alpha})^2 + 4\sin^2\frac{k\omega}{2}e^{-k\alpha}\right]^{1/2} \tag{4.9}$$

Choosing the greatest of the two terms appearing in the square bracket of (4.9) and using diophantine estimate, (for simplicity we quote the results for $\eta = 1$) we have

$$D(k) \ge \begin{cases} \dfrac{2}{\gamma k} & \text{for } k \le k_{\max} = \dfrac{2e}{\gamma\alpha} \\[2ex] \dfrac{\alpha}{e} & \text{for } k > k_{\max} \end{cases} \tag{4.10}$$

Finally the estimate for the norm of $d\phi/dz$ reads

$$||d\phi/dz||_{r(1-\theta)} \le \epsilon e^{-\alpha}h \tag{4.11}$$

where

$$h = \max \begin{cases} 2^{-1}\gamma n(1-\theta)^n & \text{for } n \leq k_{\max} \\ e\alpha^{-1}(1-\theta)^n & \text{for } n > k_{\max} \end{cases} \leq \begin{cases} \gamma/(2\theta) & \theta \geq \alpha\gamma/(2e) \\ e/\alpha & \theta < \alpha\gamma/(2e) \end{cases} \tag{4.12}$$

since $\theta < 1$ and for $x > 0, y > 0$ one has $\min\{x, y\} \leq 2/(x^{-1} + y^{-1})$ then

$$h \leq \frac{e}{\alpha\theta} \min\left\{\theta, \frac{\alpha\gamma}{2e}\right\} \leq \frac{1}{\theta} \frac{1}{\dfrac{1}{\gamma} + \dfrac{\alpha}{2e}} \tag{4.13}$$

The difference with the usual case is that $\gamma^{-1}$ is replaced with $\Gamma^{-1} = \gamma^{-1} + \alpha/(2e)$. as a consequence recalling the result of [16] and assuming $f_2 \neq 0$ and $\xi = (1 - 1/\mu)$ we have

**Theorem** If $f(z)$ is analytic and $\|df/dz\|_R \leq C R$ in $|z| < R$, then the map is conjugated with its linear part in a disc of radius at least $R_s$ where, letting $c$ be a numerical constant

$$R_s = \frac{1}{c\Gamma} \left\{ (c\Gamma R)^{1/\eta} [1 - (c\Gamma R)^{1/\eta}] \right\}^{\eta} \tag{4.14}$$

If $f$ is holomorphic so that $R$ can be arbitrary large then the square bracket has a maximum $4^{-\eta}$ and

$$R_s \propto \Gamma^{-1} = \frac{1}{\gamma} + \frac{\alpha}{2e}$$

The effect of adding an imaginary part is evident. If $\alpha$ goes to zero we recover the usual estimate for the Siegel problem, while if the frequency is rational the limit goes to zero.

## 5. ANALYTICITY BY KAM METHOD

In order to investigate the analyticity of the normalizing transformation we need to conjugate the map (2.9) which is analytic in a polydisc of radius $R_N$ with the integrable map $\mathcal{M}$

$$\mathcal{M} = \begin{cases} \theta' = \theta + \Omega(r^2) \\ r' = r \end{cases}$$

by using a KAM technique, which allows to detect analyticity domains different from polydiscs. To this end we use the polar representation (2.27) and observe that this map preserves a measure $r + \mu(r, \theta)$ where $\mu$ is analytic in $|\text{Im}\,\theta| \leq \Delta/2$ and $|r| \leq R_N e^{-\Delta}/4$ with a bound there given by $2^{-N}/\Delta^2$. Following Moser's scheme and introducing a cutoff and letting $\gamma_0$ and $\eta_0$ be the diophantine parameters of the linear frequency of the map (2.1) we denote now with $\omega_0$, we prove the following

**Theorem** The map $M$ given by (2.27) is conjugated analytically with $\mathcal{M}$ in the disc $|r| \leq e^{-\Delta} R_N/8$ excluding two angular sectors around the real and imaginary axis defined by straigt lines of aperture $\chi$ provided that $C\, 2^{-N} |\tan \chi|^{-3} (\gamma_0/\Delta^{1+\eta_0})^2 < 1$. In term of the radius $R_N$ the bound on $\chi$ reads

$$|\tan \chi| \geq C_2 \exp\left(-\left[\frac{1}{c\gamma_0 R_N}\right]^{\frac{1}{1+\eta_0}}\right)$$

Taking the envelope of the angular sectors for $N \geq 3$ we obtain as analyticity domain the complement of the disc $|r| \leq R_3$ with respect to neighborhoods of the real and imaginary

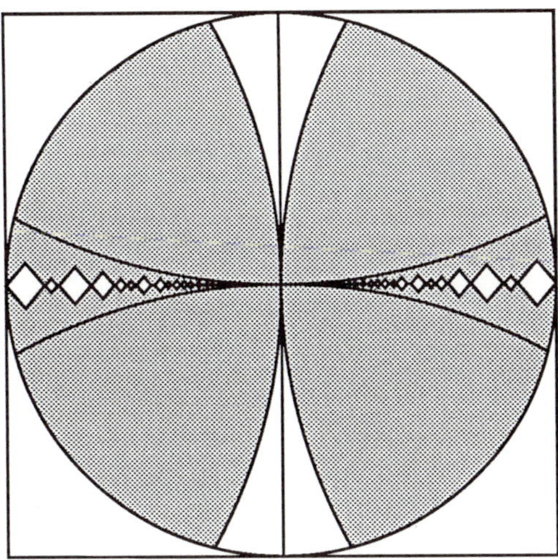

**Fig. 2** Plot of the analyticity region, dashed, in the complex $r$ plane for the function $\Phi$

axis delimited by curves having an exponential tangency order at the origin with the real and imaginary axis (see figure 2).

In the proof the invariance with respect to the measure $\mu(r)$ and the presence of a fixed point at the origin play an essential role to replace the intersection condition used by Moser which has no evident complex extension.

Starting from this initial analyticity domain we can further continue to the real axis with domains which reach the real axis at points where the frequency is diophantine. We exclude in this way neigborhoods of the points of the real axis where the frequency is resonant.

**Theorem**  The map $M_N$ can be analytically conjugated with the map $M$ in $|\mathrm{Im}\theta| < \Delta/4$ and in the images by $\rho = r^2$ of the discs in the $\rho$ plane defined by: $|\rho - \rho_c| \leq |\alpha|/2$ with $\rho_c = \tau + i\alpha$ where $\Omega(\tau) = \omega$ provided that $C' \, 2^{-N}(\Gamma/\Delta^{1+\eta})^7 \leq 1$ where $\Gamma^{-1} = \gamma^{-1} + \nu|\alpha|$ where $\nu = \min\{1, |\Omega_2|/2\pi\}$ and $\gamma, \eta$ are the diophantine constants of the frequency $\omega$.

The consequence of this theorem is that the allowed discs are given by

$$\gamma^{-1} + \nu|\alpha| \geq E_2(R_N) = C_2 \exp\left(-\left[\frac{1}{c_2\gamma_0 R_N}\right]^{\frac{1}{1+\eta_0}}\right)$$

At any diophantine point the conjugacy is possible provided that $\gamma^{-1} \geq E_2(R_N)$. In this case the analyticity domain, envelope of discs, is a cone of aperture $\pi/3$ with vertex at $\rho = \tau$ and axis parallel to the imaginary axis.

At any point $\tau_*$ such that $\Omega(\tau_*) = 2\pi p/q$ is resonant the conjugation is possible for $\nu|\alpha| > E_2$. One can also show that the non-analyticity region in this case is a rombus with the center at $\rho = \tau_*$ and the ratio of vertical and horizontal diagonals is $\sqrt{3}$ the length of the horizontal diagonal being $E_2/(\nu q^2)$. In figure 2 a scketch of the non-analyticity domains is given.

## Conclusions

We have shown that the dynamical behaviour of an area preserving map can be better understood by considering its complexification. In this case the conjugation with an integrable map has singularities in a neighborhood of the real and imaginary axis of the complexified radial coordinate corresponding to the topology changes of the orbits, while it is analytic in a strip

in the plane of the complexified angle. The leading polar singularities allow to understand in great detail the behaviour of the perturbation expansion. The rigorous KAM procedure bounds the region where the singularities are confined but the determination of their nature and their precise location is a hard problem. The techniques of holomorphic dynamics could certainly be helpful in this respect and the connection between the complex time singularity analysis of Panlevé type for an interpolating flow and the Liouville integrability whose obstructions are revealed by the complex phase space singularities will also be explored.
We wish finally to stress that the choice of area preserving maps is dictated not only by its physical relevance but also by the dramatic simplifications it allows from the geometric and analytic point of view.


REFERENCES


1. H. Poincaré *Les Méthodes Nouvelles de la Méchanique Céleste* vol. 3 Gautier-Villars Paris (1899)
   G. D. Birkhoff **Acta Mathematica 43**, 1 (1920)

2. A. D. Bruno *Trans. Moscow Mat. Soc.* **25**, 131 (1971)

3. A. Bazzani, G. Turchetti, P. Mazzanti, G. Servizi *Normal forms for hamiltonian maps and nonlinear effects in particle accelerators* Il Nuovo Cimento $\underline{B102}$,51 (1988)

4. A. Bazzani, M. Malavasi, S. Siboni, C. Pellacani, S. Rambaldi, G. Turchetti *Poincaré map and anomalous transport in a magnetically confined plasma* IL Nuovo Cimento $\underline{B103}$, 659 (1989)

5. A. N. Kolmogorov Dokl. Akad. Nauk. SSSR **98**, 527 (1954)
   V. I. Arnold *Russ. Math. Surv.* **18**, 9 (1963
   J. Moser *Nachr. Akad. Wiss. Göttingen Math. Phys. K1* **1**, 1 (1962)

6. G. Turchetti *Nekhoroshev stability estimates for symplectic maps and physical applications* in *Number theory in Physics* ed. P. Moussa, Springer pag. 223,(1990)
   A Bazzani, S. Marmi, G. Turchetti *Nekhoroshev estimates for non resonant symplectic maps* Celestial Mechanics **47**, 333 (1990)

7. G. Servizi, G. Turchetti *Phys. Lett* **A151**, 485 (1990)

8. A. Bazzani *Normal form theory for volume preserving maps* submitted to ZAMP

9. C. L. Siegel, J. Moser *Lectures in Celestial Mechanics* pag 166 Springer Verlag (1971)

10. A. I. Neishtadt PMM S.S.S.R. **45**, 776 (1982)

11. L. Siegel *Annals of Mathematics* **42**, 806 (1941)

12. N. N. Nekhoroshev *Russ. Math. Surv.* **32**, 1 (1977)

13. G. Benettin, A. Giorgilli, G. Servizi, G. Turchetti, *Resonances and asymptotic behaviour of Birkhoff series*, Phys. Letters **A95**, 11 (1983).

14. G. Servizi, G. Turchetti, *Perturbative expansion for area preserving maps* Il Nuovo Cimento **B95**, 121 (1986)

15. A. Bazzani, M. Giovannozzi, G. Servizi, E. Todesco, G. Turchetti *Resonant normal forms and interpolating hamiltonians for area preserving maps* subm. to Physica D.

16. C. Liverani, G. Turchetti *Improuved KAM estimates for the Siegel radius* Journ. of Stat. Phys. **45**, 1071 (1986)

NOTE ON A COMPLEX ECKHAUS EQUATION

Michael F. Jørgensen and Peter L. Christiansen

Laboratory of Applied Mathematical Physics
The Technical University of Denmark
DK-2800 Lyngby, Denmark

Silvana de Lillo

Department of Physics, University of Perugia and National
Institute of Nuclear Physics, Section Perugia
I-06100 Perugia, Italy

Leonor Cruzeiro-Hansson

Department of Crystallography
Birkbeck College
London WC1E 7HX, UK

ABSTRACT

A complex Eckhaus equation which is easily Bose quantized possesses an infinity of local conservation laws. Thus this equation is S-integrable as well as C-integrable.

In a recent paper Calogero[1] made a heuristic distinction between C-integrable and S-integrable nonlinear partial differential equations (PDE's) namely, equations that are linearizable by an appropriate <u>Change of variables</u> and equations that are integrable via the <u>Spectral transform technique</u>. One example of a C-integrable PDE is the Eckhaus equation[2]

$$i\psi_t + \psi_{xx} + [2(|\psi|^2)_x + |\psi|^4]\psi = 0 \qquad (1)$$

which can be obtained by introducing the transformation

$$\phi(x,t) = \psi(x,t) \exp\left\{\int^x dx' |\psi(x',t)|^2\right\} \qquad (2)$$

into the linear Schrödinger equation

$$i\phi_t + \phi_{xx} = 0 . \qquad (3)$$

In contrast to the nonlinear Schrödinger equation,

*Chaotic Dynamics: Theory and Practice*
Edited by T. Bountis, Plenum Press, New York, 1992

$$i\psi_t + \psi_{xx} + |\psi|^2\psi = 0 \ , \tag{4}$$

Eq. (1) is <u>not</u> S-integrable since the infinity of conservation laws which can be derived from the infinity of conservation laws for the linear Schrödinger equation (3) by means of the transformation (2) are all non-local in character.

Replacing Eq. (2) by the complex transformation

$$\phi(x,t) = \psi(x,t) \exp\left\{i \int^x dx' |\psi(x',t)|^2\right\} \tag{5}$$

one arrives at the complex Eckhaus equation

$$i\psi_t + \psi_{xx} + [2i(|\psi|^2)_x + |\psi|^4]\psi = 0 \tag{6}$$

(Eq. (2.12) of Ref. 1 with $s = 1$, $L_1 = 2$, $L_2 = 0$. Thus conditions (2.14) and 15) for S- and C- integrability are satisfied). Application of Eq. (5) transforms the first three conserved quantities of the linear Schrödinger equation (3), norm, momentum, and energy, in the following manner

$$\int_{-\infty}^{\infty} \phi\phi^* \, dx \ \Rightarrow \ \int_{-\infty}^{\infty} \psi\psi^* \, dx \ , \tag{7a}$$

$$i\int_{-\infty}^{\infty} (\phi^*\phi_x - \phi\phi_x^*)dx \ \Rightarrow \ \int_{-\infty}^{\infty} \left\{i(\psi^*\psi_x - \psi\psi_x^*) - 2|\psi|^4\right\} \, dx \ , \tag{7b}$$

and

$$\int_{-\infty}^{\infty} \phi_x\phi_x^* \, dx \ \Rightarrow \ \int_{-\infty}^{\infty} \left\{|\psi_x|^2 + |\psi|^6 - i(\psi^*\psi_x - \psi\psi_x^*)|\psi|^2\right\} \, dx \ . \tag{7c}$$

Thus we see that the complex transformation (5) provides local conservation laws for Eq. (6). Indeed, it can be shown that the infinity of the conserved quantities for the linear Schrödinger equation,

$$\int_{-\infty}^{\infty} \phi^* \, \frac{\partial^n}{\partial x^n} \, \phi \, dx \ , \quad n = 0,1,\cdots,$$

leads to an infinity of local conservation laws for Eq. (6). Thus, in this sense the complex Eckhaus equation is S-integrable as well as, of course, C-integrable. We are currently trying to find a Lax pair for this equation.

It seems possible to quantize the complex Eckhaus equation in the following simple manner: Replacing $\phi$ and $\phi^*$ by annihilation and creation operators for boson fields $\hat{\phi}$ and $\hat{\phi}^\dagger$ with commutation relations

$$[\hat{\phi}(x),\hat{\phi}(y)] = 0, \quad [\hat{\phi}^\dagger(x),\hat{\phi}^\dagger(y)] = 0 \ , \tag{8a}, (8b)$$

and

$$[\hat{\phi}(x),\hat{\phi}^\dagger(y)] = \delta(x-y) \ , \tag{8c}$$

the inverse of transformation (5)

$$\psi(x,t) = \phi(x,t) \, \exp\left\{- i\int^{x} dx' |\phi(x',t)|^2\right\} , \qquad (5a)$$

written in operatorform

$$\hat{\psi}(x,t) = \hat{\phi}(x,t) \, \exp\left\{- \frac{i}{2}\int^{x} (\hat{\phi}(x',t)\hat{\phi}^{\dagger}(x',t) + \hat{\phi}^{\dagger}(x',t)\hat{\phi}(x',t))dx \right\} , \qquad (9)$$

leads to commutation relations for the operators $\hat{\psi}$ and $\hat{\psi}^{\dagger}$ for the $\psi$-field of the same simple form as Eqs. (8).


## Acknowledgements

Alwyn Scott is thanked for helpful discussions. Financial support from the Danish Natural Science Research Council (under grant 11-9026) and from the EEC Science Programme under grant no. (89 1000 79/JU 1) is acknowledged.


## REFERENCES

1.    F. Calogero, "Why Are Certain Nonlinear PDEs Both Widely Applicable and Integrable?", "What is Integrability?" (Ed. V.E. Zakharov), Springer-Verlag, Berlin 1991, pp. 1-62.
2.    F. Calogero and W. Eckhaus, "Nonlinear Evolution Equations, Rescalings, Model PDEs and Their Integrability. I", Inverse Problems $\underline{3}$, 229-262 (1987).

# INTEGRABILITY OF DISCRETE-TIME SYSTEMS


B. Grammaticos*, G. Karra*, V. Papageorgiou#
and A. Ramani°

*LPN, Univ. Paris VII
Tour 24-14, 5ᵉ étage
75251 Paris, France
#Dept. of Mathematics and Computer Science
and Institute for Nonlinear Studies
Clarkson University
Potsdam NY 13699-5815, USA
°CPT, Ecole Polytechnique
91128 Palaiseau, France


KEYWORDS/ABSTRACT


Integrability / Discrete-time systems / Mappings / Painlevé
equations / Singularity confinement / Linearizable mappings

A new integrability criterion for discrete-time systems, based
on the notion of the confinement of the singularities that may
appear in rational mappings is presented. Discrete analogues
of the Painlevé equations are derived as second-order,
nonautonomous mappings, with the help of this integrability
detector. Moreover a parallel between continuous and discrete
systems is established by showing that to each kind of
"continuous" integrability there exists a "discrete" analogue.


## 1. INTRODUCTION

The study of dynamical systems was focused, till recently,
mainly on continuous systems. Discrete systems have been
considered as oversimplified, "toy", approximations of the
physical reality and their study was limited to simple,
exploratory models. This is, of course, quite unfair and the
situation is now changing rapidly. To start with, our
knowledge of the physical world through simulations is based
always on discrete systems. One can even go one step further
and ask the question of the continuous or discrete nature of

space-time. However this is a philosophical question and what a physicist can do about it is to obtain bounds on the time and space lattice constants[1]. Discrete systems are also fundamental in the sense that they contain, in the appropriate limits, a multitude of continuous ones[2]. Finally, discrete-time systems are "next of kin" to quantum systems. Although almost no systematic study exists in this direction one can, in some sense, consider the discretization as a quantum deformation of a continuous-time (classical) system[3].

Discrete systems have been used extensively for the modelling and understanding of chaotic phenomena[4]. In fact the evidence for the existence of the various routes to chaos was obtained through the study of simple mappings. On the contrary, little has been realized in the domain of integrability of discrete systems. Although integrable mappings have been known for years[5], few systematic studies were undertaken in this direction. Ablowitz and Ladik have studied integrable discretization schemes for nonlinear partial differential equations[6]. The advantage of such an approach is evident: the integration scheme is *nonlinearly* stable. Hirota[7] has produced an amazing amount of results on nonlinear discrete systems. (As was also the case for Hirota's bilinear formalism, the nonlinear community was astonishingly slow to realize the importance of his results on discrete systems). Hirota's approach was based on a discrete variation of his bilinear formalism and dealt mainly with the construction of explicit solutions to his (integrable) equations. We do not intend to review here the results of the past twenty years on integrable mappings. A more detailed guide to the bibliography can be found in Ref. 3. To put it in a nutshell very few results were produced till recently when the situation started to change dramatically. The reason for this can be found in the fact that techniques similar to those used in the study of continuous PDE's were introduced in the study of discrete systems. Thus whole families of integrable mappings and lattices (multidimensional discrete systems) where obtained. Moreover the study of partition functions in some models in statistical physics led to recursion relations that are simply integrable mappings. An important result of these approaches was the discovery of the discrete analogues of the Painlevé I and II transcendental equations. Thus Brezin and Kazakov[8], studying a model of 2-D quantum gravity obtained the nonautonomous mapping:

$$x_{n+1} + x_{n-1} + x_n = \frac{an+b}{x_n} + c \tag{1}$$

which, besides being integrable, leads to $P_I$ at the continuous limit. Integrability in this context is synonymous to the existence of a Lax pair or, equivalently, of a Zakharov-Shabat linearization. The actual integration of $d-P_I$ through the isomonodromy method was given by Fokas and collaborators[9]. A restricted form of the discrete $P_{II}$ was obtained by Periwal and Shevitz[10] in a conformal field theory model and its general form by Nijhoff and one of us (V.P.)[11]. The latter method started from a discrete lattice of KdV type and performed a discrete similarity reduction. Just as in the continuous case one obtains the continuous $P_{II}$ in the same way this procedure led to the d-$P_{II}$:

$$x_{n+1} + x_{n-1} = \frac{x_n(an+b)+c}{1-x_n^2} \qquad (2)$$

Quispel and collaborators[12] obtained the discrete analogues of elliptic functions as rational mappings, incorporating and generalizing previously known discrete systems. Integrability of mappings in the sense of Liouville, was investigated by Bruschi et al[13].

All these results have raised again the question of a criterion of integrability for discrete systems. The question is not new. It has been addressed several times in the past without success. The existence of a successful integrability detector for continuous systems (the Painlevé method[14]) was the necessary motivation, but it was not clear how one could transpose an essentially local approach to systems that are nonlocal by construction. The situation changed recently when we introduced[15] the singularity confinement method. The present paper is devoted to a presentation of the method and of its first applications. Among them a most important result is the derivation of the discrete forms of the Painlevé equations $P_{III}$, $P_{IV}$ and $P_V$.


## 2. THE SINGULARITY CONFINEMENT METHOD

How did this detector of integrability of discrete systems come to be? From our experience on continuous systems it was clear that no such criterion can be developped unless a "critical mass" of known integrable systems is attained. This was realized only recently. Moreover it turned out that these integrable mappings or lattices possessed denominators. So the question that arose naturally was: what happens to the mapping when the denominator vanishes?

Let us illustrate this point with the KdV lattice[16]:

$$x_j^{i+1} = x_{j+1}^{i-1} + \frac{1}{x_j^i} - \frac{1}{x_{j+1}^i} \qquad (3)$$

The structure of the lattice associated to the evolution equation (3) can be easily assessed in Fig.1. Now let us assume that during the successive applications of (3) the value of x at $(i-2,j+1)$ becomes zero. This is not at all impossible and the point where this occurs depends on the initial data i.e. the singularity induced is movable. From (3) it is clear that x diverges at both sites $(i-1,j)$ and $(i-1,j+1)$ and that it vanishes again at $(i,j)$. Now the crucial question is what happens at the sites $(i+1,j-1)$ and $(i+1,j)$. It turns out that, due to the precise form of (3), there exists a fine cancellation leading to finite values at both sites. Thus the singularity is perfectly confined.

This would not been true if the evolution did not have the form (3). Indeed if we consider a lattice of the form:

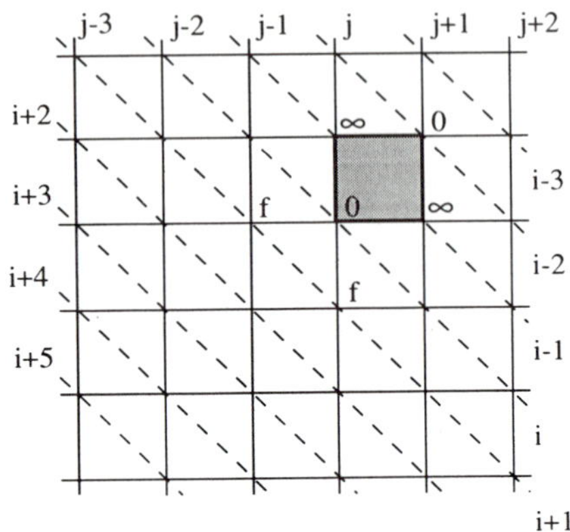



Fig 1.The index j runs over the vertical lines while i
labels the slanted (dashed) lines. Evolution can be
understood as taking place towards increasing i's.

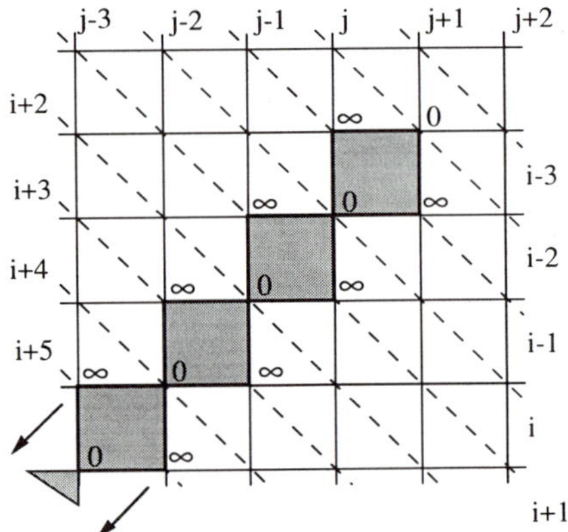



Fig 2.Same as in Fig 1. The singularities extend here all
the way to infinity.

$$x_j^{i+1} = x_{j+1}^{i-1} + \frac{1}{x_j^i} - \frac{\lambda}{x_{j+1}^i} \qquad (4)$$

with $\lambda \neq 1$ (which, presumably, suffices in order to destroy integrability) leads to the singularity pattern of Fig. 2 with singularities extending all the way to infinity. The situation is reminiscent of the difference between the singularity structures of integrable and nonintegrable continuous-time systems. Integrable systems have the Painlevé property: their singularities are isolated and single-valued; thus one can make a loop around each of them and come back to the starting point. In nonintegrable systems the singularities condense to natural boundaries[17] that one cannot cross. Thus we can see from this first example that integrability in discrete-time systems is related to confined (movable) singularities.

Let us present some more examples before formulating any conjecture. We start with a completely solvable mapping: the discretized anharmonic oscillator. Hirota[7] has presented the (integrable) mapping:

$$(x_{n+1} - 2x_n + x_{n-1})/\delta^2 = -\alpha x_n - \beta x_n^2 (x_{n+1} + x_{n-1})/2 \qquad (5)$$

that corresponds to a discretization of a quartic oscillator. Introducing $z_n = (\beta \delta^2/2)^{1/2} x_n$ and $\mu = 1 - \alpha \delta^2/2$ we remark that (5) can be written as:

$$z_{n+1} + z_{n-1} = \frac{2\mu z_n}{1 + z_n^2} \qquad (6)$$

i.e. the well-known McMillan mapping[5]. Hirota produced the complete solution of (5) in terms of elliptic functions: $x(t) = x_o \mathrm{cn}(\kappa; \Omega(t - t_o))$ where $\Omega$ and $\kappa$ are given by $1 - \mathrm{cn}(\delta\Omega)/\mathrm{dn}^2(\delta\Omega) = \alpha\delta^2/2$ and $2\kappa^2 = x_o^2 \beta \delta^2 / [\mathrm{sn}(\delta\Omega)/\mathrm{dn}(\delta\Omega)]^2$ and the time variable is discretized $t = n\delta + t_o$. Now let us assume that for a given $n$ $x(n)$ diverges. Using the addition formulae for the elliptic cosine we can verify easily that $x(n\pm 1) = \pm i (\beta\delta^2/2)^{-1/2}$ and, also, $x(n+2) = -x(n-2)$. Thus $x(n-1)$ has precisely the value that guarantees a divergence for $x(n)$ and $x(n+1)$ has the value that compensates this divergence. Moreover, the memory of the initial condition, that has propagated up to $x(n-2)$, survives past the singularity in $x(n+2)$.

The mapping:

$$x_{n+1} = x_n + \frac{1}{y_n}$$

$$y_{n+1} = y_n - x_n - \frac{1}{x_n} \qquad (7)$$

introduced by Devaney[18] is well known for its chaotic

behaviour. So, let us study the structure of its singularities. Suppose that, for some n, $x_n$ vanishes while $y_n$ is finite. Then at the next iteration we obtain $x_{n+1}=1/y_n$ while $y_{n+1}$ diverges. In fact for all the subsequent iterations $x_k$ keeps the same value while $y_k$ is infinite. In this case the singularity is *not* confined.

We can now formulate our conjecture concerning the integrability of mappings in the following (intuitive) way: the movable singularities of integrable mappings are confined i.e. they are cancelled out after a finite number of steps. Moreover the memory of the initial condition is not lost whenever a singularity is crossed. One must be careful as to whether singularities are possible, within a given discrete-time system, without assuming divergent initial conditions. In the latter case the singularity is not a movable one.

The implementation of the singularity confinement method is quite simple. Given a mapping, one must first find all possible ways a singularity can emerge (this step follows closely the first step of the algorithm for ODE's where one looks for all possible leading singular behaviours). The system is said to have passed the test (and is thus a candidate for integrability) if this divergence does not propagate in (discrete) time, i.e. that it remains confined. The second step is therefore to find how far it has to propagate before it has a chance to leave room for a regular behaviour (this is somewhat reminiscent of the "search for resonances" in the ARS algorithm), and finally one has to verify that indeed the singularity does *not* propagate beyond that (this last step is the equivalent of the "resonance condition").

As an illustration of the singularity confinement algorithm used as integrability detector we will examine a mapping of the Quispel type:

$$x_{n+1} + x_{n-1} = -\frac{dx_n^2+ex_n+f}{ax_n^2+bx_n+c} \qquad (8)$$

that generalizes the Hirota–MacMillan one. Applying our criterion we will deduce the values of the parameters a, b, …, f for the mapping to be integrable. Let us assume that at a certain time step the denominator vanishes: $ax_n^2+bx_n+c=0$. We obtain then $x_{n+1}=\infty$ and $x_{n+2}=-d/a-x_n$. The problems arise at the next step: $x_{n+3}$ will diverge because of the presence of $x_{n+1}$ unless the r.h.s. diverges also so as to cancel the divergence of $x_{n+1}$. This can happen only if $ax_{n+2}^2+bx_{n+2}+c=0$. Substituting $x_{n+2}$ we obtain an equation for $x_n$ and demand that it be proportional to $ax_n^2+bx_n+c=0$. We obtain thus as (only) constraint for the confinement of the singularity d=b. This is precisely the only relation between the parameters a,b,…,f that makes the mapping a member of the integrable Quispel family as it was already found by MacMillan [5]. It goes without saying that not all mappings will be covered by our conjecture: strictly polynomial mappings do not have movable

singularities at finite distance.  The Hénon map[19], a paradigm
of chaotic system:

$$x_{n+1} + x_{n-1} = \alpha + \beta x_n^2 \tag{9}$$

is such an example. The singularity confinement approach
cannot be applied here. Still a very indirect indication
concerning the nonintegrability of the mapping is obtained by
comparing (9) to the Quispel-type mappings (8). We can always
take a=0 in the denominator but *not* b=0 since in this case (as
d=b) the mapping becomes a linear one. Thus the Hénon mapping
cannot belong to the Quispel family. This is, admitedly, no
proof of nonintegrability but still an indication in the right
direction.


3. DISCRETE PAINLEVE TRANSCENDENTS

   The Painlevé-Gambier transcendental equations were
discovered at the beginning of the century[20]. The term
"transcendental" refers to the fact that the dependence of the
solution on the integration constants is not algebraic nor
does the equation admit a first integral involving the
constant algebraically. The method used for the derivation of
these transcendental equations is related to what came to be
known in the past decade as singularity analysis[14]. These
equations have the so-called Painlevé property, i.e. their
solutions are meromorphic functions of the independent
variable, or, equivalently, their (movable) singularities are
just poles[21]. Their solutions were given only in the past few
years. Following the pioneering work of Ablowitz and Segur[22],
it was shown that the Painlevé equations can be linearized in
terms of integrodifferential equations, using the Inverse
Scattering Transform scheme.

   Discrete forms of the Painlevé equations appeared recently
in relation to physical problems. Their derivation has so far
been fortuitous, since no systematic method existed. One
important application of the singularity confinement method is
to systematically derive the discrete forms of the Painlevé
equations I to V (recovering the known forms for equations I
and II)[23].

   The starting point for the application of the singularity
confinement method is the Quispel family of mappings:

$$x_{n+1} = \frac{f_1(x_n) - x_{n-1}f_2(x_n)}{f_2(x_n) - x_{n-1}f_3(x_n)} \tag{10}$$

The reason for this choice is that this mapping is integrable
in terms of elliptic functions. Since the autonomous limits of
the continuous Painlevé equations are integrable in terms of
elliptic functions one would expect to find the discrete forms
by "de-autonomizing" the Quispel map.

In order to gain some insight into the choice of the
we rewrite the Quispel map as:

$$f_3(x_n)\Pi - f_2(x_n)\Sigma + f_1(x_n) = 0 \tag{11}$$

where $\Sigma = x_{n-1}+x_{n+1}$, $\Pi = x_{n-1}\,x_{n+1}$ and the $f_i$ are quartic
polynomials, and ask that this equation go over to the
continuous Painlevé under consideration at the continuous
limit. We introduce a lattice parameter $\delta$ and obtain:

$$\Sigma = 2x + \delta^2 x'' + \mathcal{O}(\delta^4) \tag{12}$$
$$\Pi = x^2 + \delta^2(xx''-x'^2) + \mathcal{O}(\delta^4)$$

and when we extract from Eq.12 the part involving derivatives
we obtain a continuous limit ($\delta\to 0$) of the form:

$$x'' = \frac{f_3(x)}{xf_3(x)-f_2(x)} x'^2 + g(x). \tag{13}$$

So if we are aiming at a specific Painlevé equation, the first
thing to do is to choose $f_2$, $f_3$ in such a way as to get
$f_3(x)/(xf_3(x)-f_2(x))$ to coincide with the factor multiplying $x'^2$
in that equation.

For $P_I$ and $P_{II}$ we have clearly $f_3=0$. The form we choose for
$d-P_I$ is:

$$x_{n+1} + x_{n-1} = -x_n + B(n) + \frac{C(n)}{x_n} \tag{14}$$

A first condition for singularity confinement is $B(n+1)-B(n)=0$. Thus $B$ is constant. Once this is implemented we find
a second (and sufficient) condition: $C(n+3)-C(n+2)-C(n+1)+C(n)=0$. The general solution of this equation is
$C(n)=\alpha n+\beta+\gamma(-1)^n$. We can remark here that the last term of $C(n)$
will disappear at the continuous limit. Here again (for $\gamma=0$ and
$B=b$) we obtain:

$$x_{n+1} + x_{n-1} + x_n = b + \frac{\alpha n+\beta}{x_n} \tag{15}$$

that is precisely the discrete $P_I$.

For $d-P_{II}$ we start from the form (8) after taking $a=e=1$ by
rescaling the variables:

$$x_{n+1} + x_{n-1} = -\frac{x_n^2+B(n)x_n+C(n)}{x_n(x_n+1)} \tag{16}$$

As in the autonomous case we assume that $x_n=0$, which leads to
the following condition for singularity confinement: $C(n+1)-C(n-1)-B(n+1)+B(n)=0$. Similarly starting from the second root
of the denominator $x_n=-1$, we find: $C(n+1)-C(n-1)+B(n-1)-B(n)=0$.

Combining the two equations we obtain: $B(n+1)-2B(n)+B(n-1)=0$ and $B(n)=\lambda n+\mu$. Substituting back we obtain for C: $C(n+1)-C(n-1)=\lambda$ and thus $C(n)=\lambda n/2+\nu$. With these expressions of B and C and with $z_n=2x_n+1$ we find:

$$z_{n+1} + z_{n-1} = \frac{z_n(\alpha n+\beta)+\gamma}{1-z_n^2} \tag{17}$$

i.e. the dicrete $P_{II}$.

In the case of $P_{III}$ we have $x''= x'^2/x + g(x)$. First of all we should point out that the continuous form of $P_{III}$ we are going to work with is:

$$w'' = \frac{w'^2}{w} + e^z(aw^2+b) + e^{2z}\left(cw^3+\frac{d}{w}\right) \tag{18}$$

obtained from the usual one[21] through the transformation $z\to e^z$ that absorbs the $w'/z$ term. This form agrees with (13) if we simply take $f_2=0$. In that case, in Quispel's approach, $f_1$ and $f_3$ have one quadratic common factor and assuming that this remains true when the coefficients become n-dependent, the mapping takes the form:

$$x_{n-1}\,x_{n+1} = \frac{\kappa(n)x_n^2 + \zeta(n)x_n + \mu(n)}{x_n^2 + \beta(n)x_n + \gamma(n)} \tag{19}$$

To fix the n-dependent coefficients we will study the singularity behavior as described before. When one solves for $x_{n+1}$ there are two possible sources of singularity for this mapping. Either $x_n$ is a zero of the denominator $x_n^2 + \beta(n)x_n + \gamma(n)$ or $x_{n-1}$ becomes zero. In the first case, the singularity sequence is the following: $x_{n+1}$ diverges, $x_{n+2}$ has a finite value $\kappa(n+1)/x_n$ and $x_{n+3}$ would in principle be proportional to $1/x_{n+1}$ and thus zero. This would lead to a new divergence. The only way out is to ask that $x_{n+2}$ also be a zero of the appropriate denominator, so that $x_{n+3}$ does not vanish. Expressing $x_{n+2}$ in terms of $x_n$ and taking into account that this must be true for both zeros $x_n$ of $x_n^2 + \beta(n)x_n + \gamma(n)$, we obtain $\beta(n) = \beta(n+2)\kappa(n+1)/\gamma(n+2)$ and $\gamma(n) = \kappa^2(n+1)/\gamma(n+2)$. Multiplying $x_n$ by an arbitrary function of n does not change the form of Eq.20 but only affects the coefficients. This scaling freedom allows us to take a constant value $\beta$ for $\beta(n)$, resulting to $\kappa(n+1)=\gamma(n+2)$, $\gamma(n) = \gamma(n+2)$. Thus the $\gamma$'s and $\kappa$'s must be constants within a given parity: $\gamma(\text{even})=\kappa(\text{odd})=\gamma_+$, $\gamma(\text{odd})=\kappa(\text{even})=\gamma_-$.

In order to study the second kind of singularity, we start with $x_n$ such that $x_{n+1}$ vanishes (i.e. $\kappa(n)x_n^2 + \zeta(n)x_n + \mu(n)=0$). We find then that $x_{n+2}$ has a finite value $\mu(n+1)/\gamma(n+1)x_n$ and this would lead to a divergent $x_{n+3}$ unless the numerator also vanishes. Substituting the expression for $x_{n+2}$ and using the fact that again this must be true for both zeros of $\kappa(n)x_n^2 + \zeta(n)x_n + \mu(n)$, we obtain $\mu(n) = \zeta(n)\mu(n+1)/\zeta(n+2) = \mu^2(n+1)/\mu(n+2)$. The solution to these equations is straightforward: $\mu(n)=\mu_0\lambda^{2n}$ and $\zeta(n)=\zeta_{0,\pm}\lambda^n$, where $\mu_0$, $\zeta_{0,\pm}$ are constants, the $\pm$ sign being related to the parity of n. Note that, in that case, there is *no second kind* of singularity at all! Indeed $x_{n+3}$ is not allowed to diverge even though $x_{n+1}=0$. (This is reminiscent of the case of continuous equations where, if a denominator appears, one must consider the values of the dependent variable that makes this denominator vanish to ascertain that this does *not* generate a singularity).

In order to go to the continuous limit, we start with a change of the mapping variable $y_n=\lambda^{n/2}x_n$. Moreover, at the continuous limit the distinction between even and odd must disappear. We thus write d-$P_{III}$ as:

$$y_{n-1}\,y_{n+1} = \frac{\gamma y_n^2 + \zeta_0\lambda^{n/2}y_n + \mu_0\lambda^n}{\lambda^n y_n^2 + \beta\lambda^{n/2}y_n + \gamma} \tag{20}$$

The continuum limit is obtained by letting the lattice parameter $\delta$ go to zero, while $\gamma \approx -1/c\delta^2$ and all the other constants are of order unity: $\beta\approx a/c$, $\mu_0\approx -b/c$, $\zeta_0\approx -d/c$. Simultaneously, one must take $\lambda \approx 1+2p\delta$, leading to Eq. (18) with $e^z$ replaced by $e^{pz}$. But p can be absorbed by rescaling z and redefining a, b, c and d, thus we recover $P_{III}$.

We will not give the precise derivation of d-$P_{IV}$ and d-$P_V$. It can be found in Ref. 23. Here are the final forms for d-$P_{IV}$:

$$x_{n+1}\,x_{n-1} + x_n(x_{n+1} + x_{n-1}) = \frac{-(an+b)x_n^3 + (\varepsilon_0-\tfrac{1}{4}(an+b)^2)x_n^2 + \mu}{x_n^2 + (an+b)x_n + (\gamma_0+\tfrac{1}{4}(an+b)^2)} \tag{21}$$

and for d-$P_V$:

$$(2x_n-1)x_{n+1}x_{n-1}-x_n(x_{n+1}+x_{n-1})$$
$$= \frac{\tfrac{1}{2}(\sigma-\alpha_0\lambda^{2n})x_n^3 + \left\{\theta+\tfrac{1}{4}(\sigma+\alpha_0\lambda^{2n}-2\rho_0\lambda^n)\right\}x_n^2 - 2\mu x_n + \mu}{\alpha_0\lambda^{2n}x_n^2 + (\rho_0\lambda^n-\alpha_0\lambda^{2n})x_n + \tfrac{1}{4}(\sigma+\alpha_0\lambda^{2n}-2\rho_0\lambda^n)} \tag{22}$$

the latter corresponding to the "symmetric" form of $P_V$:

$$w''=\frac{1}{2}\left(\frac{1}{w} + \frac{1}{w-1}\right)w'^2 + a\frac{w}{w-1} + b\frac{w-1}{w} + ce^z w(w-1) + de^{2z}w(w-1)(2w-1) \tag{23}$$

As in the case of d-$P_{III}$ the appropriate limits of d-$P_{IV}$ and d-$P_V$ lead to the continuous $P_{IV}$ and $P_V$.

One further property of the Painlevé equations[24] that is verified by the discrete ones is the parameter coalescence that gives the following reduction chain: $P_V \rightarrow \{P_{IV}, P_{III}\} \rightarrow P_{II} \rightarrow P_I$. This is a further indication that the d-P's are indeed the discrete Painlevé equations.

Some questions remain open at this point. While the linearization of d-$P_I$ and d-$P_{II}$ has been given, no Lax pair is known yet for d-$P_{III}$, d-$P_{IV}$ and d-$P_V$. We are convinced that this will be taken care of in the near future. Moreover no discrete form has been obtained yet for $P_{VI}$. This is a problem presenting technical difficulties that, although considerable, should not be insuperable.

## 4. RICCATI MAPS

The mappings we have considered in the previous sections were integrable either thanks to the existence of invariants or through the existence of a Lax pair leading to a Zakharov-Shabat linearization. However the study of continuous-time systems reveals the existence of an "intermediate" class of systems, namely those linearizable to differential equations, often through a Cole-Hopf transformation. It is clear that something analogous exists for mappings.

The best-known linearizable ODE is the Riccati equation ($x' = \frac{dx}{dt}$):

$$x' = \alpha x^2 + \beta x + \gamma \tag{24}$$

where $\alpha, \beta$ and $\gamma$ are functions of t. Its linearization is obtained by the Cole-Hopf transformation:

$$x = -v'/\alpha v \tag{25}$$

that transforms (24) to:

$$v'' - (\beta + \alpha'/\alpha)v' + \alpha\gamma v = 0 \tag{26}$$

The discrete equivalent of (24) is just the homographic map:

$$x_{n+1} = \frac{bx_n + c}{ax_n + d} \tag{27}$$

As $a \neq 0$ (otherwise (27) is affine), we can always have $d \neq 0$ (by translation) in which case we can take $d=1$. Equation (27) goes over to the Riccati at the continuous limit. By writing $x_{n+1} = x_n + \varepsilon x' + \mathcal{O}(\varepsilon^2)$ and taking $a = -\alpha\varepsilon$, $c = \gamma\varepsilon$ and $b = 1 + \beta\varepsilon$ we recover exactly equation (24) in the limit $\varepsilon \rightarrow 0$.

The discretized Riccati was initially studied by Hirota[7], who obtained a discrete form equivalent to (27) starting from his bilinear formalism. This approach offers also a natural framework for the final linearization of the Riccati. In order to linearize (27) it suffices to put $x=u/v$ and equate separately numerator and denominator. We find thus:

$$u_{n+1}=bu_n+cv_n \qquad (28a)$$

$$v_{n+1}=au_n+v_n \qquad (28b)$$

The linear system (28) allows one to construct the solution $x_n$ at every (discrete) time step, starting from a given initial condition $x_0$ at $n=0$. It is interesting to remark that equation (28b) is the one defining the Cole-Hopf relation at the limit $\varepsilon \to 0$: $u_n=(v_{n+1}-v_n)/a$ goes over to $u=-v'/\alpha$.

Higher order linearizable mappins exist as well. Second order mappings were studied from the point of view of linearizability[25]. We have shown that these mappings have the general form:

$$x_{n+1} = \frac{f_1(x_n) - x_{n-1}\, f_2(x_n)}{f_4(x_n) - x_{n-1}\, f_3(x_n)} \qquad (29)$$

where $f_k(x_n)=a_k x_n+b_k$. This form (29) is a generalisation of the Quispel map with $f_4 \neq f_2$ but with linear $f_k$'s. However not all such mappings are integrable. We can thus ask the question: under which conditions can the mappings (29) be integrable? The natural framework for the investigation of this question is the "singularity confinement" method. A glance to eq. (29) suffices to convince us that an infinite value for $x_i$ ($i=n$, $n\pm1$) does not play any particular role. In fact, the relation between the $x_i$ is "bi-homographic" and thus infinity can be taken to any finite value by a simple homographic transformation of variables that only changes the values of the $a_k$ and $b_k$. Still, (29) may pose a subtler problem. It may turn out that for a certain $n$ the mapping (apparently) loses one degree of freedom. This occurs when $x_{n+1}$ is defined independently of $x_{n-1}$, and this happens whenever:

$$f_1(x_n)f_3(x_n)-f_2(x_n)f_4(x_n) = 0 \qquad (30)$$

Once $x_n$ is obtained from (30), one can compute $x_{n+1}$ simply as:

$$x_{n+1} = \frac{f_1(x_n)}{f_4(x_n)} = \frac{f_2(x_n)}{f_3(x_n)} \qquad (31)$$

*unless* $x_{n-1}$ was such that both the numerator and the denominator of the fraction defining $x_{n+1}$ vanished, i.e.

$$x_{n-1} = \frac{f_1(x_n)}{f_2(x_n)} = \frac{f_4(x_n)}{f_3(x_n)} \; . \tag{32}$$

Thus one sees the two ways in which the "singularity confinement" can be preserved: either (32) is satisfied, or it is not, in which case $x_{n+1}$ is determined and independent of $x_{n-1}$. In this latter case one degree of freedom will be definitely lost as $x_{n+2}$ will be determined in terms of $x_n$ only, unless both numerator and denominator of the fraction that define it vanish, i.e.

$$x_n = \frac{f_1(x_{n+1})}{f_2(x_{n+1})} = \frac{f_4(x_{n+1})}{f_3(x_{n+1})} \tag{33}$$

(note that the coefficients $a_k$ and $b_k$ of the $f_i(x_{n+1})$ are *not* in general the same as those of the $f_i(x_n)$: they are indeed functions of n). Equation (33) means that $x_n$ must satisfy two relations in terms of the $a_k$ and $b_k$ at times n and n+1.

In the first case where (32) is satisfied, on the other hand, it would seem that a degree of freedom suddenly *appears* at time n+1. The only way out is to demand that $x_n$ was determined by $x_{n-1}$ only, independent of $x_{n-2}$, i.e. that one already had at the previous time step:

$$x_n = \frac{f_1(x_{n-1})}{f_4(x_{n-1})} = \frac{f_2(x_{n-1})}{f_3(x_{n-1})} \; . \tag{34}$$

This, in turn, means that $x_n$ must satisfy two relations in terms of the $a_k$ and $b_k$ at times n and n-1.

In summary, for the "singularity confinement" condition to be verified, $x_n$ must satisfy either (31) and (33) or (32) and (34), when $x_n$ is a solution of (30). Four conditions for integrability are thus obtained. Now we can come back to the initial question on the conditions for linearizability. To make a long story short, the conditions for linearizability turn out to be *precisely* the same as the ones for integrability.

Equation (29) is the generic form of the second-order linearizable mapping. However when some of the $f_i$'s vanish the analysis of the generic case cannot be transposed in a straightforward way. This point will be illustrated with the following example. Let us consider the case of (29) with $f_3 \equiv 0$ i.e. the mapping:

$$x_{n+1} = \frac{f_1(x_n) - x_{n-1} f_2(x_n)}{f_4(x_n)} \tag{35}$$

In order to simplify the computations we take $a_4 = 1$ and $a_2 = -1$. This can be realized through a n-dependent rescaling of the variable x, and a suitable redefinition of the remaining $a_k$, $b_k$. What are the possible singular behaviours of the mapping (35)? First let us take $f_2=0$ at some time n, i.e. $x_n=b_2$. This leads to $x_{n+1}=(a_1 b_2+b_1)/(b_2+b_4)$ and the mapping will lose one degree of freedom unless $x_{n+2}$ becomes indeterminate of the form $0/0$. These two constraints lead to: $a_1(n)=-(b_4(n+1)+b_2(n-1))$ and $b_1(n)=b_2(n)b_2(n-1)-b_4(n)b_4(n+1)$. Thus the mapping takes the following form:

$$(x_{n+1}+b_4(n+1))(x_n+b_4(n))=(x_{n-1}-b_2(n-1))(x_n-b_2(n)) \qquad (36)$$

In order to integrate the mapping we introduce the quantity:

$$\Omega_n = \frac{x_{n+1}+b_4(n+1)}{x_n-b_2(n)} \qquad (37)$$

whereupon (36) reduces to:

$$\Omega_n \Omega_{n-1}=1 \qquad (38)$$

with obvious solution $\Omega_n=\Omega_0^{(-1)^n}$. Injecting this solution in (37) we obtain a linear equation for $x_n$.

The "singularity" $f_2=0$ is not the only one that can appear in the mapping (35). Another form of singularity exists whenever $f_4=0$. However, if we ask that $x_n+b_4(n)=0$ at some time n this means, from (36), that at time $(n-1)$ we must have $x_{n-1}-b_2(n-1)=0$, i.e. $f_2(n-1)=0$. Thus, the singularity $f_4=0$ cannot precede the $f_2=0$ one in mapping (36). In fact, if we ask that $f_4=0$ without being preceded by $f_2=0$ then we obtain a totally new branch of the mapping (35). So we start with $x_n=-b_4(n)$, leading to $x_{n+1}=\infty$ and $x_{n+2}=a_1(n+1)-b_4(n)$. For $x_{n+3}$ to be finite we must have $f_2(n+2)=0$ and, using the value of $x_{n+2}$ we obtain the relation $a_1(n+1)=b_4(n)+b_2(n+2)$. Before writing the resulting mapping explicitly, we introduce the auxiliary function $c_1$ (discrete primitive of $b_1$): $b_1(n)=c_1(n)-c_1(n-1)$. We have thus:

$$(x_n x_{n+1}+b_4(n)x_{n+1}-b_2(n+1)x_n-c_1(n))-(x_n x_{n-1}+b_4(n-1)x_n-b_2(n)x_{n-1}-c_1(n-1))=0$$
$$(39)$$

Thus, in this case, the mapping (35) is just the (discrete) derivative of a Riccati mapping. Equation (39) can be integrated once to $(x_n x_{n+1}+b_4(n)x_{n+1}-b_2(n+1)x_n-c_1(n))=K$ and the resulting Riccati linearized through the usual Cole-Hopf transformation.

# 5. CONCLUSIONS

In the preceding sections we have presented our recent findings on the integrability of discrete-time systems. As a synopsis we will stress here the most important points.

First, new integrable mappings that are the discrete equivalents of the continuous Painlevé equations have been obtained. Thus the list of the discrete-Painlevé is now almost complete.

Second, a new method (the singularity confinement) has been devised for the detection of integrability in discrete-time systems. The power of the method is undeniable given the richness of the results already obtained. Still the practical implementation of the method is not algorithmic to the point that it can be fully automatized. This is not an altogether negative point as it promotes reflection instead of thoughtless production of results.

The last point and most important of the three is that it has been made clear that a close parallel exists between discrete and continuous systems. To each type of "continuous" integrability we have been able to assign a "discrete" analogue. Moreover the main criterion used for its detection, the Painlevé property, presents considerable similitudes in the two domains. Singularity singlevaluedness for continuous systems becomes singularity confinement for discrete ones but still singularity is the key word.

Discrete systems are attracting more and more attention and their integrability is at the center of intensive research. We are convinced that our new method will constitute a precious tool in this domain and that it will lead to a host of interesting results.


## ACKNOWLEDGMENTS


The authors are greatly indebted to the following colleagues who have participated through direct collaboration or stimulating discussions to the elaboration of these novel ideas on discrete systems: M.J. Ablowitz, F. Bureau, H. W. Capel, B. Dorizzi, A. Fokas, J. Hietarinta, R. Hirota, M.D. Kruskal, F. W. Nijhoff, J. Satsuma and J.M. Strelcyn.


## REFERENCES


1.  R.P. Feynman, Intl. Jour. Theor. Phys. 21 (1982) 467.
2.  F.W. Nijhoff, V. Papageorgiou and H. Capel, "Integrable time-discrete systems: lattices and mappings", talk at the International Workshop on Quantum Groups, Leningrad 1990, preprint INS #166/90, and references therein.
3.  R. Hirota, J. Phys. Soc. Japan, 50 (1981) 3785.
4.  A.J. Lichtenberg and M.A. Lieberman, "Regular and Stochastic Motion", Springer-Verlag, New-York (1983).
5.  E.M. McMillan, in *Topics in Modern Physics*, eds. W.E. Brittin and H. Odabasi, (Colorado Associated Univ. Press, Boulder, 1971), p. 219.
    E. Date, M. Jimbo and T. Miwa, J. Phys. Soc. Jpn 51 (1982) 4125, 52 (1983) 388,766.

F.W. Nijhoff, G.R.W. Quispel and H. Capel, Phys Lett 97A (1983) 125.
A.P. Veselov, Funct. Anal. Appl. 22 (1988) 83.
P.A. Deift and L.C. Li, Comm. Pure Appl. Math. 42 (1989) 963.
Yu. B. Suris, Phys. Lett. 145A (1990) 113.
H. Capel, F.W. Nijhoff and V. Papageorgiou, Phys. Lett. 155A (1991) 337.

6.   M.J. Ablowitz and F.J. Ladik, Stud. Appl. Math. 55 (1976) 213, 57 (1977) 1.

7.   R. Hirota, J. Phys. Soc. Jpn 43 (1977) 1424, 2074, 2079, 45 (1978) 321, 46 (1979) 312, 56 (1987) 4285.

8.   E. Brezin and V. Kazakov, Phys. Lett. B236 (1990) 144.

9.   A.R. Its, A.V. Kitaev and A.S. Fokas, Usp. Math. Nauk 45 (1990) 135.

10.  V. Periwal and D. Shevitz, Phys. Rev. Lett. 64 (1990) 1326.

11.  F.W. Nijhoff and V. Papageorgiou, Phys. Lett. 153A (1991) 337.

12.  G.R.W. Quispel, J.A.G. Roberts and C.J. Thompson, Phys. Lett. 126A (1988) 419, Physica D34 (1989) 183.

13.  M. Bruschi, O. Ragnisco, P. M. Santini and Tu Gui- Zhang, Physica D49 (1991) 273.

14.  M.J. Ablowitz, A. Ramani and H. Segur, Lett. Nuov. Cim. 23 (1978) 333.
A. Ramani, B. Grammaticos and A. Bountis, Phys. Rep. 180 (1989) 159.

15.  B. Grammaticos, A. Ramani and V. Papageorgiou, Phys. Rev. Lett. 67 (1991) 1825.

16.  V. Papageorgiou, F.W. Nijhoff and H. Capel, Phys. Lett. 147A (1990) 106.

17.  Y.F. Chang, J. M. Greene, M. Tabor, J. Weiss, Physica D8 (1983) 3183.

18.  R. Devaney, Comm. Math. Phys. 80 (1981) 465.

19.  M. Hénon, Quart. J. Appl. Math. 27 (1969) 291.

20.  P. Painlevé, Acta Math. 25 (1902) 1.
B. Gambier, Acta Math. 33 (1910) 1.

21.  E. L. Ince, "Ordinary differential equations", Dover, New York, 1956.

22.  M.J. Ablowitz and H. Segur, Phys. Rev. Lett. 38 (1977) 1103.

23.  A. Ramani, B. Grammaticos, and J. Hietarinta, Phys. Rev. Lett. 67 (1991) 1829.

24.  P. Painlevé, C. R. Acad. Sci. Paris 143 (1906) 1111.

25.  A. Ramani, B. Grammaticos, and G. Karra, Physica A 180 (1992) 115.

# FROM WEAK TO FULL PAINLEVÉ PROPERTY

# VIA TIME SINGULARITIES TRANSFORMATIONS

Alain Goriely

Service de Physique Statistique
Université Libre de Bruxelles, Campus Plaine CP231
Boulevard du Triomphe, 1050 Brussels, Belgium

## 1. INTRODUCTION

In modern nonlinear science, integrability theories play a modest but important rôle. Indeed, in many areas, the existence of global solutions seem necessary for itself or as the first step of perturbation schemes. Amongst these integrability theories for differential equations, the singularity analysis method (the so-called Painlevé theory) seems one of the most succesful, especially for partial differential equations.[1] The aim of these methods is to link the local behaviour of the solution near movable singularities in the complex plane to the global concept of integrability. More precisely, we define the P-integrability for a system of differential equations as the Painlevé property, that is, the solutions exhibit no worst singularities than movable poles. Despite its success, problems remain, essentially due to the existence of integrable systems (in the Liouville sense for example) which do not possess the Painlevé property. The underlying hope is that these systems possess the Painlevé property in some hidden set of variables, ensuring therefore the existence of global integrals of motion.

In this paper, we show that many of these systems can also be detected by singularity analysis and even explicitly linked to the Painlevé property thanks to two sets of transformations. We also define an extended Painlevé test which is coordinate invariant in order to obtain a global test taking into account the transformation of singularities in the complex plane.

## 2. THE PAINLEVÉ TEST

We give here a complete definition of the Painlevé test. The Painlevé test is an algorithmic procedure to detect the Painlevé Property for a system of ODEs.[2] The main idea is to check the existence of a formal Laurent series, checking therefore the occurence of algebraic and transcendent branch points.

A system of ODE's with rational vector field

*Chaotic Dynamics: Theory and Practice*
Edited by T. Bountis, Plenum Press, New York, 1992

$$\frac{dx_i}{dz} = f_i(x_1, \ldots, x_n) \qquad\qquad i = 1, \ldots, n \qquad\qquad (2.1)$$

with $z \in \mathbb{C}$, will satisfy the Painlevé test in the variables $(x_i, z)$, if all its solutions (general and singular) can be expanded in terms of formal Laurent series:

$$x_i(z) = \alpha_i(z - z_0)^{p_i} \sum_{j=0}^{\infty} a_{ij}(z - z_0)^j \qquad\qquad i = 1, \ldots, n \qquad\qquad (2.2)$$

## 2.1 Definitions

### 1. The dominant behaviour

In order to study the properties of a system near a movable singular point, we have to take into account two different contributions of the vector field. Therefore, we split it in two parts:

$$f_i = \widehat{f}_i + \check{f}_i \qquad\qquad (2.3)$$

The first part $\widehat{f}$ is the *dominant part* determining the behaviour of the system at the singularity. This behaviour is given by

$$x_i = \alpha_i(z - z_0)^{p_i} \qquad\qquad i = 1, \ldots, n \qquad\qquad (2.4)$$

The set $(\alpha, \mathbf{p})$ defines a *balance* (of *order* equals to the number of non-arbitrary $\alpha_i$), for which one has to prove the existence of the full Laurent series (2.2).

The second part $\check{f}$ is the *non-dominant part* characterized by a singular behaviour that can be neglected at the singularity:

$$\check{f}_i(\alpha_j(z - z_0)^{p_j}) \underset{z \to z_0}{\sim} \beta_i(z - z_0)^{p_i + q_i} \qquad\qquad i = 1, \ldots, n \qquad\qquad (2.5)$$

with $q_i > -1 \ \forall \ i = 1, \ldots, n$.

### 2. The resonances

The *resonances* of a balance are the indices in the Laurent series where arbitrary constants first appear. For instance, for a balance of order $l$, $(n - l)$ resonances vanish due to the arbitrariness of the first coefficients $\alpha_i$.

### 3. The Painlevé test

The conditions on a formal solution to be a Laurent series are the following:

1) All $p_i$ are integers.
2) All $q_i$ are positive integers.
3) All resonances are integers.
4) The arbitrariness of the coefficients at the resonances does not introduce incompatible conditions on the coefficients $a_{ij}$.

## 2.2 The Extended Painlevé Test

The aim of the Painlevé test is to give an algorithmic way to detect integrable systems. Indeed, it appears that whenever a system possesses the Painlevé property it can be, in some way, integrated. It means that the regularity of the solutions in the complex plane might ensure the system to have global integrals of motion. This hope, hidden behind the Painlevé conjecture, presents some major problems. Indeed, the Painlevé property is not conserved under simple coordinate transformations. Therefore, it can not provide us with a useful definition for integrability which should be coordinate invariant. This is why, we introduce a coordinate invariant test, the *extended Painlevé test*. A system satisfies the test, when there exists a system of coordinates in which it satisfies the usual Painlevé test as described above. The problem lies then in the choice of the transformations in order to find the suitable variables for which the system exhibits the Painlevé property. Thus it seems necessary to know how the solutions will be transformed in order to select the proper variables. In the next sections, we show that the behaviour of the solutions near the singularities can be controlled in such a way that this selection can be easily computed whenever it is possible. This allows for understanding many problems still present in the usual approaches such as the quasi Painlevé property,[3] the weak Painlevé conjecture[4] or the problem of negative resonances.[5]

## 3. A MATRIX REPRESENTATION AND TWO SETS OF TRANSFORMATIONS

In the latter, we will focus on the following class of equations:[6]

$$\frac{dx_i}{dt} = x_i \sum_{j=1}^{m} A_{ij} \prod_{k=1}^{n} x_k^{B_{jk}} \qquad i = 1,\ldots,n \quad m \geq n \qquad (3.1)$$

with $A_{ij}$, $B_{ij} \in \mathbb{R}$

This class includes, for example, all systems with polynomial vector fields. The main advantadge of this matrix form is its form-invariance under two sets of transformations. The first one, *the quasimonomial transformations* (QMT), acts only on the dependent variables:

$$x_i = \prod_{k=1}^{n} x_k'^{C_{ik}} \qquad i = 1,\ldots,n \qquad (3.2)$$

and transforms (3.1):

$$\frac{dx_i'}{dt} = x_i' \sum_{j=1}^{m} A_{ij}' \prod_{k=1}^{n} x_k'^{B_{jk}'} \qquad i = 1,\ldots,n \qquad (3.3)$$

$$\text{where} \qquad \begin{cases} A' = C^{-1}A \\ B' = BC \end{cases}$$

The second set of transformations, *the new-time transformations* (NTT),[7] acts on the independent variables:

$$dt = (\prod_{i=1}^{n} x_i^{\beta_i})\, d\tilde{t} \qquad (3.4)$$

This new time parametrization maps system (3.1) on

$$\frac{dx_i}{d\tilde{t}} = x_i \sum_{j=1}^{m} \tilde{A}_{ij} \prod_{k=1}^{n} x_k^{\tilde{B}_{jk}} \qquad i = 1, \ldots, n \tag{3.5}$$

$$\text{where} \qquad \begin{cases} \tilde{A} = A \\ \tilde{B}_{ij} = B_{ij} + \beta_j \end{cases} \tag{3.6}$$

## 4. THE EXTENDED PAINLEVÉ TEST

In order to define a coordinate invariant test, we study the variance of the various elements of the test (dominant and non dominant behaviours, resonances and compatibility conditions) under the sets of transformations defined above. This is necessary to pick up the right transformation when one has to prove the integrability of a system using singularity analysis methods.

### 4.1  Action of a QMT on a Balance

Let us assume that in a set of variables $(x_i, t)$, we have a balance of order $l$ for which the dominant behaviour $(\alpha, \mathbf{p})$, the non-dominant behaviour $\mathbf{q}$ and the resonances $\mathbf{r}$ have been determined. We now look for the transformed balance, that is, the corresponding solution after the transformations (3.2) in the set $(\mathbf{x}', t)$ in terms of the old one and the matrix C:

$$\begin{cases} p' = C^{-1}p \\ \alpha'_i = \prod_{j=1}^{n} \alpha_j^{C_{ij}} \\ q' = q \\ r' = r \end{cases} \tag{4.1}$$

The $p_i$'s are not invariant under the QMT. Moreover for a given balance, they can be transformed at will (for instance, matrix $C$ can be chosen in such a way that $p'_i = -1$). Therefore, we clearly see that no condition should be required on the dominant exponent $\mathbf{p}$. At first sight, it might seem in contradiction with the Painlevé test defined above (requiring integers $p_i$) but the extended Painlevé test ensures that whatever the exponents $p_i$ are, they always can be transformed to integers by an adequate QMT. Let us already note that by using a simple set of transformations as (3.2) we can detect new solutions.

### 4.1  Action of a NTT on a Balance

Here again, we assume that we know the elements of a balance of order $l$ in the $(x_i, t)$ variables and express them in the new $(x_i, t')$ variables in terms of the vector $\beta$ introduced by the NTT:

$$\tilde{p}_i = \frac{p_i}{1+c} \qquad i = 1, \ldots, l \tag{4.2}$$

$$\tilde{q}_i = \frac{q_i - c}{1+c} \qquad i = 1, \ldots, l \tag{4.3}$$

$$\widetilde{r} = \begin{cases} -1 & \text{for the resonance } r = -1 \text{ associated with the pole position} \\ \dfrac{r}{1+c} & \text{for the other resonances} \end{cases} \qquad (4.4)$$

with $c = -\sum_{i=1}^{n} \beta_i p_i$.

New features appear here. The resonances (and henceforth, the kind of singularities) are transformed by a NTT. It is thus possible to map some systems exhibiting solutions with algebraic branch points (or even transcendent singularities) on systems with no other singularities than poles. On the other hand, a NTT may introduce new branches of solutions which will not be of Painlevé type. Therefore, it is necessary to be careful with such kind of transformations. These transformations of singularities can be used to understand some problems arising in the usual singularity analysis methods such as the weak-Painlevé conjecture and the problem of negative resonances.

## 5. APPLICATIONS

### 5.1 An Useful Lemma

The last step of the Painlevé test requires the resolution of a recursion relation. Indeed, in order to ensure the arbitrariness of some coefficients of the Laurent series (the one whose indices are resonances), one has to compute all the coefficients up to the highest resonance and check that no logarthmic terms should be added to render the series generic. We now show a usefull lemma giving sufficient conditions for the Laurent series to exist once the resonances and the non-dominant behaviours $q_i$ are known:

**Lemma:**
Let $(\alpha, \mathbf{p})$ be a balance of order $l$ with non-dominant behaviour $q_i$ and consider the set of strictly positive resonance $R$. Let $r_{\max} = \max(R)$ and assume that

1) $R$ is an incomensurate set, that is

$$r_i \neq \sum_{j=1}^{n-1} n_j r_j \quad \forall\, n_j \in \mathbb{N},\ \forall\, r_i, r_j \in R$$

2) Either
    2.a) $q_i > r_{\max}$
    2.b) $(q_i - 1)$ is comensurate with $R$

then the balance $(\alpha, \mathbf{p})$ satisfies the Painlevé test.

We will see that this lemma is in many cases powerful even if it only provides sufficient conditions.

### 5.1 The Weak Painlevé Conjecture

The discovery of two dimensional integrable Hamiltonians with algebraic branch points lead to the conjecture that certain peculiar branch points are compatible with integrability.[4] This weak Painlevé conjecture can be stated as follows:[8]

If all solutions can be expanded in Puisseux series:

$$x_i(z) = \alpha_i (z - z_0)^{e_i/g} \sum_{j=0}^{\infty} a_{ij}(z - z_0)^{j/g} \qquad i = 1, \ldots, n \qquad (5.1)$$

then the system is integrable.

It means that both the resonances and the leading behaviours are multiple of the same rational $1/g$.

Let us focus on a balance. If this balance is of weak Painlevé type, it is possible to map it on a full Painlevé balance by choosing a NTT with $1 + c = 1/g$. In this process, the rational branch points are mapped on poles. If in the same process, all balances can be transformed in Laurent solutions, the system satisfies the Painlevé test in the new set of variables. Let us see on an example how the transformation acts:

### 5.1.1  The Fokas-Langerstrom potential

The Hamiltonian reads:

$$H = 1/2(p_x^2 + p_y^2) + 3/2(xy)^{-2/3} \qquad (5.2)$$

The regularization of this Hamiltonian has already been presented by Hietarinta $et$ $al.$[7] using a "coupling constant metamorphosis" to show the duality beetween (5.2) and another integrable full-Painlevé Hamiltonian. Let us compute the regularizing NTT for this system. In variables $(x_1, x_2, x_3, x_4) = (x, y, p_x, p_y)$, Hamilton's equation reads:

$$\begin{cases} \dot{x}_1 = x_3 \\ \dot{x}_2 = x_4 \\ \dot{x}_3 = x_1^{-5/3} x_2^{-2/3} \\ \dot{x}_4 = x_1^{-2/3} x_2^{-5/3} \end{cases} \qquad (5.3)$$

Written in form (3.1) this system is defined by:

$$A = \mathrm{Diag}(1, 1, -1, -1) \quad B = \begin{pmatrix} -1 & 0 & 1 & 0 \\ 0 & -1 & 0 & 1 \\ -5/3 & -2/3 & -1 & 0 \\ -2/3 & -5/3 & 0 & -1 \end{pmatrix} \qquad (5.4)$$

The singularity analysis gives ($\mathbf{p} = (p_1, p_2, p_3, p_4)$ $R = (r_1, r_2, r_3, r_4)$):

$$\begin{cases} \mathbf{p} = (3/4, 0, -1/4, 0) & \mathbf{q} = (0, -1/2), & R = (-1, 0, 0, 1/2) \\ \mathbf{p} = (0, 3/4, 0, -1/4) & \mathbf{q} = (0, -1/2), & R = (-1, 0, 0, 1/2) \\ \mathbf{p} = (3/5, 3/5, -2/5, -2/5) & \mathbf{q} = (), & R = (-1, 4/5, 3/5, -4/5) \end{cases}$$
$$(5.5)$$

An obvious choice of parameter $c$ for the balance of order 4 is $1 + c = 1/5$, that is $c = -4/5$ and $\beta = (2/3, 2/3, 0, 0)$. After the NTT (3.4) one obtains a new dynamical systems:

$$\begin{cases} \dfrac{dx_1}{\widetilde{dt}} = (x_1 x_2)^{2/3} x_3 \\[2ex] \dfrac{dx_2}{\widetilde{dt}} = (x_1 x_2)^{2/3} x_4 \\[2ex] \dfrac{dx_3}{\widetilde{dt}} = x_1^{-1} \\[2ex] \dfrac{dx_4}{\widetilde{dt}} = x_2^{-1} \end{cases} \tag{5.6}$$

with new balances:

$$\begin{cases} \mathbf{p} = (3/2, 0, -1/2, 0) & \mathbf{q} = (0, -1), & R = (-1, 0, 0, 1) \\ \dot{\mathbf{p}} = (0, 3/2, 0, -1/2) & \mathbf{q} = (0, -1), & R = (-1, 0, 0, 1) \\ \mathbf{p} = (3, 3, -2, -2) & \mathbf{q} = (\,), & R = (-1, 4, 3, -4) \end{cases} \tag{5.7}$$

In the light of the previous results, it is obvious that this last sytem satisfies the extended Painlevé test (all resonances are incomensurate and the $p_i$'s can be turned into integers by a QMT of matrix $C = \mathrm{Diag}(1/2, 1/2, 1/2, 1/2)$).

## 5.3  The Problem of Negative Resonances

Certain systems of equations have only negative resonances, these resonances can not be tested and their integrability properties can not be decided by using the usual Painlevé test. Indeed some of these systems are known to have the Painlevé property while the others do not have it. In order to test these resonances, new tests have been devised such as the Fuchs-Painlevé test.[9] We now show that the extended Painlevé test takes also into account the negative resonances. Indeed, if we choose a NTT with a vector $\beta$ such that $c = -1$, the negative resonances will be mapped onto the positive ones. The latter can be tested using the usual Painlevé test.

### 5.3.1  The Chazy equation

The system reads:[10]

$$\frac{d^3 x}{dt^3} = -3\left(\frac{dx}{dt}\right)^2 + 2x\frac{d^2 x}{dt^2} \tag{5.8}$$

The Chazy equation possesses the Painlevé property but does not have positive resonances: $R = (-1, -2, -3)$. In order to test these resonances, we perform a NTT with $c = -2$, followed by a QMT with $C = \mathrm{Diag}(-1, -1, -1)$.

Let us first write system (5.8) in matrix representation (3.1):

$$\begin{cases} \dfrac{dx_1}{dt} = x_2 \\[2ex] \dfrac{dx_2}{dt} = x_2^{-1} x_3 \\[2ex] \dfrac{dx_3}{dt} = -3x_2^2 + 2x_1 x_3 \end{cases} \tag{5.9}$$

$$A = \begin{pmatrix} 1 & 0 & 0 & 0 \\ 0 & 1 & 0 & 0 \\ 0 & 0 & -3 & 2 \end{pmatrix} \qquad B = \begin{pmatrix} -1 & 1 & 0 \\ 0 & -1 & 1 \\ 0 & 2 & -1 \\ 1 & 1 & -1 \end{pmatrix} \tag{5.10}$$

The NTT of vector $\beta = (0,-1,0)$ followed by the QMT of matrix $C = \mathrm{Diag}(-1,-1,-1)$ gives:

$$\left\{ \begin{array}{l} \dfrac{dx_1}{d\tilde{t}} = 1 \\[2mm] \dfrac{dx_2}{d\tilde{t}} = -x_2^{-1}x_3 \\[2mm] \dfrac{dx_3}{d\tilde{t}} = -3x_2 + 2x_1 x_2^{-1} x_3 \end{array} \right. \tag{5.11}$$

This system has a balance of order 3 with resonances $R = (-1,2,3)$. For this balance, the compatibility conditions are satisfied due to incomensurate relations between the resonances.

Hence, the Painlevé test property allows to distinguish systems which possess the Painlevé property by checking explicitly the negative resonances.


## 5.4 New Integrable Systems

We study here a simplified 3-dimensional LV system.[11]

$$\dot{x}_i = \lambda_i x_i + x_i M_{ij} x_j \qquad i = 1,2,3 \tag{5.12}$$

with $M = \begin{pmatrix} 0 & a & 1 \\ 1 & 0 & b \\ c & 1 & 0 \end{pmatrix}$

Let us see how new particular solutions can be detected by using the extended Painlevé test. We consider the case $abc \neq 0$. In order to find new solutions, we take $\lambda_1 = \lambda_2 = \lambda_3 = \lambda$. If $a = 1$, $b > 0, c > 1$, and we obtain a unique balance of order 3:

$$p = (-1,-1,-1) \quad q = (\,) \quad r = \left(-1, \frac{c}{\alpha}, \frac{1-c+bc}{\alpha}\right) \tag{5.13}$$

where $\alpha = (1 + bc)$

If $b, c \in \mathbb{N}$, then, one can regularize this system by a NTT with $\beta = (-bc/\alpha, 0, 0)$. The QMT of matrix

$$C = \begin{pmatrix} \alpha & 0 & 0 \\ \alpha - 1 & 1 & 0 \\ \alpha - 1 & 0 & 1 \end{pmatrix}$$

gives a new 3-dimensional LV system which satisfy the Painlevé test. The first integrals can be explicitely computed:[12]

$$\left\{ \begin{array}{l} I_1 = \left( c\, x_1^{-1/\alpha} x_2^{1/\alpha} x_3^{1/\alpha} - x_1^{bc/\alpha} x_2^{-bc/\alpha} x_3^{c/\alpha} \right) e^{-\frac{c\lambda}{\alpha} t} \\[4mm] I_2 = e^{\lambda t} x_1^{1/\alpha} x_2^{-bc/\alpha} x_3^{-c\alpha} - \displaystyle\int^{x_1^{b/\alpha} x_2^{-b/\alpha} x_3^{1/\alpha} e^{\lambda t}} dv \left( \frac{v^c}{c} + I_1 \right)^{b-1} \end{array} \right. \tag{5.14}$$

where $I_1$ and $I_2$ are integration constants determined by initial conditions

Following the same procedures, other integrable cases can be detected.

# 6. CONCLUSIONS

The extended Painlevé test is not a new test or a new conjecture. It is just a way of taking explicitly into account the variance of solution behaviours near movable singularities under time transformations. It points out the difference between the Painlevé property in a given set of variables and the P-integrability which should be defined independently of the set of coordinates. In physics, one needs all possible solutions even the worst ones. For instance, we saw that even solutions with algebraic branch point in the complex plane can be turned into regular solution. The quasimonomial transformation and the new-time transformation are just two possible sets of transformations, but it is clear that other transformations could also be devised (polynomial ones for example) and used to regularize non-Painlevé systems once their effect on singularity behaviour is explicitly known. This will provide us with a complete set of tools for dealing with solutions of dynamical systems in the complex plane and finally investigate the largely unexplored zone of nonintegrable systems.

## BIBLIOGRAPHY


1. M.J. Ablowitz, A. Ramani and H. Segur, J. Math. Phys **21** (1980) 715, and 1006.
2. A. Ramani, B. Grammaticos and T. Bountis, Phys. Rep. **180** (1990) 159.
3. W.H. Steeb, M. Kloke, B.M. Spieker and A. Kunnick, Foundation of Physics (1984) 637.
4. A. Ramani, B Dorizzi and B. Grammaticos, Phys. Rev. Lett. **49** (1982) 1538.
5. M. Kruskal *in Painlevé Transcendents, their asymptotics and physical applications*, edited by P. Winternitz and D. Levi (Plenum Publishing Corp, New York. 1992).
6. L. Brenig and A. Goriely, Phys. Rev. A **40** (1989) 4119.
7. J. Hietarinta, B. Grammaticos, B. Dorizzi and A. Ramani, Phys. Rev. Lett. **53** (1984) 1707.
8. B. Grammaticos, B. Dorizzi and A. Ramani, J. Math. Phys. **25** (1984) 3470.
9. P. Fordy and A. Pickering, (1991) *Analysing Negative Resonances in the Painlevé Test* (preprint).
10. R. Conte (1991) *Unification of PDE and ODE versions of Painlevé analysis into a single invariant version,in Painlevé Transcendents, their asymptotics and physical applications*, edited by P. Winternitz and D. Levi (Plenum Publishing Corp, New York, 1992).
11. B. Grammaticos, J. Moulin-Ollagnier, A. Ramani, J-M. Strelcyn and S. Wojciechowski, Physica A **163**, 683.
12. A. Goriely (1988), *Transformations Quasi-Monomiales et Intégrabilité*, Mémoire de licence.

# THE ROLE OF NEGATIVE RESONANCES IN THE PAINLEVÉ TEST


Allan Fordy and Andrew Pickering

Department of Applied Mathematical Studies and
Centre for Nonlinear Studies
University of Leeds
Leeds, LS2 9JT, UK



## Abstract

We present a recent improvement [1] to the Painlevé test such that
negative resonances can be treated.  To this end we demand that the general
solution of  both the given nonlinear equation <u>and</u> its linearisation be
<u>single</u> valued.  This gives rise to compatibility conditions for <u>every</u>
integer resonance, whether positive or negative.  We generalise this
approach further by considering the singularity structure of higher order
perturbation equations.  We present 3 examples which illustrate the need
for these generalisations.


## 1.  Introduction

Since the discovery of the connection between soliton equations and
Painlevé transcendents [2] and the subsequent introduction of the Painlevé
test [3], there has been a booming industry, developing and applying the
'Painlevé method'.  The Painlevé test has been an extremely successful
method of testing both ODEs and PDEs for 'complete integrability'  and has
been the subject of a large number of papers (see the reviews [4,5], the
conference proceedings [6]  and references therein).  Nevertheless, this
method (as currently used) is <u>incapable</u> of decisively testing certain
equations, such as those possessing several <u>negative</u> <u>resonances</u>.  One such
equation is that of Chazy which has <u>only</u> negative resonances.  This
equation has a movable natural boundary, but its general solution can be
written down in terms of hypergeometric functions [7] and is single valued
in its domain of definition.  Even  integrable equations, such as the
members of the KdV hierarchy, have 'secondary' Painlevé branches which
possess several negative resonances.  There has been some discussion of
negative resonances [8,9], but they were largely ignored as being harmless.

In this paper (following [1]) we maintain that negative resonances can contain important (sometimes decisive) information regarding the integrability of an equation. To this end we give a method of deriving compatibility conditions for negative resonances and show that it can be extremely hazardous to ignore them!  Furthermore, we show that 'secondary' branches are by no means secondary; for many equations our method extracts the _same_ information from _all_ branches, even though the conventional Painlevé analysis could only derive _full_ information from the principal branch.

Our approach is to _simultaneously_ test a nonlinear equation _and_ its linearisation, treated as a (rather weakly) coupled system.  Applied to the linearised equation, the Painlevé analysis reduces to a Fuchsian analysis about a regular singularity.  The roots of the indicial equation of the linearisation are just (up to a constant integer shift) the resonances of the nonlinear equation. We demand that the solutions to both the nonlinear and linear equations, be single-valued.  We therefore require that all roots of the indicial equation (and thus all resonances of the nonlinear equation) be _distinct_ _integers_, whether positive or negative. Since the roots of the indicial equation differ by integers, compatibility conditions arise.  We call this the "Fuchs-Painlevé test".

Whilst this method greatly strengthens the power and scope of singularity analysis in testing equations for integrability, it turns out to be not enough.  We thus extend the Fuchs-Painlevé method (testing the nonlinear equations and its _first_ _order_ perturbation) to a test which investigates the singularity structure of a nonlinear equation and its higher order perturbation equations.

We illustrate the need for both the Fuchs-Painlevé and the 'perturbative Painlevé' tests by three examples in section 3.

In section 4 we change our perspective from testing an equation for complete integrability to analysing hierarchies of integrable nonlinear evolution equations.  We show that there is a direct correspondence between the presence of negative resonances and the existence of lower order commuting flows.  We also show how to use the Miura map between the KdV and MKdV hierarchies to deduce the relationship between the resonance polynomials of the two hierarchies.  This is a general procedure, which could be used with any Miura map between integrable hierarchies.


## 2.  The Fuchs-Painleve Test and Beyond

For simplicity, we restrict attention to nonlinear evolution equations of the form:

$$u_t = K[u] \qquad\qquad (2.1a)$$

where $K[u]$ is a polynomial function of $u, u_x, \ldots, u_{Nx}$. The case of ODEs is then easily incorporated by taking $u_t = 0$. The linearisation of (2.1a) is:

$$w_t = K'[u]w \qquad\qquad (2.1b)$$

where $K'[u]w = \dfrac{d}{d\varepsilon} K[u + \varepsilon w]\big|_{\varepsilon = 0}$. Equation (2.1b) is the equation satisfied by generalised symmetries (commuting flows) of (2.1a), when they exist.

<u>Remark</u>

Our examples of section 3 are, in fact, all ODEs, but the method is applicable to all cases. In this paper PDEs arise in section 4.

We first carry out a standard Painlevé expansion of the nonlinear equation (2.1a), using the modification [10] of the WTC method (see [5]). This 'invariant' approach simplifies many of the expressions, hiding the complications in the definition of $S, C$ and $\chi$ (see the appendix). We seek a solution of the form:

$$u^{(p)} = \chi^{-\alpha} \sum_{i=0}^{\infty} u_i \chi^i . \qquad\qquad (2.2)$$

A leading order analysis gives a number of possible choices of $\alpha$, depending upon the nonlinearities. Each $\alpha$ corresponds to a possible choice of dominant terms $\hat{K}[u]$ of $K[u]$ and this, in turn, leads to a number of possible starting terms $u_0$ as solutions of the <u>algebraic</u> equation:

$$\hat{K}(u_0 \chi^{-\alpha})\big|_{\chi' - 1 = u_0' = 0} = P_\alpha(u_0)\chi^\beta = 0 , \qquad\qquad (2.3a)$$

where $\beta$ is the <u>weight</u> of the dominant expression $\hat{K}$ and $P_\alpha$ a polynomial. The coefficients $u_i$ are determined recursively by:

$$\left[ \hat{K}'[u_0\chi^{-\alpha}]\chi^{i-\alpha}\Big|_{\chi' - 1 = u_0' = 0} \right] u_i = \text{expressions involving } u_0, \ldots, u_{i-1}, S, C .$$

$$(2.3b)$$

The coefficient of $u_i$ on the left vanishes for certain values of i, called resonances, which must be integer for an 'integrable' equation:

$$\mathcal{R} = \{r_1, \ldots, r_n\} , \quad r_1 \leq r_2 \leq \ldots \leq r_n . \qquad\qquad (2.3c)$$

The number n depends upon the highest order derivative in the dominant expression $\hat{K}[u]$, not on the order of $K[u]$. The right hand side of (2.3b) must also vanish for these values of i, giving rise to compatibility conditions whenever i reaches a <u>positive</u> $r \in \mathcal{R}$. No conditions arises at <u>negative</u> resonances.

To pass the Painlevé part of our test an equation should have the following properties:

(PI) Each possible choice of $\alpha$ and all corresponding $r_i$ must be integer, with $r_i$ being distinct.

(PII) all branches should be such that at any <u>positive</u> resonance, the compatibility conditions are identically satisfied.

(PIII) there exists a branch with the number of resonances n=N, the order of the <u>full</u> operator K[u].

<u>Remarks</u>

(i) We have dropped the requirement of a <u>principal</u> branch ($r_1 = -1$, $r_2 \geq 0$) with the consequence that it may not be possible to build the general solution as a finite pole Laurent expansion. However, we still require the existence of a branch with the 'correct' number of resonances.

(ii) Negative resonances of 'secondary' branches are allowed since they give no conditions which <u>contradict</u> integrability.

This weakened condition (PIII) allows some 'bad' equations through our net, but these are caught at step 2 below.

Our second step is to consider the linearised equation (2.1b), with $u = u^{(p)}$, the Painlevé expansion (2.2), for <u>each</u> of the branches. For each branch we can write (2.1b) as:

$$\hat{K}'[u^{(p)}]w = I'[u^{(p)}]w , \qquad (2.4)$$

where $\hat{K}'$ and $I'$ are respectively the linearisations of the dominant and inferior parts of $K - u_t$. $\hat{K}'[u^{(p)}]$ is scaled in such a way that $\chi = 0$ is a <u>regular singularity</u> in the Fuchsian sense. We seek an expansion:

$$w = \chi^\sigma \sum_{i=0}^{\infty} w_i \chi^i , \quad w_0 \neq 0 , \qquad (2.5a)$$

where $\sigma \in \{\sigma_1, \ldots, \sigma_n\}$ is a root of the indicial equation:

$$\hat{K}'[u_0 \chi^{-\alpha}]\chi^\sigma \Big|_{\chi'-1 = u_0' = 0} = 0 . \qquad (2.5b)$$

Comparing this with (2.3), we see that:

$$\sigma_i = r_i - \alpha \text{ for } r_i \in \mathcal{R} , \quad i = 1, \ldots, n . \qquad (2.5c)$$

In order that the general solution of the <u>linear</u> equation (2.4) be single valued we require, in addition to PI-III,

FI that the indicial equation has n <u>distinct</u>, <u>integer</u> solutions $(\sigma_1, \ldots, \sigma_n)$ ,

FII that the compatibility conditions arising at each $\sigma_i$, $1 \leq i \leq n$ be identically satisfied.

<u>Remarks</u>

(i)     FII gives compatibility conditions for $r_i$, regardless of whether they are positive or negative.

(ii)    It should be noticed that since the position of the singularity is determined by the nonlinear equation, even this <u>linear</u> equation has moveable poles.

We are demanding much more than (2.4) being just Fuchsian when we ask for w to be single valued. This is a strong constraint which, together with the above weakened Painlevé property of (2.1a) enables us to distinguish integrable cases in a wide variety of equations hitherto untestable by the Painlevé method. A feature of equations which pass our test, but have no principal branch, is the presence of an essential singularity [1].

Sometimes it is not enough to consider just the given nonlinear equation and its linearisation. In [11,12] we generalise this approach to include higher order perturbations:

Let $u^{(0)}$ be a Painlevé expansion for equation (2.1a). Let:

$$u = u^{(0)} + \varepsilon u^{(1)} + \varepsilon^2 u^{(2)} + \ldots \tag{2.6}$$

be a 'nearby' solution. Then the coefficients of $\varepsilon^n$, n=0,1,..., of (2.1a), are given by:

$$u_t^{(0)} = K[u^{(0)}] , \tag{2.7a}$$

$$u_t^{(1)} = K'[u^{(0)}]u^{(1)} , \tag{2.7b}$$

$$u_t^{(n)} = K'[u^{(0)}]u^{(n)} + F^{(n)}[u^{(0)},u^{(1)},\ldots,u^{(n-1)}], \quad n=2,\ldots . \tag{2.7c}$$

For n≥2, a new 'inhomogeneous' term $F^{(n)}$ arises from nonlinear combinations of lower order terms in the series (2.6). The Fuchs–Painlevé method stops at (2.7b). As previously discussed the first two components have leading order behaviour:

$$u^{(0)} \sim \chi^{-\alpha} , \quad u^{(1)} \sim \chi^{r_1-\alpha} ,$$

where $r_1$ is the lowest root of the indicial equation for (2.7b). Our generalisation is to study the singularity structure of the higher order perturbation equations (2.7c). Now the leading order behaviour is no longer determined by the indicial equation but by the inhomogeneity. The method will be illustrated in example (3.3) for which it is necessary to go to n=2 in order to extract <u>full</u> information from the second branch.

### 3. Examples

Here we present just three examples. More can be found in [1, 11, 12].

<u>Example 1</u> Chazy's equation [7]

$$K[u] = u_{xxx} - 2uu_{xx} + 3u_x^2 = 0 ,  \tag{3.1a}$$

$$K'[u]w = w_{xxx} - 2uw_{xx} + 6u_x w_x - 2u_{xx} w = 0 .  \tag{3.1b}$$

There are two branches:

(i)  $\alpha = 1$, $u_0 = -6$ , $\hat{K}[u] = K[u]$ , $\beta = -4$ , $\mathcal{R} = \{-3, -2, -1\}$ ,  (3.2a)

(ii)  $\alpha = 2$, $u_0$ arbitrary , $\hat{K}[u] = -2uu_{xx} + 3u_x^2$ , $\beta = -6$ , $\mathcal{R} = \{-1, 0\}$.  (3.2b)

Neither branch is principal and no resonance can be tested by the Painlevé method. However, branch (i) has 3 resonances as required by PIII. Corresponding to each branch we write the linearised equation (2.4):

(i)  $w_{xxx} - 2uw_{xx} + 6u_x w_x - 2u_{xx} w = 0$ ,  (3.3a)

(ii)  $-2uw_{xx} + 6u_x w_x - 2u_{xx} w = -w_{xxx}$ .  (3.3b)

For branch (ii) we have put the 'inferior' part on the right. In each case the corresponding branch of the Painlevé expansion must be inserted for u so, noting the respective leading order behaviour given in (3.2), we see that $\chi=0$ is a regular singularity in each case. The Fuchsian analysis of branch (ii) treats equation (3.3b) as <u>effectively</u> second order, since the third order part is not dominant. Thus, the indicial equation of this branch is only quadratic, giving rise to only two resonances. In fact, the indicial equations are respectively:

(i)  $(\sigma+4)(\sigma+3)(\sigma+2) = 0$ ,  (3.4a)

(ii)  $(\sigma+3)(\sigma+2) = 0$ .  (3.4b)

The expansions for u and w take the form:

(i)  $u = \chi^{-1} \sum_{i=0}^{\infty} u_i \chi^i$ , $u_0 \neq 0$ , $w = \chi^{-4} \sum_{i=0}^{\infty} w_i \chi^i$ , $w_0$ arbitrary ,  (3.5a)

(ii)  $u = \chi^{-2} \sum_{i=0}^{\infty} u_i \chi^i$ , $w = \chi^{-3} \sum_{i=0}^{\infty} w_i \chi^i$ , $w_0$ arbitrary .  (3.5b)

Substituting u of (3.5a) into (3.1) gives, for branch (i):

$$K[u] \equiv \chi^{-4} \sum_{i=0}^{\infty} K_i \chi^i = 0 , \quad K'[u]w \equiv \chi^{-7} \sum_{i=0}^{\infty} \ell_i \chi^i = 0 ,  \tag{3.6a}$$

where

$$K_0 = -u_0(u_0+6) , \quad K_1\Big|_{u_0=-6} = 24u_1 , \quad K_2\Big|_{u_0+6=u_1=0} = 60(u_2+S) .  \tag{3.6b}$$

Since there are no positive resonances, no obstructions arise, so that all the $u_i$ can be recursively constructed, the first three being (with $u_0 \neq 0$):

$$u_0 = -6 \ , \quad u_1 = 0 \ , \quad u_2 = -S \ . \tag{3.6c}$$

With this expansion for u inserted into (3.3a) we can try to construct the expansion (3.5a) for w in the same way. However, this time there <u>are</u> possible obstructions. The higher roots of the indicial equation are <u>positive</u> resonances (as $s_i = \sigma_i + 4$) for this expansion, but correspond to <u>negative</u> resonances for u. The following compatibility conditions arise $(w_0 \neq 0)$:

$$s_1 = 0 \ : \ \ell_0 \ = \ -20w_0(u_0 + 6) \ = \ 0 \ ,$$

$$s_2 = 1 \ : \ \ell_1 \Big|_{u_0 = -6} \ = \ -40w_0 u_1 \ = \ 0 \ , \tag{3.6d}$$

$$s_3 = 2 \ : \ \ell_2 \Big|_{u_0 + 6 = u_1 = 0} \ = \ -64w_0(u_2 + S) \ = \ 0 \ ,$$

which are <u>identically satisfied</u> when (3.6c) hold. No further conditions arise.

Similarly, for branch (ii), substituting u of (3.5b) into (3.1) leads to:

$$u_0 \text{ arbitrary} \ , \ u_1 = -6 - u_{0x} \ , \tag{3.7a}$$

while the compatibility conditions for w at $s_i = \sigma_i + 3$ are:

$$w_0 \text{ arbitrary} \ ; \ s_2 = 1 \ : \ -10w_0(u_1 + u_{0x} + 6) = 0 \ , \tag{3.7b}$$

which are identically satisfied.

Thus we see that in both cases the compatibility conditions at resonances of the Fuchsian expansion are identically satisfied as a consequence of the definitions of $u_i$, given by the corresponding Painlevé expansions. Thus, Chazy's equation passes our test, as it should.

In [1] we represent the general solution of Chazy's equation as an infinite expansion in <u>negative</u> powers of x (not $\chi$). The two branches of the (classical) Painlevé expansion arise as <u>particular</u> solutions of this general solution.

<u>Example 2</u>  Bureau's equation (20.4)

This equation was studied by Bureau [13]:

$$K[u] = u_{xxx} + 3uu_{xx} + 3u_x^2 + 3u^2 u_x - c_0 u_x - d_3(u^3 + 2uu_x) - d_2 u^2 - d_1 u - d_0 = 0 \ , \tag{3.8a}$$

$$K'[u]w = w_{xxx} + 3uw_{xx} + (6u_x + 3u^2 - c_0 - 2d_3 u)w_x \tag{3.8b}$$
$$+ (3u_{xx} - 3d_3 u^2 + 6uu_x - 2d_3 u_x - 2d_2 u - d_1)w = 0 \ .$$

Here, we have used Bureau's notation for the coefficients $c_0, d_i$ , which are functions of $x$. There are two branches:

$$\text{(i)} \quad \alpha = 1 \ , \ u_0 = 1 \ , \ \hat{K}[u] = u_{xxx} + 3uu_{xx} + 3u_x^2 + 3u^2 u_x \ , \ \beta = -4 \ , \ \mathcal{R} = \{-1, 1, 3\}, \tag{3.9a}$$

$$\text{(ii)} \quad \alpha = 1 \ , \ u_0 = 2 \ , \ \hat{K}[u] = u_{xxx} + 3uu_{xx} + 3u_x^2 + 3u^2 u_x \ , \ \beta = -4 \ , \ \mathcal{R} = \{-2, -1, 3\}. \tag{3.9b}$$

A standard Painlevé analysis on branch (i):

$$u = \chi^{-1} \sum_{i=0}^{\infty} u_i \chi^i \ , \ K[u] = \chi^{-4} \sum_{i=0}^{\infty} K_i \chi^i \tag{3.10a}$$

leads to the following compatibility conditions at the positive resonances:

$$r_2 = 1 : d_3 = 0 \ ; \ r_3 = 3 : d_2 = d_1 - c_{0x} = 0 \ . \tag{3.10b}$$

Thus, the equation:

$$(u_{xx} + 3uu_x + u^3 - c_0 u)_x - d_0 = 0 \tag{3.11}$$

has the Painlevé property, with branch (i) being principal. The compatibility conditions at the resonances of the linearised equation (3.8b) are identically satisfied.

Branch (ii) only has one positive resonance, so a standard Painlevé analysis only gives one condition:

$$d_1 + c_0 d_3 - 2d_2 d_3 - c_{0x} + 2d_{2x} = 0 \ , \tag{3.12}$$

which is identically satisfied if (3.10b) hold. However, it is seen that a Painlevé analysis of this branch does not provide enough information to _imply_ (3.10b). On the other hand, a Fuchsian analysis of the linear equation (3.8b) does give rise to _two_ nontrivial compatibility conditions which, when taken with (3.12), give precisely (3.10b)!

Thus, we see that the secondary branch, whose Painlevé expansion does _not_ give the general solution and does _not_ give enough conditions to determine integrability, _does_ contain the same amount of information as the principal branch. However, this can only be extracted through the Fuchsian part of our analysis.

The integrable equation (3.11) is just the stationary, third order Burgers equation when $c_0 = d_0 = 0$.

Example 3  Bureau's equation (20.5)

This equation is also discussed in [13]. In order to study both branches we need to use (2.7c) with n=2 (second order perturbation). For simplicity of notation, we use $u^{(0)}=u$, $u^{(1)}=w$, $u^{(2)}=v$. Then:

$$K[u] \equiv u_{xxx}+2uu_{xx}+4u_x^2+2u^2u_x-c_0u_x-d_3u^3-d_2u^2-d_1u-d_0 = 0 , \qquad (3.13a)$$

$$K'[u]w \equiv w_{xxx}+2uw_{xx}+(8u_x+2u^2-c_0)w_x+(2u_{xx}+4uu_x-3d_3u^2-2d_2u-d_1)w = 0, \qquad (3.13b)$$

$$K'[u]v \equiv v_{xxx}+2uv_{xx}+(8u_x+2u^2-c_0)v_x+(2u_{xx}+4uu_x-3d_3u^2-2d_2u-d_1)v$$

$$= -2ww_{xx}-4w_x^2-2w^2u_x-4uww_x+3d_3uw^2+d_2w^2 . \qquad (3.13c)$$

There are two branches:

(i) $\quad \alpha = 1$ , $u_0 = 1$ , $\hat{K}[u] = u_{xxx}+2uu_{xx}+4u_x^2+2u^2u_x$, $\beta = -4$ , $\mathcal{R}=\{-1,1,4\}$ ,

$$(3.14a)$$

(ii) $\quad \alpha = 1$ , $u_0 = 3$ , $\hat{K}[u] = u_{xxx}+2uu_{xx}+4u_x^2+2u^2u_x$, $\beta = -4$ , $\mathcal{R}=\{-3,-1,4\}$ .

$$(3.14b)$$

A standard Painlevé analysis on the principal branch (i) yields the following compatibility conditions at the positive resonances:

$$r_2 = 1 : d_3 = 0 \;;\; r_3 = 4 : d_2 = d_1-\tfrac{1}{2}c_{0x} = d_0 = 0 . \qquad (3.15)$$

The compatibility conditions at the resonances of the linearised equation (3.13b) are then identically satisfied, as are those of (3.13c).

Branch (ii) has only one positive resonance, so a standard Painlevé analysis gives rise to only one compatibility condition. The Fuchsian part of our analysis gives rise to two more resonance conditions. Taken together, these define $d_0, d_1$ and $d_2$ in terms of $d_3$ and $c_0$, which must satisfy one constraint:

$$d_0 = -\frac{27}{16}c_0 d_3^2 + \frac{729}{128}d_3^4 + \frac{3}{4}(d_3c_{0x}+c_0d_{3x}) - \frac{1377}{64}d_3^2d_{3x} + \frac{99}{16}d_{3x}^2 - \frac{63}{16}d_3d_{3xx} + \frac{9}{2}d_{3xxx} ,$$

$$(3.16a)$$

$$d_1 = -c_0d_3 + \frac{405}{64}d_3^3 + \frac{1}{2}c_{0x} - \frac{141}{16}d_3d_{3x} - \frac{3}{4}d_{3xx} , \qquad (3.16b)$$

$$d_2 = \frac{27}{8}d_3^2 - \frac{3}{2}d_{3x} , \qquad (3.16c)$$

$$f(c_0,d_3) = 0 . \qquad (3.16d)$$

The function f is rather long, but satisfies $f(c_0,0)=0$. When $d_3=0$, the conditions (3.15) are recovered. However, in order to extract this information it is necessary to consider equation (3.13c).

<u>Remark</u>

Whilst a detailed description and analysis of this method is left to [11,12], we note the following: the leading order behaviour of v is determined by the <u>inhomogeneous</u> right hand side of (3.13c):

$$v_{xxx} \sim w w_{xx} \;,\;\; w \sim \chi^{-4} \;\Rightarrow\; v \sim \chi^{-7} \;. \tag{3.17a}$$

In general, the leading order behaviour of $u^{(n)}$ is given by:

$$u^{(n)} \sim \chi^{nr_1 - \alpha} \;. \tag{3.17b}$$

However, the resonances are still determined by the operator $\hat{K}'[u]$ and the arbitrary $v_j$ are $v_3, v_5$ and $v_{10}$ .

Thus, for this example, branch (ii) contains the same information (regarding integrability) as the principal branch, but we need to go to the <u>second order perturbation</u> to extract this.


## 4.    Integrable Hierarchies

In this section we are not concerned with <u>testing</u> equations for integrability, but assume that we have a hierarchy of integrable equations, meaning that we have a family of commuting flows. Our aim is to briefly describe the relationship between negative resonances and lower order commuting flows in these circumstances. For simplicity we use the KdV hierarchy to illustrate this point, leaving the details and further examples to [14].

The KdV hierarchy:

$$u_{t_{2n+1}} = R^n u_x = K_{2n+1}[u] \;,\;\; R = \partial^2 + 4u + 2u_x \partial^{-1} \;,\;\; n = 0,1,\dots \;, \tag{4.1}$$

where R is the recursion operator, has recently been analysed from the Painlevé point of view (see [5]). The $n^{th}$ flow (4.1) possesses n branches, with $u_0$ given by:

$$u_0 = -k(k+1) \;,\;\; k = 1,\dots,n \;. \tag{4.2a}$$

The resonances for the kth branch are solutions of the polynomial $P(r;n,k)$, given by:

$$P(r;n,k) = (r-2n-2) \prod_{i=1}^{k} (r+2i-1)(r-2i-2n-2) \prod_{j=k+1}^{n} (r-2j-1)(r+2j-2n-2). \tag{4.2b}$$

The first k of these are negative:

$$r_i = -2(k-i)-1 \;,\;\; i = 1,\dots,k \;. \tag{4.2c}$$

<u>Remark</u>

In [5] a recursion relation is given for $P(r;n,k)$ but not the closed form solution (4.2b), which appears to be new.   This is proved in [11,14].

The flows (4.1) mutually commute, so that:

$$K'_{2n+1}[u] \, K_{2m+1}[u] \, - \, K'_{2m+1}[u] \, K_{2n+1}[u] \, = \, 0 \quad \forall \, n,m \, . \tag{4.3a}$$

This is equivalent to saying that for fixed n, $w = K_{2m+1}[u]$ is a solution of the symmetry equation (linearisation) of (4.1):

$$w_{t_{2n+1}} \, = \, K'_{2n+1}[u] \, w \, . \tag{4.3b}$$

We note that the leading order part of (4.3a) is:

$$\left\{ K'_{2n+1}[u_0\chi^{-2}] \, K_{2m+1}[u_0\chi^{-2}] \, - \, K'_{2m+1}[u_0\chi^{-2}] \, K_{2n+1}[u_0\chi^{-2}] \right\} \Big|_{\chi'=1} \, = \, 0 \, . \tag{4.4}$$

For a given branch with:

$$u_0 \, = \, -k(k+1) \quad \text{for } \underline{\text{fixed}} \, k \, , \quad 1 \leq k \leq n \, , \tag{4.5}$$

the operators $K_{2m+1}[u]$ fall into two categories:

(i)      $m \geq k$, for which $K_{2m+1}[u_0\chi^{-2}]\Big|_{\chi'=1} \, = \, 0$ , \hfill (4.6a)

(ii)     $0 \leq m \leq k-1$, for which $K_{2m+1}[u_0\chi^{-2}]\Big|_{\chi'=1} \, = \, \gamma_{km}\chi^{-2m-3}$ , \hfill (4.6b)

where $\gamma_{km}$ is a <u>non-zero</u> constant.   It should be noticed that $K_{2n+1}[u]$ itself falls into category (i).   For type (i), equation (4.4) is trivially satisfied, while type (ii) leads to:

$$K'_{2n+1}[u_0\chi^{-2}] \, \gamma_{km}\chi^{-2m-3}\Big|_{\chi'=1} \, = \, 0 \, , \quad 0 \leq m \leq k-1 \, . \tag{4.7a}$$

Comparing (4.7a) with (2.3b), with $\alpha=2$, and the definition of a resonance, we see that the k negative resonances:

$$r_{k-m} \, = \, -2m-1 \, , \quad 0 \leq m \leq k-1 \tag{4.7b}$$

(see (4.2c)) arise as a direct result of the first k flows in the hierarchy (4.1).

We can construct the  general solution of (4.3b) by Fuchsian expansion, with u given by one of the Painlevé branches.   Particular solutions of (4.3b) are the commuting flows (4.1), and expressions for these can be obtained  by  special  choices  of  arbitrary coefficients. For  a  given integrable hierarchy it is fairly straightforward to find associated Miura maps and corresponding modified hierarchies (see appendix B of [15]).   We can use the Miura map to 'pull back' the 'Painlevé information' from the given hierarchy to its modification.   We illustrate this with the KdV/MKdV hierarchies, directing the reader to [11,14] for a full discussion.

The MKdV hierarchy, given by:

$$v_{t_{2n+1}} \, = \, \bar{R}^n v_x \, = \, \bar{K}_{2n+1}[v] \, , \tag{4.8}$$

is related to the KdV hierarchy (4.1) through the Miura Map:

$$u = M[v] \equiv -v_x - v^2 ,$$ (4.9)

where $\bar{R}$ denotes the recursion operator:

$$\bar{R} = \partial^2 - 4v^2 - 4v_x \partial^{-1} v ,$$ (4.10a)

related to the KdV recursion operator by:

$$RM' = M'\bar{R} , \quad M' = -\partial - 2v ,$$ (4.10b)

where $M'$ is the Fréchet derivative of (4.9). It follows from (4.9) that:

$$v = \chi^{-1} \sum_{0}^{\infty} v_i \chi^i ,$$ (4.11a)

where $u_0 = -v_0 - v_0^2$. Corresponding to (4.2a) we have:

$$v_0^2 - v_0 - k(k+1) = 0 \quad \Rightarrow \quad v_0 = -k, k+1, \quad k=0,\ldots,n .$$ (4.11b)

Two of these solutions are spurious: $v_0 = 0$ contradicts the leading order behaviour and $v_0 = n+1$ arises from the action of $(\partial + 2v)$ in:

$$\left. K_{2n+1} \right|_{u=M[v]} = -(\partial + 2v)\bar{K}_{2n+1} ,$$ (4.12)

obtained by differentiating (4.9) with respect to $t_{2n+1}$. We are thus left with the 2n solutions:

$$v_0 = \pm k , \quad k = 1,\ldots,n ,$$ (4.13)

but there are only n branches, the k and $-k$ branches being related by $v \to -v$, a symmetry of the MKdV hierarchy. We take the negative roots below. We next take the Fréchet derivative of (4.12) with respect to $v$ (with $v \to v + \varepsilon\varphi$)

$$\frac{DK_{2n+1}}{Du} \frac{DM}{Dv} \varphi = -\left\{ (\partial + 2v)\frac{D\bar{K}_{2n+1}}{Dv} + 2\bar{K}_{2n+1} \right\} \varphi ,$$ (4.14a)

where we have used $\frac{D}{Du}$ and $\frac{D}{Dv}$ instead of primes, to avoid confusion. To obtain the resonance polynomials, we put $\varphi = \chi^{r-1}$ and $v = k\chi^{-1}$ and use the left hand side of (2.3b):

$$\frac{DK_{2n+1}}{Du}[-k(k+1)\chi^{-2}](\partial - 2k\chi^{-1})\chi^{r-1}\Big|_{\chi'=1} = (\partial - 2k\chi^{-1})\frac{D\bar{K}_{2n+1}}{Dv}[-k\chi^{-1}]\chi^{r-1}\Big|_{\chi'=1} .$$ (4.14b)

The second term on the right of (4.14a) vanishes as a consequence of (2.3a). Noting that:

$$\frac{DK_{2n+1}}{Du}[-k(k+1)\chi^{-2}]\chi^{r-2}\Big|_{\chi'=1} = P(r;n,k)\chi^{r-2n-3} ,$$ (4.15a)

$$\frac{D\bar{K}_{2n+1}}{Dv}\,[-k\chi^{-1}]\chi^{r-1}\Big|_{\chi'=1} = \bar{P}(r;n,k)\chi^{r-2n-2}\,, \tag{4.15b}$$

where $P$ and $\bar{P}$ are the resonance polynomials of the KdV and MKdV hierarchies respectively, then

$$\bar{P}(r;n,k) = \frac{(r-2k-1)}{(r-2n-2k-2)}\,P(r;n,k)$$

$$= (r-2n-1)\prod_{i=1}^{k}(r+2i-1)(r-2i-2n)\prod_{j=k}^{n-1}(r-2j-1)(r+2j-2n)\,, \tag{4.15c}$$

where $P$ is given by (4.2b). Thus, given that we have $P$ for the KdV hierachy, it is a simple matter to calculate $\bar{P}$ for the MKdV hierarchy. The formula (4.2b) for $P$ takes a little longer to derive, so is left to [14].

We recently learnt of the paper [16] in which Strampp also discusses the relationship between negative resenances and commuting flows. However, the interesting formulae (4.2b) and (4.15c) are not presented.

## Acknowledgements

We thank Martin Kruskal and Robert Conte for many useful discussions and the SERC for their Earmarked Studentship. We also thank W. Strampp for sending us a copy of his paper [16].

## References

[1]    A. P. Fordy and A. Pickering, Analysing negative resonances in the Painlevé test, Phys. Letts. A, 1991 (in press).

[2]    M. J. Ablowitz and H. Segur, Exact linearisation of a Painlevé transcendent, Phys. Rev. Lett. **38**, 1103–6 (1977).

[3]    M. J. Ablowitz, A. Ramani and H. Segur, Nonlinear evolution equations and ordinary differential equations of Painlevé type. Lett. Nuovo Cimento **23**, 333–8 (1978).

[4]    A. Ramani, B. Grammaticos and T. Bountis, The Painlevé property and singularity analysis of integrable and non-integrable systems, Phys. Rep. **180**, 159 (1989).

[5]    M. Tabor, Painlevé property for partial diffrential equations, pp. 427–46, Soliton theory: a survey of results, ed. A. P. Fordy, MUP, Manchester 1990.

[6]    D. Levi and P. Winternitz (eds), Painlevé transcendents, their asymptotics and physical applications, Plenum, NY, 1991.

[7]    J. Chazy, Sur les équations différentielles dont l'intégrale générale est uniforme et admet des singularités essentielles mobiles, C. R. Acad. Sc. Paris **149**, 563–5 (1909).

[8]     M. D. Kruskal, Flexibility in applying the Painlevé test, published in [6].

[9]     M. D. Kruskal and P. A. Clarkson, The Painlevé-Kowalevski and Poly-Painlevé Tests for Integrability, preprint (1991).

[10]    R. Conte, Invariant Painlevé analysis of partial differential equations, Phys. Letts. A, <u>140</u>, 383-90 (1989).

[11]    A. Pickering, "Testing nonlinear evolution equations for complete integrability", Ph.D. Thesis, Leeds University, in preparation.

[12]    R. Conte, A. P. Fordy and A. Pickering, A perturbative Painlevé approach to nonlinear partial differential equations, in preparation.

[13]    F. J. Bureau, Differential equations with fixed critical points, Annali di Matematica pura ed applicata, <u>LXVI</u>, 1-116, (1964).

[14]    A. P. Fordy and A. Pickering, Integrable hierarchies and negative resonances, in preparation.

[15]    M. Antonowicz and A. P. Fordy, Hamiltonian structure of nonlinear evolution equations, pp. 273-312 in: "Soliton theory: a survey of results", (ed) A. P. Fordy, MUP, Manchester, 1990.

[16]    W. Strampp, On the resonance pattern of integrable equaitons, Prog. Theor. Phys. <u>80</u>, 384-96 (1988).

## Appendix

In this appendix we state some formulae from Conte's invariant analysis [10]. The functions $\chi$, in this paper, and the usual WTC $\varphi$ are related by:

$$\chi = (\varphi_x/\varphi - \varphi_{xx}/2\varphi_x)^{-1} . \tag{A1}$$

This $\chi$ satisfies the two Riccati equations:

$$\chi_x = 1 + \frac{1}{2}S\chi^2 , \quad \chi_t = -C + C_x\chi - \frac{1}{2}(C_{xx} + CS)\chi^2 , \tag{A2}$$

with integrability condition:

$$S_t + C_{xxx} + 2C_x S + CS_x = 0 , \tag{A3}$$

where S and C are the unique differential invariants of the Möbius group, given by:

$$S = \varphi_{xxx}/\varphi_x - \frac{3}{2}(\varphi_{xx}/\varphi_x)^2 , \quad C = -\varphi_t/\varphi_x . \tag{A4}$$

The <u>practical</u> advantage of using this approach is to reduce the size of many expressions and thus simplify many of the calculations.

# ON THE CONVERGENCE OF SERIES SOLUTIONS
# OF NONINTEGRABLE SYSTEMS WITH
# ALGEBRAIC SINGULARITIES

**L.B. Drossos and T.C. Bountis**

**Department of Mathematics**
**University of Patras**
**26110 Patras, Greece**

## ABSTRACT

It has recently been shown that there exist nonintegrable 2-degree-of-freedom Hamiltonian systems with only algebraic singularities in complex time, which do not cluster on the same Riemann sheet in the $t-$plane. The general solution $x(t)$, $y(t)$ of these systems around any one of these singularities at $t = t_*$ can be written in the form of series expansions

$$x(t) = \sum_{n \geq n_1} a_n (t - t_*)^{n\,p/q} \quad , \quad y(t) = \sum_{n \geq n_2} a_n (t - t_*)^{n\,r/q}$$

with $n_1$, $n_2 \in Z$ and $p$, $q$, $r \in N$. In this paper we prove, for a class of such systems, that these series converge within a finite (non-zero) radius of convergence around $t = t_*$ . We also demonstrate numerically that this radius extends all the way to the singularity nearest to $t$ in the complex $t-$plane.

## 1. INTRODUCTION

Recently, it has been shown that there exist nonintegrable dynamical systems whose singularities in complex time are of purely algebraic type and do not cluster on the same Riemann sheet in the complex $t-$plane [1,2].

In particular, in the case of 2-degree-of-freedom Hamiltonian systems of the form

$$H(x,y,p_x,p_y) = \frac{1}{2}(p_x^2 + p_y^2) + V(x,y) \tag{1.1}$$

with

$$V(x,y) = \frac{1}{2}(Ax^2 + By^2) + \frac{k}{[f(x,y)]^2} \tag{1.2}$$

$$f(x,y) = \begin{cases} x^2 + y^2 - 1 & \text{(1.3a)} \\[2em] x - y & \text{(1.3b)} \end{cases}$$

and A, B, k arbitrary real parameters, we have found that the general solution about one such singularity, at $t = t_*$, can be written in the form of series expansions as follows :

$$x(t) = \sum_{n \geq n_1} a_n(t - t_*)^{n/2} \quad , \quad y(t) = \sum_{n \geq n_2} a_n(t - t_*)^{n/2} \tag{1.4}$$

with $p_x = \dot{x}$, $p_x = \dot{y}$.

Although this result shows that all the singularities of these systems are algebraic, the question of the convergence of the series (1.4) remained open. In this paper we prove that these series indeed converge absolutely within a finite (non-zero) radius of convergence around $t = t_*$.

In section 2, we present the main result of this work, based on an extension and appropriate application of the method of majorants [3,4] to the second order o.d.es satisfied by $x(t)$, $y(x)$

$$\ddot{x} = -\frac{\partial V}{\partial x} \quad , \quad \ddot{y} = -\frac{\partial V}{\partial y} \tag{1.5}$$

and illustrate our result in detail on the example of the soft-billiards potential (1.3a).

In section 3, we examine other potentials to which this method applies, including some with more than 2-sheeted branching. Finally, in section 4, we present numerical results demonstrating that the coefficients in (1.4) typically grow exponentially with $n$. Hence, we are able to estimate the radius of convergence numerically and find, as expected, that it equals the distance to the nearest singularity of our series solutions.

## 2. CONVERGENCE AND UNIQUENESS OF SOLUTIONS

In order to prove our main result concerning the convergence and uniqueness of our series solutions (1.4) we first need to establish the following 2 Lemmas:

**Lemma 1**

Let $f(s, u, v)$ be an analytic function where $s$, $u$, $v$ are complex variables lying within circles $C_a$, $C_b$ and $C_d$ of radii $a$, $b$ and $d$, respectively, drawn about the origin of the $s-$, $u-$ and $v-$planes. Further, let $f(s, u, v)$ be continuous on the circumferences of $C_a$, $C_b$, and $C_d$. Under these conditions, $|f(s, u, v)|$ is bounded within this domain. Let $M > 0$ be its lowest upper bound, i.e. $|f| < M$ when $|s| < a$, $|u| < b$, $|v| < d$. Then

$$\left| \frac{\partial^{p+q+r} f(s, u, v)}{\partial s^p \partial u^q \partial v^r} \right|_{(0,0,0)} < \frac{p!q!r!}{a^p b^q d^r} M' \tag{2.1}$$

where

$$M' = M|A_{\bar{p},\bar{q},\bar{r}}| \tag{2.1a}$$

$$|A_{\bar{p},\bar{q},\bar{r}}| = \max_{0 \leq p,q,r \leq \infty} \left\{ |A_{p,q,r}| \right\} \tag{2.1b}$$

and $A_{p,q,r}$ are the coefficients in Taylor series expansion of $f/M$ about $(0,0,0)$.

<u>Proof.</u> Let

$$f_1 = \frac{1}{M}f, \quad s_1 = \frac{1}{a}s, \quad u_1 = \frac{1}{b}u, \quad v_1 = \frac{1}{d}v \tag{2.2}$$

whence

$$|f_1| \leq 1 \quad \text{when} \quad |s_1| \leq 1 \ , \ |u_1| \leq 1 \ , \ |v_1| \leq 1 \tag{2.3}$$

and

$$\frac{\partial^{p+q+r} f_1}{\partial s_1^p \partial u_1^q \partial v_1^r} = \frac{a^p b^q d^r}{M} \frac{\partial^{p+q+r} f}{\partial s^p \partial u^q \partial v^r} \tag{2.4}$$

Now, by analyticity, $f_1$ can also, be expanded in a Taylor series about the origin, which converges for all $|s_1| < 1$, $|u_1| < 1$, $|v_1| < 1$, i.e.

$$f_1(s_1, u_1, v_1) = \sum_{p,q,r \geq 0} A_{p,q,r} s_1^p u_1^q v_1^r \tag{2.5a}$$

with

$$A_{p,q,r} = \frac{1}{p!q!r!} \left| \frac{\partial^{p+q+r} f_1}{\partial s_1^p \partial u_1^q \partial v_1^r} \right|_{(0,0,0)} \tag{2.5b}$$

Examining the convergence of (2.5) with respect to every one of its variables separately we get, for the $s_1$ variable, that there exists a number $p_0$ such that, for any $u_1$, $v_1$

$$\left| \frac{A_{p+1,q,r} \, s_1^{p+1} u_1^q v_1^r}{A_{p,q,r} \, s_1^p u_1^q v_1^r} \right| < 1 \tag{2.6}$$

for any $p > p_0$ with $p_0$ large enough. Since the above relation holds for all $s_1$ with $|s_1| < 1$, $|s_1| = 1$, we get from (2.6) that

$$|A_{p,q,r}| < |A_{p_0,q,r}| \qquad \forall p > p_0 \tag{2.7a}$$

Similarly, for the other variables $u_1$, $v_1$ there exist, $q_0$ and $r_0$ large enough, such that

$$|A_{p,q,r}| < |A_{p,q_0,r}| \qquad \forall q > q_0 \tag{2.7b}$$

and

$$|A_{p,q,r}| < |A_{p,q,r_0}| \qquad \forall r > r_0 \tag{2.7c}$$

Now, since for every pair of variables relations like (2.7) are valid, it follows that, for $p$, $q$, $r$ large enough, $|A_{p,q,r}|$ decrease monotonically. Thus, there exists only a finite number of possibilities $p < p_0$, $q < q_0$, $r < r_0$ for which this monotonicity will generally break down. We can, therefore, always find an integer triplet $(\bar{p}, \bar{q}, \bar{r})$ such that

$$|A_{\bar{p},\bar{q},\bar{r}}| = \max_{0 \leq p,q,r \leq p_0,q_0,r_0} \left\{ |A_{p,q,r}| \right\}$$

whence

$$|A_{p,q,r}| < |A_{\bar{p},\bar{q},\bar{r}}| \qquad \forall p, q, r$$

and using this result together with (2.4), (2.5b) we conclude

$$\left|\frac{\partial^{p+q+r} f(s,u,v)}{\partial s^p \partial u^q \partial v^r}\right|_{(0,0,0)} < \frac{p!q!r!}{a^p b^q d^r} M |A_{\bar{p},\bar{q},\bar{r}}| \tag{2.8}$$

which was to be proved.

**Lemma 2**

Let the function $f(s,u,v)$ satisfy all the hypotheses of Lemma 1. Then the Taylor series about $(0,0,0)$ of the function,

$$\varphi(s,u,v) = \frac{M'}{(1-\frac{s}{a})(1-\frac{u}{b})(1-\frac{v}{d})} - M'\left(1 + \frac{u}{b} + \frac{v}{d}\right) \tag{2.9}$$

with $M'$ as defined in (2.1a), is a dominant series for the corresponding Taylor expansion of the function

$$h(s,u,v) = f(s,u,v) - f(0,0,0) - C_{0,1,0}u - C_{0,0,1}v \tag{2.10}$$

<u>Proof</u>. According to Lemma 1, all the coefficients of the function f,

$$C_{p,q,r} = \frac{1}{p!q!r!}\left|\frac{\partial^{p+q+r} f}{\partial s^p \partial u^q \partial v^r}\right|_{(0,0,0)}$$

are bounded above by the quantity

$$|C_{p,q,r}| = \frac{M'}{a^p b^q d^r} \tag{2.11}$$

cf. (2.1), while the Taylor expansion of (2.9) about $(0,0,0)$ is :

$$\varphi(s,u,v) = \sum_{p,q,r\leq 0} M' \left(\frac{s}{a}\right)^p \left(\frac{u}{b}\right)^q \left(\frac{v}{d}\right)^r - M'\left(1 + \frac{u}{b} + \frac{v}{d}\right)$$

$$= \sum_{p\geq 1, q,r\geq 2} \frac{M'}{a^p b^q d^r} s^p u^q v^r \tag{2.12}$$

It is clear from (2.11) that the series (2.12) is dominant for the function $h$ of (2.10), which is what we wanted to show.

We now prove our main theorem for the equations of motion (1.5), for $V(x,y)$ of the form (1.2), with (1.3a) and indicate in section 3 how it can be applied to other potentials with algebraic singularities.

**Theorem**

Given the system

$$\ddot{x} = -Ax + \frac{4kx}{(x^2 + y^2 - 1)^3} \quad , \quad \ddot{y} = -By + \frac{4ky}{(x^2 + y^2 - 1)^3} \tag{2.13}$$

there exists a unique, convergent series solution of the form

$$x = x_* + \sum_{n \geq 0} a_n \tau^{n/2} \quad , \qquad y = y_* + \sum_{n \geq 0} a_n \tau^{n/2} \tag{2.14}$$

with

$$x_*^2 + y_*^2 = 1, \qquad \tau = t - t_*.$$

Proof. Since, on the singularity $t = t_*$ the r.h.s of equations (2.13) becomes infinite, we apply the transformation

$$s = \tau^{1/2}, \quad x = x_* + s(u_0 + u), \quad y = y_* + s(v_0 + v) \tag{2.15}$$

where $u_0$, $v_0$ are constants to be specified below. Thus (2.13) becomes, after multiplication by $4s^3$,

$$s^2 \frac{d^2u}{ds^2} + s\frac{du}{ds} - u - u_0 = 4As^3(x_* + su_0 + su)$$

$$+ \frac{16k(x_* + su_0 + su)}{\left[2u_0 x_* + 2v_0 y_* + 2u x_* + 2v y_* + (u + u_0)^2 s + (v + v_0)^2 s\right]^3}$$

$$= f_1(s, u, v) \tag{2.16a}$$

$$s^2 \frac{d^2v}{ds^2} + s\frac{dv}{ds} - v - v_0 = 4Bs^3(y_* + sv_0 + sv)$$

$$+ \frac{16k(y_* + sv_0 + sv)}{\left[2u_0 x_* + 2v_0 y_* + 2u x_* + 2v y_* + (u + u_0)^2 s + (v + v_0)^2 s\right]^3}$$

$$= f_2(s, u, v) \tag{2.16b}$$

We now seek solutions u(s), v(s) of (2.16), with u(0)=v(0)=0, in the form

$$u(s) = u_1 s + u_2 s^2 + u_3 s^3 + \cdots$$

$$\tag{2.17}$$

$$v(s) = v_1 s + v_2 s^2 + v_3 s^3 + \cdots$$

Inserting (2.17) in (2.16) and equating like powers of $s$, we get at lowest order

$$u_0 = (-2k)^{1/4}x_* \quad , \quad v_0 = (-2k)^{1/4}y_*$$

whence

$$u_0 x_* + v_0 y_* = (-2k)^{1/4} \neq 0$$

So both, $f_1$ and $f_2$, can be expanded in Taylor series about $(0,0,0)$ and (2.16) become

$$s^2\frac{d^2u}{ds^2} + s\frac{du}{ds} - u - \lambda_1 u - \lambda_2 v = \sum_{p\geq 1, q, r\geq 2} A_{p,q,r}s^p u^q v^r \tag{2.18a}$$

$$s^2\frac{d^2v}{ds^2} + s\frac{dv}{ds} - v - \lambda_2' u - \lambda_1' v = \sum_{p\geq 1, q, r\geq 2} B_{p,q,r}s^p u^q v^r \tag{2.18b}$$

with

$$\lambda_1 = \left.\frac{\partial f_1}{\partial u}\right|_{(0,0,0)} = 3x_*^2 \qquad \lambda_2 = \left.\frac{\partial f_1}{\partial v}\right|_{(0,0,0)} = 3x_* y_*$$

$$\tag{2.19}$$

$$\lambda_1' = \left.\frac{\partial f_2}{\partial u}\right|_{(0,0,0)} = 3x_* y_* \qquad \lambda_2' = \left.\frac{\partial f_2}{\partial v}\right|_{(0,0,0)} = 3y_*^2$$

Now inserting (2.17) into (2.18) and equating coefficients of $s^n$ we get

$$(n^2 - 1 - \lambda_1)u_n - \lambda_2 v_n = P_n(A_{p,q,r}, u_0, u_1, \cdots, u_{n-1}, v_0, v_1, \cdots, v_{n-1})$$

$$\tag{2.20}$$

$$-\lambda_1' u_n + (n^2 - 1 - \lambda_1')v_n = Q_n(B_{p,q,r}, u_0, u_1, \cdots, u_{n-1}, v_0, v_1, \cdots, v_{n-1})$$

where $P_n$, $Q_n$ are polynomials in their arguments. Also, computing the determinant

$$\Delta = \begin{vmatrix} n^2 - 1 - \lambda_1 & -\lambda_2 \\ -\lambda_1' & n^2 - 1 - \lambda_2' \end{vmatrix}$$

$$= (n^2 - 1)\left[n^2 - 1 - (\lambda_1 + \lambda_2')\right] + \lambda_1 \lambda_2' - \lambda_1' \lambda_2$$

with the aid of (2.19) and $x_*^2 + y_*^2 = 1$, we finally find

$$\Delta = (n+1)(n+2)(n-1)(n-2) \tag{2.21}$$

It is clear from (2.21) that $\Delta = 0$ only when $n = 1$ and $n = 2$. In the former case, we have

$$-\lambda_1 u_1 - \lambda_2 v_1 = A_{1,0,0} \quad , \quad -\lambda_1' u_1 - \lambda_2' v_1 = B_{1,0,0}$$

and since

$$\frac{\lambda_1}{\lambda_1'} = \frac{\lambda_2}{\lambda_2'} = \frac{x_*}{y_*} = \frac{A_{1,0,0}}{B_{1,0,0}}$$

it follows that one of the coefficients $u_1$ (or $v_1$) is arbitrary [1,2]. Similarly, in the case $n = 2$, we find that one of the $u_2$ (or $v_2$) is arbitrary.

Since $\Delta \neq 0$, we can solve system (2.20) for all $n > 2$, and obtain:

$$u_n = \frac{n^2 - 1 - \lambda_1'}{(n^2 - 1)(n^2 - 4)} P_n + \frac{\lambda_2}{(n^2 - 1)(n^2 - 4)} Q_n$$

$$v_n = \frac{\lambda_2'}{(n^2 - 1)(n^2 - 4)} P_n + \frac{n^2 - 1 - \lambda_1}{(n^2 - 1)(n^2 - 4)} Q_n$$

$$\text{(2.22)}$$

Now, the functions $f_1$, $f_2$ of (2.16) fulfill the hypotheses of Lemmas 1 and 2 and satisfy $|f_i| \leq M_i$ when $|s| \leq \rho_i$, $|u| \leq \sigma_i$, $|v| \leq \mu_i$, $i = 1, 2$. Then, by virtue of Lemma 2, for $i = 1, 2$, the functions

$$\varphi_i(s, u, v) = \frac{M_i'}{(1 - \frac{s}{\rho_i})(1 - \frac{u}{\sigma_i})(1 - \frac{v}{\mu_i})} - M_i'\left(1 + \frac{u}{\sigma_i} + \frac{v}{\mu_i}\right) \qquad \text{(2.23)}$$

are dominant for the series expansion of

$$f_1'(s, u, v) = f_1(s, u, v) - f_1(0, 0, 0) - \lambda_1 u - \lambda_2 v \qquad \text{(2.24a)}$$

and

$$f_2'(s, u, v) = f_2(s, u, v) - f_2(0, 0, 0) - \lambda_1' u - \lambda_2' v \qquad \text{(2.24b)}$$

about $(0,0,0)$ and $M_1'$, $M_2'$ are as defined in Lemma 1. These functions, (2.24a,b) represent exactly the r.h.s of relations (2.18a,b) respectively.

Coming back to (2.22), since the coefficients of $P_n$, $Q_n$ tend to zero as $n \to \infty$ and $n > 2$, we can always find upper bounds $A$, $B$, $C$, $D$ such that

$$A \geq \left| \frac{n^2 - 1 - \lambda_1'}{(n^2 - 1)(n^2 - 4)} \right| \qquad B \geq \left| \frac{\lambda_2}{(n^2 - 1)(n^2 - 4)} \right|$$

$$C \geq \left| \frac{\lambda_2'}{(n^2 - 1)(n^2 - 4)} \right| \qquad D \geq \left| \frac{n^2 - 1 - \lambda_1}{(n^2 - 1)(n^2 - 4)} \right|$$

$$\text{(2.25)}$$

with $AD - BC \neq 0$. Consider now the algebraic system:

$$\frac{D}{AD - BC} U + \frac{-B}{AD - BC} V = \varphi_1(s, U, V)$$

$$\frac{-C}{AD - BC} U + \frac{A}{AD - BC} V = \varphi_2(s, U, V)$$

$$\text{(2.26)}$$

with $\varphi_1$, $\varphi_2$ given by (2.23) and $U$, $V$ new majorizing variables for $|u|$, $|v|$.

Note first that the system (2.26) with (2.23) possesses the solution $(U,V) = (0,0)$. It is also not difficult to find the analytical solution $U(s)$, $V(s)$ of the above system, which when expnded about $(0,0,0)$, has the form

$$U = U_1 s + U_2 s^2 + U_3 s^3 + \cdots \quad , \quad V = V_1 s + V_2 s^2 + V_3 s^3 + \cdots \qquad (2.27)$$

with a finite (non-zero) radius of convergence. Substituting (2.24) into (2.26) and taking into account the fact that

$$\varphi_i(s,U,V) = \sum_{p\geq 1, q, r\geq 2} C^{(i)}_{p,q,r} s^p U^q V^r \quad , \quad C^{(i)}_{p,q,r} = \frac{M'_i}{\rho_i^p \sigma_i^q \mu_i^r} \quad , \quad i = 1, 2$$

cf. (2.12), we obtain for $n > 2$:

$$\frac{D}{AD-BC} U_n + \frac{-B}{AD-BC} V_n = P_n(C^{(1)}_{p,q,r}, U_0, U_1, \cdots, U_{n-1}, V_0, V_1, \cdots, V_{n-1}) = P'_n$$

$$(2.28)$$

$$\frac{-C}{AD-BC} U_n + \frac{A}{AD-BC} V_n = Q_n(C^{(2)}_{p,q,r}, U_0, U_1, \cdots, U_{n-1}, V_0, V_1, \cdots, V_{n-1}) = Q'_n$$

hence

$$U_n = A P'_n + B Q'_n \quad , \quad V_n = C P'_n + D Q'_n \qquad (2.29)$$

Note that for a given set of initial conditions $u_i$ (or $v_i$), $i=0$, 1, 2, we can always choose $A$, $B$, $C$ and $D$ in (2.26) such that (2.29) gives

$$U_i \geq |u_i| \quad , \quad V_i \geq |v_i| \qquad (2.30)$$

$i=0$, 1, 2. Let us assume that (2.30) holds for $i=0, 1, 2, 3, \cdots, k - 1$. We will prove that it also holds for $i = k$:

From (2.22) and (2.25) we have that

$$|u_k| \leq A|P_k| + B|Q_k| \leq A P'_k + B Q'_k = U_k \qquad (2.31)$$

cf. (2.20) and (2.28), since

$$|A_{p,q,r}| < C^{(1)}_{p,q,r} \quad , \quad |B_{p,q,r}| < C^{(2)}_{p,q,r}$$

and

$$|u_i| < U_i \quad , \quad |v_i| < V_i \quad , \quad i = 0, 1, 2, \cdots, k - 1.$$

Similarly, one shows that $|v_k| \leq V_k$ and the absolute convergence of series (2.17) has been proved.

That these solutions are unique can now be easily established by the fact that they correspond to a unique choice of initial conditions and that they are represented by convergent series expansions.

## 3. APPLICATIONS TO OTHER POTENTIALS

In this section, we briefly demonstrate how the approach of section 2 can be used to prove the convergence of series solutions of other potentials with purely algebraic singularities. In particular, we will consider here:

(a) The Calogero-Moser potential [5]

$$V(x,y) = \frac{1}{2}(Ax^2 + By^2) + \frac{k}{(x-y)^2} \qquad A, B, k > 0 \qquad (3.1)$$

(b) The Weak-Painleve potential [6].

$$V(x,y) = y^5 + y^3 x^2 + \frac{3}{16} y x^4 \qquad (3.2)$$

(c) The Grammaticos-Dorizzi-Ramani potential [6].

$$V(x,y) = \frac{1}{8} x^4 + \frac{3}{4} x^2 y^2 + y^4 + \lambda x \qquad (3.3)$$

The equations of motion for the potential (3.1),

$$\ddot{x} = -Ax + \frac{2k}{(x-y)^3} \quad , \quad \ddot{y} = -By - \frac{2k}{(x-y)^3} \qquad (3.4)$$

have been shown to have solutions of the form [1,2]

$$x = a + \sum_{n \geq 0} a_n \tau^{n/2} \quad , \quad y = a + \sum_{n \geq 0} b_n \tau^{n/2} \quad , \quad \tau = t - t_* \qquad (3.5)$$

near a singularity $t = t_*$, for all $A$, $B$, $k$ ($a$, $t_*$ and two more coefficients in (3.5) are arbitrary to be specified by the 4 initial conditions of the problem).

Hence, in order to regularize the r.h.s. of eqs. (3.4) we introduce the transformation of variables

$$s = \tau^{1/2}, \quad x = a + s(u_0 + u), \quad y = a + s(v_0 + v) \qquad (3.6)$$

cf. (2.15), with the aid of which the equations of motion (3.4) can be written in the form

$$s^2 \frac{d^2 u}{ds^2} + s \frac{du}{ds} - u - u_0 = 4As^3(x_* + su_0 + su) + \frac{8k}{[(u_0 - v_0) + (u - v)]^3} \qquad (3.7)$$

and a similar equation for $v$, cf. (2.16). Again, we seek solutions of these equations, $u(s)$, $v(s)$ as power series in $s$, with $u(0) = v(0) = 0$, cf. (2.17), and find at leading order that the constants $u_0$, $v_0$ satisfy:

$$-u_0 = \frac{8k}{(u_0 - v_0)^3} \quad , \quad v_0 = \frac{8k}{(u_0 - v_0)^3}$$

whence

$$u_0 = (-k)^{1/4} \quad , \quad u_0 = (-k)^{1/4} \tag{3.8}$$

From (3.8) we conclude that we can write eq. (3.7) as follows:

$$s^2 \frac{d^2u}{ds^2} + s\frac{du}{ds} - u - \lambda_1 u - \lambda_2 v = \sum_{p\geq 0, q, r \geq 1} A_{p,q,r} s^p u^q v^r \tag{3.9}$$

with a similar equation for $v$, cf. (2.18). Proceeding in the same way as in section 2, we substitute for $u$, $v$ in these equations the series

$$u(s) = u_1 s + u_2 s^2 + u_3 s^3 + \cdots \quad , \quad v(s) = v_1 s + v_2 s^2 + v_3 s^3 + \cdots \tag{3.10}$$

and solve for the coefficients $u_n$ and $v_n$, equating like powers of $s$. We subsequently introduce the majorizing functions $\varphi_1$, $\varphi_2$ cf. (2.24), and show that there exist converging series expansions

$$U = U_1 s + U_2 s^2 + U_3 s^3 + \cdots \quad , \quad V = V_1 s + V_2 s^2 + V_3 s^3 + \cdots \tag{3.11}$$

such that $U_n \geq |u_n|$ and $V_n \geq |v_n|$. Thus, the series (3.10), and, consequently, also $x(t)$, $y(t)$ converge, cf. (3.6) and (3.5).

Turning to the Weak-Painleve potential (b) in (3.2) we note that the leading order behavior of the solutions near a movable branch point is of the form $x, y \sim (t - t_*)^{-2/3}$. Hence, upon transforming to new variables

$$s = \tau^{1/3}, \quad x = s^{-2}(u_0 + u), \quad y = s^{-2}(v_0 + v) \tag{3.12}$$

the equations of motion for (3.2) become

$$s^2 \frac{d^2u}{ds^2} - s\frac{du}{ds} + 10(u + u_0) = g(s, u, v) \tag{3.13}$$

with a similar equations for $v$. Here also $g(s, u, v)$ has a convergent power series expansion about $(0,0,0)$ of the same form as the r.h.s of (3.9). Therefore, the same procedure as before applies and the series solutions (3.10) in (3.13) are seen to be absolutely convergent. In fact, by an appropriate change of variables, one can obtain convergence results for the solutions of the family of Weak-Painleve potentials [6]

$$V_n = \sum_{k=0}^{[n/2]} 2^{n-2k} \frac{(n-k)!}{k!(n-2k)!} x^{2k} y^{n-2k} \tag{3.14}$$

whose leading order behavior near a singularity is $x, y \sim (t - t_*)^{-2/(n-2)}$, $t \to t_*$ (For $n = 2p + 1$, we would need to set $s = \tau^{1/2}$, and for $n = 2k$, $s = \tau^{1/(p-1)}$).

Finally, in the case of potential (c) in (3.3), the singularities are again of square root type and transforming to new variables

$$s = \tau^{1/2}, \quad x = fs^{-1} + s(u_0 + u), \quad y = \frac{2}{\sqrt{2}}s^{-2} + s(v_0 + v)$$

we use the analysis of section 2 to prove convergence of the series $u = u_1 s + u_2 s^2 + \cdots$ and $v = v_1 s + v_2 s^2 + \cdots$, in exactly the same way.

## 4. NUMERICAL CONVERGENCE RESULTS

In this section we obtain numerical results on the convergence of series solutions of a potential with only algebraic singularities. In particular we examine here the Calogero-Moser case (b), cf. (3.1) whose equations of motion (3.4) yield for the coefficients $a_n$, $b_n$ of (3.5) the following recursion relations.

$$\left[ \frac{n}{2}\left(\frac{n}{2} - 1\right) - \frac{3}{8}\right] a_n + \frac{3}{8} b_n = -A a_{n-4} + S_n \tag{4.1}$$

$$\frac{3}{8} a_n + \left[\frac{n}{2}\left(\frac{n}{2} - 1\right) - \frac{3}{8}\right] b_n = -B b_{n-4} - S_n$$

for $n \geq 3$, and

$$S_n = \frac{k}{8a_1^3}\left[ \sum_{p=2}^{[(n-1)/2]} (-1)^p (p+1)(p+2) C_{n-2p-1,p}\right]$$

where $a_0$, $a_2 = b_2$, $a_3 = -b_3$ (and $t_*$) are the 4 free parameters of the solution, $a_1 = -b_1 = (-k)^{1/4}$, and the coefficients $C_{m,p}$, satisfy [7]

$$C_{m,p} = \frac{1}{mA_0}\sum_{k=1}^{m}(kp - m + k)A_k C_{m-k,p} \qquad m \geq 1$$

$$\tag{4.2}$$

$$A_k = \frac{a_{k+3} - b_{k+3}}{2a_1}, \quad k \geq 0, \quad C_{0,p} = \left(\frac{a_3}{a_1}\right)^p$$

Using (4.1) and (4.2) we numerically compute the coefficients $a_n$, $b_n$ and found that they depend exponentially on $n$, i.e.

$$|a_n| \propto e^{\alpha n}, \quad |b_n| \propto e^{\beta n}$$

Plotting $\log|a_n|$ vs. n in Fig.1, we see that these curves can be very accurately fitted by straight lines with slope $\alpha$ (In this example, it turns out that $\alpha = \beta$). On the other hand, plotting $\log|a_n|$ vs. $\log(n!)$ one finds, by comparison a discernible deviation from a straight line.

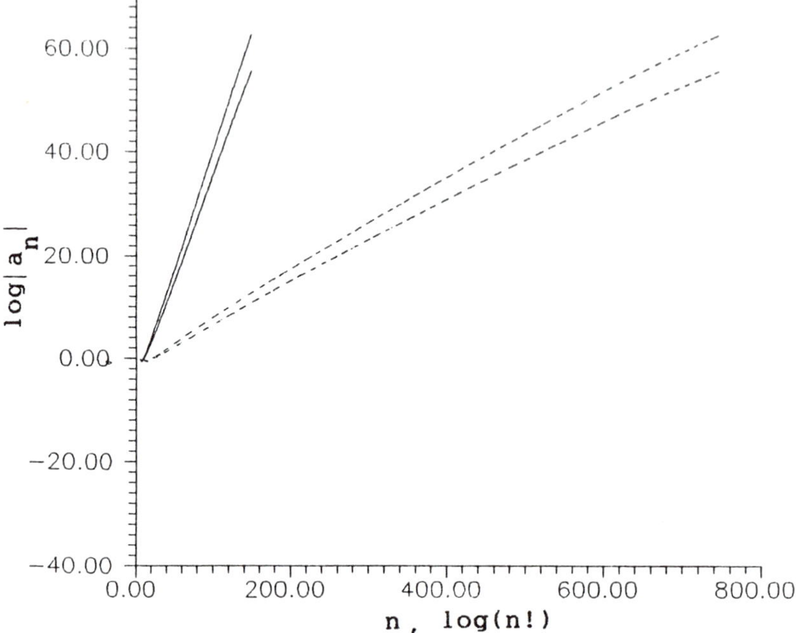


**Fig.1.** Plots of $\log|a_n|$ vs. $\log(n)$ (solid lines) and vs. $\log(n!)$ (dashed curves) for the solution $x(t)$ of eqs. (3.4), cf. (3.5), for $A=1$ and different values of $B$.

Finally computing the radius of convergence of our series, $R_\alpha$, $R_\beta$, using the root test,

$$R_\alpha = \lim_{n\to\infty} |a_n|^{1/n} \quad , \quad R_\beta = \lim_{n\to\infty} |b_n|^{1/n} \tag{4.3}$$

we obtain, in Fig.2 a clear indication that the limit (4.3) exists and that $R = \min(R_\alpha, R_\beta)$ (here, $R = R_\alpha = R_\beta$) equals the distance between $t_*$ and its nearest singularity in the complex $t$–plane. For example using the initial conditions

$$x = 1.623282 + i2.2383 \cdot 10^{-2} \ , \ x = 12.112234 + i11.213834,$$

$$y = 1.578516 - i2.2383 \cdot 10^{-2} \ , \ y = -10.3154348 - i11.213834$$

we find $t_* = -10^{-3}$ and a distance $R$ to the nearest singularity $t'_* = -0.4646$ is

$$R = |t - t'_*|^{1/2} = 0.6823$$

which well approximated by the limiting value of $R_n$ in Fig.2.

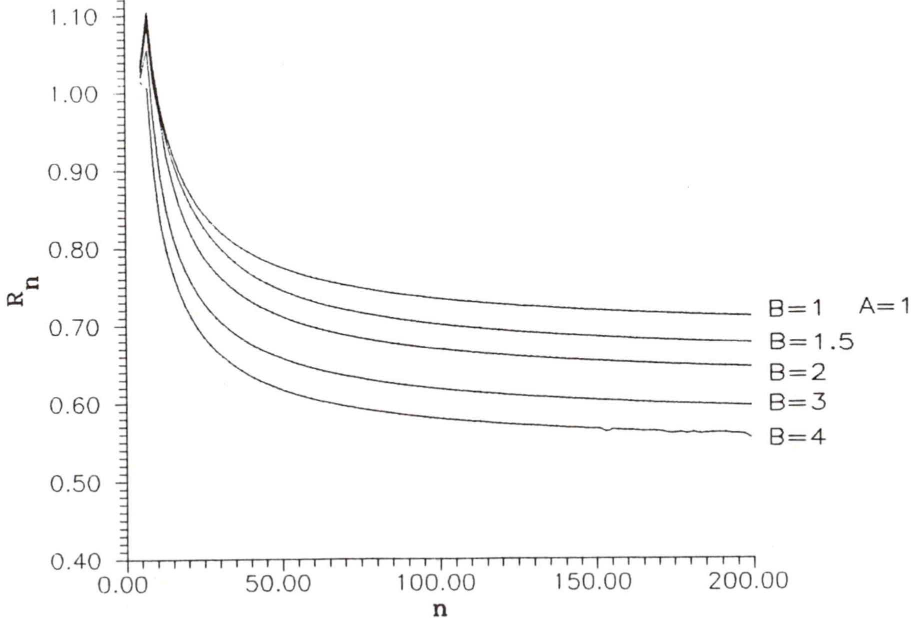

**Fig.2.** Convergence of the root test for the radius of convergence of series (3.5) for $x(t)$. Plotted here is $R=|a_n|$ vs. $n$ for different values of $B$ ($A=1$) in (3.4).

## 5. ACKNOWLEDGEMENTS

We acknowledge many useful conversations with Ian Percival, Mark Ablowitz and Martin Kruskal on topics related to the main theme of this paper. We also wish to thank the EEC Science / Stimulation Program for partially supporting this work under contract SCI-0156.

## 6. REFERENCES

[1]  T. Bountis, L. Drossos, I. C. Percival, "On Nonintegrable Systems With Square Root Singularities", Phys. Let. A159 (1991).

[2]  T. Bountis, L. Drossos, I.C. Percival, "Nonintegrable Systems With Algebraic Singularities in Complex Time", J. Phys. A24(1991).

[3]  E. L. Ince, "Ordinary Differential Equations" Dover edition, New York (1956).

[4]  G. Birkhoff, G. C. Rota, "Ordinary Differential Equations" 2nd Edition Blaisdell, New York (1969).

[5]  M. A. Olshanetsky, A. M. Perelomov, Phys. Rep. 71 (1981) 313.

[6]   A. Ramani, B. Grammaticos, T. Bountis Phys. Rep. 180 (1989) 160.

[7]   I. S. Gradshteyn, I. M. Ryzhik, "Table of Integrals, Series, and Products, Corrected and Enlarged edition" p.14(0.314) Academic Press (1980).

# PHASE TRANSITIONS WITHIN THE

# FULLY DEVELOPED CHAOTIC REGIME

R. Kluiving and H.W. Capel

Institute for Theoretical Physics
Valckenierstraat 65
1018 XE Amsterdam, The Netherlands

and

R.A. Pasmanter

Koninklijk Nederlands Meteorologisch Instituut
Postbus 201
3730 AE De Bilt, The Netherlands

## INTRODUCTION

In this contribution we report on a special type of phase transitions, which arises
in a particular one-dimensional fully developed chaotic (FDC) map, the bungalow-
tent map $f_a(x)$, depending on a control variable $a$. Phase-transition-like behaviour
can be observed in the Lyapounov spectrum $\lambda(a)$ for a sequence of critical values of
the parameter $a$.[1] The phase-transition-like phenomena can be understood in terms
of a particular symbolic dynamics description of the chaotic process, which is equivalent to a statistical description of a phase transition in a half-infinite spin chain. The
(long-range) interactions between the spins are implicitly given by the probabilities
of the spin configurations. Exact expressions for these probabilities can be derived
due to the special choice of the symbolic dynamics. This enables us to find exact
analytical expressions for the correlation function and the critical exponents near
the critical value of $a$. It thus turns out that the bungalow-tent map provides an
exactly solvable model for a phase transition in one dimension. More details about
the (rather lengthy) derivations will be published soon.[2]

*Chaotic Dynamics: Theory and Practice*
Edited by T. Bountis, Plenum Press, New York, 1992

The bungalow-tent map [1] is defined on the interval $[-1, 1]$ as follows :

$$f_a(x) = \begin{cases} 1 + 2a + 2(a+1)x & \text{for} & -1 \le x \le -\frac{1}{2} \\ 1 + 2(1-a)x & \text{for} & -\frac{1}{2} \le x \le 0 \\ 1 + 2(a-1)x & \text{for} & 0 \le x \le \frac{1}{2} \\ 1 + 2a - 2(a+1)x & \text{for} & \frac{1}{2} \le x \le 1 \end{cases} \tag{1}$$

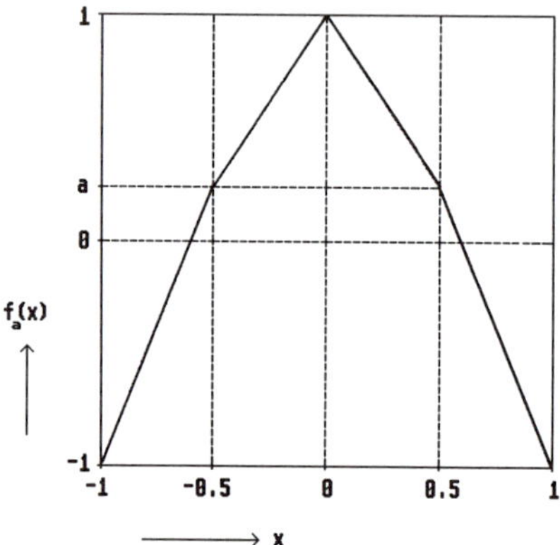


Figure 1. The bungalow-tent map.

cf. fig. 1. It is a symmetric piece-wise-linear one-hump map. The height of the extreme left and right kink is tuned by the control parameter $a$. The iterated system

$$x_{n+1} = f_a(x_n) \tag{2}$$

is FDC [3] (i.e. chaotic and ergodic on the whole interval $[-1, 1]$) for $-\frac{1}{2} < a < 1$, whereas for $-1 < a < -\frac{1}{2}$ the fixed point $\hat{x} = -1$ is stable.

In order to calculate analytically the Lyapounov exponent

$$\lambda_a = \int_{-1}^{1} \rho_a(x) \ln |f_a'(x)| \, dx \tag{3}$$

one needs an exact expression for the probability density $\rho_a(x)$ satisfying the Frobenius-Perron equation of the map. Since this is only possible for isolated values of $a$ [1], we used numerical techniques to plot $\lambda_a$ against $a$ for $-\frac{1}{2} < a < 1$, cf. fig. 2.

The most remarkable thing to be seen in this spectrum is the occurrence of several discontinuities, the first one being located at $a_1 = \frac{1}{2}$, the second one at $a_2 = 0,866\cdots$,

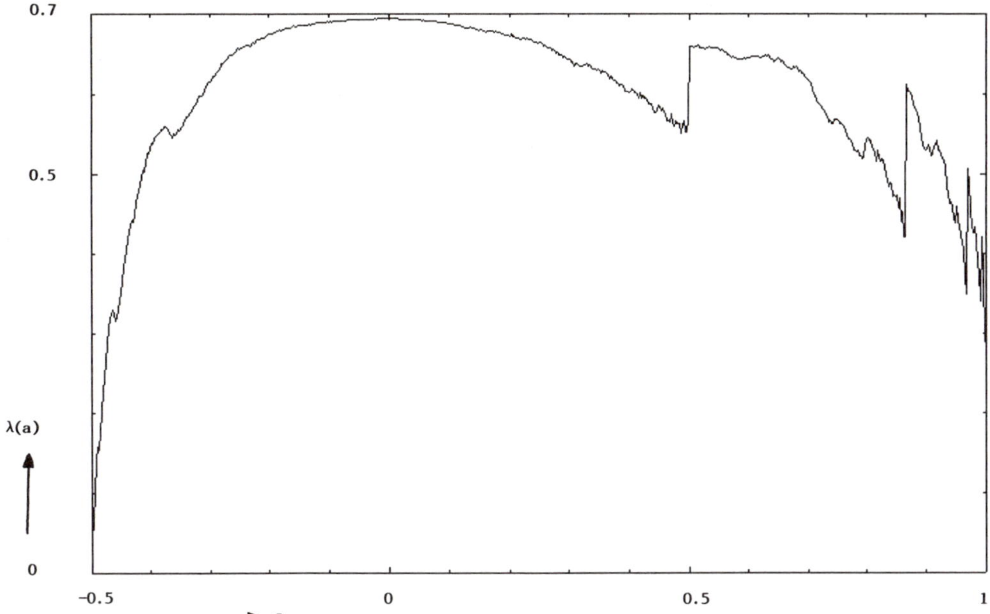

Figure 2. The Lyapounov spectrum of the bungalow-tent map.

etc. It turns out that there is an infinite number of discontinuities, their locations $a_k$ being determined by the equation

$$[2\,(a_k + 1)]^{k-1}\,(1 - a_k) = \frac{1}{2}\,, \qquad k = 1, 2, \ldots \tag{4}$$

This equation was derived using an exact renormalization procedure which showed that the jump phenomena at $a_2, a_3, \ldots$ are rescaled copies of the phenomenon at the principal value $a_1 = \frac{1}{2}$.[1]

Therefore it is sufficient to concentrate on the dynamical behaviour of the system (2) near the value $a = \frac{1}{2}$, in order to understand what is going on.

Closer analysis, see fig. 3, reveals that a laminar interval consists of a slow spiralling away from the (unstable) fixed point $x_F$.

The closer the parameter $a$ is set near $\frac{1}{2}$ from below, the longer the orbit keeps spiralling around $x_F$. Eventually the orbit escapes, giving rise to a chaotic burst, until it returns again to the close neighbourhood of $x_F$ (remember: the map is FDC, i.e. ergodic on the entire interval $[-1, 1]$, thus the orbit will certainly return to this neighbourhood), giving birth to a new laminar interval.

In a previous paper [1] we showed, using qualitative arguments, that

$$\langle n_{\mathrm{lam}} \rangle = \frac{1}{2\varepsilon} + O(1) \tag{5}$$

with $\langle n_{\mathrm{lam}} \rangle$ the mean spiralling time and $\varepsilon = \frac{1}{2} - a$. Numerical observations were in perfect agreement with eq. (5).

Solving the Frobenius-Perron equation of the bungalow-tent map for $a = \lim_{\varepsilon \downarrow 0} \frac{1}{2} - \varepsilon$ and $a = \frac{1}{2}$ yielded the following expressions for the probability

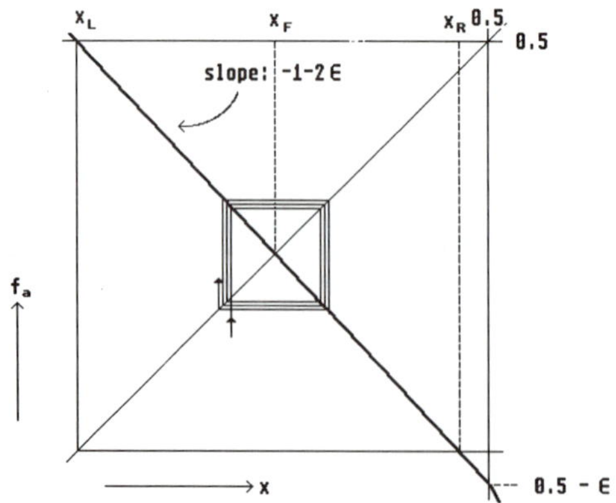


Figure 3. Magnified picture of the bungalow-tent map for fixed $a = \frac{1}{2} - \varepsilon$ and for $x$-values close to $\frac{1}{2}$.

density:

$$\lim_{\varepsilon \downarrow 0} \rho_{\frac{1}{2}-\varepsilon}(x) = \frac{1}{3}\theta\left(\frac{1}{2} - x\right)$$

$$+\frac{2}{3}\theta\left(x - \frac{1}{2}\right) + \frac{1}{6}\delta\left(x - \frac{1}{2}\right) \tag{6}$$

$$\rho_{\frac{1}{2}}(x) = \frac{2}{5}\theta\left(\frac{1}{2} - x\right) + \frac{4}{5}\theta\left(x - \frac{1}{2}\right) \tag{7}$$

So the probability density changes *shockwise* at the transition $(a \uparrow \frac{1}{2}) \to (a = \frac{1}{2})$. This explains quantitatively [1] the discontinuity in the Lyapounov spectrum at $a = \frac{1}{2}$ (cf. eq. (3)). In fact, any spectrum

$$A_a \equiv \int_{-1}^{1} \rho_a(x) A(x) dx \tag{8}$$

with $A(x)$ an arbitrary (but integrable) function of $x$, shows a discontinuity at $a = \frac{1}{2}$, as well as at the other critical values $a_2, a_3, \ldots$ .

The phenomenon is generic for all control-variable-tuned FDC maps which have a kink in such a way that the slope at one side of the kink grows towards $-1$ as the fixed point approaches the kink.

## THE $(+,-)$-SYMBOLIC DYNAMICS

Let

$$\vec{x} = x_0, x_1, x_2, \cdots \tag{9}$$

be an orbit or trajectory of a FDC one-hump map. We define the half-infinite symbol
sequence $\vec{s}$ as follows:

$$\vec{s} = [s_1, s_2, \ldots] \quad , \quad s_i = \begin{cases} + & \text{if } x_i > x_{i-1} \\ - & \text{if } x_i < x_{i-1} \end{cases} \tag{10}$$

An equivalent way of forming $\vec{s}$ is:

$$\vec{s} = [s_1, s_2, \ldots] \quad , \quad s_i = \begin{cases} + & \text{if } x_{i-1} < x_F \\ - & \text{if } x_{i-1} > x_F \end{cases} \tag{11}$$

where $x_F$ is the extreme right fixed point of the one-hump map, cf. fig. 4.

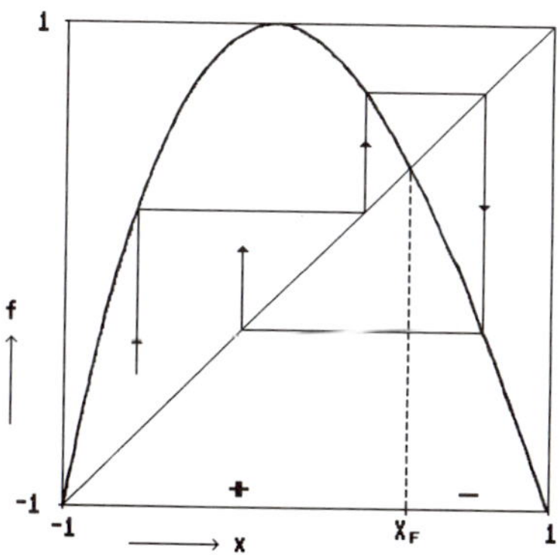



Figure 4. The $(+-)$ - partition.

The probability of finding a particular *word* or subsequence

$$\vec{w}^k = [w_1, w_2, \ldots, w_k] \ , \ w_i = + \text{ or } - \ , \quad k = 1, 2, \ldots \tag{12}$$

of $k$ successive symbols $w_1, w_2, \ldots, w_k$ in $\vec{s}$ is given by

$$P_{\vec{s}}\left(\vec{w}^k\right) = \lim_{N \to \infty} \frac{1}{N - k + 1} \sum_{i=1}^{N-k+1} \delta_{w_1 s_i} \delta_{w_2 s_{i+1}} \cdots \delta_{w_k s_{i+k-1}} \tag{13}$$

This probability can be shown to be equal to

$$P_{\vec{s}}\left(\vec{w}^k\right) = \int_{-1}^{1} dx \, \rho(x) \theta \left[\text{sgn}\left(w_1\right)\left[f(x) - x\right]\right]$$

$$\times \theta \left[\text{sgn}\left(w_2\right)\left[f^{(2)}(x) - f(x)\right]\right] \cdots \theta \left[\text{sgn}\left(w_k\right)\left[f^{(k)}(x) - f^{(k-1)}(x)\right]\right] \tag{14}$$

where $\rho(x)$ is the probability density of the map and

$$\theta(y) = \begin{cases} 1 & \text{if } y > 0 \\ 0 & \text{if } y < 0 \end{cases} \tag{15}$$

From fig. 4 it can be seen that a minus in the $(+\,-)$ - sequence $\vec{s}$ is always followed by a plus, or, equivalently, one cannot find words $\vec{w}^k$ which possess one or more pairs of neighbouring minusses in $\vec{s}$. This can also be deduced from eq. (14). At this point we would like to stress that the $(+\,-)$ - symbol sequence $\vec{s}$ is *not* intended to give a complete description of the chaotic orbit $\vec{x}$, since it can be shown that no one-to-one correspondence exists between $\vec{s}$ and $\vec{x}$. The symbol sequence $\vec{s}$ only gives a description of the going up and going down behaviour of $\vec{x}$, and, due to the FDC aspect of the iterated map, the characteristics of $\vec{s}$ are independent of the starting position $x_0$ (as long as $x_0$ does not belong to an unstable cycle), as follows from eq. (14).


APPLICATION TO THE BUNGALOW-TENT MAP


In this section we present the results of the application of the $(+, -)$-symbolic dynamics to the bungalow-tent map for values of the control parameter $a$ near the critical value $\frac{1}{2}$. More details will be published elsewhere [2].

If $a = \frac{1}{2}$, i.e. just after the phase transition, one finds

$$P_{\vec{s}}\left(\vec{w}^k\right) = \frac{3}{5} \left[\prod_{i=1}^{k-1} Q\left(w_i, w_{i+1}\right)\right] \left(\frac{2}{3}\right)^{\sum\limits_{i=1}^{k} \delta_{-,w_i}}$$

$$\times \left(\frac{1}{3}\right)^{\delta_{-,w_k} - \delta_{+,w_1} + k - 2 \sum\limits_{i=1}^{k} \delta_{-,w_i}} \tag{16}$$

for the probability of finding a particular word $\vec{w}^k$ in the $(+, -)$-symbol sequence $\vec{s}$ associated with the bungalow-tent map with $a = 1/2$. Here $Q$, being a function of two neighbouring symbols $w_i$ and $w_{i+1}$ is defined by

$$Q\left(w_i, w_{i+1}\right) = \begin{cases} 0 & \text{if } w_i = - \text{ and } w_{i+1} = - \\ 1 & \text{otherwise} \end{cases} \tag{17}$$

For the conditional probability

$$P_{\vec{s}}\left([w_1, \ldots, w_k] \rightarrow [w_{k+1}]\right) \equiv \frac{P_{\vec{s}}\left([w_1, \ldots, w_{k+1}]\right)}{P_{\vec{s}}\left([w_1, \ldots, w_k]\right)} \tag{18}$$

one finds

$$P_{\vec{s}}\left([w_1, \ldots, w_k] \rightarrow [w_{k+1}]\right) = Q\left(w_k, w_{k+1}\right) \left(\frac{2}{3}\right)^{\delta_{-,w_{k+1}}}$$

$$\times \left(\frac{1}{3}\right)^{\delta_{+,w_{k+1}} - \delta_{-,w_k}}$$

$$= P_{\vec{s}}\left([w_k] \rightarrow [w_{k+1}]\right) \tag{19}$$

Thus the $(+\,-)$ - symbol sequence $\vec{s}$ of the $(a = \frac{1}{2})$ - bungalow-tent map is *order-1 Markovian*, i.e. the memory extends only over one time step.

Introducing

$$P_{\vec{s}}\left([w_1,(n),w_{n+2}]\right) \equiv \sum_{w_2=\pm} \cdots \sum_{w_{n+1}=\pm} P_{\vec{s}}(w_1,w_2,\ldots,w_{n+1},w_{n+2}) \qquad (20)$$

one can define four correlation functions:

$$\begin{aligned}
\gamma_{++}(n) &\equiv P_{\vec{s}}([+,(n),+]) - (P_{\vec{s}}([+]))^2 &\qquad (21)\\
\gamma_{+-}(n) &\equiv P_{\vec{s}}([+,(n),-]) - P_{\vec{s}}([+])P_{\vec{s}}([-]) &\qquad (22)\\
\gamma_{-+}(n) &\equiv P_{\vec{s}}([-,(n),+]) - P_{\vec{s}}([-])P_{\vec{s}}([+]) &\qquad (23)\\
\gamma_{--}(n) &\equiv P_{\vec{s}}([-,(n),-]) - (P_{\vec{s}}([-]))^2 &\qquad (24)
\end{aligned}$$

These four correlation functions are related as follows:[4]

$$\gamma_{++}(n) = \gamma_{--}(n) = -\gamma_{+-}(n) = -\gamma_{-+}(n) \qquad (25)$$

Therefore we need only to consider one correlation function, say $\gamma_{++}(n)$. From eq. (16) one derives straightforwardly:

$$\gamma_{++}(n) = (-1)^{n+1}\frac{9}{25}\left(\frac{2}{3}\right)^{n+2}$$

cf. fig. 5, in which the behaviour of $\gamma_{++}(n)$ is shown. Notice that the correlations die out rapidly.

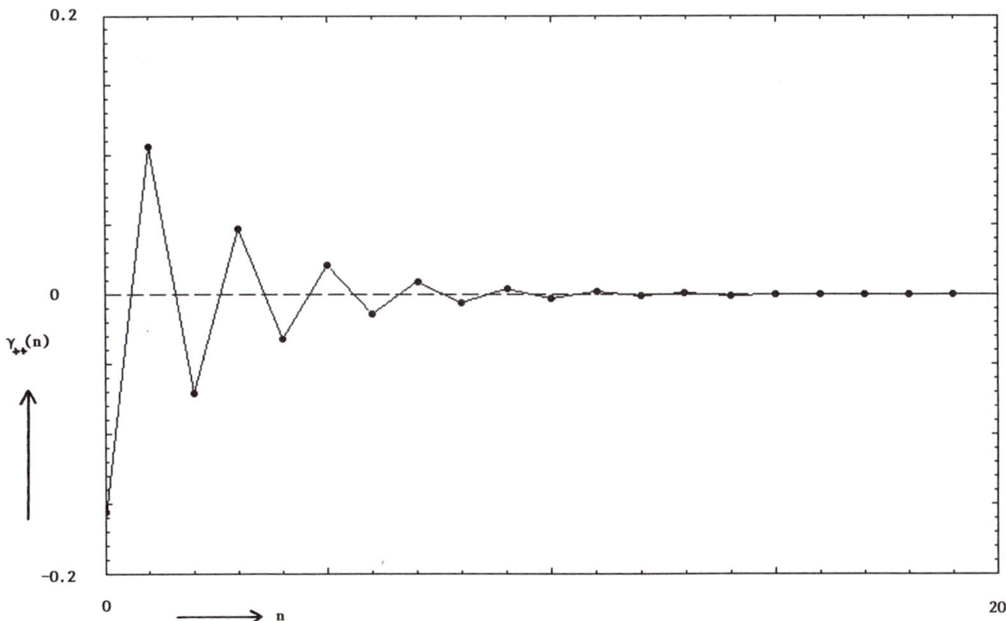

Figure 5. The correlation function $\gamma_{++}(n)$ of the $+$ - $-$ symbol sequence $\vec{s}$ associated with the $(a = \frac{1}{2})$ - bungalow-tent map.

For $a = \frac{1}{2} - \epsilon$, i.e. just before the phase transition, one finds the recursion relation

$$P_{\vec{s}}\left([w_1, w_2, \cdots, w_k]\right) = Q\left(w_1, w_2\right)\Big[(3 - 2\varepsilon)^{-\delta_{+,w_2}}\Big\{(2 - 2\varepsilon)^{\delta_{-,w_1}\delta_{+,w_2}} \times$$
$$P_{\vec{s}}\left([w_2, w_3, \cdots, w_k]\right) + \delta_{+,w_2}(-1)^{\delta_{-,w_1}}\varepsilon I_1^{[w_2,\cdots,w_k]}\Big\} +$$
$$(-1)^{\delta_{-,w_1}}\varepsilon\left[\prod_{i=1}^{k}\delta_{(-)^i,w_i}\right]\left\{\frac{\frac{1}{2}A(\varepsilon)}{(1 + 2\varepsilon)^{k-2}} - I_2^{[w_2,\cdots,w_k]}\right\}\Big]$$
$$+O(\varepsilon^2) \qquad (k \geq 3) \tag{26}$$

with

$$A(\varepsilon) = \frac{1}{3}\frac{1}{1 + 2\varepsilon} - \frac{1}{3 - 2\varepsilon}\frac{1}{6\varepsilon} = -\frac{1}{\varepsilon}\frac{1}{18} + \frac{8}{27} + O(\varepsilon) \tag{27}$$

In eq. (26), $I_1^{[w_2,\cdots,w_k]}$ and $I_2^{[w_2,\cdots,w_k]}$ are unknown integrals, which turn out to be irrelevant for the derivation of critical exponents and correlation functions.

From eq. (26) one can deduce that the $(+, -)$-sequence $\vec{s}$ has an infinite memory for $a = \lim_{\epsilon \to 0}\frac{1}{2} - \epsilon$, i.e. there does not exist a finite integer $m$ such that

$$P_{\vec{s}}([w_1, \ldots, w_k] \to [w_{k+1}]) = P_{\vec{s}}([w_{k-m+1} \ldots, w_k] \to [w_{k+1}]) \tag{28}$$

for all $k > m$ and all allowed words $w^{\to k+1}$. We were able to prove eq. (5) rigorously from eq. (26).

Also we derived an exact expression for the correlation function $\gamma_{+,+}(n)$ for $a = \frac{1}{2} - \epsilon$:

$$\gamma_{++}(n) = \frac{7}{720} + O(\varepsilon)$$
$$+ \left(\frac{-36}{720} + O(\varepsilon)\right)(1 + 2\varepsilon)^{-n}$$
$$+ \left(\frac{-2}{15} + O(\varepsilon)\right)\left[\frac{3}{2} + O(\varepsilon)\right]^{-n} \tag{29}$$
$$\text{for } n = 2, 4, 6, \ldots ,$$

and

$$\gamma_{++}(n) = \frac{7}{720} + O(\varepsilon)$$
$$+ \left(\frac{24}{720} + O(\varepsilon)\right)(1 + 2\varepsilon)^{-(n-1)}$$
$$+ \left(\frac{4}{45} + O(\varepsilon)\right)\left[\frac{3}{2} + O(\varepsilon)\right]^{-(n-1)} \tag{30}$$
$$\text{for } n = 1, 3, 5, \ldots .$$

From the above expressions one observes the following critical behaviour: After a very short 'set-in' time (which is of the order $1/\ln\frac{3}{2} \sim 3$) the correlation function starts to jump between the value $31/720 + O(\varepsilon)$ (if $n$ is odd) and the value $-29/720 + O(\varepsilon)$ (if $n$ is even). This up-and-down-jumping behaviour will continue, until $n$ reaches the critical slowing-down time $n_c = 1/2\varepsilon$. (Notice that this critical time is equal to the mean spiralling time $\langle n_{\text{lam}}\rangle$.) For $n > n_c$ the correlation function eventually saturates on the value $7/720 + O(\varepsilon)$, cf. fig. 6.

Thus the correlations do *not* die out, indicating the existence of longe-range ordering just before the critical value.

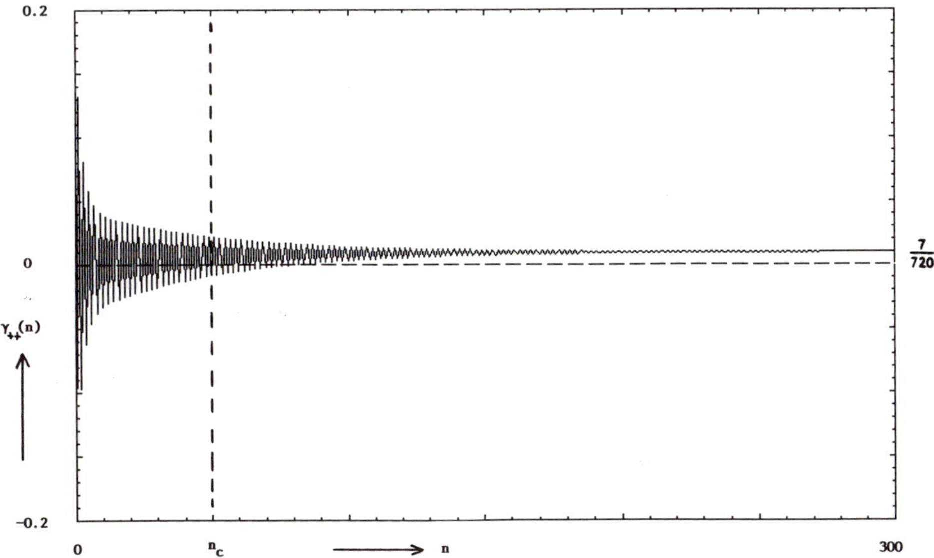


Figure 6. The correlation function $\gamma_{++}$ for $a = \frac{1}{2} - \varepsilon$ with $\varepsilon = 0.01$.

## CONCLUDING REMARKS

In this contribution we described the phase-transition near the critical value $a = \frac{1}{2}$ in the bungalow-tent map. The $(+,-)$-symbolic dynamics proved to be very successful for the analysis of this phenomenon. Moreover, this analysis cannot be given in terms of the more commonly used $LR$-symbolic dynamics, since a $(+,-)$-sequence cannot be translated unambiguously into a $LR$-sequence.[5]

This $(+,-)$-symbolic dynamics of the phase-transition in the bungalow-tent map can be reformulated in terms of an Ising model in the ensemble with given probabilities. It might be of interest to investigate which effective Hamiltonians could give rise to such probabilities and related phase transition phenomena.

## ACKNOWLEDGEMENT


This investigation is part of the research program of the Stichting voor Fundamenteel Onderzoek der Materie (FOM) which is financially supported by the Nederlandse Organisatie voor Wetenschappelijk Onderzoek (NWO).


## REFERENCES


1. R. Kluiving, H.W. Capel and R.A. Pasmanter,
   Physica A 164 (1990) 2593.
2. R. Kluiving, H.W. Capel and R.A. Pasmanter,
   Symbolic dynamics of fully developed chaos, part 3:
   Infinite-memory sequences and phase transitions,
   in preparation, to appear in Physica A.

3.  G. Györyi and P. Szépfalusy, J. Stat. Phys. 34 (1984) 451.

4.  R. Kluiving, H.W. Capel and R.A. Pasmanter,
    Symbolic dynamics of fully developed chaos, part 1:
    Statistics and characteristics of two-symbol sequences,
    preprint, to appear in Physica A.

5.  R. Kluiving, H.W. Capel and R.A. Pasmanter,
    Symbolic dynamics of fully developed chaos, part 2:
    Random and order-1 Markovian sequences,
    preprint, to appear in Physica A.

# FROM DYNAMICAL SYSTEMS

# TO LOCAL DIFFUSION PROCESSES


Armando Bazzani, Stefano Siboni°,  and  Giorgio Turchetti[‡]

Department of Physics, University of Bologna
Via Irnerio 46, 40126 Bologna Italy

Sandro Vaienti[*]

Centre de Physique Théorique - C.N.R.S. Luminy
Case 907 13288 Marseille Cedex 9 France


## 1. INTRODUCTION

The transport in chaotic regions of phase space of Hamiltonian systems is quite relevant for many physical systems (a confined plasma, the beam of a particle accelerator, a spinning planet or a galaxy). A theory of transport is still missing due to its extreme complexity even for systems with a small number of degrees of freedom [1,2]. Indeed if we consider an integrable system sufficiently perturbed then a variety of topological structures appear such as KAM curves, Cantori issued from their break up, chains of islands, chaotic regions issued from homoclinic and etheroclinic intersections of hyperbolic fixed points which replicate continuing ad infinitum under scale changes [3,4]. The picture simplifies somehow when the perturbation strength is increased since the measure of the "chaotic regions" increases, but unlikely the limit of vanishing perturbation where the measure of the tori equals the measure of all orbits, one cannot show that the measure of surviving islands of regular motion converges to zero. The description of transport seems to be possible only on a local base and any "transport coefficient" we could define, will surely be strongly varying in phase space. If the system is described by action and angle variabes $j \in \mathbb{R}^d$ and $\phi \in \mathbb{T}^d$ it is usually assumed that the angles are fast variables randomizing on the torus in a time much shorter with respect to the time scale on which the actions show an appreciable diffusion. As a consequence assuming a Markhov process for the actions and an invariance principle, a Fokker-Planck equation for the actions distribution function is written: for example this program is well accomplished for some statistical systems of billiard type [5]. In spite of the efforts devoted to prove the random phase approximation for Hamiltonian maps such as the Greene-Chirikov model, no rigorous and positive result was obtained up to now: partial results are only available for almost hyperbolic mappings with singularities [6]. In this work we describe a simple mathematical model corresponding to an integrable map where the frequency is modulated by a random or a deterministic process. Since the integrable model describes the dynamics on a KAM curve, the modulated map can give some information on the local properties of the diffusion of a realistic model. The modulated diffusion seems to be the basic diffusion mechanism for particles in hadronic accelerators and in plasmas heated by electromagnetic waves. In the

---


°  and Centre de Physique Théorique - C.N.R.S. Luminy, Marseille, France
‡  Work partially supported with NATO GRANT 890383
*  Supported by Contract CEE SC1*0281 and partially by Ministero Pubblica Istruzione - Italy



*Chaotic Dynamics: Theory and Practice*
Edited by T. Bountis, Plenum Press, New York, 1992

present paper we prove the existence and evaluate the standard diffusion coefficient related to the second moment of the unperturbed integral of motion. When the modulation is an independent identically distributed stochastic process, with probability measure absolutely continuous with respect to the Lebesgue measure on $\mathbb{R}$ and of amplitude $\varepsilon$, we show that for large values of $\varepsilon$ the diffusion coefficient is the quasi-linear one according to the random phase approximation. As a second case the modulation of a map by a deterministic process isomorphic to a Bernoulli shift ($x' = 2x \bmod 1$) has been studied to simulate the coupling with the hyperbolic motion on a separatrix. The existence of the diffusion coefficient is already a non-trivial problem and has been proved if the unperturbed frequency satisfies a diophantine condition. Even in this case for $\varepsilon \gg 1$, the diffusion coefficient tends to the quasi-linear estimate. As a last case we show that for a periodic modulation of the frequency incommensurate with respect to the frequency of the unperturbed model, no diffusion occurs since with an additional diophantine condition the extended phase space is foliated into two-dimensional tori acting as topological barriers to diffusion. The hyperbolic and the periodic modulation described above have a natural higher dimensional extension when our phase space is $\mathbb{R} \times \mathbf{T}^n$ and the modulation is given either by the linear automorphism or by a quasi-periodic motion on $\mathbf{T}^n$. The behaviour of a hyperbolic automorphism on a torus is by itself a nontrivial problem. It is of interest also in view of understanding diffusion processes on a compact manifold; examples can be found when we consider the diffusion process between two invariant curves after excluding the eventual invariant submanifolds. For the linear automorphism of the torus $\mathbf{T}^2$ the existence of the diffusion coefficient defined in terms of the moments of arbitrary order(*), has been partially proved and confirmed by numerical simulations. We remark that the numerical simulations are nontrivial since the accuracy limits to a very defined value the number of iterations. In conclusion: the simple model studied in this paper suggests that the local diffusive behaviour of an Hamiltonian system arises when the system is coupled with a hyperbolic degree of freedom. Even though the behaviour of the second moment is almost the same as for a stochastic modulation, we expect that the full statistics of the process could sharpy discriminate these two cases. When considering the diffusion on a compact manifold where the probability and the phase space can be identified, the final goal is to characterize completely the limit process and to derive a Fokker-Planck equation on the manifold itself. If such a program can be achieved one will obtain the diffusion coefficient for any dynamical variable.


## 2. THE MODEL

Given a quasi-integrable area-preserving map the KAM theory assures the existence of invariant curves characterized by the rotation number $\omega$. The motion on a single KAM curve is described by the mapping

$$
M : \begin{cases} \theta' = \theta + \omega \quad \bmod 2\pi \\ \\ j' = j + V(\theta) \end{cases}
\tag{2.1}
$$

for a fixed value of $j$. Indeed if $\omega/2\pi$ is a diophantine irrational and $V(\theta)$ an analytic periodic function in $\theta$ with period $2\pi$ and vanishing mean, the map (2.1) is an isochronous integrable map whose phase space is completely foliated in invariant curves. As a consequence if we introduce the invariant $J$

$$
J = j - F(\theta) = j - \sum_{k=-\infty}^{+\infty}{}' \frac{V_k}{e^{ik\omega} - 1} e^{ik\theta}
\tag{2.2}
$$

where $V_k$ are the Fourier coefficients of $V(\theta)$, the mapping (2.1) can be conjugated with the normal form

---

(*) and this in analogy with the solution of a one dimensional Fokker-Planck equation (see Sect. 4)

140

$$
N : \begin{cases} \Theta' \;=\; \Theta \;+\; \omega \quad \mathrm{mod}\, 2\pi \\[2ex] J' \;=\; J \end{cases}
\tag{2.3}
$$

By taking in mind the initial picture, we say that the dynamics of the map (2.1) gives a local description of the dynamics of a quasi-integrable map.

We now introduce a perturbation to the linear frequency, which leads to the modulated map:

$$
\begin{cases} \theta_{n+1} \;=\; \theta_n \;+\; \omega \;+\; \alpha_n \varepsilon \quad \mathrm{mod}\, 2\pi \\[2ex] j_{n+1} \;=\; j_n \;+\; V(\theta_n) \qquad\qquad\qquad \forall\, n \in \mathbb{N} \cup \{0\} \end{cases}
\tag{2.4}
$$

where che process $\alpha_n$ is chosen in one of the following forms:

(1) $\alpha_n \;=\; 2\,\alpha_{n-1} \,\mathrm{mod}\,[0,1[\,-\,1/2$, a Markov map invariant with respect to the Lebesgue measure;

(2) $\alpha_n$ is an independent identically distributed (i.i.d.) process on $\mathbb{R}$ with probability measure absolutely continuous with respect to the Lebesgue measure;

(3) $\alpha_n \;=\; \cos(\Omega n)$, where $\Omega\,/\,2\pi$ is irrational.

Such a perturbation simulates both the effect of an external noise or modulation on a dynamical system and the effect of a linear coupling between two degrees of freedom in a time independent system. We observe that in the case (2) the perturbation is uncorrelated at each time, whereas in the case (1) the correlation decays exponentially fast in $n$; finally in the case (3) we have no decaying of the correlation at all. Due to the perturbation, the invariant (2.2) will depend on time according to:

$$
J_{n+1} \;=\; J_0 \;+\; \sum_{m=0}^{n} \sideset{}{'}\sum_{k=-\infty}^{+\infty} \frac{V_k}{e^{ik\omega} - 1} \cdot e^{ik[\theta_0 + (m+1)\omega]}.
\tag{2.5}
$$

$$
\cdot \left[ e^{ik\varepsilon(\alpha_{-1} + \alpha_0 + \ldots + \alpha_{m-1})} \;-\; e^{ik\varepsilon(\alpha_{-1} + \alpha_0 + \ldots + \alpha_m)} \right]
$$

where we posed $\alpha_{-1} = 0$. In order to see if the perturbed system produces a diffusive behaviour we evaluate the diffusion coefficient $D(J_0)$ by taking the variance of the process $J_{n+1}$ averaged on the angular variable $\theta_0$ and the realization of the process $\alpha_n$ (we denote such a mean by $\langle\,\cdot\,\rangle$). Thus $D(J_0)$ is given by:

$$
D(J_0) \;\overset{def}{=}\; \lim_{n\to+\infty} \frac{1}{2n} \cdot \langle (J_{n+1} - J_0)^2 \rangle
\tag{2.6}
$$

We agree to call the coefficient $D$ a local diffusion coefficient, since we look to the dynamics of (2.1) as a local description of the dynamics of a quasi-integrable system; in such a case the coefficient $D$ is an indication of the diffusion of trajectories in a neighborhood of a KAM curve.


## 3. STATEMENT OF THE RESULTS

The main results we will discuss in this paper are summirized by the following

**Theorem[7]**

1) In the case of process (1) and owing to the diophantine condition on $\dfrac{\omega}{2\pi}$ the diffusion coefficient exists and can be written as:

$$D(J_0) \;=\; \frac{1}{2} \cdot \sum_{k=-\infty}^{+\infty}{}' |V_k|^2 \;+$$

$$+ \sum_{k=-\infty}^{+\infty}{}' |V_k|^2 \cdot \sum_{r=0}^{+\infty} \cos\left[k\omega(r+1)\right] \cdot \frac{\sin k\frac{\varepsilon}{2}\left(1 - \frac{1}{2^{r+1}}\right)}{k\frac{\varepsilon}{2}\left(1 - \frac{1}{2^{r+1}}\right)} \cdot \prod_{j=1}^{r+1} \cos\frac{\varepsilon}{2}k\left(1 - \frac{1}{2^j}\right) \tag{3.1}$$

2) In the case (2) the diffusion coefficient exists again and takes the form:

$$D(J_0) \;=\; \sum_{k=-\infty}^{+\infty}{}' \frac{|V_k|^2}{\left|e^{ik\omega} - 1\right|^2} \cdot \Re\left[1 - \mathcal{X}(\varepsilon k) - e^{ik\omega}\frac{\left[1 - \mathcal{X}(\varepsilon k)\right]^2}{1 - e^{ik\omega}\mathcal{X}(\varepsilon k)}\right] \tag{3.2}$$

where $\mathcal{X}(z)$ is the characteristic function of the distribution, and one can prove that the diffusion coefficient is the same without averaging on the angle.

3) In the case (3), if the frequencies $(\omega, \Omega)$ satisfy a diophantine condition, the diffusion is forbidden by the presence of topological barriers in the phase space. $\qquad\square$

The theorem can be extended [7] to include other types of perturbations, which give rise to different kinds of diffusive processes. For example another interesting perturbation is when $\alpha_n$ is an i.i.d. stochastic process with discrete distribution (Bernoulli shift). In this case one can perform an analysis according to the proof of theorem 1, but with an important difference: for a particular choice of the structural parameters of the mapping, the diffusion coefficient becomes infinite and this can be interpreted as a ballistic variation of the action. In the case (3) the inhibition of the diffusion can be proved by observing that the perturbed map (2.4) is equivalent to a volume preserving map defined in $T^2 \times \mathbb{R}$ according to

$$\begin{cases} \phi' &= \phi + \Omega \qquad \mathrm{mod}\,2\pi \\[2mm] \theta' &= \theta + \omega + \varepsilon\cos\phi \qquad \mathrm{mod}\,2\pi \\[2mm] j' &= j + V(\theta) \end{cases} \tag{3.3}$$

Then if we performe an analytic change of variables of the form

$$\begin{cases} \phi &= \Phi \\[2mm] \theta &= \Theta + \epsilon h(\Phi) \\[2mm] j &= J + F(\Theta, \Phi) \end{cases} \tag{3.4}$$

it is possible to reduce the map (3.3) into the normal form

$$\begin{cases} \Phi' &= \Phi + \Omega \qquad \mathrm{mod}\,2\pi \\[2mm] \Theta' &= \Theta + \omega \quad \mathrm{mod}\,2\pi \\[2mm] J' &= J \end{cases} \tag{3.5}$$

by solving the following homological equations

142

$$h(\Phi + \Omega) \quad - \quad h(\Phi) \quad = \quad \cos \Phi$$

$$\tag{3.6}$$

$$F(\Theta + \omega, \Phi + \Omega) \quad - \quad F(\Theta, \Phi) \quad = \quad V(\Theta + \epsilon h(\Phi))$$

Since by hypothesis $V(\theta)$ is analytic in a strip $|\mathrm{Im}\theta| \leq \Delta$, one can choose two constants $\Delta_\Phi$ and $\Delta_\Theta$ such that $V(\Theta + \epsilon h(\Phi))$ is analytic in the strip $|\mathrm{Im}\Theta| \leq \Delta_\Theta$ and $|\mathrm{Im}\Phi| \leq \Delta_\Phi$; as a consequence it is always possible to construct an analytic solution of the system (3.6) by using the Fourier expansions in the strip $|\mathrm{Im}\Theta| \leq \Delta_\Theta/2$ and $|\mathrm{Im}\Phi| \leq \Delta_\Phi/2$ and the diophantine condition in order to bound the small divisors. Therefore the map (2.4) presents barriers to the diffusion which are the projections on the initial phase space $T^2 \times \mathbb{R}$ of the invariant tori $J = const.$ of the extended map (3.3). An extension of this result to modulated symplectic maps is under consideration [8].

We now return to formulae (3.1) and (3.2) expressing the diffusion coefficient as a function of the parameter of perturbation $\varepsilon$. As briefly sketched in the Introduction, we distinguish two regimes:

(i) for $\varepsilon \gg 1$, the diffusion coefficient tends to its quasi-linear part, both oscillating as for the perturbation (1), or monotonically as for some choices of the parameters in the perturbation (2). These behaviours are usually suggested by the Random Phase Approximation [1].

(ii) in the intermediate region of values of $\varepsilon$, the diffusion coefficient turns out to be larger than its quasi-linear estimate and this effect has been really observed in plasma physics.

In figs. 1-2-3 we illustrate the graph of $D(J_0)$ versus $\varepsilon$ for the perturbations $\alpha_n$ of type (1) and (2).

## 4. RELAXATION TO EQUILIBRIUM ON THE TORUS

In this final section we consider the problem of the diffusion on a compact manifold. This could be understood either as a local diffusion process describing the motion in a bounded region of the phase space, or as relaxation to equilibrium of the correlation of any two observables defined on the phase space.

A formal analogy with the solution of the diffusion equation on the 1-torus leads to the following definition for the diffusion coefficient:

$$D = \lim_{n \to +\infty} -\frac{1}{n} \log \left| \langle j^2 \rangle_\Omega(n) - \langle j^2 \rangle(+\infty) \right| \tag{4.1}$$

where $\langle j^2 \rangle_\Omega(n)$ is the 2-moment of the action computed with respect to an initial Lebesgue-measurable domain $\Omega$ of the torus $\mathbf{T}^2 = [-1/2, 1/2[ \times [-1/2, 1/2[$, with measure $\mu_L(\Omega) > 0$, at the n-th iteration of the map:

$$\langle j^2 \rangle_\Omega(n) = \frac{1}{\mu_L(\Omega)} \int_{\mathbf{T}^2} X_{M^n(\Omega)}(\theta, j) \, j^2 \, d\theta \, dj \tag{4.2}$$

and $\langle j^2 \rangle(+\infty)$ denotes the corresponding moment for the equilibrium uniform distribution, whose numerical value is simply $1/12$. The diffusion coefficient can also be rewritten in the form of a thermodynamic limit of correlations:

$$D = \lim_{n\to+\infty} -\frac{1}{n}\log\left| \int_{\mathbf{T}^2} j^2\, X_\Omega\left(M^{-n}(\theta,j)\right) d\theta dj \;-\; \int_{\mathbf{T}^2} j^2\, d\theta dj \cdot \int_{\mathbf{T}^2} X_\Omega(\theta,j)\, d\theta dj \right| \qquad (4.3)$$

The existence of the thermodynamic limit (4.3) <u>for all the moments</u> could have an important statistical implication; in fact it could also be related to the existence of the thermodynamic limit for characteristic functions of sets and this is equivalent to a strong mixing property — the $\varphi$-mixing condition — which allows to compute all the statistical properties of the systems: decay of correlations, central limit theorem and invariance principle.
For an initial domain $\Omega$ given by a parallelogram we have the following:

**Theorem[9,10]**
$\forall \delta \in \,]0,1[, \ \forall \overline{x}_2 \in \mathbf{T}^2$ and $\forall \delta' \in \,]0,1[$ small enough, $\exists$ a L-measurable set $B(\delta) \subseteq \mathbf{T}^2$, of measure $m_L(B(\delta)) \geq 1 - \delta$, and a L-measurable set $B'(\delta,\overline{x}_2,\delta') \subseteq \mathbf{T}^2$, of measure $m_L(B'(\delta,\overline{x}_2,\delta')) \geq 1 - \delta'$, such that $\forall\, \overline{\Delta x} \in B(\delta)$ and $\forall \overline{x}_1 \in B'(\delta,\overline{x}_2,\delta')$ the parallelogram with vertices:

$$\overline{x}_1, \qquad \overline{x}_1 + \overline{\Delta x}, \qquad \overline{x}_2 + \overline{\Delta x}, \qquad \overline{x}_2$$

satisfies the thermodynamic limit defining the diffusion coefficient:

$$\exists \quad \lim_{n\to+\infty} -\frac{1}{n}\log\left|\langle j^2\rangle_\Omega(n) - \langle j^2\rangle(+\infty)\right| \;=\; 2\log\lambda \qquad (4.4)$$

where $\log\lambda$ is the positive Liapunov exponent of the automorphism. $\qquad\square$

The previous statement shows that for a random choice of the parameters $\overline{\Delta x}, \overline{x}_2$ and $\overline{x}_1$, the probability of getting a parallelogram for which the diffusion coefficient defined as above exists is arbitrarily close to 1.
A very nontrivial guess is that the previous limit (4.4) exists and takes the same value for most initial domains $\Omega$. A few numerical investigations confirm these conjectures (see for example Fig. 4, where the usual Cat Map is applied to an initial domain $\Omega$ consisting in a pentagon).


REFERENCES

[1] A.J. Lichtenberg, M.A. Lieberman, "Regular and Stochastic Motion", Springer-Verlag, New York, N.Y., (1983)
[2] B.V. Chirikov, A Universal Instability of Many-dimensional Oscillator Systems, *Phys. Rep.* **52**, 263-379, (1979)
[3] J. Moser, "Lectures on Hamiltonian Systems", Mem. Amer. Math. Soc., (1968)
[4] R.S. MacKay, J.D. Meiss, I.C. Percival, Transport in Hamiltonian Systems, *PhysicaD*, 55-81, (1984)
[5] Ya G. Sinai, L. Bunimovich, *Comm.Math.Phys.*, **78**, 479, (1981)
[6] J. Bellissard, S. Vaienti, Rigorous Diffusion Properties for the Sawtooth Map, Preprint CPT-90/P.2427 (1990), to appear in Comm. Math. Phys.
[7] A. Bazzani, S. Siboni, S. Vaienti, G. Turchetti, in preparation
[8] A. Bazzani, in preparation
[9] S. Siboni, G. Turchetti, S. Vaienti, in preparation
[10] S. Siboni, S. Vaienti, in preparation

# NONINTEGRABILITY, SEPARATRICES CROSSING AND HOMOCLINIC ORBITS IN THE PROBLEM OF ROTATIONAL MOTION OF A SATELLITE

Andrzej J. Maciejewski and Krzysztof Goździewski

Institute of Astronomy, Nicolaus Copernicus University
87-100 Toruń, Chopina 12/18, Poland

## 1.  INTRODUCTION

It is well known that in hamiltonian systems the existence of transversal homoclinic orbit to a hyperbolic periodic solution leads to complicated behavior of phase trajectories and nonintegrability[1,2]. However, if we take an unstable equilibrium instead of a periodic solution, the situation is much more complicated. Devaney[3] gave an example (the Neumann problem) of integrable system with transversal homoclinic orbit to the saddle equilibrium point.

The first result, connecting the existence of transversal homoclinic orbit to an unstable equilibrium with nonintegrability, is due to Devaney[4] (for a hamiltonian system with two degrees of freedom and the saddle-focus type equilibrium). For a practical use of Devaney's theorem one should prove, for a given system, the existence and transversality of a homoclinic orbit. As far as we know, it is possible to do this, in general, for an almost integrable system by means of the Melnikov-like method[5]. However, it should be noted, that there are many examples (e.g. the restricted three body problem) where this approach cannot be applied.

Recently Bolotin[6] proved a theorem similar to that of Devaney for the case of a saddle equilibrium. This not only explains the 'miracle' of the Neumann system but, what is more important, gives another tool for proving nonintegrability. In order to apply this result, as with Devaney's theorem, one should prove the existence, transversality and additional properties of the homoclinic orbit. Thus, our remarks concerning Devaney theorem can be repeated here. (In the cited paper Bolotin used his theorem and proved nonintegrability of the perturbed Lagrange top i.e. an almost integrable system.)

Both these theorems are of special importance in the context of chaotic dynamics. They allow us to state not only nonintegrability—as a consequence of them, we can describe behavior of phase trajectories on the equilibrium energy level in terms of symbolic dynamics (the case of the saddle-node) or in terms of similar formalism (the case of the saddle) as it was done by Turaev and Shilnikov[7] (see also literature cited in this paper).

In many studies of dynamical systems, especially those concerning chaotic behavior, numerical methods are intensively used. For example, numerically calculated Lyapunov exponents[8] are used as the basic indicator of nonintegrability. Here we propose to apply numerically the theorems of Devaney and Bolotin for the same purposes.

The aim of this paper is to present our algorithm and some results of its application which allow us to investigate the problem of rotational motion of a rigid satellite on a circular orbit.

## 2. THEORY

Let us consider an analytic hamiltonian system with $n$ degrees of freedom with the hamiltonian function $H = H(\mathbf{z})$, $\mathbf{z} = (q_1, \ldots, q_n, p_1, \ldots, p_n)$. Let $\mathbf{z} = \mathbf{z}_0$ be an equilibrium position for this system and $(\boldsymbol{\lambda}, -\boldsymbol{\lambda})$, $\boldsymbol{\lambda} = (\lambda_1, \ldots, \lambda_n)$ are eigenvalues of linearized system. We assume that none of them is imaginary. Then, there exist two $n$-dimensional analytic invariant manifolds $W^s$ and $W^u$ immersed into the phase space, consisting of trajectories asymptotic to $\mathbf{z}_0$ for $t \to +\infty$ and $t \to -\infty$, respectively. The intersection $W^s \cap W^u$ consists of double-asymptotic (homoclinic) trajectories $t \to \mathbf{z}(t)$, such that $\mathbf{z}(t) \to \mathbf{z}_0$ as $t \to \pm\infty$. If the intersection of the manifolds $W^s$ and $W^u$ on the $(2n-1)$-dimensional energy level is transversal along the double-asymptotic trajectory $\Gamma$, i.e. for $z \in \Gamma$ the intersection $T_z W^s \cap T_z W^u$ is a straight line tangent to $\Gamma$, then the double-asymptotic trajectory is called transversal[6].

For the case of $n = 2$ and the saddle-focus equilibrium:

$$\lambda_k = \alpha_k + i\beta_k, \quad \alpha_k \beta_k \neq 0, \quad k = 1,2$$

Devaney[4] proved the following theorem.

**Theorem 1** *Let $\mathbf{z}_0$ be a saddle-focus equilibrium for a hamiltonian system with two degrees of freedom and suppose $\Gamma$ is a transverse homoclinic orbit at this equilibrium. Then, to any local transverse section $\Sigma$ to $\Gamma$, and for any positive integer $N$, there is a compact, invariant, hyperbolic set $\Lambda_N \subset \Sigma$ on which the Poincaré map is topologically conjugate to the Bernoulli shift on $N$ symbols.*

If $\mathbf{z}_0$ is a saddle equilibrium then the transversality of a homoclinic orbit is not sufficient for nonintegrability. The additional conditions were formulated by Bolotin [6] in terms of the Birkhoff normal form.

It is assumed that $\mathbf{z}_0 = \mathbf{0}$ is the saddle equilibrium and all its characteristic exponents are different ($\lambda_i > 0$). It is possible that there are resonances:

$$0 = \,<\mathbf{k}, \boldsymbol{\lambda}> = \sum_{i=1}^{n} k_i \lambda_i; \quad \mathbf{k} \in \mathbf{Z}^n, \quad \mathbf{k} \neq 0 \tag{1}$$

but we exclude multiple resonances among (1):

$$\lambda_i = \,<\mathbf{m}, \boldsymbol{\lambda}>, \quad \lambda_i + \lambda_j = \,<\mathbf{m}, \boldsymbol{\lambda}>, \tag{2}$$

$$i, j = 1, \ldots, n; \quad \mathbf{m} \in \mathbf{Z}_+^n, \quad |\mathbf{m}| = \sum_{i=1}^{n} |m_i| > 1.$$

When there are no multiple resonances, the Hamiltonian function can be transformed in a neighborhood of equilibrium $\mathbf{z}_0$ to the Birkhoff normal from:

$$H = \sum_{i=1}^{n} \lambda_i \sigma_i + \sum_{i,j=1}^{n} a_{ij} \sigma_i \sigma_j + O(|p|^5 + |q|^5), \quad H(\mathbf{z}_0) = 0 \tag{3}$$

where $\sigma_i = q_i p_i$. An equilibrium position will be called *weakly non-resonant* if among resonances (1) there are no multiple ones (2), and matrix $\mathbf{A} = (a_{ij})$ is not degenerated:

$$\mathbf{A}\mathbf{k} \neq \mathbf{0}; \quad \mathbf{k} \in \mathbf{Z}^n, \quad \mathbf{k} \neq \mathbf{0}, \quad < \mathbf{k}, \boldsymbol{\lambda} > = 0. \tag{4}$$

Characteristic exponents of the hamiltonian system restricted to the $W^u$ have the same sign, thus the transformation to the Poincaré normal form converges on $W^u$. When there are no multiple resonances (2) then this normal form is linear—there exist analytical coordinates $q_1, \ldots, q_n$ on $W^u$ in a neighborhood of $\mathbf{z}_0$, such that the hamiltonian system on $W^u$ has the form:

$$\dot{q}_i = \lambda_i q_i; \quad i = 1, \ldots, n. \tag{5}$$

A trajectory asymptotic to $\mathbf{0}$ for $t \to -\infty$ will be called *main* if along it one of the normal coordinates $q_i$ is equal zero. Main asymptotic trajectories for $t \to +\infty$ are defined similarly (taking $W^s$ instead of $W^u$).

**Theorem 2** (Bolotin[6]) *Let $\mathbf{z}_0$ be an weakly nonresonant equilibrium position with real characteristic exponents. If there exists a transversal double asymptotic to $\mathbf{z}_0$ trajectory $\Gamma$, and it is not main for $t \to \pm\infty$, then the hamiltonian system does not have first integrals that are analytic in a neighborhood of the set $\Gamma \cup \{\mathbf{z}_0\}$ except those dependent on $H$.*

This theorem is applicable to a case of two equilibria and heteroclinic trajectories and to a case when $T_z W^u \cap T_z W^s$ , $z \in \Gamma$, is $k$-dimensional (see the cited paper of Bolotin for details and proof).

## 3. PROBLEM OF ROTATIONAL MOTION OF A SYMMETRIC SATELLITE

The problem of two rigid bodies leads to such difficulties, that its restricted versions are considered. In the restricted problem of rotational motion it is assumed that the motion of the bodies mass centers is given (e.g. keplerian orbits). This formulation is appropriate in a case when we have a spherically symmetric central body and a small rigid satellite. In most cases the potential of the torque contains only the second order harmonics[9]. The popularity of this problem is caused by its richness—comparable, in our opinion, with the restricted three body problem.

Here we assume that the satellite is symmetric. Markeev[10,11] was the first who started to study systematically such a system in the framework of hamiltonian mechanics. His investigations (continued[12,13] till now) were developed by Sokolsky[14], Barkin[15], Demin[16] and many others.

Let the center of mass of the satellite move on a circular orbit. We introduce two right handed orthonormal reference frames: the orbital frame ($x$-axis is directed along the radius vector, $z$-axis has direction of the orbital angular momentum) and the principal axes reference frame. As the generalized coordinates $(q_1, q_2, q_3)$ we will use the Euler angles of the type 1-2-1 that parametrize an orientation of the principal axes with respect to the orbital frame. Using the standard technique, the hamiltonian of the problem is derived in the form:

$$H = \frac{1}{2}\left\{\frac{p_1 - \gamma\cos(q_2)}{\sin(q_2)}\right\}^2 + \frac{p_2^2}{2} - \cos(q_1)\cot(q_2)p_1 \tag{6}$$

$$- \sin(q_1)p_2 + \gamma\frac{\cos(q_1)}{\sin(q_2)} + \frac{3}{2}(\alpha - 1)\cos^2(q_2)$$

where $p_1, p_2, p_3$ are canonical momenta ($q_3$ is cyclic, so we put $\gamma = p_3(0) = \alpha\omega_1/\omega_o$; $\omega_1$ is the projection of the total angular velocity of the body onto the symmetry axis, $\omega_o$ is the orbital angular velocity), $\alpha = A/B$; $A, B = C$ are the principal moments of inertia.

Two parametric hamiltonian system given by (6) possesses three families of isolated equilibria:

$$q_1^0 = 0 \text{ (or } \pi), \quad q_2^0 = \frac{\pi}{2}, \quad p_1^0 = p_2^0 = 0, \tag{7}$$

$$q_1^0 = 0 \text{ (or } \pi), \quad \sin(q_2^0) = \mp\frac{\gamma}{4 - 3\alpha}, \quad p_1^0 = \pm\frac{3(a-1)}{2}\sin(2q_2^0), \quad p_2^0 = 0, \tag{8}$$

$$\cos(q_1^0) = \gamma, \quad q_2^0 = \frac{\pi}{2}, \quad p_1^0 = 0, \quad p_2^0 = \sin(q_1^0), \tag{9}$$

called the cylindric, conic and hyperbolic precessions, respectively.

The stability of these equilibria was investigated almost completely[14]. The hyperbolic precession cannot be of saddle or saddle-focus type. Depending on the parameters values characteristic exponents of the cylindric as well as the conic precession can be all real or complex. Figure 1 shows these regions on the $(\alpha, \gamma)$ plane.

Coordinates chosen in Eq.(6) are local—they are good for investigation of cylindric precession neighborhood. However, for some values of parameters for conic precession (or a motion close to it) values of $q_2$ can be small. The presence of terms proportional to negative powers of $\sin(q_2)$ in the hamiltonian can cause numerical difficulties. In order to omit these problems we derived the hamiltonian function choosing as the generalized coordinate the Euler angle of the type 3-2-1. In these coordinates the Hamiltonian has the form:

$$H = \frac{1}{2}\left\{\frac{p_1 + \gamma\sin(q_2)}{\cos(q_2)}\right\}^2 + \frac{p_2^2}{2} - p_1 + \frac{3}{2}(\alpha - 1)\cos^2(q_1)\cos^2(q_2) \tag{10}$$

(we used the same symbols for coordinates and momenta). In new variables the conic precession is given by:

$$q_1^0 = 0 \text{ (or } \pi), \quad \sin(q_2^0) = -\frac{\gamma}{4 - 3\alpha}, \quad p_1^0 = \cos^2(q_2^0) - \gamma\sin(q_2^0), \quad p_2^0 = 0. \tag{11}$$

## 4.  ALGORITHMS

The only information we have about the asymptotic surfaces is that obtained from the local theory. Thus, every method for obtaining asymptotic surfaces, that does not suppose any special feature of the system, should use a continuation process. Examples of such a procedure can be found in the papers of Danby[17,18] (for a case of periodic solution and saddle-focus equilibrium) and Dovbysh[19] (for a case of periodic solution). In these examples, the procedure starts with approximation of asymptotic surfaces by means of *linear* theory. Next, a small interval of the asymptotic surfaces on the cross section plane is continued by means of the Poincaré map.

In a case of a saddle-focus equilibrium our algorithm is similar but we made some changes and improvements.

In a neighborhood of the equilibrium point we approximate asymptotic surfaces by means of *nonlinear* theory i.e., we use nonlinear normal form of the hamiltonian and nonlinear transformation to this normal form. For this purposes we used our

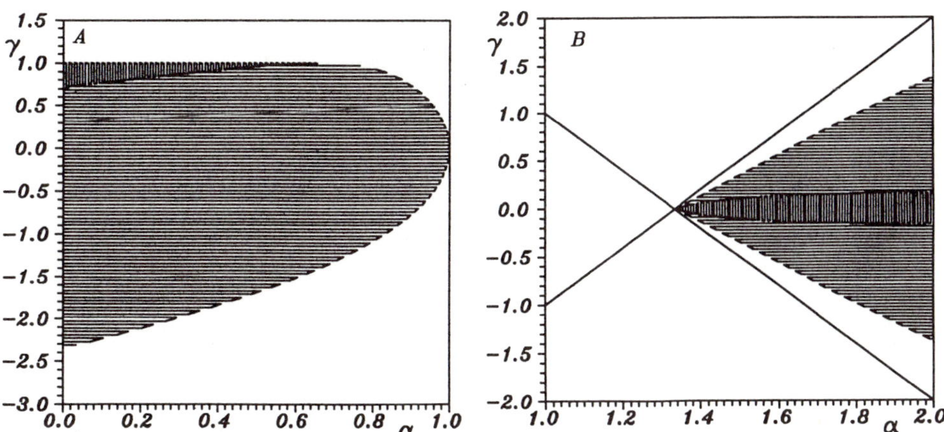

Figure 1. Regions on the $(\alpha, \gamma)$ plane where regular precessions are of saddle type (vertical lines) and saddle-focus type (horizontal lines); $A$ for cylindric precession, $B$ for conic precession.

algorithm[20,21]. This allows to reduce number of iterations needed for obtaining the maximal continuation. Practice shows that, in many cases, linear theory fails—only small pieces of asymptotic surfaces can be obtained.

Instead of obtaining successive images of the initial interval under the Poincaré map we used an algorithm that we called 'self-stepping'. In our algorithm we also obtain an approximation of a curve that is intersection of asymptotic surface with a chosen cross section plane. The beginning of this curve can be generated with prescribed accuracy by means of the local nonlinear theory. The algorithm consists of the following steps.

(1). For a given initial set of points, which lie on the curve fit the cubic spline (parametrized by $s \in [0, 1]$).

(2). Take a point from the spline and obtain its image under Poincaré map.

(3). If the distance of the obtained point from the end of the curve is bigger then desired, then take smaller value of $s$; if it is not, take the next point from the spline.

(4). If the end of the spline is reached, then take points from the curve which continue the spline and repeat the whole procedure.

This procedure ends when the increment of the parameter $s$ is smaller than the prescribed limit.

The basis of our algorithm for numerical reconstruction of saddle equilibrium asymptotic surfaces is the local flow on this surfaces (see Fig. 2). We always choose the cross section plane in such a way that it passes through the equilibrium and transversely crosses the local asymptotic surface. As in the previous case, we will approximate the curve that is an intersection of the asymptotic surface with the cross section plane. However now, because of the flow nature, it is generally impossible to obtain successive points on the curve taking an image under the Poincaré map of a point from this curve. Our algorithm consists of the following steps (see Fig. 2):

(1). Choose a cross section plane and determine the beginning of the curve, that is the intersection of the asymptotic surface with the plane, by means of local theory.

(2). Determine the position of the cross section plane with respect to the axes of the normal coordinates on an asymptotic surface.

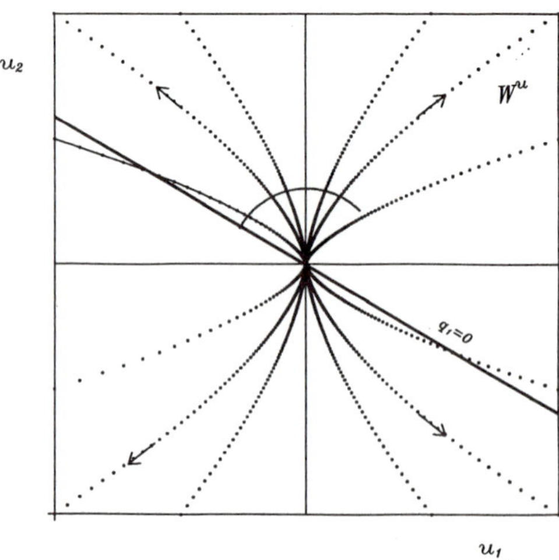



Figure 2. Local phase flow on unstable surface $W^u$ of conic precession. $(u_1, u_2)$ are normal coordinates. Continuous line is the intersection of $W^u$ with the chosen cross section plane $q_1 = 0$ (see text for further explanation).

(3). Fix a radius of a circle on the asymptotic surface (the value of which is limited by the precision of the transformation to the normal variables).

(4). Start from a point that lies on the circle and is close to the cross section plane (coordinates of the point are parametrized by the polar angle $\varphi$). Choose an arc of the circle such that the phase trajectories pass first the circle and next the cross section plane as $t \to \infty$, if the asymptotic surface is $W^u$ (for $W^s$ the same but as $t \to -\infty$).

(5). Transform the normal coordinates of the chosen point to the original ones and generate the phase curve described by these coordinates. Determine the first intersection point of the phase curve with the cross section plane.

(6). If the distance of the obtained point from the end of the curve is bigger than the assumed limit, change the value of $\varphi$ and repeat the procedure starting from (5); if it is not, take the next point from the circle.

The procedure ends if the increment (or decrement) of $\varphi$ is smaller than the prescribed limit.

We tested this procedure in many different cases (changing cross section plane, parameters of the problem etc.) and always found good results—the curve obtained is maximal—it ends on the lines of discontinuity[17,18] of the cross section or on the limits of the variables domain.

If the asymptotic surfaces cross, then the curves on the cross section plane also cross, and initial conditions for the homoclinic orbit can be determined. We did this by taking small arcs of curves around the crossing point and fitting splines to them. Coordinates of the crossing point are obtained by means of a simplex minimalization procedure.

All algorithms presented here were implemented in Pascal (Turbo Pascal v. 6.0 of Borland International) and calculations were done on an IBM 386 personal computer.

150

# 5. RESULTS

Let us start our investigation from the cylindric precession for such values of the parameters that this point is of the saddle-focus type. We do not present here the systematic overview of our results, only some characteristic features are shown (detailed discussion is under preparation).

Generally, asymptotic surfaces cross transversely and their shapes on the cross section plane depend strongly on parameters values. This is illustrated in Figs.3–5 where the cross section plane is chosen at $q_1 = 0$ (as local coordinates on the equilibrium energy level we take $(q_1, q_2, p_2)$). One can observe that the number of visible intersections of $W^u$ with $W^s$ decreases with the value of $\alpha$. As an example we show homoclinic orbits obtained for the parameters $\alpha = 0.815, \gamma = 0.9$ (see Fig. 7). The Fig. 6 and Fig. 8 show the Poincarè map in a neighborhood of the homoclinic loops and the global picture, respectively.

For values of $\alpha$ small enough, the curves on the cross section plane obtained by our algorithm do not have common points. They end on the $q_2 = \pm\pi/2$ line and do not cross the $q_2$-axis. This does not mean that they do not cross at all. For investigation of such a case it is necessary to introduce other coordinates.

A special example is that for $\alpha = 0.827, \gamma = 0.9$ (see Figs. 9–10), when it seems that asymptotic surfaces coincide. We observe that this takes place when characteristic dimensions of the images of asymptotic surfaces on the cross section plane are small enough. This effect is connected with local integrability of the system—if a homoclinic loop lies in a domain of integrability it must be degenerated. We test if it is the case computing maximal Lyapunov characteristic exponent (see Fig. 12), for several different orbits starting close to the homoclinic loop. It seems to agree with our suggestion, however it can hardly be expected that the system is integrable on the whole energy level. This is shown by the global Poincarè map (Fig. 11).

For values of parameters when the cylindric precession is of saddle type a typical shape of asymptotic solutions is shown in Fig. 13 (we note here that a single curve in this figure *was not* obtained from cross section points of one orbit; every point of the curve belongs to a different orbit). For acceptable set of parameters we did not find a case when they cross. However, it should be mentioned that we stopped our algorithm at the first discontinuity point on the cross section plane. Thus, our result does not exclude that the asymptotic surfaces intersect.

We obtained much more interesting results for real characteristic exponents case for the conic precession. For all tested values of parameters, asymptotic surfaces have very similar features as is illustrated in Fig. 14. It is visible that they cross (homoclinic orbit is shown in Fig. 15). Figure 16 shows that the homoclinic orbit is not main. Since for these values of parameters there are no resonances, it is not necessary to check the nondegeneracy condition Eq.(4).

For both investigated cases conclusions arise from Bolotin's and Devaney's theorems. For most values of parameters, the application of our algorithms allows one to state that the system is not integrable and the nonintegrability is caused by the existence of a transversal homoclinic orbit to the equilibrium.

Till now we have not tested our algorithms for any other system but, we believe, they are general enough to be applied to arbitrary hamiltonian system with two degrees of freedom.

Let us conclude with the remark that the methodology of our investigation is the same as that of Dovbysh[19] who gave an example of a 'computer proof' of separatrices crossing and the Kolmogorov stability in the problem of heavy top.

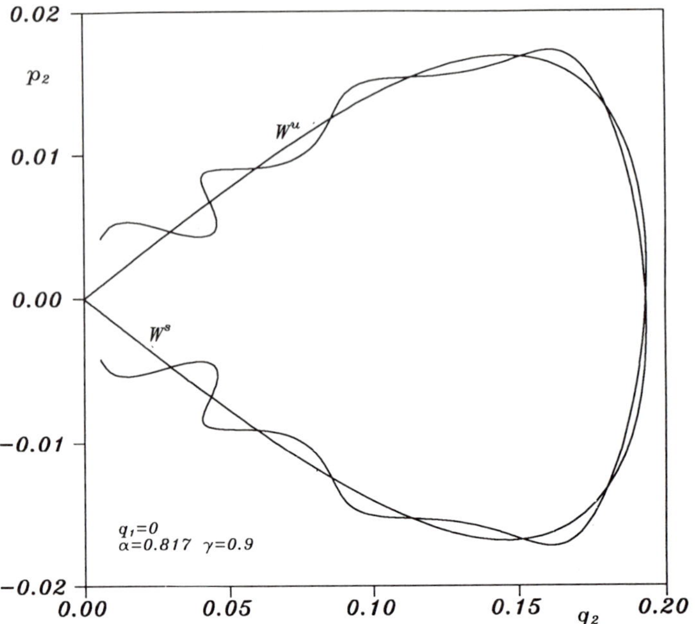



Figure 3. Asymptotic surfaces of the cylindric precession on the cross section plane.

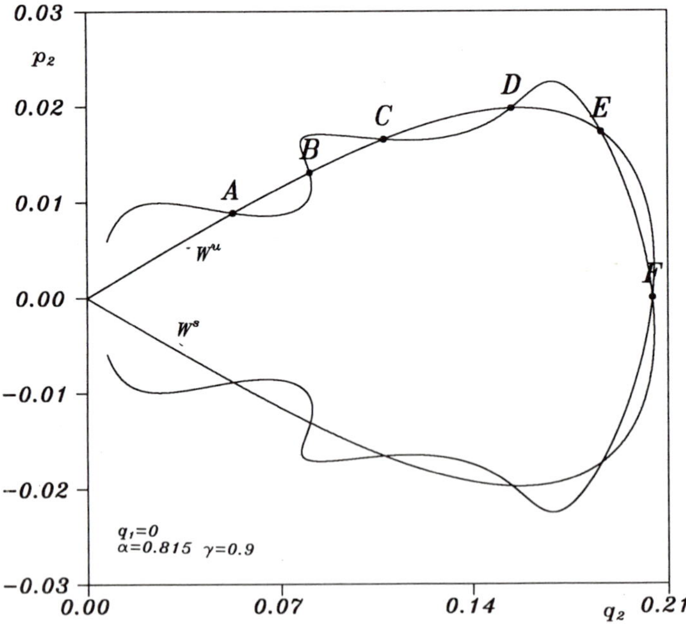



Figure 4. Asymptotic surfaces of the cylindric precession on the cross section plane.

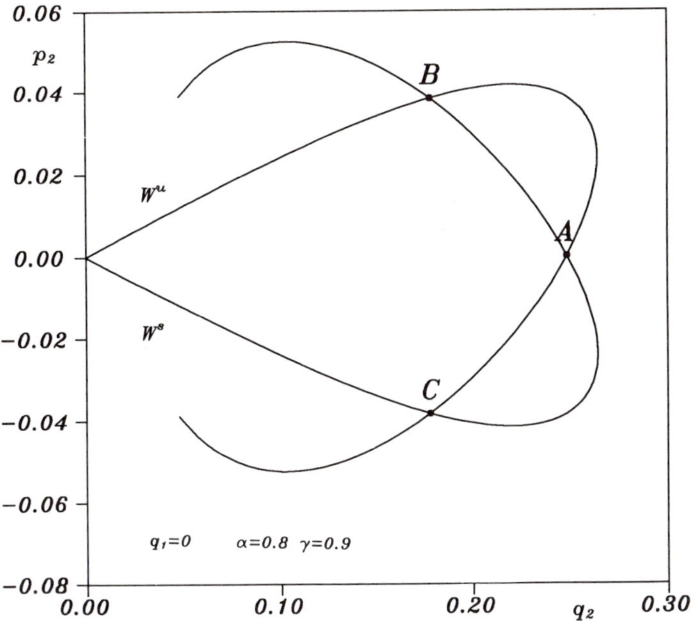

Figure 5. Asymptotic surfaces of the cylindric precession on the cross section plane.

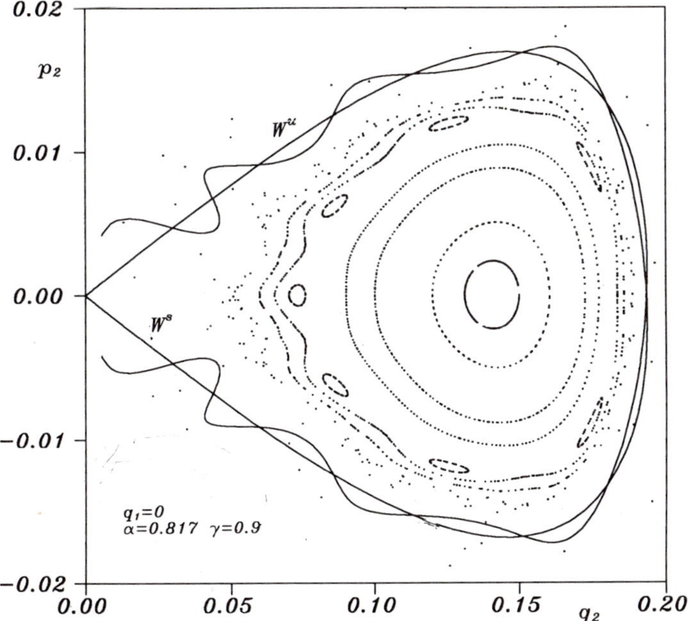

Figure 6. Poincaré map in a neighborhood of asymptotic surfaces.

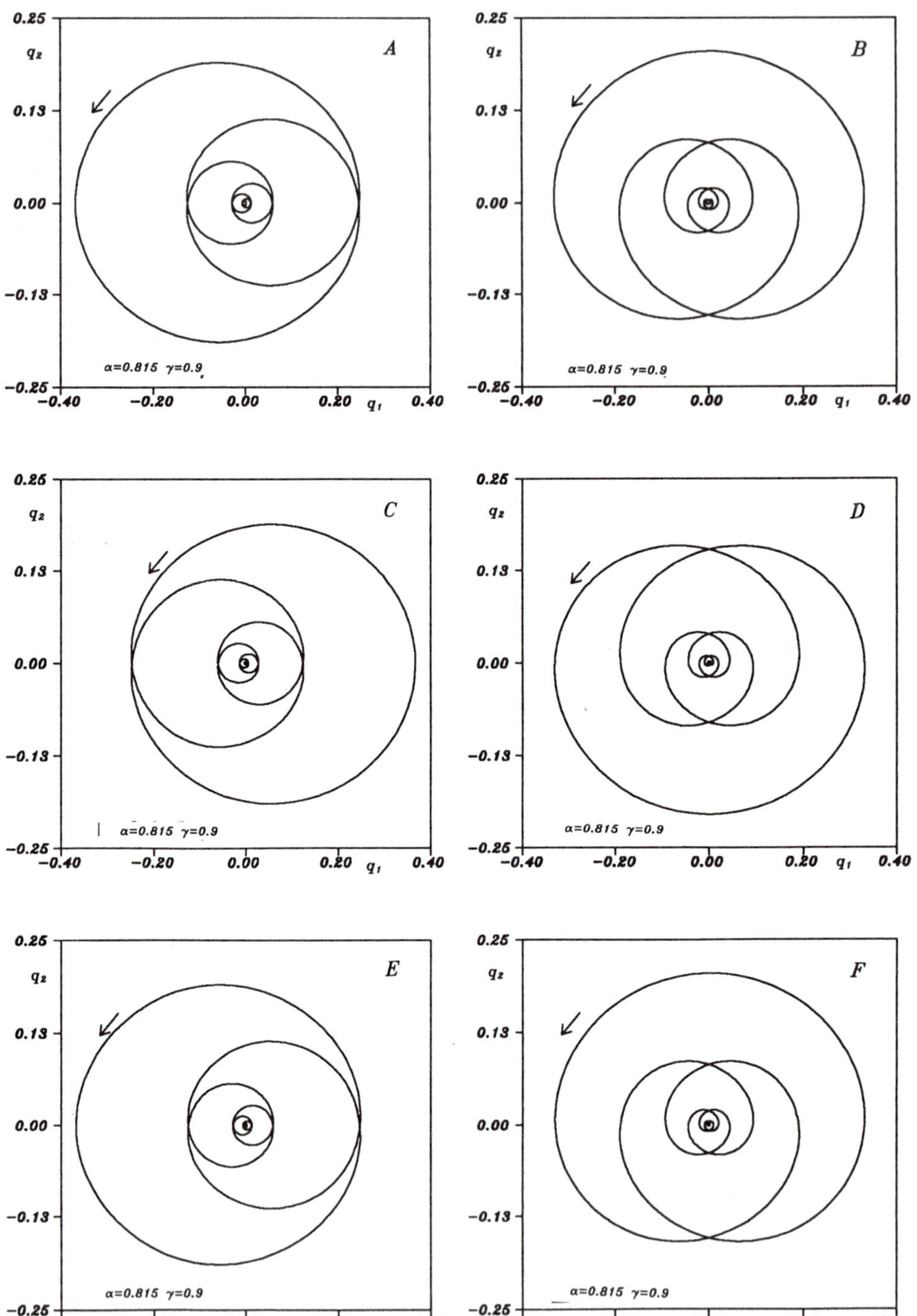

Figure 7. Examples of of orbits homoclinic to the cylindric precession (see Fig. 4).

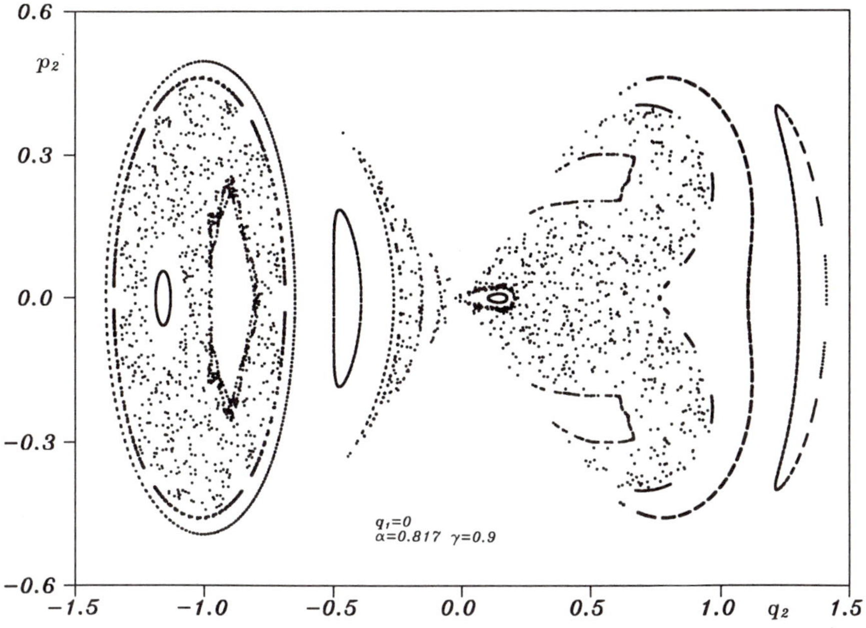



Figure 8. Global Poincaré map.

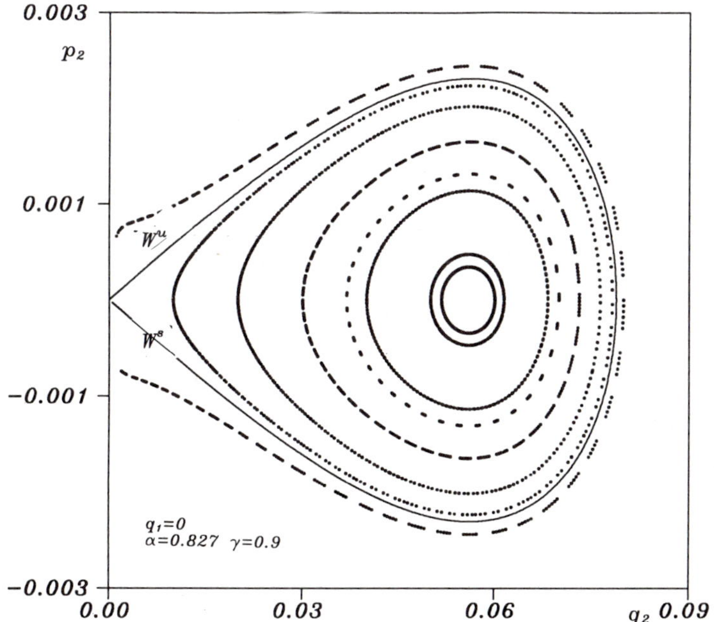



Figure 9. Poincaré map and doubled asymptotic surfaces.

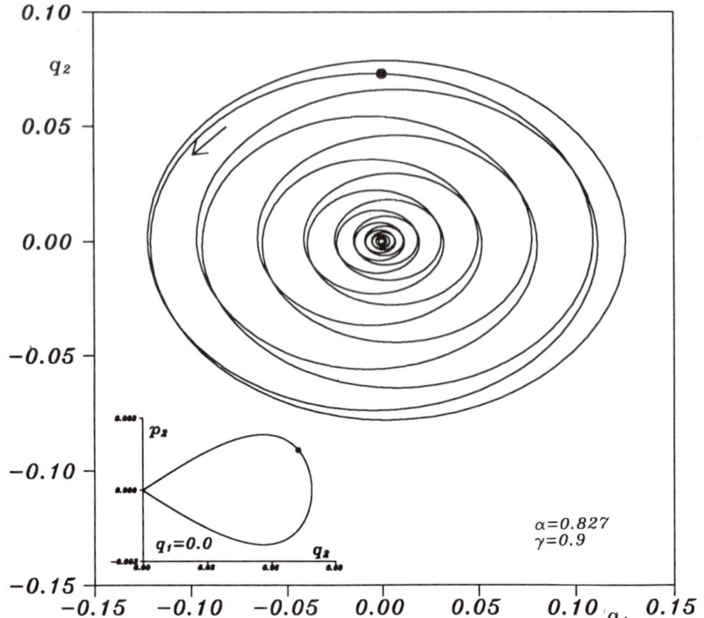



Figure 10. Degenerated orbit homoclinic to the cylindric precession.

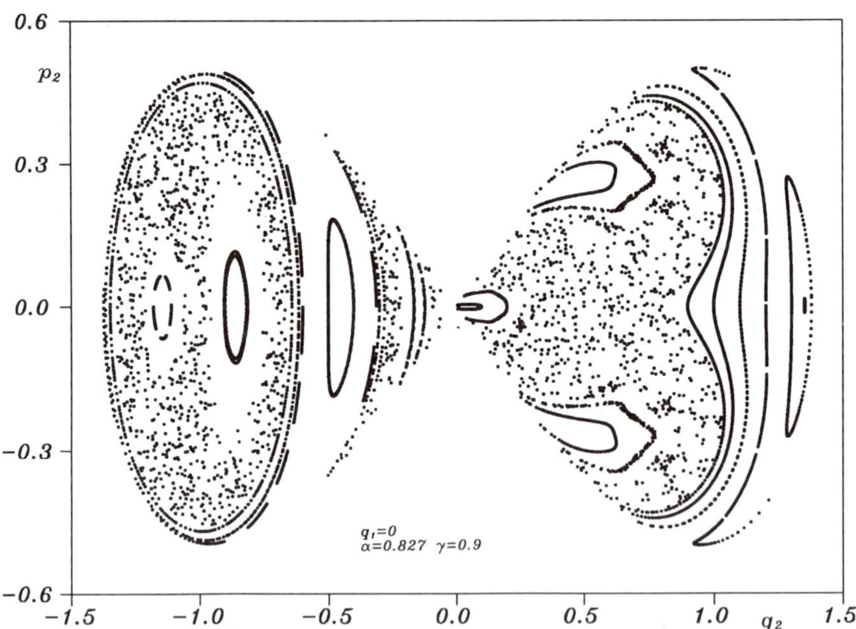



Figure 11. Global Poincaré map for a case with degenerated homoclinic orbit.

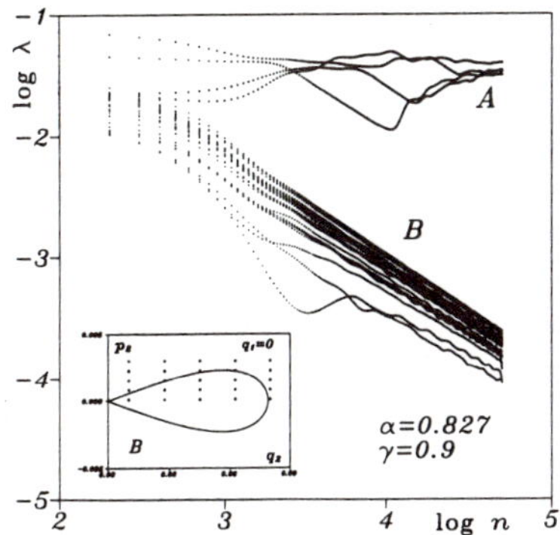

Figure 12. The maximal Lyapunov's exponent for orbits from a vincinity of homoclinic orbit (group B) and for some orbits from chaotic region (see Fig.11)—group A; $n$ denotes the number of iterations (with time step equal 1).

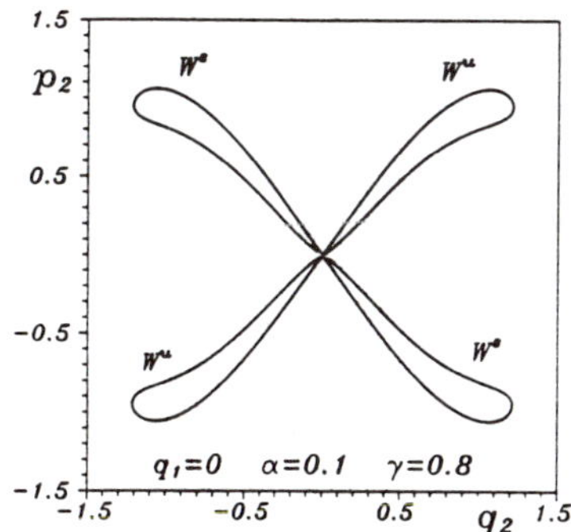

Figure 13. Asymptotic surfaces of the cylindric precession when it is of the saddle type.

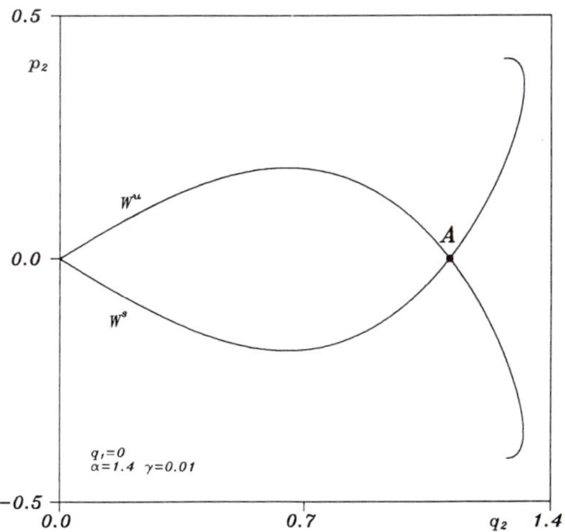

Figure 14. Asymptotic surfaces of the conic precession when it is of the saddle type.

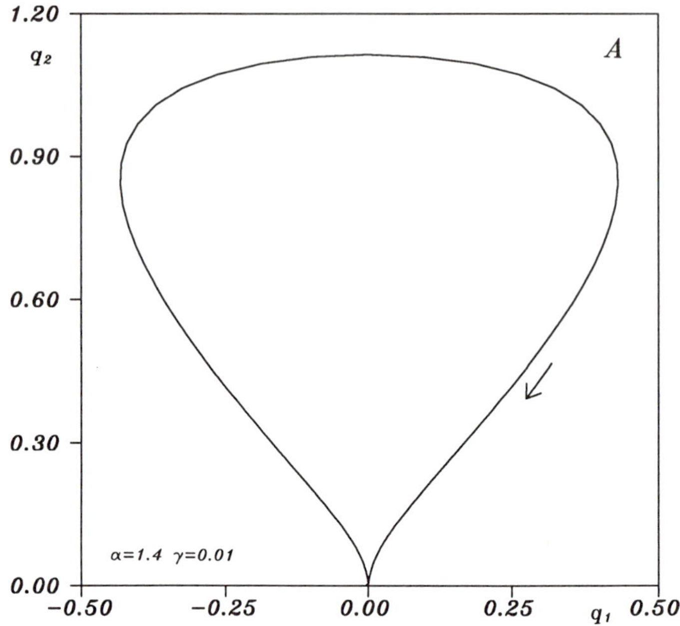


Figure 15. Orbit homoclinic to the conic precession when it is of the saddle type.

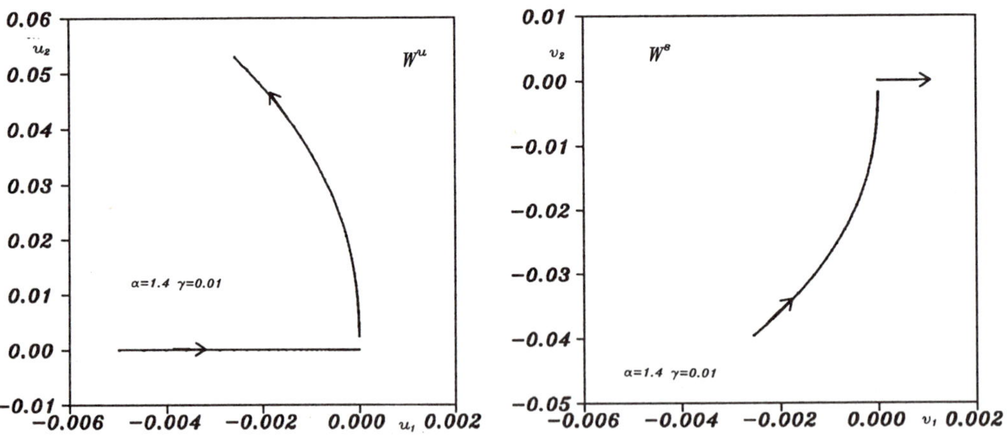


Figure 16. Demonstration that the homoclinic orbit from Fig. 15 is not main for $t \to \pm\infty$.

ACKNOWLEDGEMENTS

This work was supported by the Grant UMK 513A.

REFERENCES

1. S. V. Bolotin, Liouville nonitegrability condition for a hamiltonian system, Vest. Mosc.Univ. Ser. Mat. Mech., 3:58 (1986).
2. S. L. Ziglin, Separatrices splitting and first integrals nonexistence in Hamiltonian system with two degrees of freedom, Izv. Akad. Nauk SSSR Ser. Mat., 51:1088 (1987).
3. R. L. Devaney, Homoclinic orbits in Hamiltonian systems, American J. Math., 100:631 (1978).
4. R. L. Devaney, Transversal homoclinic orbits in an integrable system, J. Diff. Eqs., 21:431 (1976).
5. L. M. Lerman and Ya. L. Umanskii, About separatices loop existence in four dimensional systems close to integrable hamiltonian ones, Prikh. Math. Mech., 47:395 (1983).
6. S. V. Bolotin , Double-asymptotic solutions and nonitegrability conditions of a hamiltonian system, Vest.Mosc.Univ.Ser.Mat.Mech., 1:55 (1990).
7. D. W. Turaev and L. P. Shilnikov, On Hamiltonian systems with homoclinic saddle curves, Dokl.Acad.Nauk SSSR, 304:811 (1989).
8. G. Benttin, L. Galgani, A. Giorgilli and J.-M. Strelcyn, Lyapunov characteristic exponents for smooth dynamical systems, Meccanica, March:9 (1980).
9. V. V. Beletskii, "The motion of an artificial satellite around the center of mass", Nauka, Moscow, (1965).
10. A. P. Markeev, Resonance effects and stability of a satellite stationary rotations, Kosm. Issled., 5:365 (1967).
11. A. P. Markeev, Rotations of a satellite on elliptic orbit, Kosm. Issled., 5:530 (1967).
12. A. P. Markeev, On periodic motions of a satellite, Kosm. Issled., 23:323 (1985).
13. A. P. Markeev, Asymptotic trajectories and stability of periodic motions of an autonomous hamiltonian system, Prikh. Mat. Mech., 52:363 (1988).
14. A. G. Sokolsky , Stability of regular precessions, Kosm. Issled., 28:698 (1980).
15. Yu. W. Barkin, Skewed regular motions of a satellite and some small effects in the motion of the Moon and Phobos, Kosm. Issled., 23:26 (1985).
16. M. W. Demin, Plane periodical motion of the satellite around the mass center in the vincinity of the trigonal libration point, Kossm. Issled., 27:347 (1989).
17. J. M. A. Danby, Two notes on the Copenhagen problem, Celest. Mech., 33:251 (1984).
18. J. M. A. Danby, The evolution of periodic orbits close to heteroclinic points, Celest. Mech., 33:26 (1984).
19. S. A. Dovbysh, Numerical investigations of the transversal separatrices crossing and the Kolmogorov stability, in: "Numerical analyzis, mathematical modeling in mechanics", B. E. Poberdi, ed., Moscow State Univ., Moscow (1988).
20. K. Goździewski and A. J. Maciejewski, System for normalization of a Hamiltonian function, Celest. Mech., 49:1 (1990).
21. A. J. Maciejewski and K. Goździewski, Normalization algorithms of Hamiltonian near an equilibrium point, Astoph. Space Sci., 179:1 (1990).

# RELATIVISTIC CHAOS IN ROBERTSON-WALKER COSMOLOGIES: THE

# TOPOLOGICAL STRUCTURE OF SPACE-TIME AND THE MICROSCOPIC DYNAMICS


Roman Tomaschitz

Dipartimento di Matematica Pura ed Applicata
dell'Università degli Studi di Padova
Via Belzoni 7, I-35131 Padova, Italy and
Service de Physique Théorique de Saclay
F-91191 Gif-sur-Yvette, France


## I. INTRODUCTION

Among the general relativistic models that attempt to describe the rough overall structure of the universe and its origins , Robertson-Walker geometries are today regarded as the most likely candidates. Appealing to the principles of homogeneity and isotropy, a R.-W. cosmology is locally described by a line element $ds^2 = -dt^2 + a^2(t)\, d\sigma^2$, where $d\sigma^2$ is the line element of of a 3-space of constant curvature. The cosmological expansion factor $a(t)$ determines the time dependence of the frequencies of light in the universe, and is according to the observed red-shifts an increasing function of time. The second fundamental property of $a(t)$ is that it determines via Einstein's equations together with the Gaussian curvature of 3-space the energy density $\varepsilon$ and the pressure p of the light-matter content of the universe.

Though Einstein's equations give the relation between $\varepsilon$, p and $a(t)$, they do not determine the global topological structure of the universe, in particular they say nothing about the topology of the spacelike 3-sections at a fixed instant of time. Homogeneity and isotropy demand that these sections are 3-manifolds of constant curvature, but nothing prevents them from having different geometries and even topologies at two different instants of time.

Traditionally three different geometries are considered: in the case of zero curvature the 3-sections are Euclidean 3-space , for positive curvature one considers the 3-sphere, for negative curvature a shell of the Minkowski hyperboloid. The topology of 4-space is then the product of the real line or a semi-infinite interval with one of these topologies. Moreover different spacelike sections at different instants of time are isometric after a simple rescaling. The spacelike slices in the case of positive curvature are compact and such cosmologies are called closed, the other two cases correspondingly open.

These three examples do by far not exhaust all possible topologies of 3-manifolds of constant curvature, and in [14] we started to investigate the influence of a possible non-trivial topology of the spacelike slices on the microscopic dynamics. Which topologies come in question? Three-dimensional manifolds of constant zero or positive curvature are rather exceptional, cf.[9], similar to positively curved or flat Riemann surfaces. The generic case are manifolds of negative curvature, called hyperbolic. Such manifolds are best imagined in hyperbolic space as polyhedra with their faces identified in pairs, cf.[5], analogous to the 2-torus as a square in the Euclidean plane. As a representation of hyperbolic space we may choose the Poincaré ball $B^3$, $|\vec{x}| < R$, $d\sigma^2 = (1 - |\vec{x}|^2/R^2)^{-2}\, d\vec{x}^2$. The geodesics in this


*Chaotic Dynamics: Theory and Practice*
Edited by T. Bountis, Plenum Press, New York, 1992

ball are arcs of circles, and the the totally geodesic planes are spherical caps, both orthogonal to the sphere $S_\infty$, the boundary of $B^3$. The polyhedral faces lie on geodesic planes and are identified in pairs by glueing mappings, elements of the group of isometries of $B^3$, which happens to be the Lorentz group, cf.[1]. The polyhedra may also have faces lying on $S_\infty$ which are not identified in pairs, and which constitute the boundaries of the manifold. Such polyhedra have then infinite volume if measured by the Poincaré metric $d\sigma^2$ in $B^3$.

The glueing transformations generate a discrete subgroup $\Gamma$ of the Lorentz group, and the images of the polyhedron F, $\Gamma(F)$ will create a tessellation of the Poincaré ball. There are accumulation points of this tiling on the boundary of $B^3$ (the fractal curves in Figs. 1,2) This limit set plays a large role in the spectral theory of the Laplace-Beltrami operator on the spacelike sections, for example its Hausdorff dimension determines the ground state eigenvalue, cf.[3, 8].

There are the compact hyperbolic manifolds, whose polyhedra do not touch the boundary of $B^3$, providing closed models of negative curvature. The five Platonic solids, regular polyhedra in $B^3$, are typical examples of them, cf.[5]. The most important feature of these finite-volume manifolds is that they are rigid [11], a given topology, defined by the face-pairing can carry only one metric of constant negative curvature. Therefore it is not possible to deform a Platonic solid a little, keeping a chosen face-identification, so that the deformed polyhedron with the deformed glueing mappings generate again a tiling of $B^3$. Thus finite-volume hyperbolic manifolds are also rather exceptional, being too rigid to be proper candidates for the spacelike slices.

The generic case of constant curvature 3-manifolds are thus hyperbolic manifolds of infinite volume, their spacelike slices being open, infinite. Here the topological structure does not fix the metric at all. The polyhedra, having free faces on $S_\infty$, are together with the covering group $\Gamma$ deformable without destroying their tiling property. Such deformations can be parametrized by a certain number (depending on the topology) of variables which characterize geometrically the polyhedron, cf.[2, 4, 13], and one obtains so an explicit realization of the deformation space of non-equivalent metrics on the topological manifold. The metric of $B^3$ is of course always induced on F. Thus a R.-W. cosmology is determined by the choice of an expansion factor $a(t)$ in the line element $ds^2$, and by a path $(\Gamma(t), F(t))$, generically time-dependent, in the deformation space of an open hyperbolic 3-manifold, cf.[13, 15].

In [14] we started to analyse R.-W. cosmologies whose spacelike slices have a non-trivial topological structure, and the consequences that arise from the topology, both for classical world lines and for scalar quantum fields. In [16] we discussed de Sitter space, i.e. an expansion factor of the form $a(t) = \sinh(\Lambda t)$, and spacelike slices of the form $I \times S$, I an open finite interval, S a Riemann surface ('thickened surfaces'). We calculated the time evolution of the energy of the corresponding wave fields, and the bearing of the spectrum of the L.-B. operator of the 3-slices on this evolution.

In this paper we will give further examples, in particular we will figure out expansion factors $a(t)$ that are compatible with the conditions of positive pressure and energy, $\varepsilon, p > 0$, ( in de Sitter space we have $\varepsilon = p = 0$ ), and we will discuss the energy asymptotics of the scalar wave fields in the asymptotically flat regime, for $t \to \infty$, and towards the initial singularity, for $t \to 0$.


## II. THE ENERGY OF SCALAR WAVE FIELDS AND THE SPECTRUM OF THE LAPLACE - BELTRAMI OPERATOR OF THE SPACE SECTIONS

As pointed out in Section 1 the negative curvature (but not the finite or infinite volume

Fig.1. Tiling induced on the boundary $S_\infty$ of the Poincaré ball $B^3$ by the universal cover of the 3-manifold $I \times S$. The convex hull of the fractal Jordan curve $\Lambda(\Gamma)$ determines a compact region $C(\Lambda)\backslash\Gamma$ (see Sec.3) in infinite 3-space F, where the chaotic trajectories lie. $g(S) = 19$, $\delta = 1.402\pm0.001$ ($\delta$ has been calculated by the method of characteristic curves, cf.[15]).

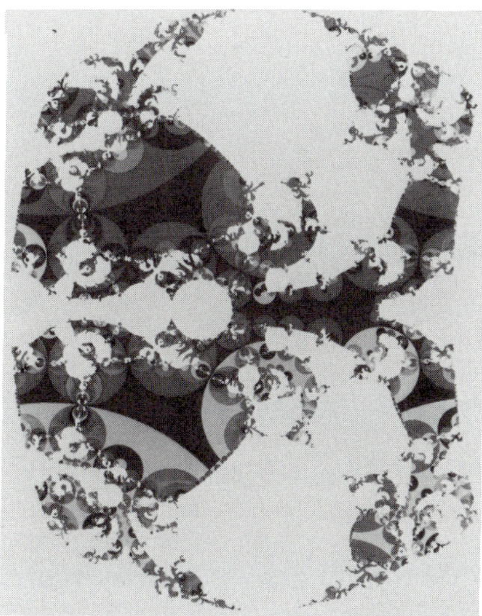

Fig.2. As Fig.1, covering of $S_\infty$ stemming from a spacelike section F of the 4-manifold. Fig.1(b) in [14] - Fig.1 - Fig.2 - Fig.2 in [16] represent a sequence of non-isometric points on a path $(F(t), \Gamma(t))$ in the deformation space of the topological manifold $I \times S$. $g(S) = 19$, $\delta= 1.423$.

of the space sections ) restricts crucially the possibilities of the asymptotic behaviour of the cosmological expansion factor a(t).

The energy density $\varepsilon$ and the pressure p of the light-mass content of the universe are simple functions of a(t) and its derivatives, cf.[7],

$$\varepsilon = 3 \frac{c^2}{8\pi k}\left[\frac{-\Lambda^2}{a^2}+\frac{\dot{a}^2}{a^2}-\frac{\hat{\lambda}c^2}{3}\right], \quad p = \frac{c^2}{8\pi k}\left[-2\frac{\ddot{a}}{a}-\frac{\dot{a}^2}{a^2}+\frac{\Lambda^2}{a^2}+\hat{\lambda}c^2\right], \tag{2.1}$$

and they have to be positive. In the following we put $c^2/8\pi k$ equal 1, $\hat{\lambda}$ is the cosmological constant.

We assume in this paper that a(t) is strictly increasing. Then, in the case of a positive $\hat{\lambda}$ the conditions $\varepsilon \geq 0$, $p \geq 0$ require exponential increase of a(t) for $t \to \infty$ as it happens in de Sitter space. If $\hat{\lambda}$ is negative a(t) cannot diverge for $t \to \infty$, in fact in this case we have an oscillating universe $a(t) = \sin(\Lambda t)$ which may be treated analogously to the examples given later on. From now on we assume $\hat{\lambda} = 0$. Then the positivity conditions require an asymptotically linear expansion factor $a(t) \sim \Lambda t$ for $t \to \infty$. $\Lambda$ is a constant setting the time scale, a(t) is dimensionless. The case $a(t) = \Lambda t$ exactly has been sketched in [14], it leads to $\varepsilon = p = 0$, but correction terms to this linear behaviour may change this situation considerably. Luckily the conditions $\varepsilon > 0$, $p > 0$ restrict the form of these correction terms too.

A.  The asymptotically flat regime: time asymptotics for $t \to \infty$

We discuss a form of the expansion factor that exhausts qualitatively all possibilities of the asymptotic behaviour of the solutions of the wave equation for $t \to \infty$.

*Example 1*

$$a(t) = \Lambda t + c(\log\Lambda t)^{\alpha}. \tag{2.2}$$

From (2.1) we have

$$\varepsilon \sim 6\alpha c \Lambda^{-1}t^{-3}\left(\log\Lambda t\right)^{\alpha-1}, \quad p \sim 2c\alpha(1-\alpha)\Lambda^{-1}t^{-3}\left(\log\Lambda t\right)^{\alpha-2}, \text{ if } \alpha =1: \quad p \sim 2c^2\Lambda^{-2}t^{-4}\log\Lambda t. \tag{2.3}$$

Thus to ensure $\varepsilon > 0$, $p > 0$ we have to require $0 < \alpha \leq 1, c > 0$, or $\alpha < 0$, $c < 0$. The case $\alpha = 1$, $c = 1$ appears in [6, p. 222].

With the a(t) in (2.2) we will now calculate solutions of the wave equation and the asymptotic behaviour of the energy evolution of these wave fields. The way to do this can be found in [14] in some detail, and we sketch it here very shortly to keep track with self-containedness. If we make a separation ansatz $\psi(t, \vec{x}) = \tilde{\psi}(\vec{x})\,\varphi(t)$ in the Klein-Gordon equation

$$\left[\Box - \xi R - (mc/\hbar)^2\right]\psi = 0, \tag{2.4}$$

we obtain for the time dependence of the wave fields

$$\ddot{\varphi} + 3\frac{\dot{a}}{a}\dot{\varphi} + \left[m^2 + \frac{\Lambda^2\lambda}{a^2} + c^2\xi R\right]\varphi = 0 \quad . \tag{2.5}$$

We write in (2.5) m for $mc^2/\hbar$ ; $\square$ denotes the wave operator on the 4-manifold and R is its curvature scalar,

$$R = \frac{6}{c^2}\left[\frac{-\Lambda^2}{a^2} + \frac{\ddot{a}}{a} + \frac{\dot{a}^2}{a^2}\right] . \tag{2.6}$$

$\xi$ is a dimensionless coupling constant of the field to the scalar curvature, and $\Lambda$ appears in the expansion factor. The radius of the Poincaré ball in Sec.1 we fix as $c/\Lambda$. Finally $\lambda$ is the spectral parameter for the space part of the wave equation, and thus it varies over the spectrum of the Laplace-Beltrami operator of the spacelike slices of the 4-manifold. In the example of the manifold in Figs.1, 2 $\lambda$ can admit a discrete value $\delta(2-\delta)$, $\delta$ the Hausdorff dimension of the fractal limit set $\Lambda(\Gamma)$, $1 \le \delta < 2$, which corresponds to the ground state eigenvalue of the L.-B. operator, and it can vary in $[1, \infty)$, corresponding to the continuous spectrum. We also assume that the 3-manifold (F, $\Gamma$) does not vary in time.

For the asymptotic expansions that we will carry out it is useful to eliminate the first derivative in (2.5) by introducing a new dependent variable $\psi := a^{3/2}\varphi$,

$$\ddot{\psi} + \left[m^2 + \frac{\Lambda^2}{a^2}(\lambda-6\xi) + \frac{\ddot{a}}{a}(6\xi-3/2) + \frac{\dot{a}^2}{a^2}(6\xi-3/4)\right]\psi = 0 \quad . \tag{2.7}$$

To obtain the right asymptotic behaviour of the energy , $E \sim mc^2$ in the limit $t \to \infty$, we have to impose the normalization condition

$$\dot{\psi}\overline{\psi} - \overline{\dot{\psi}}\psi = \pm 2i \tag{2.8}$$

on the solutions of (2.7).

Inserting a(t) of (2.2) into (2.5, 2.7) we calculate in the asymptotic order we need

$$\varphi = Am^{-1/2}\left(\Lambda t\right)^{-3/2} e^{imt}\left[1 - iBt^{-1} + Ct^{-2} + i\varepsilon Dt^{-2}(\log\Lambda t)^\alpha + O(t^{-2}(\log\Lambda t)^{\alpha-1})\right] \tag{2.9}$$

with

$$A = 1 - \frac{3}{2}\frac{\varepsilon}{\Lambda t}(\log\Lambda t)^\alpha + \frac{15}{8}\frac{\varepsilon^2}{(\Lambda t)^2}(\log\Lambda t)^{2\alpha} + O((\log\Lambda t)^{3\alpha}/(\Lambda t)^3) \quad ,$$

and $B = (\lambda-3/4)/2m$ , $C = -(\lambda^2 + \lambda/2 - 15/16)/8m^2$ . D is a real constant that does not enter in the final result for the energy.

In [16] the following formula for the energy of a wave field has been derived:

$$E = \frac{1}{2}\hbar a^3\left\{\dot{\varphi}\overline{\dot{\varphi}} + \varphi\overline{\varphi}\left[m^2 + (\lambda-6\xi)\Lambda^2 a^{-2} + 6\xi\frac{\dot{a}^2}{a^2}\right] + 6\xi\frac{\dot{a}}{a}(\dot{\varphi}\overline{\varphi} + \overline{\dot{\varphi}}\varphi)\right\} \quad , \tag{2.10}$$

it is always positive definite for $0 \le 6\xi \le \min[1, \delta(2-\delta)]$, which can be seen easily by completing the first and the last term to a square. Inserting (2.9) into( 2.10) we arrive at $(m \to mc^2/\hbar)$

$$E(\lambda) = mc^2A^2 + \frac{1}{2}\frac{\hbar^2}{mc^2}\frac{1}{t^2}(\lambda + \frac{9}{4} - 18\xi) + O((\log\Lambda t)^\alpha/t^3) \quad . \tag{2.11}$$

From (2.11) we have in 'universal' quantities

$$E(\lambda_1) - E(\lambda_2) \sim \frac{1}{2} \frac{1}{a^2(t)} \frac{\Lambda^2 \hbar^2}{mc^2} (\lambda_1 - \lambda_2) \quad . \tag{2.12}$$

If m=0 the solution (2.9) has to be replaced by

$$\varphi = \Lambda^{-1/2}(\lambda -1)^{-1/4}(\Lambda t)^{-1+i\sqrt{\lambda-1}} \left[ 1 + O((\log\Lambda t)^{\alpha}/t) \right] , \tag{2.13}$$

with the energy

$$E = \frac{\hbar}{t}\left[ \sqrt{\lambda-1} + (1-6\xi)/\sqrt{\lambda-1} \right] + O((\log\Lambda t)^{\alpha}/t^2) \quad . \tag{2.14}$$

The frequency of the oscillation of (2.13) is

$$\nu = (e^{2\pi/\sqrt{\lambda-1}} -1)^{-1} t^{-1} , \tag{2.15}$$

and thus we may write (2.14) as

$$E \sim h(\lambda)\nu , \tag{2.16}$$

with $h(\lambda\to\infty) \to h$, $h(\lambda\to1) \to \infty$ .The asymptotic expansions are carried out for a fixed $\lambda$. In the massive case (2.9) we have of course $\nu = mc^2/h$.

An expansion factor of the form

$$a(t) = \Lambda t + c(\Lambda t)^{-\beta} \tag{2.17}$$

can be treated completely analogously to (2.2): we have for (2.17)

$$\varepsilon \sim - 6c\beta\Lambda^2(\Lambda t)^{-\beta-3} , \; p \sim - 2c\beta^2\Lambda^2(\Lambda t)^{-\beta-3} , \tag{2.18}$$

positivity is insured for $\beta > 0$, $c < 0$. One has to replace in the formulae of Ex.1 $(\log\Lambda t)^{\alpha}$ by $(\Lambda t)^{-\beta}$. In particular the formulae for the energy and the frequency remain in the order given exactly the same. Mixtures of powers and logarithms in $a(t)$ do not alter anything qualitatively.


### B.  The approach to the initial singularity:  time asymptotics of the fields and their energies for $t \to 0$

In the limit $t \to 0$ the expansion factor may go either to zero or approach a finite value. Exponential decay of $a(t)$ violates $\varepsilon > 0$, $p > 0$, power law decay $a(t) \sim (\Lambda t)^{\alpha}$, $\alpha > 0$ leads to a positive energy density and pressure in the range $0 < \alpha \le 2/3$ ,

$$\varepsilon \sim 3\alpha^2 t^{-2}, \; p \sim \alpha(2-3\alpha)t^{-2} \quad . \tag{2.19}$$

There is also the possibility $a(t) = \Lambda t$ strictly , without any corrections for $t \to 0$, that has been discussed in [14], and the two mentioned de Sitter examples. All these cases give $\varepsilon=p=0$.

Logarithmic decay , $a(t) = (\log 1/\Lambda t)^{-\beta}$ , $\beta > 0$ , leads to

$$\varepsilon \sim 3\beta^2 t^{-2}(\log 1/\Lambda t)^{-2} \; , \; p \sim 2\beta t^{-2}(\log 1/\Lambda t)^{-1} \; . \tag{2.20}$$

Models with a two-sided infinite time scale, $a(t \to -\infty)$ strictly decreasing, are incompatible with negative curvature, violating $\varepsilon > 0, p > 0$.

*Example 2*

$$a(t) = (\Lambda t)^{\alpha}, \; 0 < \alpha \le 3/2, \; t \to 0 ; \tag{2.21}$$

the following discussion holds even true for $0 < \alpha < 1$, we will comment on this in Ex.7.

The wave equation reads according to (2.7)

$$\ddot{\psi} + \left[ m^2 + \hat{A} \, t^{-2} + \hat{B} \, \Lambda^2 (\Lambda t)^{-2\alpha} \right] \psi = 0 \; , \tag{2.22}$$

with $\hat{A} = 3\alpha(1 - 3\alpha/2)/2 + 6\xi\alpha(2\alpha - 1) \; , \; \hat{B} = \lambda - 6\xi$ .

In the lowest order asymptotic expansion that we will use we can drop the $m^2$-term in (2.22), a fundamental system is then $t^{1/2 \pm v}$ , $v := \sqrt{1/4 - \hat{A}} \ge 0$ ; $v^2 \ge 0$ is always satisfied for $0 < \alpha < 1, \; 0 \le 6\xi \le 1$.

The normalized (see (2.8)) general solution of (2.5) is

$$\varphi = \left[ \Lambda^{-1/2} A(\Lambda t)^{1/2 - 3\alpha/2 - v} + \Lambda^{-1/2} B(\Lambda t)^{1/2 - 3\alpha/2 + v} \right]\left[ 1 + O(t^{2(1-\alpha)}) \right] \tag{2.23}$$

with

$$A = a e^{i\varphi/2} , \; B = b e^{-i\varphi/2} , \; \sin\varphi = \pm 1/2vab \; . \tag{2.24}$$

In choosing A, B as we did we have fixed a constant overall phase factor which always drops out in terms like $\overline{\varphi}\dot{\varphi}, \; \varphi\dot{\overline{\varphi}}$ , and so it does not affect the energy.

The solution (2.9) of Eq. (2.5 ) for $t \to \infty$ we fixed by imposing the end-value condition that for $t \to \infty$ $\varphi$ should approach as closely as possible the Minkowski space solution. This fixes in principle also a and b in (2.24), which are functionals of the expansion factor a(t) via (2.5). The problem is of course that we do not know the function a(t) in the intermediate regime, only its asymptotic limits. Therefore we will discuss the time asymptotics of the energy leaving the numerical values of a, b undetermined. In the de Sitter example in Ref.[16] we knew the expansion factor in the intermediate region, and we could determine a, b by solving Eq.(2.5).

From (2.23) and (2.10 ) we get

$$E \sim \frac{1}{2} \hbar\Lambda |A|^2 (\Lambda t)^{-1-2v}\left[ (-3\alpha/2 + 1/2 - v + 6\xi\alpha)^2 + \alpha^2 6\xi(1 - 6\xi) \right] \; . \tag{2.25}$$

We assume as always $0 \le 6\xi \le \min[1, \delta(2-\delta)]$ to have the functional (2.10 ) positive definite. For $\xi = 1/6$ this expression vanishes, and we have to calculate higher orders. For $\xi = 0$ and $0 < \alpha \le 1/3$ it vanishes likewise.

<u>(a) $\xi = 1/6$, i. e. $\nu = (1 - \alpha)/2$</u>

Instead of (2.23) we have

$$\varphi = \Lambda^{-1/2} A(\Lambda t)^{-\alpha}\left[1 + c(\Lambda t)^{2(1-\alpha)} + O((\Lambda t)^{4(1-\alpha)})\right] + \Lambda^{-1/2} B(\Lambda t)^{1-2\alpha}\left[1 + O((\Lambda t)^{2(1-\alpha)})\right] ; \tag{2.26}$$

c is a real constant, its numerical value does not enter in E which reads now

$$E \sim \frac{1}{2}\hbar\Lambda(\Lambda t)^{-\alpha}\left[(1-\alpha)^2 |B|^2 + (\lambda-1)|A|^2\right] . \tag{2.27}$$

For small $\alpha$ the energy in (2.25) is a factor $t^{-2}$ stronger divergent, compare also Eqs. (2.34, 2.36 ).

<u>(b) $\xi = 0, 0 < \alpha < 1/3, \nu = 1/2 -3\alpha/2$</u>

For the wave field and the energy we get in this case

$$\varphi = \Lambda^{-1/2}\left[A + B(\Lambda t)^{1-3\alpha}\right]\left[1 + O((\Lambda t)^{2(1-\alpha)})\right] , \tag{2.28}$$

$$E \sim \frac{1}{2}\hbar\Lambda(\Lambda t)^{-3\alpha}|B|^2(1-3\alpha)^2 , \tag{2.29}$$

the power in (2.25) would be $t^{-2+3\alpha}$, again a factor of $t^{-2}$ stronger for small $\alpha$.

<u>(c) $\xi=0, \alpha = 1/3, \nu = 0$</u>

$$\varphi = \Lambda^{-1/2}\left[A + B\log\Lambda t\right]\left[1 + O((\Lambda t)^{4/3})\right] , \tag{2.30}$$

$$E \sim \frac{1}{2}\hbar t^{-1}|B|^2 , \tag{2.31}$$

A and B are connected here via $\sin\varphi = \pm 1/ab$, see (2.24 ).

Finally, from (2.25) it follows that the energy for $\xi = 0, 1/3 < \alpha < 1$ is given by (2.29) with B replaced by A.

<u>*Example 3*</u>

$$a(t) = (\log 1/\Lambda t)^{-\beta} , \beta > 0, t \rightarrow 0. \tag{2.32}$$

According to (2.20) we have $\varepsilon > 0, p > 0$. The normalized solution of (2.5) is

$$\varphi = \Lambda^{-1/2}\left[A(\log 1/\Lambda t)^{6\beta\xi} + B\Lambda t(\log 1/\Lambda t)^{-\beta(6\xi-3)}\right]\left[1 + O((\log 1/\Lambda t)^{-1})\right] , \tag{2.33}$$

with A, B as in Ex.2(c). For $\xi = 0, 1/6$ the O-term has to be replaced by $O(t^2(\log 1/\Lambda t)^\gamma)$.

168

The energy is

$$E \sim \frac{1}{2}\hbar\Lambda\frac{1}{(\Lambda t)^2}(\log 1/\Lambda t)^{-3\beta+12\beta\xi-2}|A|^2\,\beta^2 6\xi(1-6\xi) \quad,\tag{2.34}$$

a limit case of (2.25) for $\alpha \to 0$.

For $\xi = 0$ we have instead of (2.34)

$$E \sim \frac{1}{2}\hbar\Lambda\,(\log 1/\Lambda t)^{3\beta}|B|^2 \quad,\tag{2.35}$$

and for $\xi = 1/6$

$$E \sim \frac{1}{2}\hbar\Lambda\,(\log 1/\Lambda t)^{\beta}\,(\,|B|^2 + (\lambda-1)|A|^2\,)\quad.\tag{2.36}$$

In (2.36) we have to assume $\lambda \geq 1$, i. e. to exclude the ground state value $\lambda = \delta(2-\delta)$, because of the positivity condition in (2.10).

*Example 4*   the case of finite initial radius, cf.[12],

$$a(t) = b + c(\Lambda t)^{\alpha}\,,\ \alpha,\,b,\,c > 0\,.\tag{2.37}$$

Energy and pressure are positive for $\alpha < 1$, and still singular,

$$\varepsilon \sim \frac{3\Lambda^2 c^2\alpha^2}{b^2}(\Lambda t)^{2\alpha-2}\,,\ p \sim \frac{2\Lambda^2 c\alpha(1-\alpha)}{b}(\Lambda t)^{\alpha-2}\quad.\tag{2.38}$$

The solution of (2.5) is

$$\varphi = \Lambda^{-1/2}b^{-3/2}A\left[1 - 6\xi\frac{c}{b}(\Lambda t)^{\alpha}+ O(t^{2\alpha})\right] + \Lambda^{-1/2}b^{-3/2}\,B\Lambda t\left[1 + O(t^{\alpha})\right]\quad,\tag{2.39}$$

with  A, B as in Ex.2(c). For E we have

$$E \sim \frac{1}{2}\hbar\Lambda(\Lambda t)^{2\alpha-2}|A|^2\,\alpha^2\frac{c^2}{b^2}\,6\xi(1-6\xi)\quad,\tag{2.40}$$

special treatment is again needed for $\xi = 0,\,1/6$.

<u>(a) $\xi = 1/6$</u>
If we drop in (2.5, 2.7) the $m^2$ and $a^{-2}$ terms that contribute only positive powers we can solve (2.5) by

$$\varphi = \Lambda^{-1/2}Aa^{-1}(t) + \Lambda^{1/2}Ba^{-1}(t)\int^{t} a^{-1}(t)dt\tag{2.41}$$

with A, B as before. E approaches a finite value,

$$E \sim \frac{1}{2}\hbar\Lambda b^{-1}\left[|B|^2 + |A|^2\,(b^2 m^2\Lambda^{-2} + \lambda - 1)\right]\quad.\tag{2.42}$$

<u>(b) $\xi = 0$</u>
With the same approximations as in (a) we have as independent solutions $1, \int a^{-3}\,dt$.

The normalized asymptotic solution of (2.5) is

$$\varphi = \Lambda^{-1/2} A b^{-3/2} + \Lambda^{1/2} t B b^{-3/2} \left[ 1 + O(t^\alpha) \right],$$

(2.43)

with the energy

$$E \sim \frac{1}{2} \hbar \Lambda \left[ |B|^2 + |A|^2 \, (m^2 \Lambda^{-2} + \lambda b^{-2}) \right],$$

(2.44)

which goes over for b=1 and $\varphi = \omega^{-1/2}(1-i\omega t)$ in Eq. (47) of Ref.[14], likewise (2.42).

*Example 5*   a limit case of Ex.4,

$$a(t) = b + c(\log 1/\Lambda t)^{-\alpha}, \quad \alpha,\, b,\, c\, > 0.$$

(2.45)

We have here always a positive $\varepsilon$ and p,

$$\varepsilon \sim \frac{3c^2\alpha^2}{b^2} \frac{1}{t^2} (\log 1/\Lambda t)^{-2\alpha-2}, \quad p \sim \frac{2c\alpha}{b} \frac{1}{t^2} (\log 1/\Lambda t)^{-\alpha-1},$$

(2.46)

otherwise this case is completely analogous to Ex.4, if we replace in the formulae their $(\Lambda t)^\alpha$ by $(\log 1/\Lambda t)^{-\alpha}$. For the energy we obtain instead of (2.40)

$$E \sim \frac{1}{2} \hbar \Lambda \frac{1}{(\Lambda t)^2} (\log 1/\Lambda t)^{-2\alpha-2} |A|^2 \frac{\alpha^2 c^2}{b^2} 6\xi(1 - 6\xi);$$

(2.47)

formulae (2.42, 2.44) hold still true.

*Example 6*   a limit case of Ex.4,

$$a(t) = b + c\Lambda t(\log 1/\Lambda t)^\alpha, \quad \alpha > 0.$$

(2.48)

$\varepsilon$ and p are positive,

$$\varepsilon \sim \frac{3c^2\Lambda^2}{b^2} (\log 1/\Lambda t)^{2\alpha}, \quad p \sim \frac{2c\alpha}{b} \frac{\Lambda}{t} (\log 1/\Lambda t)^{\alpha-1}.$$

(2.49)

If $\xi$ does not take its two limit values 0, 1/6, we have as a fundamental system of (2.5)
$t$, $1 - 6\xi cb^{-1}\Lambda t \,(\log 1/\Lambda t)^\alpha$ , with

$$E \sim \frac{1}{2} \hbar \Lambda (\log 1/\Lambda t)^{2\alpha} |A|^2 \frac{c^2}{b^2} 6\xi(1-6\xi).$$

(2.50)

The cases $\xi = 0$, 1/6 reduce to the preceding ones.


*Example 7*

Finally we discuss

$$a(t) = (\Lambda t)^\alpha, \quad \alpha > 1, \, t \to 0.$$

(2.51)

These expansion factors violate the positivity of $\varepsilon$, but we will also see why that happens, and thus we think it is worthwhile to treat them here.

The normalized solution of (2.5) reads

$$\varphi = \Lambda^{-1/2}\cosh(r)\, e^{i\vartheta/2}\, \widetilde{\varphi} + \Lambda^{-1/2}\sinh(r)\, e^{-i\vartheta/2}\, \widetilde{\widetilde{\varphi}} , \qquad (2.52)$$

with

$$\widetilde{\varphi} = B^{-1/4}\exp\!\left( i\,\frac{\sqrt{B}}{\alpha-1}(\Lambda t)^{1-\alpha}\right)(\Lambda t)^{-\alpha}\left\{ 1+iC(\Lambda t)^{\alpha-1} + O(t^{2(\alpha-1)}) + \right.$$

$$\left. +\, im^2\Lambda^{-2}(\Lambda t)^{1+\alpha}D\left[1 + O(t^{\alpha-1})\right] + O(m^4 t^{2(1+\alpha)}) \right\}, \qquad (2.53)$$

and

$$B = \lambda - 6\xi \;,\quad C = \frac{\alpha}{2}\frac{(1-2\alpha)(1-6\xi)}{\sqrt{B}\,(1-\alpha)} \;,\quad D = \frac{-1}{2(\alpha+1)\sqrt{B}} .$$

In (2.52) $r$ and $\vartheta$ are parameters analogous to a, b, $\varphi$ in (2.24).

The energy splits in

$$E = E_{monotonic} + E_{periodic} , \qquad (2.54)$$

with

$$E_m = (\cosh^2(r) + \sinh^2(r))\,\hbar\Lambda(\Lambda t)^{-\alpha}\sqrt{B} + O(t^{\alpha-2}) , \qquad (2.55)$$

$$E_p = -2\sinh(r)\cosh(r)\hbar\, t^{-1}\,\alpha(1-6\xi)\sin\!\left(2\sqrt{B}(\alpha-1)^{-1}(\Lambda t)^{1-\alpha} + \vartheta\right) + O(t^{-2+\alpha}) . \qquad (2.56)$$

The case $\xi = 1/6$ needs again special treatment, instead of (2.53) we have

$$\widetilde{\varphi} = B^{-1/4}\exp\!\left( i\,\frac{\sqrt{B}}{\alpha-1}(\Lambda t)^{1-\alpha}\right)(\Lambda t)^{-\alpha}\left\{ 1 + m^2\Lambda^{-2}(\Lambda t)^{1+\alpha}\left[ iD + E(\Lambda t)^{\alpha-1} + \right.\right.$$

$$\left.\left. +\, iF(\Lambda t)^{2(\alpha-1)} + O(t^{3(\alpha-1)})\right] + \left(m^2\Lambda^{-2}(\Lambda t)^{1+\alpha}\right)^2 G\left[1 + O(t^{\alpha-1})\right] + O(m^6 t^{3(1+\alpha)}) \right\}$$

with
$$\qquad (2.57)$$

$$E = \frac{-1}{4B} \;,\; G = \frac{-1}{8B(\alpha+1)^2} .$$

F does not enter in the following equations that replace (2.55, 2.56),

$$E_m = (\cosh^2(r) + \sinh^2(r))\,\hbar\Lambda\left[\frac{\sqrt{B}}{(\Lambda t)^{\alpha}} + \frac{(\Lambda t)^{\alpha}\,m^2\Lambda^{-2}}{2\sqrt{B}}\right] + O(t^{3\alpha-2}) , \qquad (2.58)$$

and

$$E_p = -\sinh(r)\cosh(r)\,\hbar\Lambda\,(\Lambda t)^{2\alpha-1}\,\frac{m^2\Lambda^{-2}\alpha}{B}\,\sin\!\left(2\frac{\sqrt{B}}{\alpha-1}(\Lambda t)^{1-\alpha} + \vartheta\right) + O(t^{3\alpha-2}) . \qquad (2.59)$$

For $m = 0$ $E_p$ is identically zero.

The frequency of an oscillation $\exp\!\left(\pm i\beta(\Lambda t)^{1-\alpha}\right)$, $t \to 0$, $\alpha > 1$, $\beta > 0$ as it occurs in (2.53, 2.57) and (2.56, 2.59) is asymptotically

$$\nu = 1/\Delta t \sim \frac{\Lambda\beta(\alpha-1)}{2\pi(\Lambda t)^{\alpha}} \to \infty , \qquad (2.60)$$

where $\Delta t$ is determined by $\beta\left(\Lambda(t-\Delta t)\right)^{1-\alpha} - \beta(\Lambda t)^{1-\alpha} = 2\pi$. For $E_m$ in (2.55, 2.58) we have then with $\beta = B^{1/2} (\alpha - 1)^{-1}$,

$$E_m \sim (\cosh^2(r) + \sinh^2(r)) \, h\nu \; . \tag{2.61}$$

E in (2.54) oscillates with a frequency of $2\nu$ between the two curves

$$E_m(t) \pm 2\sinh(r) \cosh(r) \, \hbar t^{-1} \, \alpha(1 - 6\xi), \tag{2.62}$$

and for $\xi = 1/6$ between

$$E_m(t) \pm \sinh(r) \cosh(r) \, \hbar\Lambda \, (\Lambda t)^{2\alpha-1} m^2 \Lambda^{-2} \alpha/B \; . \tag{2.63}$$

The break-down of formulae (2.1) based on classical relativistic hydrodynamics is not surprising in a situation when energy density and pressure diverge to infinity, and phenomena like (2.61) get dominant. Thus the conditions $\varepsilon > 0$, $p > 0$, with $\varepsilon$ and $p$ as in (2.1) are unlikely to give a good selection criterion for expansion factors in the limit $a(t) \to 0$. Thus we have decided to give in Examples 2 and 7 a discussion of factors $a(t) \sim (\Lambda t)^\alpha$ for the whole range of values $0 < \alpha < \infty$, and their bearing on the solutions of the wave equation.


## III. DISCUSSION AND CONCLUSION

We outline at first the classical dynamics, namely geodesic motion on the space-time manifold, and compare then with the remnants of classical chaos in the energy formulae derived in Section 2.

Applying the geodesic variational principle to the line element $ds^2$ (Sec.1) on the covering space $R^+ \times B^3$ , cf.[10], we calculate readily the geometric shapes of the geodesics,

$$r^2 + 2Mr \cos\varphi + (c/\Lambda)^2 = 0 , \tag{3.1}$$

arcs of circles centred at $|\vec{M}| = M$, orthogonal to the boundary $S_\infty$ of $B^3$ ($c/\Lambda$ is the radius of $B^3$). Their time parametrization is given by

$$r^2(t) = (c/\Lambda)^2 \, \frac{\eta^2 + 1/4 - \eta\sqrt{1 - (c/\Lambda M)^2}}{\eta^2 + 1/4 + \eta\sqrt{1 - (c/\Lambda M)^2}} \; , \quad \eta(t) = C \exp\left( \pm\Lambda \int_{1/\Lambda}^{t} \frac{dt}{\sqrt{1 + \mu^{-2} \, a^2(t)} \, a(t)} \right), \tag{3.2}$$

C and $\mu$ are integration constants, $\mu$ determines the hyperbolic length, possibly infinite, of the arc that is run through during the whole evolution $0 \le t \le \infty$ .The constant C fixes the location of the arc on (3.1). The parameter $\mu$ regulates the velocity via

$$\mu/a(t) \; = \; \frac{v/c}{\sqrt{1 - v^2/c^2}} \; , \tag{3.3}$$

for the definition of v and the derivation of (3.2, 3.3) see [14]. If $\mu = 0$ the particle is at rest, $\eta(t = 0) = \eta(t = \infty) = C$; if $\mu = \infty$ then (3.2) gives the time parametrization of light rays, (3.1) holds also true for rays. For $\eta \to 0$ or $\eta \to \infty$ we have $r(t) \to c/\Lambda$, the trajectory approaches the boundary $S_\infty$. This has interesting consequences for the chaotic properties of geodesic motion in the polyhedron F, as we will see. From the positivity condition $\varepsilon, p > 0$

in Ex.1, we know that a(t) ~ $\Lambda$t for t $\to \infty$. With this asymptotic behaviour of a(t) in (3.2) we have $\eta$(t =$\infty$, $\mu$) finite for $\mu < \infty$, and $\eta(t = \infty, \mu \to \infty) \sim \mathrm{const}.\mu^{\pm 1}$ , approaching 0 or $\infty$ for v$\to$ c according to (3.3).

Finally we discuss the behaviour of $\eta$ for t $\to$ 0. With a(t) ~ $(\Lambda t)^\alpha$ , $0 < \alpha < 1$ as in Ex.2, we see easily from (3.2) that $\eta$(t = 0, $\mu$) is finite and uniformly bounded away from 0 and $\infty$ for all $\mu$. The same holds true for Exs.3-6. But for $\alpha \geq 1$ as in Ex.7 we have $\eta$(t = 0, $\mu$) $\to$ 0 or $\infty$ regardless of the value of $\mu$.

Up to now we have discussed geodesic motion in $R^+ \times B^3$, the covering manifold of our space-time manifold $R^+ \times F$ , F is the polyhedron in $B^3$ that represents with its face-identification via $\Gamma$ (cf. Sec.1) a hyperbolic 3-manifold, a spacelike section at a given instant of time. The concept of the covering space is the convenient tool to analyse the possibly very chaotic motion in F in simple terms. Every trajectory in F is constructed from an arc of a $B^3$-geodesic (3.1).This arc intersects a certain number of tiles $\gamma$(F) of the tessellation $\Gamma$(F), (cf. Sec.1 and [13, 15]). An arc piece lying in $\gamma$(F) is projected via $\gamma^{-1}$ into F. The trajectory in F consists thus of a number of arc pieces, whose initial and end points are identified by the face-pairing transformations of F, to give a smooth curve in F. The time parametrization of the $B^3$-geodesic is inherited by the F-geodesic. In this way a trajectory is realized in $R^+ \times F$.

The ergodic properties of the trajectory in F depend of course on the arc that is projected. To discuss them we cut the arc into two pieces, say $1/\Lambda \leq t \leq \infty$, and $0 \leq t \leq 1/\Lambda$, and consider the two limits t $\to \infty$, and t $\to$ 0 separately. The initial point t = $1/\Lambda$ lies always in $B^3$, the end point t = $\infty$ or t = 0 lies either in $B^3$ or on its boundary $S_\infty$, depending on the value of $\eta$($\infty$) or $\eta$(0).

If the end point lies inside $B^3$ the trajectory is bounded, i.e. lies inside a sphere of finite hyperbolic radius. Moreover, because the accumulation points of the tiling, the limit set $\Lambda$($\Gamma$) in Figs.1,2 lie on $S_\infty$, the arc intersects only finitely many polyhedra, and thus there are only finitely many arc pieces in F constituting the trajectory. Its evolution is perfectly predictable and stable, because of its finite hyperbolic length.

If the end point lies on $S_\infty$ , $\eta$ = 0 or $\infty$ , there are two cases to distinguish. If it lies outside the limit set the trajectory intersects again only finitely many polyhedra, but clearly its F-projection is now unbounded reaching at t = $\infty$ or t = 0 a boundary of F on $S_\infty$. There is also a positive Lyapounov exponent, but the propagation of the error in the initial conditions is only proportional to the hyperbolic distance that is run through. If the end point lies in $\Lambda$($\Gamma$), the arc intersects infinitely many tiles and that gives rise to chaotic behaviour of its projection. The F-trajectory is bounded, lying in a finite compact domain $C(\Lambda)\backslash \Gamma$ of 3-space, namely the intersection of the hyperbolic convex hull of the limit set with F, cf.[15]. It is mixing there and even Bernoullian.

Finally there is the case that the end point of the arc to be projected lies in $B^3$, but that its prolongation terminates in $\Lambda$($\Gamma$). We have than a finite arc on a trajectory whose F-projection is chaotic, as it may happen with massive particles. By increasing their speed, i.e. by increasing the chaoticity parameter $\mu$ in (3.2, 3.3), the end point can come arbitrarily close to $S_\infty$, and the corresponding F-trajectory, though always regular can approximate its infinite and mixing prolongation to any wished degree.

To summarize, in the limit  t → ∞ there is a finite compact region  C(Λ)\Γ in infinite 3-space in which chaotic motion can occur: rays have the Bernoulli property, and massive particles can approximate chaotic motion for v → c arbitrarily well in the above described way. A trajectory can enter this domain C(Λ)\Γ and it may get trapped there, but it can also go through unaffectedly, depending on its lifts into the covering space $B^3$, if they end in Λ(Γ) or not. Massive particles are always bounded, whereas rays are either Bernoullian or unbounded.

The limit t → 0 : in the case of finite initial radius, Exs.4-6, particles and rays start to spread out regularly from inside the manifold, which already exists at t = 0 with a well defined metric and topology. In Exs.2,3,7  the 3-space contracts to a point, the distance between two points in the polyhedron F goes to zero. Keeping this in mind one can have nevertheless very different qualitative behaviour. Trajectories and rays in Exs.2,3 are regular and bounded for t → 0, in Ex.7 they are either unbounded and regular or bounded and Bernoullian, trapped in  C(Λ)\Γ.

The time dependence of the energy in Eq. (2.11) and the frequency in (2.15) is remarkably similar to that of a classical particle moving along a geodesic,

$$E = mc^2 \sqrt{1 + \mu^2\, a^{-2}(t)} \; = h\nu \tag{3.4}$$

with μ as in (3.3). For the wave length λ we have, using de Broglie's relation as above, $\lambda/a(t) = h/mc\mu$, cf.[14].

Concerning (2.12) we have classically the same time dependence, but there is a gap $\Delta\lambda = 1 - \delta(2 - \delta)$, δ the Hausdorff dimension of the limit set Λ(Γ), in the spectrum of the L.-B. operator of the spacelike slices between the ground state wave function and the wave fields of the continuous spectrum.

From the eikonal , cf.[14],

$$\psi(t,r,\varphi) = - \omega \int_{1/\Lambda}^{t} a^{-1}(t)\, dt + \tilde{\psi}(r,\varphi) \tag{3.5}$$

we derive easily $\nu = c/\lambda \sim t^{-1}\left(e^{2\pi\Lambda/\omega} - 1\right)^{-1}$. This and the Einstein relation is again reflected in (2.15),(2.16). The ground state is excluded for massless particles, because the spectral parameter in (2.15) must be larger than 1.

The exponents in Exs.2,3 determining the singular behaviour of φ and E depend only on the exponent in a(t) and on ξ, the geometric coupling to the curvature scalar, neither the mass nor the spectral parameter enter in them. ξ ranges in the interval [1, 1/6], cf.(2.10), the limits ξ = 0, 1/6 are discontinuous in the power laws. There is no periodicity of φ and E in the limit t → 0. The same holds true for Exs.4-6 (finite initial radius), in particular the classical ε in (2.38) and E in (2.40), have the same singular power laws, determined solely by the exponent in a(t). Only the limits ξ = 0, 1/6 have a finite energy.

Finally we discuss Ex.7. The energy in (2.55) and (2.58) has the same asymptotic behaviour as the classical E in (3.4) (which is not the case in Exs.2-6, except for ξ = 0, 1/6).The solution of the wave equation in (2.52) is periodic, and in (2.60), (2.61) we have the proportionality of E and ν like in (3.4) for m > 0. For rays we get from the eikonal in

(3.5) (with a(t) ~ $(\Lambda t)^{\alpha}$, $\alpha > 1$), and Einstein's relation the same time dependence as in the massive case, E ~ $(\Lambda t)^{-\alpha} \hbar \omega$.

In (2.60) the spectral variable $\lambda$ (it should not be mixed up with the time dependent wave lengths  in this section, we use the same notation) enters in the frequency, as it does in Ex.1 for m = 0,  and the positivity condition in (2.10) imposes again restrictions on $\lambda$ and $\xi$. For  m > 0 and $6\xi < \delta(2-\delta)$ there is a gap in the frequency (2.60) increasing in time for t $\to$ 0, corresponding to the gap $\Delta\lambda$ in the spectrum already observed in the limit t $\to \infty$ in Ex.1.


## ACKNOWLEDGMENT

The author acknowledges the support of the European Communities in their science programme under grant  B/SC1*-915078 .


## REFERENCES

[1]    Ahlfors L.V. (1981) Möbius Transformations in Several Dimensions, Lecture Notes (Univ. of Minnesota).

[2]    Bers L. (1970) On Boundaries of Teichmüller Spaces and on Kleinian Groups, Ann. Math. 91, 570-600.

[3]    Elstrodt J., F. Grunewald and J. Mennicke(1983) Discontinuous Groups on Three-Dimensional Hyperbolic Space, Russ. Math. Surv. 38, 137-168.

[4]    Hejhal D.A. (1987) Boundary-Groups, Degenerating Riemann Surfaces, and Spectral Theory, Univ. of Minnesota preprint.

[5]    Krushkal S.L., B.N. Apanasov and N.A. Grusevskii (1986) Kleinian Groups and Uniformization in Examples and Problems, Transl. of Math. Monographs vol. 62 (Amer. Math. Soc., Providence, R.I.).

[6]    McVittie G.C. (1956) General Relativity and Cosmology (Chapman and Hall,London).

[7]    Misner C.W., K.S.Thorne and J.A.Wheeler (1973) Gravitation (Freeman, N.Y.)

[8]    Patterson S.J. (1987) Lectures on Limit Sets of Kleinian Groups, in : Analytical and Geometrical Aspects of Hyperbolic Space, London Math. Soc. Lecture Notes 111, 281-323, Ed. D.B.Epstein (Cambridge Univ. Press, London).

[9]    Scott P.(1983) The Geometries of 3-manifolds, Bull. London Math. Soc. 15, 401-487.

[10]    Singer I.M. and J. A. Thorpe (1976) Lecture Notes on Elementary Topology and Geometry (Springer, N. Y.).

[11]    Thurston W. (1978) The Geometry of 3-manifolds, Lecture Notes, (Princeton Univ.)

[12]    Tolman R. C. (1934) Relativity, Thermodynamics and Cosmology (Clarendon Press, Oxford).

[13]    Tomaschitz R. (1989) On the Calculation of Quantum Mechanical Ground States from Classical Geodesic Motion on Certain Spaces of Constant Negative Curvature, Physica D 34, 42-89.

[14]    Tomaschitz R. (1991) Relativistic Quantum Chaos in Robertson-Walker Cosmologies, J. Math. Phys. 32, 2571-2579.

[15]    Tomaschitz R.(1992) An Application of Kleinian Groups in the Quantization of an Unstable Dynamical System , Int. J. Theoret. Phys. 31, no.2 (to appear).

[16]    Tomaschitz R., Quantum Chaos in de Sitter Cosmologies, to be published in the Proceedings of a NATO Advanced Research Workshop on Quantum Chaos, held May 28-June 1 1991 in Copenhagen; Ed. P. Cvitanovic (NATO-ASI Series, Plenum, NY).

# FEEDBACK CONTROL OF CHAOTIC SYSTEMS

Filipe J. Romeiras

Departamento de Mathemática
Instituto Superior Técnico, 1096 Lisboa Codex, Portugal

Celso Grebogi[a], Edward Ott[a], and W.P. Dayawansa[b]

[a]Laboratory for Plasma Research
[b]Systems Research Center
University of Maryland, College Park, MD 20742

## Abstract

A method is discussed whereby motion on a chaotic attractor can be converted to a desired attracting time-periodic motion by applying a small control. The method is illustrated numerically using a periodically driven dissipative four dimensional system.

## I.   Introduction

It is common for systems to evolve with time in a chaotic way. In practice, however, it is often desired that chaos be avoided and/or that the system be optimized with respect to some performance criterion. Given a system which behaves chaotically, one approach might be to make some large (and possibly costly) alteration in the system which completely changes its dynamics in such a way as to achieve the desired objectives. Here we assume that this avenue is not available. Thus we address the following question: Given a chaotic system, how can we obtain improved performance and achieve a desired attracting time-periodic motion by making only *small* controlling temporal perturbations in an accessible system parameter.

*Chaotic Dynamics: Theory and Practice*
Edited by T. Bountis, Plenum Press, New York, 1992

The key observation is that a chaotic attractor typically has embedded densely within it an infinite number of unstable periodic orbits [1]–[5]. In addition, chaotic attractors can also sometimes contain unstable steady states (e.g., the Lorenz attractor has such an embedded steady state). Since we wish to make only small controlling perturbations to the system, we do not envision creating new orbits with very different properties from the already existing orbits. Thus we seek to exploit the already existing unstable periodic orbits and unstable steady states. Our approach is as follows: We first determine some of the unstable low-period periodic orbits and unstable steady states that are embedded in the chaotic attractor. We then examine these orbits and choose one which yields improved system performance. Finally, we apply small controls so as to stabilize this already existing orbit.

Some comments concerning this method are the following:

1. Before settling into the desired controlled orbit the trajectory experiences a chaotic transient whose expected duration diverges as the maximum allowed size of the control approaches zero.

2. Small noise can result in occasional bursts in which the orbit wanders far from the controlled orbit.

3. Controlled chaotic systems offer an advantage in flexibility in that any one of a number of different orbits can be stabilized by the small control, and the choice can be switched from one to another depending on the current desired system performance.

For the sake of simplicity we consider a discrete time dynamical system,

$$\mathbf{Z}_{i+1} = \mathbf{F}(\mathbf{Z}_i, p), \tag{1.1}$$

where, $\mathbf{Z}_i \in \Re^n$, $p \in \Re$ and $\mathbf{F}$ is smooth. Here, $p$ is considered a real parameter which can be controlled in a small interval

$$(\bar{p} - \delta, \bar{p} + \delta)$$

around some nominal value $\bar{p}$. We assume that the nominal system (i.e., for $p = \bar{p}$) contains a chaotic attractor. Our objective is to control the parameter in such a way that for almost all initial conditions in the basin of the chaotic attractor, the dynamics of the system converge onto a desired time periodic orbit contained in the attractor. The control strategy is the following. We will find a stabilizing local feedback control law which is defined on a neighborhood of the desired periodic orbit. This is done by considering the first order approximation of the system at the chosen unstable periodic orbit. Here we assume that this approximation is stabilizable. Since stabilizability is a generic property of linear systems, this assumption is quite reasonable. The ergodic nature of the chaotic dynamics ensures that the state trajectory eventually enters into the neighborhood. Once inside, we apply the stabilizing feedback control law in order to steer the trajectory towards the desired orbit.

For simplicity we shall describe the method as applied to the stabilization of fixed points (i.e., period one orbits) of the map $\mathbf{F}$. The consideration of periodic orbits of period larger than one is straightforward. Let $\mathbf{Z}_*$ denote an unstable fixed point on the attractor. Now, the first order approximation to (1.1) near $\bar{p}$ is

$$(\Delta \mathbf{Z})_{i+1} \cong \mathbf{A}(\Delta \mathbf{Z})_i + \mathbf{B}(\Delta p)_i, \tag{1.2}$$

where

$$\mathbf{A} \;=\; D_\mathbf{z}\mathbf{F}(\mathbf{Z}_*, \bar{p}),$$

$$\mathbf{B} \;=\; D_p\mathbf{F}(\mathbf{Z}_*, \bar{p}),$$

$$(\Delta \mathbf{Z})_i \;=\; \mathbf{Z}_i - \mathbf{Z}_*,$$

$$(\Delta p)_i \;=\; p_i - \bar{p}.$$

Generically, $(\mathbf{A}, \mathbf{B})$ pair is controllable, and hence we may find a linear feedback control law,

$$(\Delta p)_i = \mathbf{K}(\Delta \mathbf{Z})_i,$$

such that the spectrum of $(\mathbf{A} + \mathbf{B}\mathbf{K})$ is in the open unit disc. Furthermore, this control law stabilizes the nonlinear system (1.1) in a neighborhood of the nominal operating point as well.

Although we describe the details only in the case of discrete time systems, this method is applicable in the continuous time case as well by considering the discrete time system obtained from the induced dynamics on a Poincaré section.

In order to illustrate the method we apply it to a periodically forced mechanical system (the kicked double rotor), which results in a four dimensional map. Amongst the examples considered, we study cases where the unstable orbit of the uncontrolled system has two unstable eigenvalues and two stable eigenvalues, and the stabilization is achieved by variation of one control parameter characterizing the strength of the periodic forcing. The present paper generalizes our previous work [6] to the case of higher dimensional systems and also includes new material illustrating the effect of the choice of stabilization on the length of the chaotic transient experienced by the orbit before control is achieved. Other relevant references on the stabilization of periodic or steady orbits embedded in chaotic attractors are the experiments of Ditto et al. [7], and Singer et al. [8], and the paper of Fowler [9].

## II.  Illustrative Example: The Kicked Double Rotor

The double rotor [10] is composed of two thin, massless rods connected as shown in Fig. 1. The first rod, of length $\ell_1$, pivots about $P_1$ (which is fixed), and the second rod, of length $2\ell_2$, pivots about $P_2$ (which moves). The angles $\theta_1(t)$, $\theta_2(t)$ measure the position at time $t$ of the first and second rods, respectively. A mass $m_1$ is attached to the first rod at $P_2$, and masses $m_2/2$ are attached to each end of the second rod ($P_3$ and $P_4$). Friction at $P_1$ (with coefficient $\nu_1$) slows the first rod at a rate proportional to its angular velocity $\dot{\theta}_1(t) \equiv d\theta_1(t)/dt$; friction at $P_2$ (with coefficient $\nu_2$) slows the second rod (and simultaneously accelerates the first rod) at a rate proportional to $\dot{\theta}_2(t) - \dot{\theta}_1(t)$. The end of the second rod marked $P_3$ receives impulse kicks at times $t = T, 2T, ...$, always from the same direction and with constant strength $f_0$. Gravity and air resistance are ignored.

We obtain the four dimensional map

$$\mathbf{Z} \to \mathbf{Z}' = \mathbf{F}(\mathbf{Z}),$$

defined by

$$\mathbf{Z} = \begin{bmatrix} \mathbf{X} \\ \mathbf{Y} \end{bmatrix} \to \begin{bmatrix} \mathbf{X}' \\ \mathbf{Y}' \end{bmatrix} = \begin{bmatrix} (\mathbf{MY} + \mathbf{X}) \quad \mod 2\pi \\ \mathbf{LY} + \mathbf{G}(\mathbf{X}') \end{bmatrix}, \tag{2.1}$$

where

$$\mathbf{X} = \begin{bmatrix} x_1 \\ x_2 \end{bmatrix} \in S^1 \times S^1, \qquad \mathbf{Y} = \begin{bmatrix} y_1 \\ y_2 \end{bmatrix} \in \Re \times \Re,$$

and

$$\mathbf{G}(\mathbf{X}') = \begin{bmatrix} c_1 \sin x_1' \\ c_2 \sin x_2' \end{bmatrix}. \tag{2.2}$$

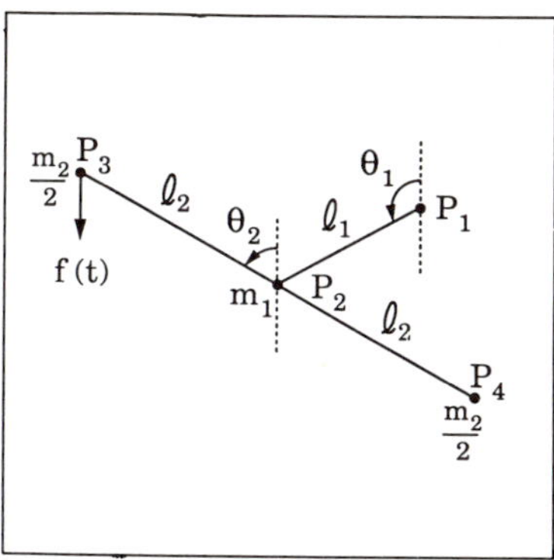

Figure 1. The double rotor.

$x_1$, $x_2$ are the positions of the rods at the instant of the $k$th kick, $x_j = \theta_j(kT)$, while $y_1, y_2$ are the angular velocities of the rods immediately after the $k$th kick, $y_j = \dot{\theta}_j(kT^+)$. $S^1$ is the circle $\Re$ [mod $2\pi$]. $\mathbf{L}$ and $\mathbf{M}$ are constant $2 \times 2$ matrices. For the sake of simplicity we assume $(m_1 + m_2)\ell_1^2 = m_2\ell_2^2 \equiv I$. We then obtain

$$\mathbf{L} = \sum_{j=1}^{2} \mathbf{W}_j e^{\lambda_j T}, \qquad \mathbf{M} = \sum_{j=1}^{2} \mathbf{W}_j \frac{e^{\lambda_j T} - 1}{\lambda_j},$$

$$\mathbf{W}_1 = \begin{bmatrix} a & b \\ b & d \end{bmatrix}, \qquad \mathbf{W}_2 = \begin{bmatrix} d & -b \\ -b & a \end{bmatrix},$$

$$a = \frac{1}{2}\left(1 + \frac{\nu_1}{\Delta}\right), \qquad d = \frac{1}{2}\left(1 - \frac{\nu_1}{\Delta}\right), \qquad b = -\frac{\nu^2}{\Delta},$$

$$\begin{matrix} \lambda_1 \\ \lambda_2 \end{matrix} = -\frac{1}{2}(\nu_1 + 2\nu_2 \pm \Delta), \qquad \Delta = (\nu_1^2 + 4\nu_2^2)^{1/2},$$

$$c_j = \frac{f_0}{I}\ell_j, \qquad (j = 1, 2).$$

The following relation between matrices $\mathbf{L}$ and $\mathbf{M}$ will be useful:

$$\mathbf{L} = \mathbf{I} + \mathbf{A}_\nu \mathbf{M}, \qquad \text{where} \qquad \mathbf{A}_\nu = \begin{bmatrix} -(\nu_1 + \nu_2) & \nu_2 \\ \nu_2 & -\nu_2 \end{bmatrix}. \tag{2.3}$$

($\lambda_1, \lambda_2$ are the eigenvalues of $\mathbf{A}_\nu$.) From now on we assume that $\nu_1 = \nu_2 \equiv \nu$. In all the numerical work described in the rest of this paper the parameters were kept fixed at the following values

$$\nu = T = I = m_1 = m_2 = \ell_2 = 1, \qquad \ell_1 = \frac{1}{\sqrt{2}}. \tag{2.4}$$

We will use as our control the forcing parameter $f_0$, and we take as its nominal value $\bar{f}_0 = 9$.

# III.    Properties of the Attractor

For chaotic attractors of an n dimensional map there are n Lyapunov exponents $L_j$ for which we choose the subscript ordering such that

$$L_1 \geq L_2 \geq \ldots \geq L_n.$$

A *chaotic* attractor is defined by the condition $L_1 > 0$.

For typical dynamical systems the Lyapunov exponents are the same for almost all initial conditions in the basin of attraction of the attractor. Thus the spectrum of Lyapunov exponents may be considered to be a property of the attractor.

From the spectrum of Lyapunov exponents we define the *Lyapunov dimension* [11,12]

$$d_L = k_L + \frac{\sum_{j=1}^{k_L} L_j}{|L_{k_L+1}|}, \tag{3.1}$$

where $1 \leq k_L \leq n - 1$ is the largest integer for which $\sum_{j=1}^{k_L} L_j \geq 0$. If $L_1 < 0$, we define $d_L = 0$; if $\sum_{j=1}^{n} L_j \geq 0$, we define $d_L = n$. It has been conjectured [11,12] that $d_L$ is equal to the information dimension of the attractor.

We have numerically calculated the Lyapunov exponents and the Lyapunov dimension of the chaotic attractor for the double rotor map using the parameters given at the end of Sec. 2. We used the method described in [13,14] to calculate the Lyapunov exponents. The result is $L_1 = 1.21$, $L_2 = 0.26$, $L_3 = -1.74$, $L_4 = -2.72$ and the Lyapunov dimension is $d_L = 2.8$.

Figure 2 shows a two dimensional cross-section of the attractor (i.e., the intersection of the attractor with a three dimensional hyperplane in the four dimensional space). Note the fractal-like structure. Numerically we obtain this picture by approximating the hyperplane by a very narrow slab

$$|\hat{\mathbf{K}}(\mathbf{Z} - \mathbf{Z}_*)| < w, \tag{3.2}$$

where $\hat{\mathbf{K}} = (0,0,1,1)$, $w = 10^{-2}$ and $\mathbf{Z}_*$ is one of the fixed points of the attractor (see Sec. 4). We then examine a very long orbit and plot in Fig. 2 only those points satisfying (3.2). The intersection of our 2.8 dimensional attractor with a three dimensional hyperplane is 1.8 dimensional. The small scale structure of this 1.8 dimensional intersection is somewhat fuzzed out in Fig. 2 due to the finite slab thickness.

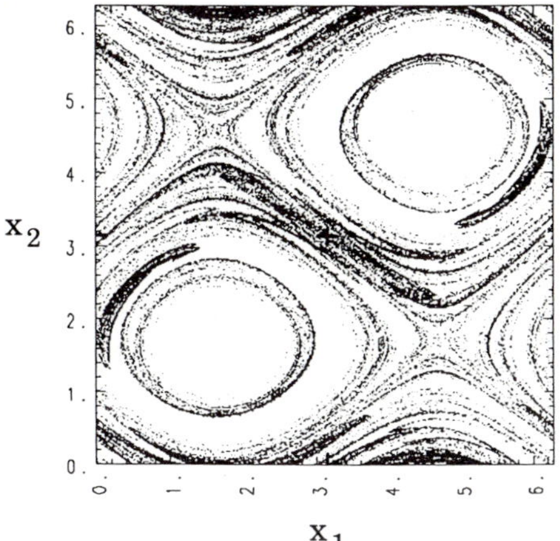

Figure 2. Double rotor map: intersection of chaotic attractor with the slab $|\hat{\mathbf{K}}(\mathbf{Z} - \mathbf{Z}_*)| < w$, $\hat{\mathbf{K}} = (0, 0, 1, 1)$, $w = 10^{-2}$, through the fixed point $\mathbf{Z}_*$ labeled $+$ in the figure. The map was iterated $10^8$ times.

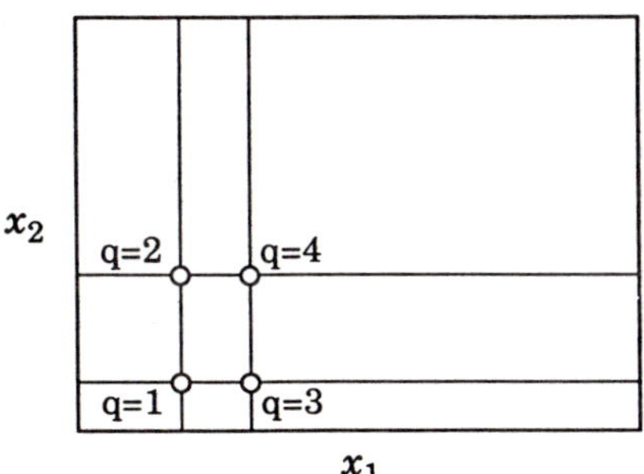


Figure 3. Labeling of fixed points.

# IV.   Fixed points of the double rotor map

The fixed points $\mathbf{Z}_* = (\mathbf{X}_*, \mathbf{Y}_*)$ of the map (2.1) are solutions of the system

$$
\begin{aligned}
\mathbf{0} &= \mathbf{M}\mathbf{Y}_* - 2\pi\mathbf{N}, \\
\mathbf{Y}_* &= \mathbf{L}\mathbf{Y}_* + \mathbf{G}(\mathbf{X}_*),
\end{aligned}
\tag{4.1}
$$

where the components of the vector $\mathbf{N} = (n_1, n_2)$ are integer and are the rotation numbers in the $x_1, x_2$ variables and $\mathbf{0}$ denotes the $2 \times 2$ matrix of zeros. The rotation numbers $n_1, n_2$ are defined as the multiples of $2\pi$ by which the angles $x_1, x_2$ are increased in one iteration of the map. From Eqs. (4.1) we obtain, using (2.3),

$$
\begin{aligned}
\mathbf{Y}_* &= 2\pi\mathbf{M}^{-1}\mathbf{N}, \\
\mathbf{G}(\mathbf{X}_*) &= -2\pi\mathbf{A}_\nu\mathbf{N}.
\end{aligned}
\tag{4.2}
$$

Using the definitions of the matrices $\mathbf{G}$ and $\mathbf{A}_\nu$, we rewrite the second of the equations (4.2) in the form

$$
\begin{bmatrix} \sin x_{1*} \\ \sin x_{2*} \end{bmatrix} = -\frac{2\pi\nu I}{f_0} \begin{bmatrix} \frac{1}{\ell_1}(-2n_1 + n_2) \\ \frac{1}{\ell_1}(n_1 - n_2) \end{bmatrix} \equiv \frac{1}{f_0} \begin{bmatrix} f_{01} \\ f_{02} \end{bmatrix},
\tag{4.3}
$$

where the identity on the right defines the two new quantities $f_{01}$ and $f_{02}$. These equations show that for each pair of rotation numbers $(n_1, n_2)$ a set of four fixed points exists if $|f_0| > \max(|f_{01}|, |f_{02}|)$. The four fixed points correspond to the four combinations of values of $(x_{1*}, x_{2*})$ that have the same pair of values of $(\sin x_{1*}, \sin x_{2*})$. We will use the notation $[(n_1, n_2); q]$, to identify the fixed points, where $q = 1, 2, 3,$ or $4$ corresponds to the ordering shown in Fig. 3. Note that $(y_{1*}, y_{2*})$ is the same for the four fixed points.

We observe, from Eq. (4.3), that as the forcing $f_0$ increases, the number of fixed points increases without bound. Not all these fixed points are necessarily embedded in the chaotic attractor, but those that are embedded in it are necessarily unstable. Furthermore, we find that the fixed points are roughly spread throughout the attractor, suggesting that there can be substantial flexibility to select among a variety of asymptotic behaviors by selecting different fixed points for control. (Even more flexibility can be achieved if we also consider periodic orbits of period greater than one.)

# V.   Time to Achieve Control

The control is activated (i.e., $p \neq \bar{p}$) only if $\mathbf{Z}_i$ falls in a narrow slab (see Sec. 6). Thus, for small $\delta$, a typical initial condition executes a chaotic orbit, unchanged from the uncontrolled case, until $\mathbf{Z}_i$ falls in this slab. Even then, because of nonlinearity not included in the linearized equation, Eq. (1.2), the control may not be able to bring the orbit to the fixed point. In this case the orbit will leave the slab and continue to wander chaotically as if there was no control. Since the orbit on the uncontrolled chaotic attractor is ergodic, at some time it will eventually be sufficiently close to the desired fixed point that control is achieved.

Thus, we create a stable orbit, but, for a typical initial condition, it is preceded in time by a chaotic transient [15-18] in which the orbit is similar to orbits on the uncontrolled chaotic attractor. The length $\tau$ of such chaotic transients depends sensitively on the initial conditions, and, for randomly chosen initial conditions in the basin of attraction of the attractor, it has an exponential probability distribution [15],

$$\phi(\tau) \cong \frac{1}{<\tau>} \exp\left(-\frac{\tau}{<\tau>}\right), \tag{5.1}$$

for large $\tau$. $<\tau>$ is the characteristic length of the chaotic transient, called in the present case the *average time to achieve control.*

We now describe the procedure used to numerically calculate the average time to achieve control, $<\tau>$. From (5.1) we obtain the fraction of chaotic transients with length smaller than some value $\tau_{\max}$ ,

$$\begin{aligned}
\rho_{\tau_{\max}} &= \int_0^{\tau_{\max}} \phi(\tau)d\tau \\
&= 1 - \exp\left(-\frac{\tau_{\max}}{<\tau>}\right),
\end{aligned}$$

and the average length of the chaotic transients with length smaller than $\tau_{\max}$,

$$\begin{aligned}
<\tau>_{\tau_{\max}} &= \int_0^{\tau_{\max}} \tau\phi(\tau)d\tau \\
&= <\tau>\left[1 - (1 + \frac{\tau_{\max}}{<\tau>})\exp(-\frac{\tau_{\max}}{<\tau>})\right].
\end{aligned}$$

Combining these two equations we obtain

$$<\tau>=<\tau>_{\tau_{\max}} \{1 - (1 - \rho_{\tau_{\max}})[1 - \log_e(1 - \rho_{\tau_{\max}})]\}^{-1}. \tag{5.2}$$

Note that $\rho_\infty = 1, <\tau>_\infty=<\tau>$.

The numerical procedure to calculate the average time to achieve control is as follows. Take a large number $N_0$ of randomly chosen initial conditions and iterate each of them with the uncontrolled map (i.e., with $\mathbf{X} \to \mathbf{F}(\mathbf{X},\bar{p})$) a sufficient number of times until they are all distributed over the attractor according to its natural measure. Then switch on the control and determine how many further iterates $\{\tau_j\}$, $j = 1, N_f$, are necessary for $N_f \leq N_0$ orbits to first fall within the control region centered at the fixed point. Define the quantities

$$\tau_{\max} = \max\{\tau_j\}_{j=1,N_f},$$

$$\rho_{\tau_{\max}} = \frac{N_f}{N_0},$$

$$<\tau>_{\tau_{\max}} = \frac{1}{N_f} \sum_{j=1}^{N_f} \tau_j.$$

Finally use Eq. (5.2) to obtain $<\tau>$. In our numerical experiments described in Section 6 we took $N_0 = 192$, $N_f = 121$, values that led to a good compromise between accuracy and computation time.

## VI.  Control

We now proceed to control the fixed points of the double rotor map with control parameter $f_0$. Let us denote by $\bar{\mathbf{Z}}_*$ the fixed points to be controlled at the nominal value $\bar{f}_0$ of the parameter. The $\mathbf{A}$ and $\mathbf{B}$ matrices of Eq. (1.2) now take the following particular form:

$$\mathbf{A} = \begin{bmatrix} \mathbf{I} & \mathbf{M} \\ \mathbf{H}(\bar{\mathbf{X}}_*) & \mathbf{L} + \mathbf{H}(\bar{\mathbf{X}}_*)\mathbf{M} \end{bmatrix},$$

$$\mathbf{H}(\bar{\mathbf{X}}_*) = \frac{\bar{f}_0}{I} \begin{bmatrix} \ell_1 \cos \bar{x}_{1*} & 0 \\ 0 & \ell_2 \cos \bar{x}_{2*} \end{bmatrix},$$

$$\mathbf{B} = \frac{1}{I} \begin{bmatrix} 0 \\ 0 \\ \ell_1 \sin \bar{x}_{1*} \\ \ell_2 \sin \bar{x}_{2*} \end{bmatrix}.$$

We use Ackermann's procedure for choosing the control vector $\mathbf{K}$ [19],

$$\mathbf{C} = [\mathbf{B}|\mathbf{AB}|\mathbf{A}^2\mathbf{B}|\mathbf{A}^3\mathbf{B}],$$

$$\mathbf{W} = \begin{bmatrix} a_3 & a_2 & a_1 & 1 \\ a_2 & a_1 & 1 & 0 \\ a_1 & 1 & 0 & 0 \\ 1 & 0 & 0 & 0 \end{bmatrix},$$

$$\mathbf{T} = \mathbf{CW},$$

$$\mathbf{K} = [\alpha_4 - a_4, \alpha_3 - a_3, \alpha_2 - a_2, \alpha_1 - a_1]\mathbf{T}^{-1}.$$

Here $\mathbf{C}$ denotes the controllability matrix which must be nonsingular for this procedure to work. The $a_j, j = 1, \ldots, 4$, are the coefficients of the characteristic polynomial of $\mathbf{A}$, while the $\alpha_j, j = 1, \ldots, 4$, are the coefficients of the desired characteristic polynomial of matrix $\mathbf{A} + \mathbf{BK}$.

One immediate conclusion is that the controllability matrix $\mathbf{C}$ is identically zero in the case of the fixed points with rotation numbers $\mathbf{N} = (0,0)$ for which $\sin \bar{x}_{1*} = \sin \bar{x}_{2*} = 0$. Hence these points are uncontrollable, at least when the control parameter is $f_0$. This set of fixed points can be controlled if we modify the double rotor map to allow for kicks with variable direction and then take as control parameter the angle the kicks make with the previously fixed direction.

The method is illustrated in Fig. 4. The control of the first fixed point was turned on at $i = 0$ with switches to control other fixed points occurring at later times. We plot the $x_1$ coordinate of an orbit as a function of (discrete) time. The parameter perturbations were programmed to control successively four different fixed points of the set with rotation numbers $\mathbf{N} = \pm(0,1)$. The times at which we switched the control from stabilizing one fixed point to stabilizing another are labeled by the arrows in the figure. The figure clearly illustrates the flexibility offered by the method in controlling different periodic motions embedded in the attractor. The figure also shows how the time to achieve control varies from case to case.

We now report the results of several numerical experiments that were carried out with the purpose of understanding the behavior of the time to achieve control.

The first experiment was intended to confirm that the time to achieve control indeed follows an exponential probability distribution as indicated in Section 5. We proceeded to control the fixed point [(0,1);4] by starting at a large number of different points on the attractor and measuring the time each orbit took to reach the stabilized neighborhood of the fixed point. We then plotted a histogram of the time to achieve control using bins of size $2 \times 10^3$. The results are presented as a semilog plot in Fig. 5 and show excellent agreement with the predicted fit to a straight line.

In our next experiment we looked at the dependence of the average time to achieve control on the size of the allowed parameter perturbations, $\delta$. The results are shown in Fig. 6, where we have used logarithmic scales in both axis. The two fixed points [(0,1);4] and [(0,1);1] were controlled. (The first of these points has two unstable eigenvalues, while the second has only one unstable eigenvalue.) We see that for the smaller values of $\delta$ the results closely follow straight lines indicating a power law dependence,

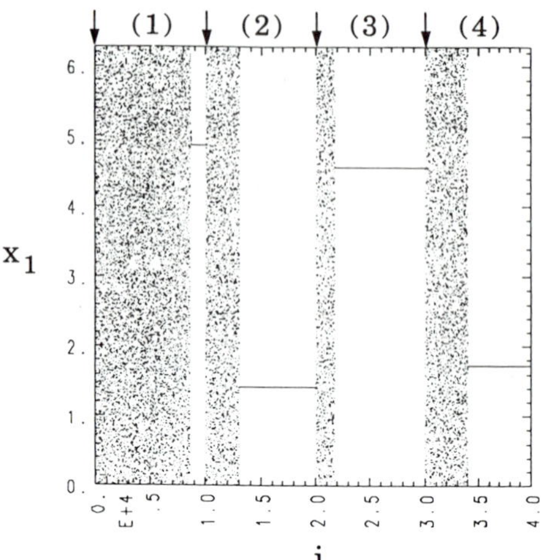

Figure 4. Double rotor map: successive control of fixed points (1) [(0,1);4], (2) [(0,-1);1], (3) [(0,1);1], (4) [(0,-1);4]. The arrows indicate the times of switching.

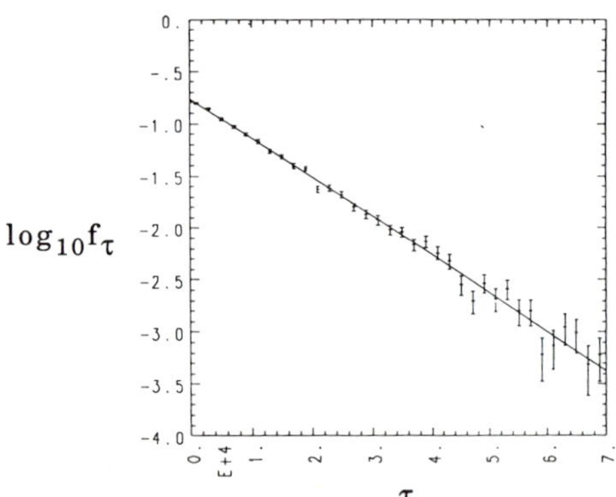

Figure 5. Double rotor map: histogram of the times $\tau$ to achieve control for a sample of 8192 orbits. The fixed point controlled was [(0,1);4]. $f_\tau$ is the frequency with which each bin is visited.

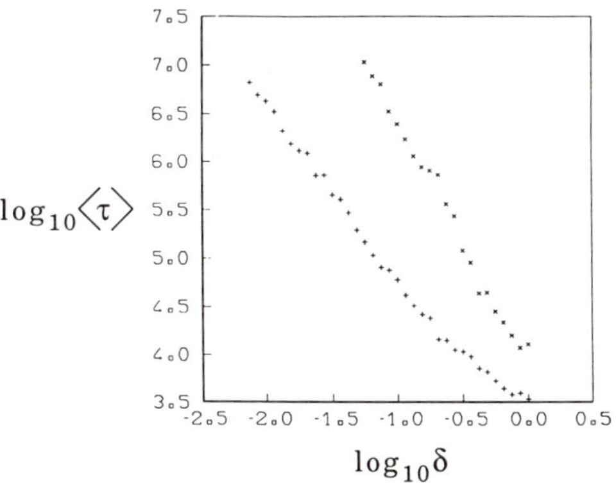


Figure 6. Double rotor map: $\log_{10} < \tau >$ versus $\log_{10} \delta$ for control of the fixed points $(+)[(0,1);1]$, $(\times)[(0,1);4]$.

in accord with the theoretical predictions of [6] for two dimensional maps. This indicates that the theory of [6] might be extended to higher dimensions.

In the experiments described until this point the choice of the regulator poles (eigenvalues of $\mathbf{A} + \mathbf{BK}$) corresponded to projection onto the stable manifold of the fixed points. That is, the stable eigenvalues of matrix $\mathbf{A}$ were left unchanged, and the unstable eigenvalues were shifted to zero.

In our next set of experiments we looked at how different choices of regulator poles affect the average time to achieve control. We considered the fixed point $[(0,1);4]$ with two unstable eigendirections and kept two of the regulator poles equal to the two stable eigenvalues of the uncontrolled fixed point. As regards the other two regulator poles, $\mu_1$ and $\mu_2$, three cases were considered:

$$(I) \quad \mu_2 = 0; \quad (II) \quad \mu_2 = \mu_1; \quad (III) \quad \mu_2 = -\mu_1.$$

$\mu_1$ was then allowed to vary in the interval $(-1,1)$. The results of the experiments are shown in Fig. 7. In cases (I) and (II) the average time to achieve control essentially increases with $\mu_1$. In case (III) the average time to achieve control passes through a broad minimum. (Note that the point $\mu_1 = \mu_2 = 0$, which is common to the three cases, corresponds to projection onto the stable manifold.)

## VII.   Discussion and Conclusion

The transient phase where the orbit wanders chaotically before locking in to a controlled orbit can be greatly shortened by applying the technique discussed by

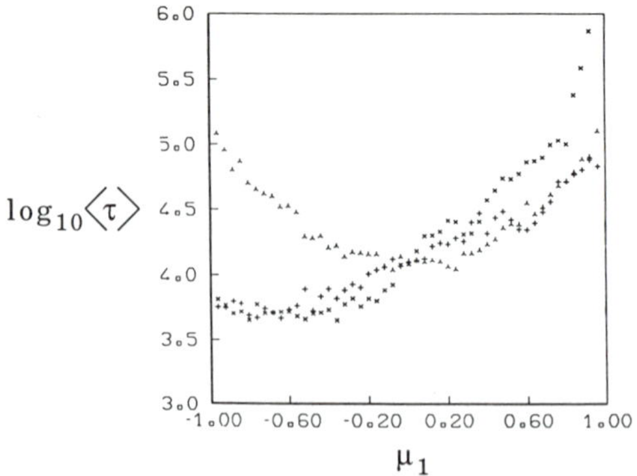


Figure 7. Double rotor map: $\log_{10} < \tau >$ versus $\mu_1$ for $(+)$ $\mu_2 = 0$, $(\times)$ $\mu_2 = \mu_1$, $(\lambda)$ $\mu_2 = -\mu_1$.

Shinbrot et al. [20]. In the latter paper it was pointed out that orbits can be rapidly brought to a target region on the attractor (in the present case the neighborhood of the periodic orbit which we wish to stabilize) by using small control perturbations. The idea was that, since chaotic systems are exponentially sensitive to perturbation, careful choice of even small control perturbations can, after some time, have a large effect on the orbit location and can be used to guide it. Thus the time to achieve control can, in principle, be greatly shortened by properly applying small controls when the orbit is far from the neighborhood of the desired periodic orbit.

One issue which we have not addressed is the effect of noise. If the noise remains small, it may not be sufficient to kick the orbit out of the neighborhood of the chosen periodic orbit where the control is activated. However, occasionally the strength of the random noise may kick the orbit far enough away from the periodic orbit that the orbit falls outside the small controlled phase space region. In this case the orbit will wander chaotically over the attractor until it falls in the controlled region again. Thus there are epochs where the orbit is kept near the desired orbit interspersed with epochs wherein the orbit wanders chaotically far from the desired orbit. If the latter are, on average, relatively much shorter than the former, then one might still regard the control as being effective. See [6] for numerical experiments on this effect. We also remark that the procedure discussed in the previous paragraph [20] can be used to greatly reduce the duration of the noise induced epochs where the orbit bursts out of the controlled phase space region.

In this paper we have considered the case where there is only a single control parameter available for adjustment. While generically a single parameter is sufficient for stabilization of a desired periodic orbit, there may be some advantage to utilizing several control variables (i.e., the single control parameter $p$ becomes a vector). In particular, the added freedom in having several control parameters might allow better means of choosing the control so as to minimize the time to achieve control.

Finally we wish to point out that full knowledge of the system dynamics is not necessary in order to apply our technique (see also [6]). In particular, we only require the location of the desired periodic orbit, the linearized dynamics about the periodic orbit, and the dependence of the location of the periodic orbit on small variation of the control parameter. Recently, the Takens embedding technique [21] has been utilized in several experimental studies (Refs. [7], [22]–[24]) to extract such information purely from observations of experimental chaotic orbits on the attractor without any *a priori* knowledge of the system of equations governing the dynamics. Hence application of our method is not limited to cases where a complete knowledge of the system is available.

In conclusion, we have demonstrated that chaotic dynamics can often be converted, by using only a small control, to motion on a desired periodic orbit. Furthermore, by switching the small control, one can switch the time asymptotic behavior from one periodic orbit to another. In some situations, where the flexibility offered by the ability to do such switching is desirable, it may be advantageous to design the system so that it is chaotic. In other situations, where one is presented with a chaotic system, the method may allow one to eliminate the chaos and achieve greatly improved behavior at relatively low cost.

## Acknowledgements

This work was supported by the U.S. Department of Energy (Scientific Computing Staff, Office of Energy Research), the Portuguese Junta Nacional de Investigação Cientifica e Tecnológica, and the National Science Foundation (Engineering Research Center Program). The computation was done at the National Energy Research Supercomputer Center.

## References

[1] C. Grebogi, E. Ott and J. A. Yorke "Unstable periodic orbits and the dimensions of multifractal chaotic attractors," Phys. Rev. A **37**, 1711-1724, 1988.

[2] D. Auerbach et al. "Exploring chaotic motion through periodic orbits," Phys. Rev. Lett. **58**, 2387-2390, 1987.

[3] T. Morita et al. "On partial dimensions and spectra of singularities of strange attractors," Prog. Theor. Phys. **78**, 511, 1987.

[4] A. Katok "Lyapunov exponents, entropy and periodic orbits for diffeomorphisms," Publ. Math. IHES **51**, 137, 1980.

[5]  R. Bowen "Periodic points and measures for axiom A diffeomorphisms," Trans. Am. Math. Soc. **154**, 377, 1971.

[6]  E. Ott, C. Grebogi and J. A. Yorke "Controlling Chaos," Phys. Rev. Lett. **64**, 1196-1199, 1990.

[7] W. L. Ditto, S. N. Rauseo and M. L. Spano "Experimental control of chaos," Phys. Rev. Lett. **65**, 3211-3214, 1990.

[8] J. Singer, Y.-Z. Wang, and H. H. Bau "Controlling a chaotic system," Phys. Rev. Lett. **66**, 1123-1125, 1991.

[9] T. B. Fowler "Application of stochastic control techniques to chaotic nonlinear systems," IEEE Trans. on Automatic Control **34**, 201-205, 1989.

[10] C. Grebogi, E. Kostelich, E. Ott and J. A. Yorke "Multi-dimensional intertwined basin boundaries: basin structure of the kicked double rotor," Physica D, **25**, 347-360, 1987; "Multi-dimensional intertwined basin boundaries and the kicked double rotor," Phys. Lett. A **118**, 448-452, 1986; E **120A**, 497, 1987.

[11] J. L. Kaplan and J. A. Yorke "Chaotic behavior of multidimensional difference equations," in *Functional Differential Equations and Approximation of Fixed Points*, edited by H.-O. Peitgen and H.-O. Walter, Lecture Notes in Mathematics **730** (Springer, Berlin, 1979) p. 204.

[12] J. D. Farmer, E. Ott and J. A. Yorke "The dimension of chaotic attractors," Physica D **7**, 153-180, 1983.

[13] G. Benettin, L. Galgani, A. Giorgilli and J. Strelcyn "Lyapunov characteristic exponents for smooth dynamical systems: A method for computing all of them. Part I: Theory," Meccanica **15**, 9-20, 1980.

[14] G. Benettin, L. Galgani, A. Giorgilli and J. Strelcyn "Lyapunov characteristic exponents for smooth dynamical systems: A method for computing all of them. Part II: Numerical application," Meccanica **15**, 21-30, 1980.

[15] C. Grebogi, E. Ott and J. A. Yorke "Crises, sudden changes in chaotic attractors, and transient chaos," Physica D **7**, 181-200, 1983.

[16] C. Grebogi, E. Ott and J. A. Yorke "Critical exponents of chaotic transients in nonlinear dynamical systems," Phys. Rev. Lett. **57**, 1284-1287, 1986.

[17] C. Grebogi, E. Ott, F. J. Romeiras, and J. A. Yorke "Critical exponents for crisis-induced intermittency," Phys. Rev. A **36**, 5365-5380, 1987.

[18] C. Grebogi, E. Ott, and J. A. Yorke "Chaotic attractors in crisis," Phys. Rev. Lett. **48**, 1507-1510, 1982.

[19] K. Ogata, *Control Engineering*, second edition (Prentice-Hall, Englewood Cliffs, NJ, 1990), pp. 782-784.

[20] T. Shinbrot, E. Ott and C. Grebogi "Using chaos to direct trajectories to targets," Phys. Rev. Lett. **65**, 3215-3218, 1990.

[21] F. Takens "Detecting strange attractors in turbulence," in *Dynamical Systems and Turbulence*, Ed. by D. A. Rand and L.-S. Young (Springer Lecture Notes in Mathematics, Vol. 898, Springer-Verlag, New York, 1980), pp. 366-381.

[22] J. C. Sommerer et al. "Experimental confirmation of the theory for critical exponents of crises," Phys. Lett. A (1991) to be published.

[23] D. P. Lathrop and E. J. Kostelich "Characterization of an experimental strange attractor by periodic orbits," Phys. Rev. A **40**, 4028, 1989.

[24] G. H. Gunaratne, P. S. Linsay and M. J. Vinson "Chaos beyond onset: A comparison of theory and experiment," Phys. Rev. Lett. **63**, 1, 1989.

# THE STRUCTURE OF BASIN BOUNDARIES

# IN A SIMPLE ADAPTIVE CONTROL SYSTEM


C.E. Frouzakis[1], R.A. Adomaitis[1,2], I.G. Kevrekidis[1], M.P. Golden[3] and B.E. Ydstie[4]

[1]Department of Chemical Engineering, Princeton University, Princeton, NJ 08544
[2]Systems Research Center, University of Maryland, College Park, MD 20742
[3]Simulation Sciences, 601 South Valencia Avenue, Suite 250, Brea, CA 92621
[4]Department of Chemical Engineering, University of Massachusetts, Amherst, MA 01003


## Abstract


We present a detailed study of the boundaries separating different basins of attraction in a discrete-time, model-reference, adaptive control system. The closed-loop system is a noninvertible map of the plane. The noninvertible nature of the map plays an important role in the shape and interactions of these basins, making them distinctly different from those of continuous-time or discrete-time invertible dynamical systems.


# 1.  Introduction

Discrete-time model reference adaptive control (MRAC) systems are known to exhibit complicated dynamics (including multistability, quasiperiodicity and chaotic behavior) in the presence of plant/model mismatch (e.g., [12, 13, 5, 4]) These systems give rise to finite dimesional maps ($R^n \mapsto R^n$), which are often *noninvertible*; this means that there exist some points in phase space which do not possess a unique preimage in time (they may have zero or more than one preimage). This property plays an important role in determining the structure of the basins of attraction and the local and global bifurcations that may occur.

In this paper, we give a detailed account of the effects of noninvertibility on the basins of attraction of a representative adaptive system. The system is designed to control a first-order plant, and is characterized by plant/model mismatch and an unmodelled additive disturbance. We focus on the analysis of the basin of attraction of the (desired) set point when it is not globally stable, since this region quantifies the finite-amplitude perturbations to which the set point is stable. We chose this particular example both because it provides a good illustration of the phenomena we want to discuss, and because a brute-force calculation of the basins of attraction for this system (performed by two of the authors, M.P.G. and B.E.Y.) has been recently reported [6].

The paper is structured as follows: we present a concise derivation of the system in Section 2; local stability analysis of the set point, an introduction of the basin picture of Golden and Ydstie, derivation of the inverse map, and the location of its critical curves in phase space are all discussed in Section 3. A detailed study of the basin boundaries and their interactions with the critical curves is contained in Section 4.

## 2.   The model

The general case of a linear, time-invariant, single-input single-output process with a time delay equal to one can be represented as [8]:

$$A(q^{-1})y(t) = q^{-1}B(q^{-1})u(t) + v(t) \tag{1}$$

where $y(t)$ is the process output, $u(t)$ the manipulated variable, and $v(t)$ the unmeasured disturbance. $A(q^{-1})$ and $B(q^{-1})$ are polynomial functions of the backward shift operator, $q^{-1}$, (defined by $q^{-1}y(t) = y(t-1)$):

$$\begin{aligned} A(q^{-1}) &= 1 + \alpha_1 q^{-1} + ...\alpha_n q^{-n} \\ B(q^{-1}) &= \beta_0 + \beta_1 q^{-1} + ...\beta_m q^{-m}. \end{aligned}$$

If we define the vectors

$$\theta^T = (-\alpha_1, ..., -\alpha_n, \beta_0, ..., \beta_m)$$

$$\phi(t)^T = (y(t), ..., y(t-n+1), u(t), ..., u(t-m))$$

equation (1) can be compactly rewritten as:

$$y(t+1) = \phi(t)^T\theta + v(t+1). \tag{2}$$

The objective is to design a controller that will track a reference signal $y^*(t)$. For the disturbance-free case, the control action can then be determined by solving

$$y^*(t+1) = \phi(t)^T\hat{\theta}(t) \tag{3}$$

for $u(t)$, with $\hat{\theta}$ the most recent estimate of the (unknown) parameters $\theta$. This estimate is obtained using input-output data available from previous time intervals. The estimation algorithm we use is the projection algorithm [8]:

$$\hat{\theta}(t) = \hat{\theta}(t-1) + P\phi(t-1)\frac{y(t) - y^*(t)}{c + \phi(t-1)^T\phi(t-1)} \tag{4}$$

where $P$ is the adaptation gain and $c$ is a small, non-negative number used to prevent division by zero. For the case of a disturbance-free process with known $n$ and $m$ the system of equations (2), (3), (4) is bounded-input bounded-output stable [7]. However, because of the inherent nonlinearity of the combined process-controller system, we must resort to numerical global stability analysis in the presence of disturbances and/or plant/model mismatch.

In the case $n = 1$, $m = 1$, and $v(t) = v =$ constant, the process becomes:

$$y(t) + \alpha_1 y(t-1) = \beta u(t-1) + v. \tag{5}$$

Let the estimate of $\beta$ be fixed to $\hat{\beta}$; the estimator will then provide the estimate, $\hat{\alpha}_1$, of $\alpha_1$ and the control algorithm becomes:

$$\hat{\alpha}_1(t) = \hat{\alpha}_1(t-1) + \frac{Py(t-1)(y(t) - y^*(t))}{c + y(t-1)^2} \tag{6}$$

$$u(t) = \frac{y^*(t) - \hat{\alpha}_1(t)y(t)}{\hat{\beta}}. \tag{7}$$

Choosing $y^*(t) =$ constant$\neq 0$, we can set $y^* = 1$ without loss of generality. Redefining the variables to be $p = \beta P/\hat{\beta}$, $\gamma = v + \beta/\hat{\beta}$, $y_n = y(t)$, and $z_n = \alpha_1 + \beta\hat{\alpha}_1(t)/\hat{\beta}$, we obtain the nondimensional form of the map $T : R^2 \to R^2$:

$$y_{n+1} = -y_n z_n + \gamma \tag{8}$$

$$z_{n+1} = z_n + \frac{py_n(-z_n y_n + \gamma - 1)}{c + y_n^2}. \tag{9}$$

## 3.  Preliminary analysis and the inverse map

The period-one fixed point of the map (the set point of the process) is located at $y = 1, z = \gamma - 1$ and its stability is determined by the eigenvalues $\lambda_{1,2}$ of its linearization evaluated at the fixed point [6]:

$$\lambda_{1,2} = \frac{1}{2}\left[ 2 - \gamma - \frac{p}{1+c} \pm \sqrt{(2 - \gamma - \frac{p}{1+c})^2 + 4(\gamma - 1)}\;\right]. \tag{10}$$

When $\gamma > 0$ the set point is stable; it loses its stability via a Hopf bifurcation when $\gamma$ crosses zero. However, global bifurcation analysis reveals that other attractors can coexist with the stable set point. Figure 1 is a one-parameter bifurcation diagram showing the coexistence of a marginally linearly stable (but attracting) set point with stable period-3 points for a wide range of values of the (scaled) estimator gain $p$. In this diagram $\gamma = 0$, $c = 0.5$ which, unless noted otherwise, will be the values used in the rest of the paper. At the same parameter values there exists a third solution to the map which is a continuous curve of degenerate fixed points at $y = 0$. These fixed points have one eigenvalue equal to unity and the other within (outside) the unit circle when $|z| < 1$ ($|z| > 1$, respectively).

The mere knowledge of the existence of multistability is not sufficient to assess the global stability characteristics of a system since the coexisting stable solutions can be so far away from the set point in phase space as to be unimportant in practice. We must quantify the stability of the various attractors to finite-amplitude perturbations by constructing their *basins of attraction* (the set of all points of the phase plane that asymptotically approach the attractor under forward iteration). In particular we are interested in the basin of attraction of the set point.

A brute-force method of constructing the basin of attraction is to set up a mesh of points in the phase plane, iterate each point forward, and mark the location of the initial point according to the attractor it asymptotically approaches. The results of this approach can be seen in Fig. 2 (reported in [6]) which provided the motivation for this work. All points in the black area belong to the basin of attraction of the stable period-3 solution; points in the shaded region approach the (period-1) set point; while all points in the white region get attracted to the degenerate solution $y = 0, |z| \leq 1$. It

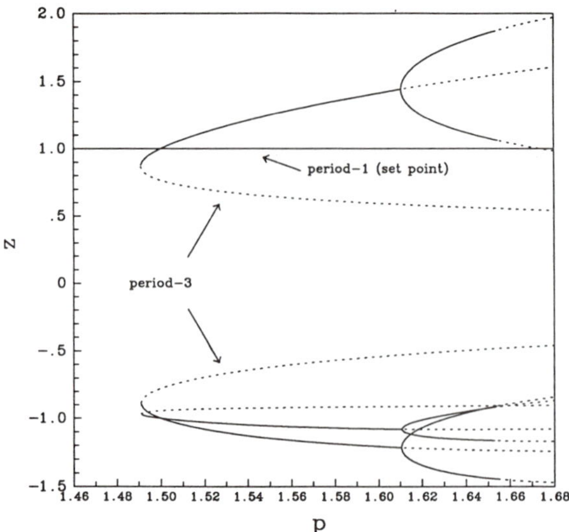



Figure 1. Partial bifurcation diagram for $c = 0.5$ and $\gamma = 0$. The period-3 solution is born at a saddle-node bifurcation and undergoes a cascade of period doublings as $p$ is increased, the first of which is shown. Broken lines indicate unstable solutions (saddle type). Branch crossings in this diagram are due to the projection and do not correspond to bifurcations.

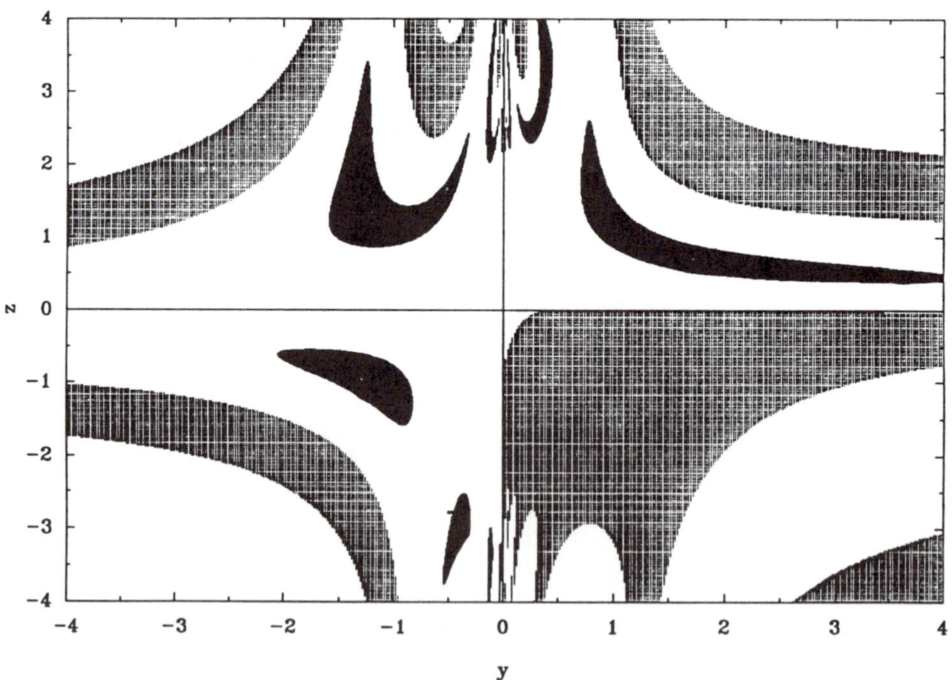



Figure 2. "Brute-force" calculation of the basins of attraction ($p = 1.5$, $c = 0.5$, $\gamma = 0$ from Golden and Ydstie). The color convention is discussed in the text.

can be seen from the figure that the basins of attraction are disconnected domains with "islands" belonging to the basin of attraction of the period-3 solution accumulating close to the upper and lower parts of the $z$-axis. This is different from what is observed for systems modeled by ODEs or for discrete-time *invertible* systems where the basin of attraction usually consists of a connected domain (which may or may not have a very complicated or even fractal structure: e.g., [14]). We will refer to a discrete-time system as being *invertible* if *all* phase space points have a single-valued inverse. All other discrete-time systems $S$, i.e., those for which the inverse map $S^{-1}$ may not exist or may not be unique, will be called *noninvertible*.

While the brute-force scheme is easy to program, considerable computational resources are required to obtain good resolution. Furthermore, it does not provide much insight into how the basins are formed. We know that for continuous-time or invertible discrete-time systems, the boundaries between the basins of attraction of coexisting attractors often consist of the stable manifolds of saddle-type, fixed or periodic points [10]. Global approximations of the stable manifolds can be computed with a two-step algorithm:

- Analytically or numerically approximate the local stable manifolds;

- Iterate the local approximations backward in time.

Invariant manifolds, or at least analytical expressions for the initial segments, can also be found as solutions of a functional equation by power series expansions (e.g., [3]).

The numerical methods of manifold construction as well as the shape and interactions of basin boundaries can become more complicated when each point of the phase plane has zero or more than one *preimage* (the preimages of a current state are the initial conditions from which the current state resulted). For these noninvertible maps, iterating the local stable manifold of a saddle point backward in time will not necessarily give rise to a (global stable) *manifold*: the results of the iteration may not lie on a continuous curve [11]. In this case we will call the set of points asymptotically attracted to the saddle forward in time the *stablemanifold* (one word as opposed to two) of the saddle (this term was proposed by R. P. McGehee). Invariant objects are invariant under both forward and backward iterations for invertible systems. Obviously this notion must be extended when considering noninvertible systems. McGehee [1991] provides a systematic framework for the study of noninvertible systems in the context of iterations of relations; the term $\star$-*invariance* is coined there to describe invariance under the inverse map.

A point $X_n \in R^2$ is called the *rank-r preimage* of the point $X_{n+r}$: i.e., $X_n = T^{-r} X_{n+r}$ is the image of $X_{n+r}$ under the period-$r$ inverse map. The difficulty in constructing the stablemanifolds lies in book-keeping, since the number of preimage branches of a local stable manifold can grow exponentially fast with $r$. Since the preimage behavior varies with the location in phase plane, our main tool will be understanding the inverse map and the "bifurcations" that mark regions of phase plane with qualitatively different preimage behavior. It must be stressed that these "bifurcations" are obtained at constant parameter values; it is the location in the phase plane that varies.

Inverting the $T$ map gives a cubic equation for $z_n$ with coefficients that are functions of the location in phase space (for fixed parameter values):

$$z_n^3 + a_2 z_n^2 + a_1 z_n + a_0 = 0 \tag{11}$$

$$a_2 = a_2(y_{n+1}, z_{n+1}; \gamma, p, c) = -z_{n+1}$$

$$a_1 = a_1(y_{n+1}, z_{n+1}; \gamma, p, c) = \frac{(\gamma - y_{n+1})(\gamma - y_{n+1} + p(y_{n+1} - 1))}{c}$$

$$a_0 = a_0(y_{n+1}, z_{n+1}; \gamma, p, c) = -\frac{(\gamma - y_{n+1})^2}{c} z_{n+1}$$

to be solved for $z_n$, and subsequently

$$y_n = \frac{\gamma - y_{n+1}}{z_n}. \tag{12}$$

The cubic equation can have one to three solutions and a "bifurcation" curve separates the set of points for which three real preimages exist ($3RP$ region) from the region where a single real preimage exists ($1RP$ region). This curve can be found explicitly by setting the discriminant of (11) to zero or by computing the image of $J_0$, the curve in phase space where the determinant of the Jacobian of the $T$ map vanishes. By taking the forward and backward images of $J_0$ we find the *critical curves* $J_n, n = 0, \pm 1, \ldots$ (in the terminology of Gumowski and Mira, [1980], who coined the term as a natural generalization of the critical points in the sense of Julia-Fatou for maps of the interval).

There are only two solutions to the cubic equation (11) on the curve $J_1$, one of which is a double, real root. This double solution (which is a double preimage of the point of interest) lies on the curve $J_0$; the third root lies on the additional preimage $J_0^*$ of $J_1$. We know that each time a stablemanifold curve enters the $3RP$ region, additional preimages of the curve appear elsewhere in the phase plane. Figure 3 shows the basin boundaries of the period-3 fixed points and illustrates their position relative to the underlying critical curves $J_0$, $J_0^*$ and $J_1$. We recall that the coefficients of the cubic equation are functions both of the parameters of the map and of the position in phase space. This means that the shape and location of the critical curves change with the system parameters.

We note that a solution to equation (12) does not exist (or can be thought of as located at infinity) when $z_n = 0$. This is a case of what we have called "bifurcation from infinity" [1]: stablemanifold segments crossing $z_n = 0$ have preimage segments that stretch to plus or minus infinity. Because no manifold crosses $z_n = 0$ in Fig. 2, this phenomenon will not be discussed further in this paper.

## 4. Basin boundaries and critical curves

### 4.1 The period three basin

To construct the basin of attraction of the stable period-3 solution, we compute the local approximation of the period-3 saddle point ($\mathcal{A}$) stable manifold and systematically compute its preimages. This is accomplished by taking initial segments along both sides of the stable eigenvector of the saddle point $\mathcal{A}$ and computing their (potentially 27 but, as we will see, actually 7) rank-3 preimages. We obtain what looks like an "island" that confines one of the stable period-3 foci. Note that in the following discussion, we will use the same letter ($\mathcal{A}$, $\mathcal{B}$, etc.) to denote both points in phase space and the islands on whose boundaries they lie.

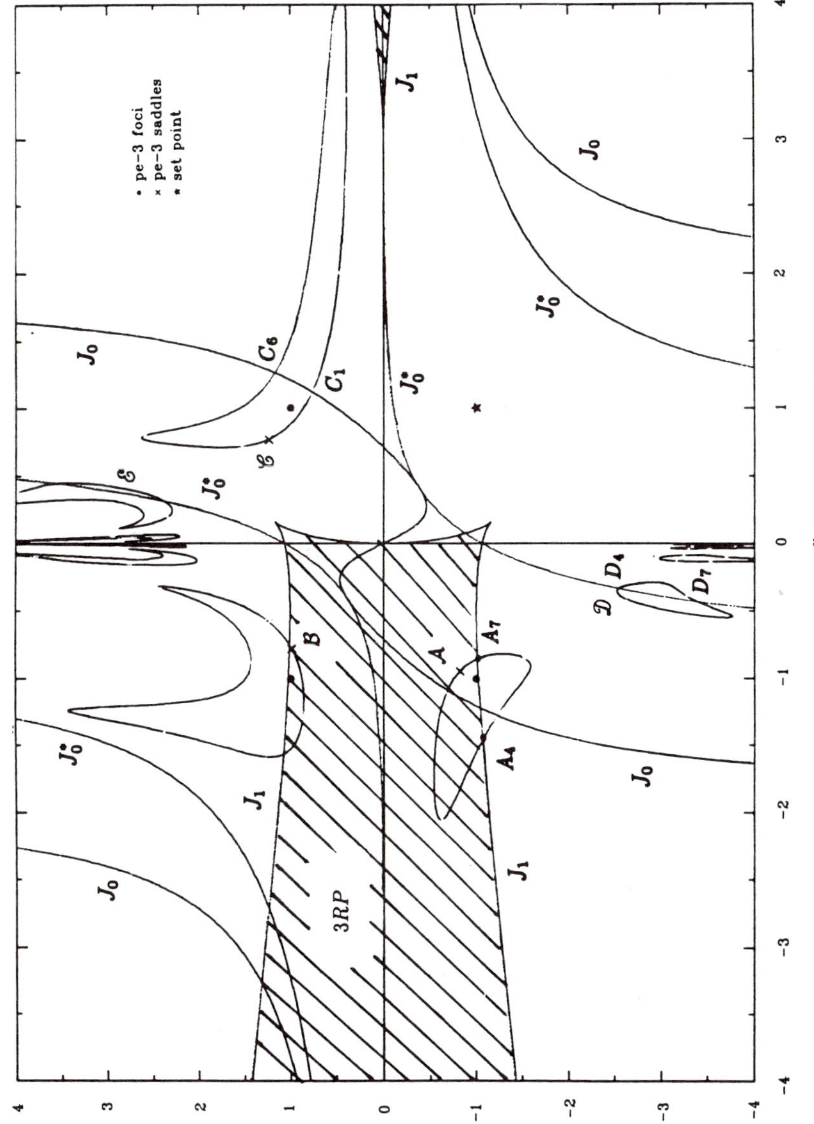


Figure 3. Critical curve $J_1$ and its preimages $J_0$ and $J_0^*$. The hatched area is the multi-inverse region. The basin boundaries of the stable period-3 solution are also included to illustrate their interaction with the critical curves. The forward map takes $\mathcal{A}$ to $\mathcal{B}$ to $\mathcal{C}$ and back to $\mathcal{A}$ again.

At first sight, it appears from fig. 3 that the boundary of island $\mathcal{A}$ is a smooth closed curve with a segment $A_4AA_7$ in the $3RP$ region. This means that if we take the rank-1 preimage of the entire island, the part in the $3RP$ region will have three preimages: two of them form the elongated island associated with saddle point $\mathcal{C}$, and the third, together with the unique preimage of the $A_4A_7$ segment, is the contracted island $\mathcal{D}$ that is not related to any of the saddle points. Two of the preimages of $A_4AA_7$ glue together at the double points $C_1$ and $C_6$ which must lie on $J_0$. The "other" preimages $D_4$ and $D_7$ of $A_4$ and $A_7$ will of course lie on the "other" preimage $J_0^*$ of the $J_1$-curve.

Both preimages $\mathcal{C}$ and $\mathcal{D}$ of the island $\mathcal{A}$ lie entirely in the $1RP$ region, and will themselves consequently have single preimages. Indeed, island $\mathcal{D}$ will be mapped by the inverse map to the even smaller island that lies next to it. This can be continued apparently forever giving an infinity of islands that accumulate close to the lower part of the $z$-axis from the left, without any of them ever reentering the $3RP$ region. The single preimage of island $\mathcal{C}$ on the other hand, is the island $\mathcal{B}$ whose boundary contains the local stable manifold of the saddle point $B$, and lies partly in the $3RP$ region. Island $\mathcal{B}$ is mapped by the inverse map to both island $\mathcal{A}$ (thus closing the period-3 loop), and to island $\mathcal{E}$ which is not associated with a saddle point. Island $\mathcal{A}$ comes from the glueing of the two preimages of the part of $\mathcal{B}$ in the $3RP$ region. The "other" preimage of this part of $\mathcal{B}$ along with the unique preimage of the remainder of $\mathcal{B}$ yields $\mathcal{E}$, which lies entirely in the $1RP$ region, and which backward in time gives rise to more islands that alternate on both sides of the upper part of the $z$-axis, all of them lying in the $1RP$ region.

This description of Fig. 3 explains the disconnected structure of the stable period-3 basin of attraction if we consider preimages of the entire $\mathcal{A}$ island. However, it is important to see how this picture develops if we start with local stable manifold approximations in the vicinity of the $\mathcal{A}$ saddle. This is shown in Fig. 4 and its enlarged details, Figs. 4a-4e.

Closer inspection of the boundary of the island $\mathcal{A}$ in Fig. 4a shows that what appears to be a smooth closed curve actually consists of different segments and their excess preimages that join smoothly at points $A_j$, $j = 1, ..., 7$. The evolution on these segments in the figure is indicated by the arrows. In the insets of Fig. 4, the negative exponents of marked points indicate the rank of the preimage, while the primes differentiate between the (up to three) preimages –roots of the cubic equation (11)– of each point.

Saddle point $A$ lies in the $3RP$ region and so does every point on its local stable manifold. There exist three rank-1 preimages of $A$: one of them is $\mathcal{D}$ (fig. 4d), while the other two, $C$ and $C_2$ lie on island $\mathcal{C}$ (fig. 4c). Since all these preimages lie in the $1RP$ region, there also exist three rank two preimages of $A$. One of them comes from $\mathcal{D}$, and its unique backward trajectory never approaches the island $\mathcal{A}$ again. The other two are the saddle point $B$ and $B_9$ (fig. 4b). There exist therefore five rank-3 preimages of $A$: one lies on the unique backward trajectory of $\mathcal{D}$; one comes from $B_9$, lies on island $\mathcal{E}$ (point $E_9$, fig. 4e) and again has a unique backward trajectory; $B$ gives rise to the remaining three rank-3 preimages of $A$. One of them also lies on island $\mathcal{E}$ (point $E$, fig. 4e) and, again, has a unique backward trajectory. The final two are the ones most relevant to island $\mathcal{A}$. One of them is, of course, the saddle $A$ itself; the other is point $A_2$ (fig. 4a).

Having discussed the backward trajectory of $A$, it becomes now easier to study the backward trajectory of its local stable manifold. Segment $AA_1$ of the stable manifold

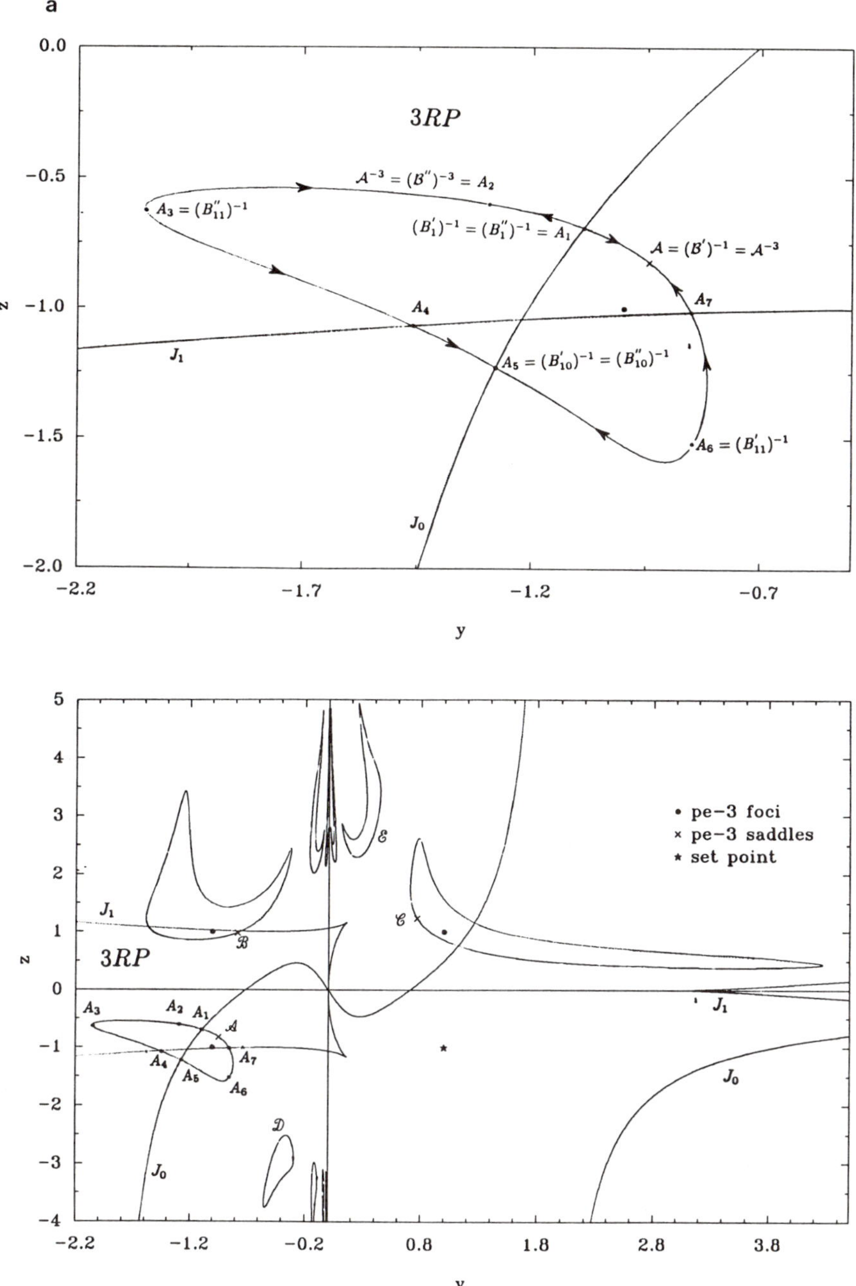



Figure 4. The detailed structure of the basin boundaries of the period-3 solution. Figures 4a-4e show enlargements of the main "islands" labelled in Fig. 4. The negative exponents of marked points indicate the rank of the preimage, while the primes differentiate between the (up to three) preimages of each point.

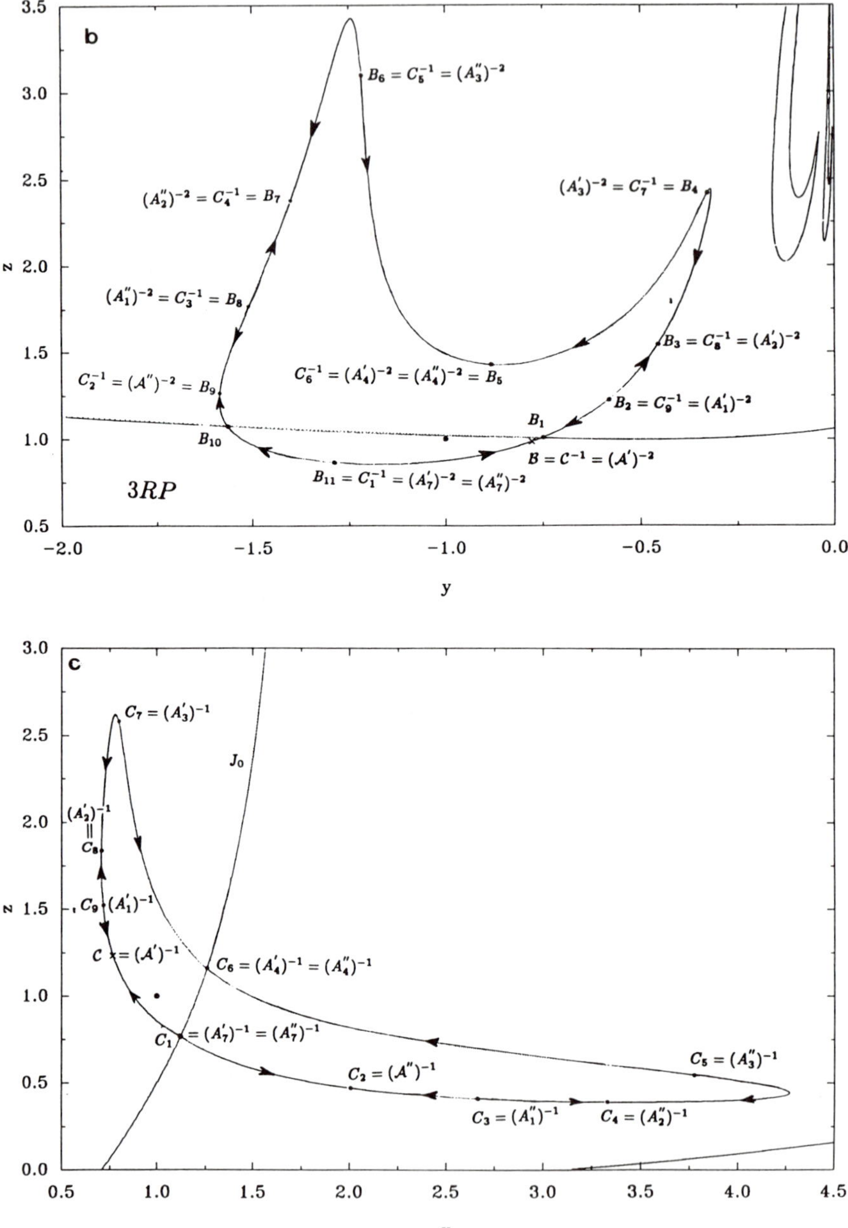

Figure 4.(continued) Islands $\mathcal{B}$, $\mathcal{C}$.

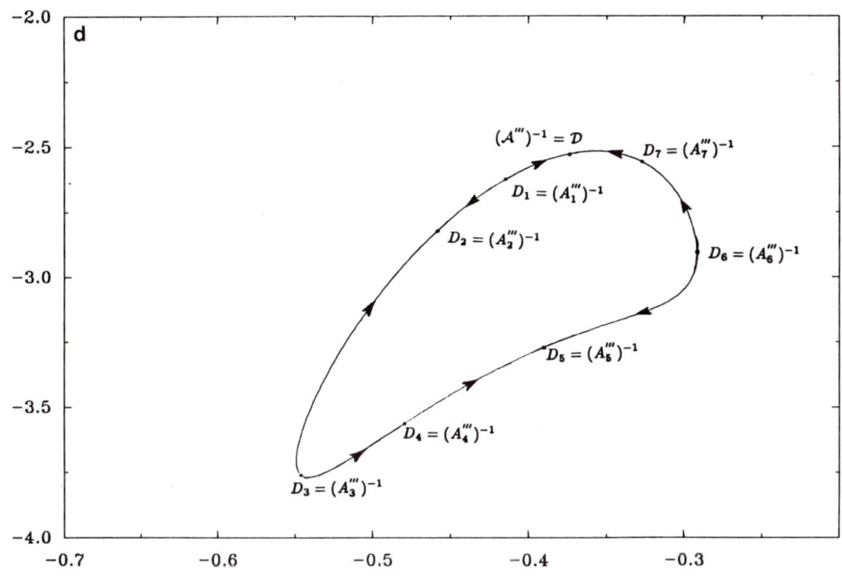

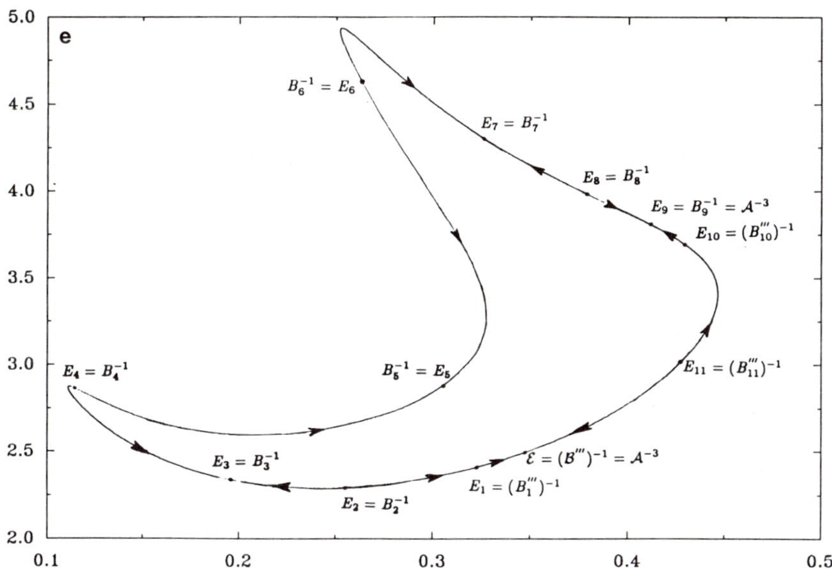

Figure 4.(continued) Islands $\mathcal{D}$, $\mathcal{E}$.

coexists with an excess rank-3 preimage, the segment $A_1 A_2$. Following a point receding along the immediate segment from $\mathcal{A}$, we see that it approaches its excess preimage and meets it at point $A_1$. Points $A_1$, $A_3$, $A_5$, and $A_6$ are boundary points of two distinct contiguous segments with different evolution directions and are called *points of alternance* [11]. They lie on critical curves (in our case $A_1$ and $A_5$ on $J_0$, $A_3$ and $A_6$ on $J_{-2}$). Preimages of points $A_j$, $j = 1, ..., 7$ lie on the other islands that form the boundary of the total basin of attraction of the stable period-3 focus and define analogous points on them. Consequently, the stablemanifold is only piecewise continuous and contains segments that join smoothly at, or start from, points of alternance. The rank-3 preimages of the $\mathcal{A} A_7$ segment "round-out" the island $\mathcal{A}$. This involves two points of alternance ($A_3$ and $A_6$, Fig. 4a) both rank-3 preimages of $A_7$.

## 4.2 The set point basin

Let us now turn to the basin boundary of the set point. For $\gamma < 0$ the set point is unstable and is surrounded by a concentric pair of invariant circles, the inner stable and the outer unstable. This pair of invariant circles appears at a "saddle-node bifurcation" (a complicated sequence of bifurcations studied by Chenciner [2]) occurring over an interval of $\gamma$ values. In Fig. 5a, the unstable invariant circle defines the basin of attraction of the stable one. At $\gamma = 0$, the stable invariant circle shrinks onto the set point during a Hopf bifurcation, rendering the set point stable; it is the basin of attraction of this marginally stable set point that constitutes the shaded region in Fig. 2. The unstable invariant circle can be computed by iterating points in the neighborhood of the set point backwards.

It is easier to compute and describe the changes in the shape of the invariant circle as $\gamma \to 0$ by studying them qualitatively for constant $\gamma$ and varying $p$. This is seen in Figs. 5a-5c, for $\gamma = -0.04$. The invariant circle grows with $p$ by forming an increasing number of folds close to the $z$-axis. As it grows, it first enters the $3RP$ region by crossing its lower tip (fig. 5b and inset, $p = 1.35$). In the inset we see that the three preimages of the segment $AB$ of the invariant circle in the $3RP$ region are glued end-to-end at the double preimages of $A$ and $B$ on the $J_0$ curve; the integrity of the invariant circle is preserved.

For $p = 1.4$ (fig. 5c) the invariant circle has also entered the part of the $3RP$ region that reenters the phase plane close to the $y$-axis on the right. The portion of the circle in this region has one preimage on the circle and two additional preimages glued together on both ends to form new islands. These islands in turn can also have parts lying in the $3RP$ region, giving birth to more and more islands, the final effect being the complicated basin illustrated in Fig. 2.

The other attracting solution that exists at these parameter values is the degenerate solution $y = 0$, $-1 \leq z \leq 1$. The eigenvalues for these fixed points are $\lambda_1 = +1$ and $\lambda_2 = -z$ and their basin of attraction is the white area in fig. 2. This solution exists only for $\gamma = 0$ and disappears for $\gamma \neq 0$. For negative values of $\gamma$ a period-2 solution is born from the upper boundary $(y, z) = (0, 1)$ (for which $\lambda_2$ was equal to -1 at $\gamma = 0$), while for $\gamma > 0$ there is a sudden expansion of the basin of attraction of the set point with the addition of the basin of the degenerate solution (fig. 6a). The first bifurcation looks like a period doubling but there is no particular period one fixed point associated with it for $\gamma \neq 0$. The basins of attraction in this case now become very intricate (much more complicated and at an even finer scale than the resolution in fig. 6b suggests, with islands belonging to the basin of attraction of one attractor appearing

206

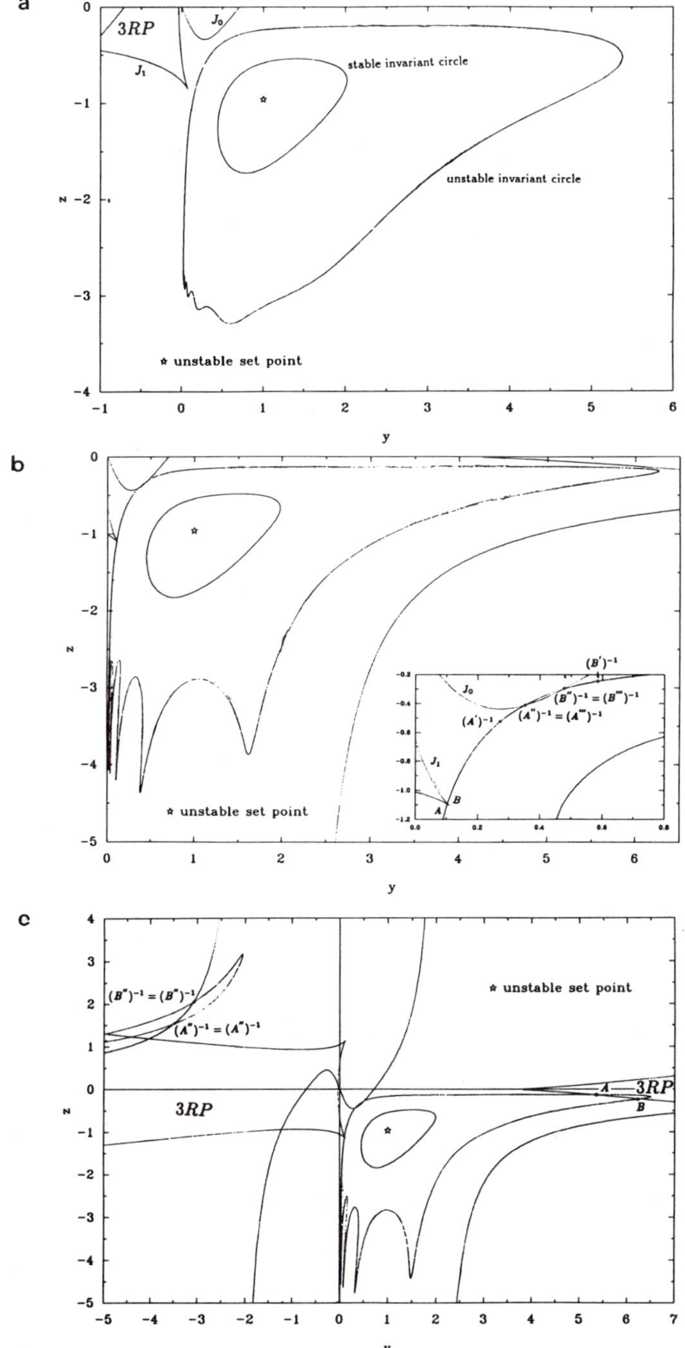



Figure 5. The basin of attraction of the set point. Figure 5a ($\gamma = -0.04, p = 1, c = 0.5)$) shows the unstable invariant circle constituting the boundary of the basin of the stable invariant circle (entirely in the $1RP$ region). Figure 5b ($\gamma = -0.04, p = 1, c = 0.5$) with the expanded inset show the initial interaction of the unstable invariant circle with the $3RP$ region. Figure 5c ($\gamma = -0.04, p = 1.4, c = 0.5$) shows a further interaction with the $3RP$ region giving rise to additional isolated basin segments.

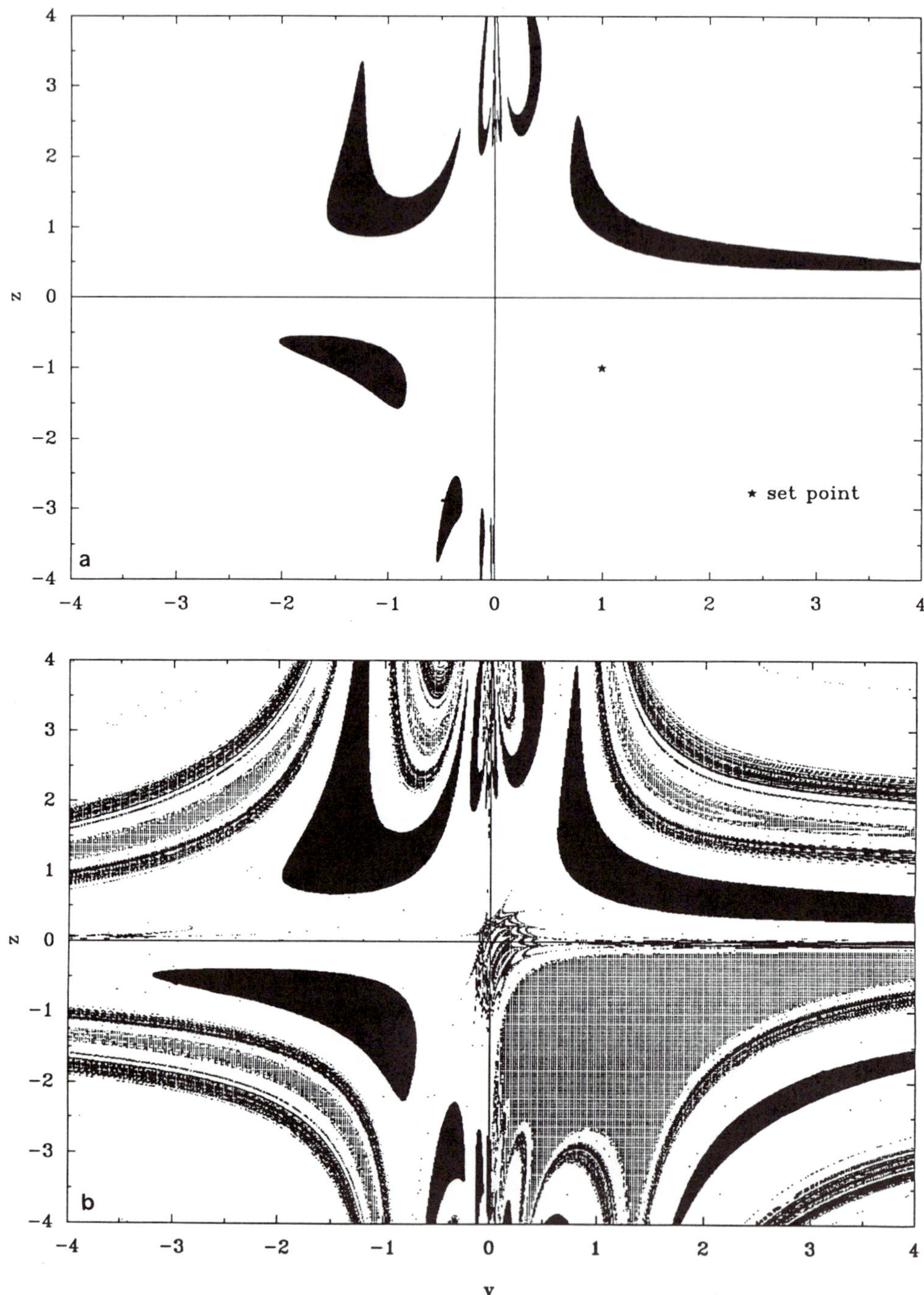

Figure 6. Representative basin structures for $\gamma \neq 0$. In figure 6a for $\gamma > 0$ ($\gamma = 0.001, p = 1.5, c = 0.5$) two attractors coexist: the period-3 and the stable set point. In figure 6b for $\gamma < 0$ the white regions are attracted to a stable period-2 ($\gamma = -0.05, p = 1.5, c = 0.5$).

*within* islands of the basin of attraction of another) since large parts of the islands and also their preimages repeatedly enter the $3RP$ region. Similarly, the sudden expansion of the basin of the set point can be considered as a non-classical global bifurcation [15] –a "crisis" in the terminology of Grebogi *et al.*[9].

# 5. Conclusions

Adaptive control systems are not only inherently nonlinear but, for the case of discrete-time control, can also be noninvertible. The first property can give rise to phenomena like multistability: coexistence of a locally stable set point with other undesirable attractors. Quantifying the stability of the set point involves the construction of its basin of attraction and, since basin boundaries often consist of stable manifolds of saddle-type solutions, the asymmetry in time that noninvertibility introduces can have a profound effect on the shape and interactions of these basins.

"Stablemanifolds" may break off to several pieces as they interact with regions with different preimage behavior, giving rise to disconnected basins. Critical curves play an important role in organizing the branching off of inverse iterates. The interaction of special orbits or families of orbits in phase space (attractors, saddles, invariant or $\star$-invariant curves) with the critical curves determines the transitions that basin boundaries can undergo as the system parameters vary.

**Acknowledgements:** The work of C.E.F. and I.G.K. was partially supported by an NSF PYI award (CTS–8957213), a David and Lucile Packard Foundation Fellowship to I.G.K. and by Shell Development Company. R.A.A acknowledges the support of the SRC throught NSF grant ECD–8803012-06.

# References

[1] R. A. Adomaitis and I. G. Kevrekidis. Noninvertibility and the structure of basins of attraction in a model adaptive control system. *J. Nonlinear Sci.*, 1:95–105, 1991.

[2] A. Chenciner. Bifurcation de points fixes elliptiques. ii. Orbites periodiques et ensembles de Cantor invariants. *Inventiones mathematicae*, 80:81–106, 1985.

[3] V. Franceschini and L. Russo. Stable and unstable manifolds of the Hénon mapping. *J. Stat. Phys.*, 35:757–769, 1981.

[4] C. E. Frouzakis, R. A. Adomaitis, and I. G. Kevrekidis. Resonance phenomena in adaptivelly controlled systems. *Int. J. Bifurcations and Chaos*, 1:83–106, 1991.

[5] M. P. Golden and B. E. Ydstie. Bifurcation in model reference adaptive control systems. *Systems & Control Letters*, 11:413–430, 1988.

[6] M. P. Golden and B. E. Ydstie. Parameter drift in adaptive control systems: Bifurcation analysis of a simple case. *IEEE Conf. Decision and Control*, Dec. 1991, (in press).

[7] G. C. Goodwin, P. J. Ramadge, and P. E. Caines. Discrete-time multivariable adaptive control. *IEEE Trans. Autom. Contr.*, 25:449–456, 1980.

[8] G. C. Goodwin and K. S. Sin. *Adaptive Filtering, Prediction and Control.* Prentice-Hall, Englewood Cliffs, NJ, 1984.

[9] C. Grebogi, E. Ott, and J. A. Yorke. Basin boundary metamorphoses: changes in the accessible boundary orbits. *Physica D*, 24:243–262, 1987.

[10] J. Guckenheimer and P. Holmes. *Nonlinear Oscillations, Dynamical Systems, and Bifurcations of Vector Fields, Applied Mathematical Sciences*, vol 42. Springer-Verlag, 1983 (Third printing).

[11] I. Gumowski and C. Mira. *Recurrences and Discrete Dynamic Systems, Lecture Notes in Mathematics*, vol 809. Springer-Verlag, 1980.

[12] I. M. Y. Mareels and R. R. Bitmead. Nonlinear dynamics in adaptive control: chaotic and periodic stabilization. *Automatica*, 22:641–655, 1986.

[13] I. M. Y. Mareels and R. R. Bitmead. Nonlinear dynamics in adaptive control: chaotic and periodic stabilization -II. *Automatica*, 24:485–497, 1988.

[14] W. McDonald, C. Grebogi, E. Ott, and J. A. Yorke. Fractal basin boundaries. *PhysicaD*, 17:125–153, 1985.

[15] C. Mira. *Chaotic Dynamics.* World Scientific Publishing Co. Pte. Ltd., 1987.

[16] R. P. McGehee. Attractors for closed relations on compact Hausdorff spaces. Preprint, University of Minnesota, 1991.

# A METHOD FOR ADAPTIVE CONTROL AND LEARNING IN CHAOTIC SYSTEMS


Dimitris Vassiliadis

Department of Physics, University of Maryland
College Park, MD 20742, USA



*Abstract:* A method is presented which can control a dynamical system through its parameters and "set" it on a desired pattern of behavior (periodic, quasiperiodic, chaotic, etc.) in spite of external perturbations. The input for the control is a comparison of the observed and the desired dynamical behaviors and does not involve the parameters explicitly. The system is brought to the desired regime exponentially fast and the length of the transient behavior is inversely proportional to the strength of the parameter control. This general method is exactly analogous to a "learning" algorithm, and this is shown to be useful for parameter identification of an unknown system.


## INTRODUCTION

The parameters of a dynamical system account for its global behavior and stability in each one of the different basins of attraction in the phase space. A change in the parameter settings can bring about a dramatic change in the observed behavior, e.g. if it takes the system across a bifurcation or a catastrophe. For physical systems a parameter fluctuation can be induced by a thermal or electric perturbation, some other form of external noise, an accident, etc, and it will affect the overall behavior and performance of the system.

This paper describes a method which controls a dynamical system by manipulating its parameters, therefore the control is called *adaptive* [1]. In this way it can suppress parametric fluctuations such as mentioned above. It will be shown below that the absolute values of the parameters need not be known, but only the relative changes brought about by the control. The only information about the system is in terms of the evolution of one or more observed dynamical variables. Once the control is activated it draws the system to the desired or *goal* pattern of behavior (periodic of a certain order, quasiperiodic, chaotic). As one would imagine one of the most interesting cases is when the goal behavior is chaotic, i.e. unstable due to sensitivity to initial conditions. In all cases investigated so far the control was effective irrespective of the goal behavior. (A less general technique [1,2] is limited to steady-state and periodic behavior.) In all these cases the system "learns" a pattern of behavior by changing its parameters until they reproduce the desired pattern of activity.


## OUTLINE OF THE METHOD

Consider a general discrete dynamical *System* described by a map (see also Fig. 1):

$$X_{n+1} = F(X_n; \mu) \tag{1}$$

(The treatment of flows, $\dot{X} = F(X; \mu)$, is analogous.) $X$ contains the dynamical variables, all of which are assumed to be observed (this assumption will be relaxed below). The scalar $\mu$ stands for the single parameter to be controlled; however, the method can be extended to control more than one parameters. The exact form of $F$ may be unknown; the System can simply be represented by a "black box" which returns a unique output given a well-defined input. For instance, in the case of a nonlinear frictionless oscillator the energy is a parameter whose value characterizes the pattern of behavior while the displacement is the variable. For a forced pendulum with dissipation the time-averaged energy is a parameter which characterizes periodic, quasiperiodic, or chaotic behavior.

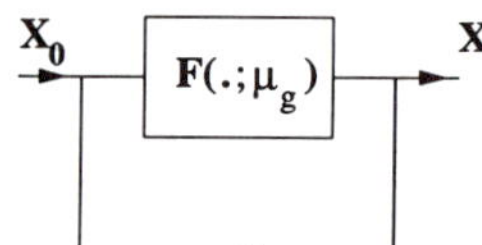


*Figure 1. A "black-box" representation of a dynamical system.*


Assuming that one or more of these parameters are accessible, their values can be changed in a controlled fashion if the dynamic variables are detected not to produce the desired pattern of behavior. For this to happen the observed output has to be compared to a reference signal, $Y_n$, which comes from a *Model*:

$$Y_{n+1} = F_M(Y_n; \mu_g) \tag{2}$$

Here the Model is supposed to be adequately accurate, i.e. it is able to mimic the System under the goal external conditions and parameter regime, $\mu_g$. The Model's functional

$\mathbf{F_M}$ need not be the same as the System's. Also the parameter $\mu_g$ does not need to appear explicitly. For example, a simple harmonic oscillator can simulate the behavior of a certain real pendulum as long as the oscillation energy is small enough. Under these conditions the functional relation for a linear Model will approximate well the nonlinear functional of the System.

One can obtain the behavior *expected* from the System by directing its observed output $\mathbf{X}$ in the Model equation:

$$\mathbf{Y_{n+1}} = \mathbf{F_M}(\mathbf{X_n}; \mu_g) \tag{3}$$

as illustrated in Fig. 2a. This is the behavior that *would be* observed at the next iteration (n+1), *were* the system already at the goal regime. Notice that, unlike the System's $\mu$, the Model parameter remains fixed at its desired value, $\mu_g$. If the System parameter $\mu = \mu_g$ then, given the same initial conditions, both will display the same behavior.

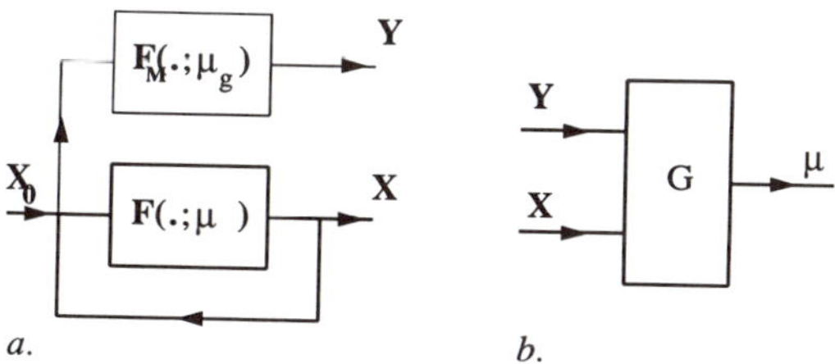


*Figure 2. a. Adaptive control of a System ($\mathbf{F}$) through its parameter $\mu$ by comparing its output to an "expected" or goal value coming from a model, $\mathbf{F_M}$. The comparison (b) is realized by a function G which supplies a correction for $\mu$.*

However, if $\mu \neq \mu_g$ then the same initial conditions in (1) and (3) will in general produce different observed behaviors. It can be shown that the System parameters can be changed towards the desired values by:

$$\mu_{n+1} = \mu_n + \alpha G(\mathbf{X_n} - \mathbf{Y_n}) \tag{4}$$

where the scalar $\alpha$ sets the strength of the parameter change (Fig. 2b). The argument of the linear or nonlinear function G is the *error signal*, $\mathbf{X_n}$-$\mathbf{Y_n}$, the difference between observed and desired output. It may be obtained as the derivative of a Lyapunov function [3] with respect to its argument, $\mathbf{X_n}$-$\mathbf{Y_n}$. It should be noted that the actual goal parameter $\mu_g$ does not enter Eq. (4) explicitly; only the observed variables are needed. Using such a control function one can show that after an external perturbation changes the parameters, they are restored to their original values in a smooth and well-defined way. At the same time the System synchronizes [4] with the Model, $X_n \to Y_n$. Once the control is activated, both the error signal and the difference $\mu$-$\mu_g$ decrease exponentially. One can prove that the control is stable, and extract scaling laws for the length of the control transient [5]. The choice of the control ($\alpha$ and G) turns out to be quite flexible. The goal value is attained independently of the magnitude and sign of the original perturbation in the parameter —as long as global motion remains stable and the initial and goal behavior share the same basin of attraction.

## NUMERICAL EXPERIMENTS

The above method was tested in cases of maps and flows which are known to exhibit chaos. In each case both the System and the Model were represented by the same dynamical system; the Model was "tuned" to the right parameter value, away from which the System was initially perturbed. Then the System parameter were controlled using (4) until the observed behavior agreed with that of the target. In contrast to the previous section where all dynamical variables were assumed to be observed, here the system observed "output" is a single variable.

It is easy to visualize the changes in the dynamics in the case of a simpler system, such as the logistic map. In Fig. 3a) the variation of both outputs can be seen as the System converges to the goal behavior. The first picture shows a "order to order" transition: the initially two-periodic System is brought in a period-16 regime. Similarly an initially chaotic System is controlled towards a period-2 regime (Fig. 3b), or to a different chaotic regime (Fig. 3c). In the second case shown (Fig. 3b) the Model also seems to have an initial "chaotic" output (which converges to the periodic behavior as well), although its parameter ($\mu_g$) is fixed to a periodic value *throughout* the run. This irregular output is only due to the chaotic input that the Model receives. For this relatively simple dynamical system, as the parameter is changed, the transient behavior reproduces parts of the familiar bifurcation diagram. Towards the end of the transient the approach to the goal behavior slows down and the increased "exposure time" magnifies some otherwise hard to distinguish intervals and bifurcations (note, for instance, the seven-fold bifurcation on the right side of Fig. 3c which is usually invisible in the bifurcation diagram).

The control is illustrated more quantitatively in the case of the Henon map. The time series of the x-coordinate constitutes the system ($X_n$) and model ($Y_n$) scalar output. One of the parameters, b (the Jacobian), is perturbed and then controlled to return to its original value ($b_g = 0.3$) using $G = X_n$-$Y_n$. Fig. 4 shows a quantitative comparison of the magnitude of the error signal, $|X_n$-$Y_n|$, plotted together with $|X_n|$. The control is on from t=300 to t=1300. The error decreases exponentially until it reaches the accuracy of the computer for single- precision reals. At this point the length of the *control transient* is recorded and the parameter can be "locked"; unless other glitches occur, it will remain set to the model value. Different functional forms of G such as $G = (X_n$-$Y_n)^3$, $(X_n$-$Y_n)+(X_{n-1}$-$Y_{n-1})$, etc. were tried successfully. The fastest and most flexible control functions turned out to be the linear ones [5].

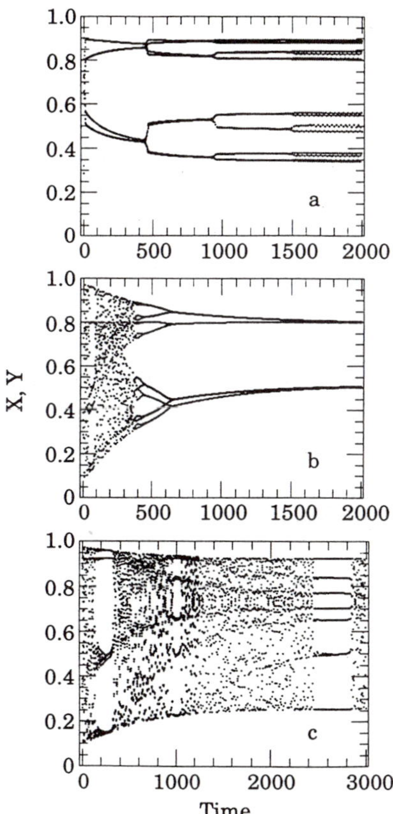

*Figure 3.  The system and model outputs are plotted together as the logistic map is controlled towards a goal activity.  At the end of the control transition the two are indistinguishable.  a. from period-2 ($\mu$=3.2) to period-16 (3.57); b. from chaos (3.9) to period-2 (3.2); c. from chaos (3.9) to chaos (3.7).*

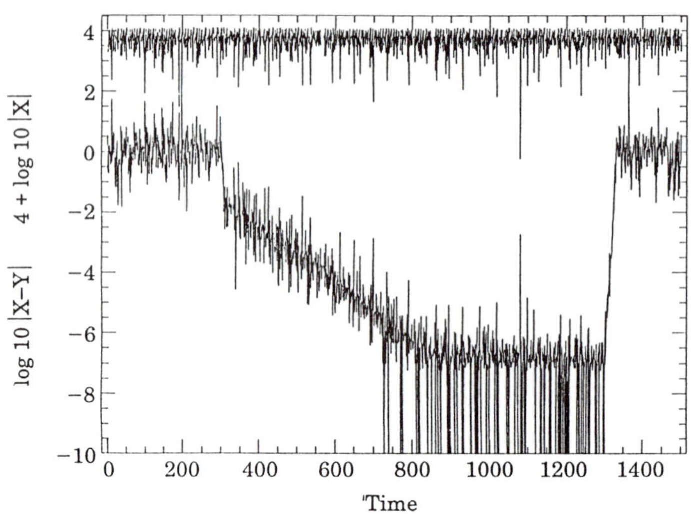

Figure 4. The Henon map parameter b is controlled (for 300<t<1300) until it reaches the model value, 0.3. The input of the linear control function is the difference between the system and model outputs, shown as a line on a logarithmic scale. The system output, X, is also plotted for reference, shifted up by 4 units for clarity.

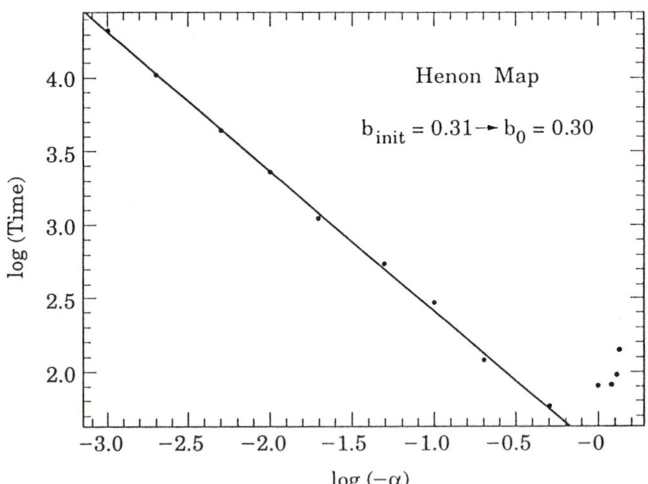



Figure 5. The transient (measured as the number of iterations) needed for the Henon map to reach the goal behavior is plotted versus the proportionality constant $\alpha$ (cf. Eq. 4).

The length of the control transient can be varied within certain limits by varying the strength $\alpha$ of the parameter change per iteration. In Fig. 5 the time is plotted as a function of $\alpha$, revealing an inverse power law. The slope is -0.95±0.17 and remains in that range for a variety of changes in the initial and other conditions. A study of the set of equations (1), (3), (4) shows that the motion is drawn to a fixed point with real negative eigenvalues, which minimizes the parameter deviation and enforces synchronization. A linearization around the fixed point allows to explain the exponential decrease of the error signal and the relation of the control transient to the strength of the parameter perturbation. As $\alpha$ is increased the control becomes faster until a bifurcation occurs at $\alpha=1/4$ where the fixed point is still stable, but the eigenvalues are now complex conjugates [5].

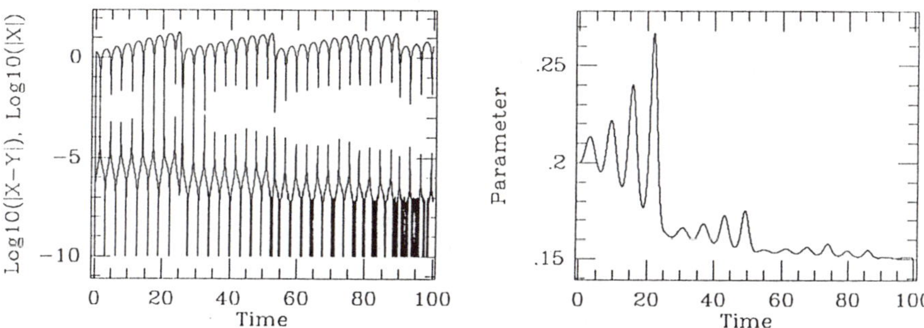

*Figure 6. Adaptive control of the Rossler system through its parameter r.*

The same technique was successfully applied in the cases of the sine map as well as the Lorenz and Rossler flows (Fig. 6). Control of more than one parameters was also tried successfully. Moreover, as outlined above the System and Model outputs were assumed to be sampled at every iteration (or at every integration timestep for continuous systems). The control was also applied at every iteration. In realistic cases, however, the sampling rate is usually much lower than that of internal dynamics. Therefore some of the numerical experiments were repeated with decreased sampling and control rates. Apart from the expected growth in the transient time scale, the method remained effective in controlling the system as desired.

Two limitations of the method were noise and inadequate model functionals, $\mathbf{F_M}$. Noise-induced fluctuations can appear not only in the parameters, but in the dynamic variables $\mathbf{X}$, as well. The comparison in (4) becomes difficult and sets a lower limit in the control accuracy. The one- and two-dimensional maps were more susceptible to the same levels of normalized additive noise than the higher dimensional flows. Additionally the model's ($\mathbf{F_M}$) accuracy will set a limit to the effectiveness of the control. The control was successful when the functional form of $\mathbf{F}$ was retained in $\mathbf{F_M}$, but model parameters (other than the control parameter $\mu_g$) were fixed to slightly different values than the system's. However, when $\mathbf{F_M}$'s functional form was distinctly different than the $\mathbf{F}$'s, the System was controlled towards a parameter value which was stable under variation

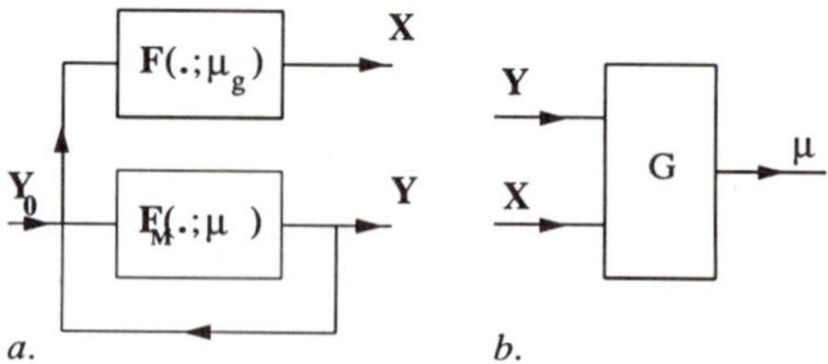


*Figure 7. a and b. (Compare to Fig. 2) By inverting the roles of the system and the model, the model parameters adjust to the settings of the system. The model "learns" how to imitate the system behavior.*

of initial perturbations, but different from $\mu_g$. Still, recent results show that models that come from knowledge of the system dynamics can successfully be replaced by models fit to time series. If the system has operated in the goal regime and a long, high S/N time series has been recorded, then the state space can be reconstructed [6] and a local model can be fit to it [7]. Linear models obtained in such a way are found to be adequate for this adaptive control and results will be reported elsewhere [5].

## PARAMETER IDENTIFICATION, OR "LEARNING"

The technique described above corresponds to a learning process. By comparing a pattern of behavior with a reference and correcting the dynamics generating the activity, the System learns to adapt itself and to perform as required. Similarly, by inverting the roles of the System and the Model (Fig. 7a), the model can be the one that adapts its activity until it imitates the behavior of the system (whose parameters are now fixed).

The same considerations apply for the control function G (Fig. 7b) as before. When the model has reached the goal behavior, its parameters have been modified to reflect those of the system. Thus knowledge is gained about the internal dynamics of the system. These cases of control and learning are strictly analogous. Each one of the numerical experiments shown above can be viewed either as a control or as a learning process.

## IN CONCLUSION

This method can be used to stabilize a system's activity in spite of external perturbations that cause parametric fluctuations. A variety of goals (periodic, chaotic, etc.) can be attained from different initial behaviors. It serves a different purpose in the control of chaos than other methods [8] which stabilize a specific trajectory which is unstable for a given parameter setting. In a real application this method can be useful in cases where a computer or other controlling device maintains a constant pattern activity of an experimental physical, chemical, or biological system by keeping its parameters at constant values, in spite of noise, thermal fluctuations, etc. The complementary procedure can be used to identify details of a system's internal dynamics, again by comparing its behavior with a model.

Moreover the method can serve as a paradigm for understanding principles of adaptive control in biological systems of higher complexity. Such systems are observed to adapt to parameter fluctuations caused by environmental noise of various origins, so that the character of their behavior is preserved [2]. This *self-regulation* is effected through internal dynamics where adaptive control and synchronization as described here may play an important role.


*Acknowledgements*:    I would like to thank T. Shinbrot for his valuable comments and encouragement.


## References


1.  B. A. Huberman and E. Luber, *IEEE Trans. Circ. Syst.*, 547 (1990)
2.  S. Sinha, R. Ramaswamy, and J. S. Rao, *Physica D 43*, 118 (1990).
3.  M. Vidyasagar, *Nonlinear Systems Analysis* (Prentice-Hall, Englewood Cliffs, NJ, 1978).
4.  L. M. Pecora and T. L. Carroll, *Phys. Rev. Lett. 64*, 821 (1990)
5.  D. Vassiliadis, *Physica D* (submitted).
6.  Packard, N. H., J. P. Crutchfield, J. D. Farmer, and R. S. Shaw, Geometry from a time series, *Phys. Rev. Lett. 45*, 712 (1980); F. Takens, Detecting Strange Attractors in Turbulence, in: A. Dold and B. Eckmann (Eds.), *Dynamical Systems and Turbulence, Warwick 1980*, Lecture Notes in Mathematics 898, Springer Verlag, Berlin, 1981
7.  J. D. Farmer and J. J. Sidorowitch, Exploiting chaos to reduce noise and predict the future, in: Y. C. Lee (ed.), *Evolution, Ecology and Cognition*, World Sci., Singapore, 1989; also references in that article
8.  E. Ott, C. Grebogi, and J. A. Yorke, *Phys. Rev. Lett. 64*, 1196 (1990); T. Shinbrot, E. Ott, C. Grebogi, and J. A. Yorke, *Phys. Rev. Lett. 65*, 3215 (1990); W. L. Ditto, S. N. Rauseo, and M. L. Spano, *Phys. Rev. Lett. 65*, 3211 (1991).

GLOBAL BIFURCATIONS, NONLINEAR AND CHAOTIC SPATIO-TEMPORAL

DYNAMICS IN SEMICONDUCTOR HETEROSTRUCTURES

R. Döttling, E. Schöll and D. Reznik

Institut für Theoretische Physik
Technische Universität Berlin
Hardenbergstr.36
D-1000 Berlin 12

## ABSTRACT

The nonlinear dynamics of charge transport parallel to the layers of a modulation-doped semiconductor heterostructure is studied theoretically. For sufficiently large dc bias limit cycle oscillations of the current in the 100 $GHz$ range, and bistability and hysteretic switching transitions between periodic attractors and fixed points attractors are predicted. We present a detailed investigation of complex bifurcation scenarios displaying subcritical Hopf bifurcations, homoclinic bifurcations of an unstable limit cycle, cyclic folds of limit cycles by condensation of paths, 'canards' and 'phantom ducks'. In a double heterostructure transient chaos associated with a chaotic repeller is found.

## INTRODUCTION

Semiconductors driven far from thermodynamic equilibrium by an external electric field represent nonlinear dynamic systems which can exhibit complex nonlinear and chaotic behavior[1]. They are particularly suitable model systems for the study of nonlinear dynamics because they are easily accessible to experimental observation through electrical current and voltage measurement, and can be taylored to specific needs by advanced materials engineering technologies. In particular, layered semiconductor heterostructures are singled out because of their current importance both in basic research and in microelectronics applications.

In this paper we consider parallel charge transport in a modulation-doped $GaAs/Al_x$-$Ga_{1-x}As$ heterostructure as schematically shown in Fig.1a. The $Al_xGa_{1-x}As$ layer is heavily n-doped with donor density $N_D$ while the $GaAs$ layer is undoped. At low bias $U_0$ the electrons reside in the $GaAs$ channel where they are separated from their parent donors in the $Al_xGa_{1-x}As$ layer. Thus the mobility $\mu_1$ of the electrons in the $GaAs$ layer is high because of the reduced impurity scattering. An electric field parallel to the layer interface induces carrier heating. If the electrons gain enough kinetic energy, thermionic emission across the barrier (Fig.1b) into the $Al_xGa_{1-x}As$ layer is possible, where the mobility $\mu_2$ is much lower due to impurity scattering. This real space transfer of electrons from the high-mobility to the low-mobility layer causes an N-shaped current-voltage characteristic with a regime of negative

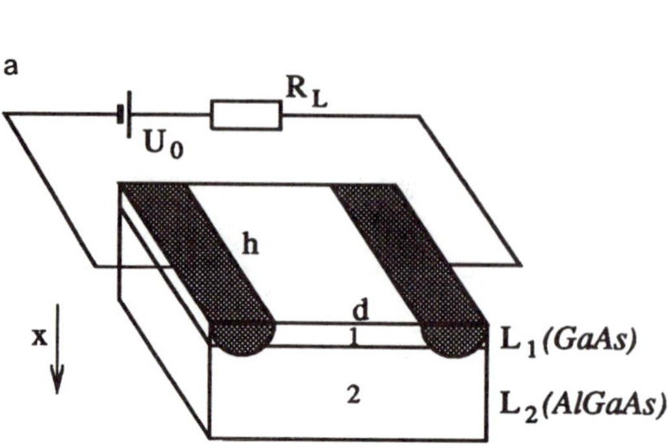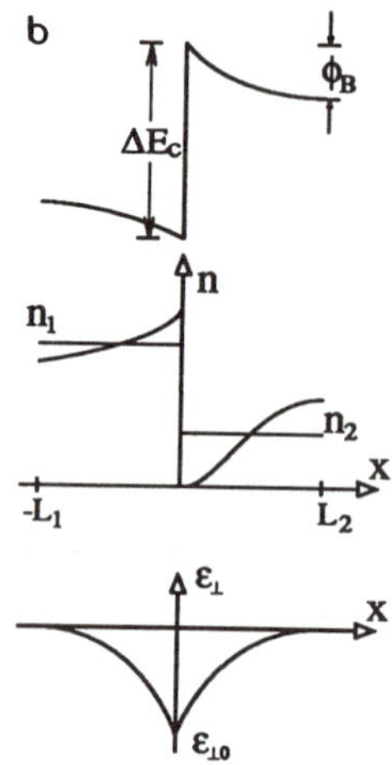



Figure 1. (a) Schematic sample and circuit configuration of a modulation doped $GaAs/Al_xGa_{1-x}As$ heterostructure with heterolayer width $L_1$ and $L_2$, respectively, and lateral dimensions $h, d$. (b) Energy-band diagram (top), carrier density (center) and perpendicular electric field (bottom) versus the perpendicular coordinate $x$ of the heterolayer (schematic).

differential conductivity (NNDC)[2–8] in analogy with intervalley transfer in the Gunn effect.

In the NNDC regime ac-driven current oscillations of $2 - 25\,MHz$ have been reported[9]. Oscillations in heterolayers were also observed under dc-conditions[10,11]. A novel physical mechanism of a real-space transfer oscillator, which gives periodic[12] and chaotic[7,13] self-generated oscillations at much *higher* frequencies ($20 - 80\,GHz$) under *dc-conditions*, has recently been proposed and numerically investigated[12]. It is based on the coupled nonlinear dynamics of the real-space electron transfer and of the space charge in the $Al_xGa_{1-x}As$ layer which controls the interface potential barrier $\Phi_B$ (Fig.1b). Real space transfer of the electrons in the $GaAs$ layer leads to an increase of the carrier density in the $Al_xGa_{1-x}As$, which diminishes the positive space charge controlling the band bending. Subsequently, the potential barrier $\Phi_B$ decreases with some delay due to the finite dielectric relaxation time. This leads to an increased backward thermionic emission current which decreases the carrier density in the $Al_xGa_{1-x}As$. Hence the space charge and $\Phi_B$ are increased. This, in turn, decreases the backward thermionic emission current, which completes the cycle.

In the present paper we analyse this physical model as a nonlinear dynamic system. We investigate in detail the resulting bifurcation scenarios as a function of bias voltage $U_0$ and load resistance $R_L$. Similar methods of nonlinear dynamics have recently been applied to different oscillation mechanisms in bulk semiconductors[14] and in heterostructures with electric fields *perpendicular* to the layer interface[15].

MODEL

The transport processes are described by the following set of nonlinear autonomous differential equations[12,16] for the carrier densities $n_1, n_2$ in the two layers, the dielectric relaxation of the applied field $\mathcal{E}_\parallel$ and the potential barrier $\Phi_B$.

The spatially averaged carrier density in the $GaAs$ layer

$$n_1 \equiv \frac{1}{L_1} \int_{-L_1}^{0} n(x,t)\, dx$$

as a function of time is governed by the equation of continuity:

$$\dot{n}_1 = \frac{1}{eL_1}\left(J_{1\to 2} - J_{2\to 1}\right)\ ,\tag{1}$$

222

where

$$J_{1\to 2} := -en_1 \sqrt{\frac{E_1}{3\pi m_1^*}} \exp\left(-\frac{3\Delta E_c}{2E_1}\right)$$

$$J_{2\to 1} := -en_2 \sqrt{\frac{E_2}{3\pi m_2^*}} \exp\left(-\frac{3\Phi_B}{2E_2}\right)$$

are the thermionic emission current densities from layer 1 to layer 2, and vice versa, respectively ($m_i^*$: effective masses) given by Bethe's theory[17], $\Delta E_c$ is the conduction band discontinuity, and $E_i = (3/2)k_B T_i$ ($i = 1,2$) are the mean carrier energies associated with the carrier temperatures $T_i$. The energy transfer between the heterolayers is described by the energy balance equations containing Joule's heat, convective, diffusive, and electron pressure induced heat flow, and energy loss due to polar optical phonon scattering[7]. It can be shown by a linear mode analysis[18] that for reasonable numerical parameters the energy relaxation occurs on a fast time scale, such that $E_1$ and $E_2$, respectively, can be eliminated adiabatically from the energy balance equations. The mean energy as a function of the applied electric field $\mathcal{E}_\parallel$ is then roughly estimated by $E_1 \approx E_L + \tau_E e\mu_1 \mathcal{E}_\parallel^2$ , $E_2 \approx E_L$ with the thermal equilibrium mean energy $E_L = (3/2)k_B T_L$ where $T_L$ is the lattice temperature, and the energy relaxation time $\tau_E$.

The dielectric relaxation of the parallel electric field is given by the equation:

$$\epsilon\dot{\mathcal{E}}_\parallel = -\sigma_L \left(\mathcal{E}_\parallel - \mathcal{E}_0\right) - \frac{en_1\mu_1 L_1 + en_2\mu_2 L_2}{L_1 + L_2} \mathcal{E}_\parallel \tag{2}$$

where $\epsilon$ is the permittivity, $\sigma_L = (h(L_1 + L_2)R_L/d)^{-1}$ is connected to the load resistance $R_L$, and $U_0 = \mathcal{E}_0 d$ is the applied bias voltage.

The interface potential barrier

$$\Phi_B \equiv -e \int_0^{L_2} \mathcal{E}_\perp(x,t)\,dx$$

is governed by the space charge dynamics in the $Al_x Ga_{1-x}As$ layer and the resulting perpendicular internal electric field $\mathcal{E}_\perp$ (Fig.1b):

$$\dot{\Phi}_B = \frac{e}{\epsilon}\left[-\mu_2 N_D \Phi_B + \mu_2 \frac{e^2}{2\epsilon}L_1^2 n_1^2 - eL_1 L_2 \dot{n}_1\right] \tag{3}$$

where we have neglected the diffusive contributions, which is appropriate if $L_2$ is less or comparable to the mean free path of the electrons.

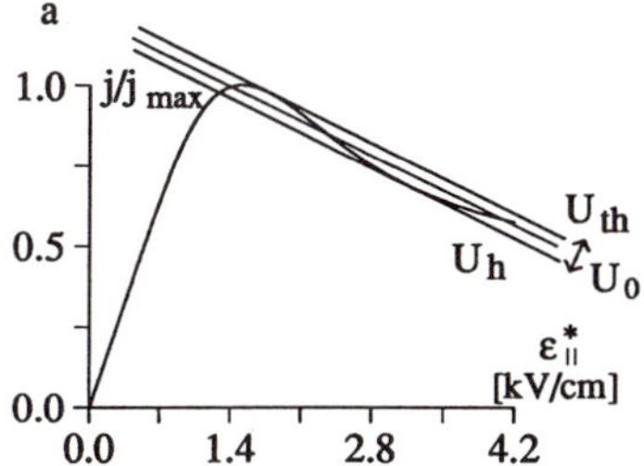


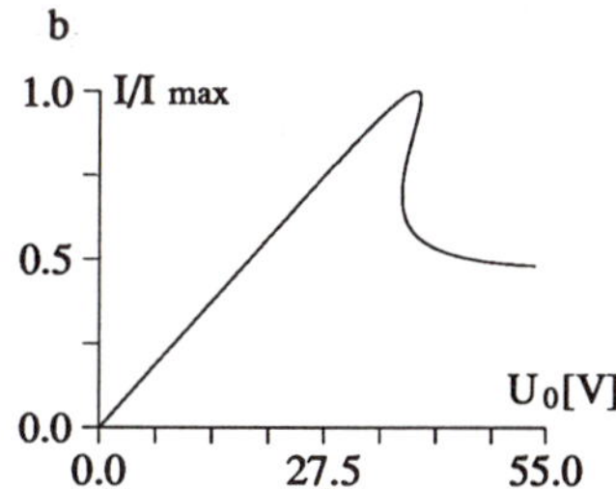


Figure 2. (a) Static current density-field characteristic as a function of the static electric field $\mathcal{E}_\parallel^*$. The load line is shown for three different bias voltages $U_0$. (b) Normalized current versus $U_0$. (Numerical parameters as in Table 1.)

Table 1. The numerical parameters used in the simulation

| | |
|---|---|
| $N_D$ | $10^{17}\ cm^{-3}$ |
| $L_1$ | $100\ \text{Å}$ |
| $L_2$ | $200\ \text{Å}$ |
| $\mu_1$ | $8000\ cm^2/V s$ |
| $\mu_2$ | $50\ cm^2/V s$ |
| $T_L$ | $300\ K$ |
| $\Delta E_c$ | $250\ meV\ (x = 0.3)$ |
| $m_1^*$ | $0.067\ m_0$ ($m_0$ free electron mass) |
| $m_2^*$ | $(0.067 + 0.083 \times x)\ m_0$ |
| $\tau_E$ | $5.0 \times 10^{-12}\ s$ |
| $h$ | $1\ mm$ |
| $d$ | $50\ \mu m$ |
| $\epsilon$ | $12\ \epsilon_0$ ($\epsilon_0$ vacuum permittivity) |
| $R_L$ | $1.144\ k\Omega$ |

Conservation of the total number of carriers requires

$$n_1 L_1 + n_2 L_2 = N_D L_2 \tag{4}$$

such that the spatially averaged carrier density in the $Al_x Ga_{1-x} As$ layer

$$n_2 \equiv \frac{1}{L_2} \int_0^{L_2} n(x,t)dx$$

is *not* an independent variable.

The autonomous nonlinear dynamic system (1)-(3) which is a reduction of the 5-variable system treated in Ref. 12 constitutes the basis of our analysis.

## FIXED POINTS

The static current density-field characteristic resulting from (1)-(3) is shown in Fig.2(a). Real space transfer of hot electrons leads to NNDC. The intersection of the load line $j \sim (\mathcal{E}_0 - \mathcal{E}_\parallel)/R_L$ with the device characteristic $j(\mathcal{E}_\parallel)$ defines the steady state operating points i.e., the fixed points of the dynamic system (1)-(3). They are denoted by the subscript *. As the applied voltage $U_0$ is varied, the load line is shifted parallel and the intersection points move along the current density-field characteristic. When the load line becomes tangential to the characteristic ($U_0 = U_{th}$ or $U_h$) two intersection points merge at $\mathcal{E}_{sn_1}^*$ and $\mathcal{E}_{sn_2}^*$, respectively, and disappear upon further variation of $U_0$. This is associated with saddle-node bifurcations of the fixed points of the dynamic system (1)-(3). In Fig.2(b) the current is plotted versus the control parameter $U_0$, exhibiting inverted SNDC (S-shaped negative differential conductivity).

## LINEAR STABILITY

The stability of the fixed points $\underline{q}^* \equiv \left(n_1^*, \mathcal{E}_\parallel^*, \Phi_B^*\right)^T$ against infinitesimal fluctuations $\delta q(t) = \delta \underline{q}(0)\, e^{\lambda t}$ is determined by linearizing the dynamical system (1)$-$(3) around its steady state and computing the eigenvalues $\lambda$ of the Jacobian matrix $J(\underline{q}^*)_{ij} = (\partial \dot{q}_i/\partial q_j)^*$ , $i,j =$

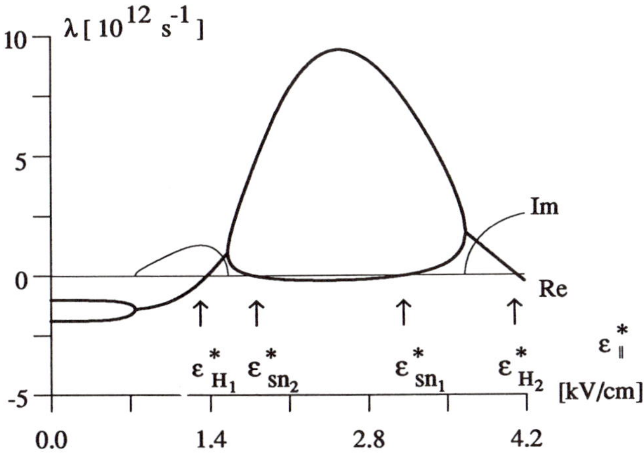


Figure 3. Eigenvalue spectrum $\lambda$ as a function of $\mathcal{E}_\parallel^*$. Bold and thin lines represent the real and the imaginary parts of $\lambda$, respectively. $\mathcal{E}_{H_1}^*$, $\mathcal{E}_{H_2}^*$ mark Hopf bifurcations and $\mathcal{E}_{sn_1}^*$, $\mathcal{E}_{sn_2}^*$ saddle-node bifurcations. (Parameters as in Table 1.)

1, 2, 3. Fig.3 shows the eigenvalues as a function of the static electric field $\mathcal{E}_\parallel^*$ which is related to the control parameter $U_0$ by (2), taken in the steady state. The third eigenvalue is always negative ($\lambda_3 < 0$ , $|\lambda_3| \gg 1$) and is not shown. It is associated with $\mathcal{E}_\parallel$, which is therefore a fast dynamic variable and can be eliminated adiabatically. The resulting two-variable system (marked by $\tilde{\ }$) has an eigenvalue spectrum $\tilde{\lambda}$ which is topologically equivalent to those in Fig.3, but more amenable to analytical treatment.

Let us now discuss the occurring bifurcations of the steady state $\underline{q}^*$ in the phase space $(n_1, \mathcal{E}_\parallel, \Phi_B)^T$. Upon increasing $\mathcal{E}_\parallel^*$ a stable node (3 real negative eigenvalues) turns into a stable focus, which becomes unstable via a Hopf bifurcation at $\mathcal{E}_{H_1}^*$ (i.e., a pair of complex conjugate eigenvalues obtains positive real parts). The unstable saddle-focus is transformed into a saddle-point with two positive and one negative eigenvalues and merges at $\mathcal{E}_{sn_2}^*$ with a saddle-point with one positive and two negative eigenvalues, corresponding to the turning point $U_{th}$ of the $I(U_0)$ characteristic (Fig.2). A reverse bifurcation scenario ($\mathcal{E}_{sn_1}^*, \mathcal{E}_{H_2}^*$) takes place upon further increase of $\mathcal{E}_\parallel^*$, corresponding to the lower branch of Fig.2(b).

## DYNAMICS

The numerical simulation of the time dependent nonlinear equations (1)-(3) reveals a more complex bifurcation scenario as a result of local bifurcations of limit cycles by a cyclic fold[19] (sometimes referred to as a condensation of paths[1]), or of a global bifurcation of a limit cycle from a separatrix[19-22]. These supplement the local bifurcations of fixed points shown in Fig.3. Fig.4 summarizes schematically the global and local bifurcations $(1-10)$. Bold lines denote stable fixed points or limit cycles (periodic attractors).

On passing through the bifurcation value $\mathcal{E}_{c_1}$ (2) a stable and an unstable limit cycle is created by a cyclic fold (saddle-node bifurcation of limit cycles). This means that two limit cycles with different stability coexist and merge at the bifurcation point. The unstable limit cycle (dotted line) shrinks and collides at $\mathcal{E}_{ho_1}$ (3) with the saddle-point $sa$ (dashed line). A saddle-to-saddle separatrix loop (homoclinic orbit) is formed surrounding the fixed points on the upper and lower branch. Upon increase of $\mathcal{E}_0$ the separatrix disappears. At $\mathcal{E}_{ho_2}$ (4) a homoclinic orbit around the fixed point on the lower branch is formed from which a limit cycle $\underline{q}(t)$ bifurcates. The stability of this periodic orbit $\underline{\tilde{q}}(t)$ in the reduced 2-variable system

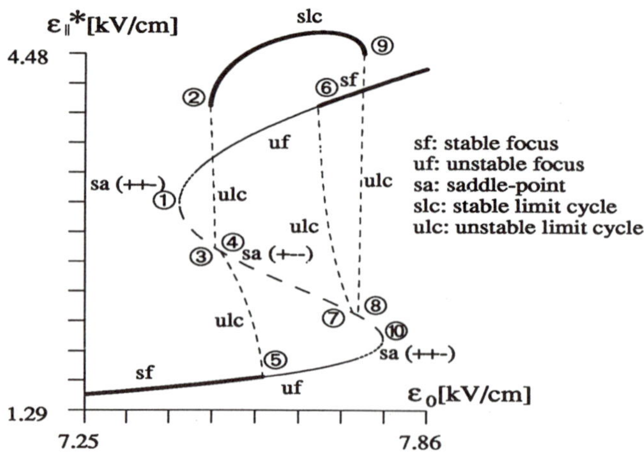



Figure 4. Bifurcation diagram of the static electric field $\mathcal{E}_\parallel^*$ versus the control parameter $\mathcal{E}_0 = U_0/d$. The local and global bifurcations are denoted by ①– ⑩; see text. (Parameters as in Table 1.)

is given in a linear approximation by the Floquet exponent $\gamma_{lc}$[21,22]

$$\gamma_{lc} = \frac{1}{T} \int_{-T/2}^{T/2} \mathrm{Tr}\,\underline{\tilde{J}}\left(\tilde{q}(t')\right) dt' \tag{5}$$

where Tr denotes the trace, and $T$ is the oscillation period. $\gamma_{lc}$ is a generalisation of the eigenvalues of the Jacobian matrix controlling the stability of *fixed points*. In the limit $\tilde{\mathcal{E}} \to \tilde{\mathcal{E}}_{ho_2}$ the limit cycle collides with the saddle $\tilde{q}^*$ of the middle branch. This yields

$$\gamma_{lc} = \mathrm{Tr}\,\underline{\tilde{J}}\left(\tilde{q}^*\right) > 0 \ , \tag{6}$$

which means that the homoclinic orbit and the bifurcating limit cycle are unstable. The unstable cycle appears immediately with a nonzero amplitude, but with frequency tending to zero. This is associated with critical slowing down according to a universal power law[19].

Upon further increase of $\mathcal{E}_0$ the unstable limit cycle shrinks and disappear via a subcritical Hopf bifurcation at $\mathcal{E}_{H_1}$ (5). When $\mathcal{E}_0$ passes $\mathcal{E}_{H_2}$ (6), an unstable limit cycle is created from a focus. As $\mathcal{E}_0$ increases this unstable cycle expands and collides at $\mathcal{E}_{ho_3}$ (7) with the saddle-point $sa$, and a saddle-to-saddle loop is formed around the fixed point on the upper branch. At the bifurcation value $\mathcal{E}_{ho_4}$ (8) another homoclinic orbit surrounding both fixed points on the upper and lower branch is created, from which an unstable limit cycle bifurcates. After that the unstable cycle expands and coalesces at $\mathcal{E}_{c_2}$ (9) with the stable limit cycle. In this way a stable limit cycle is annihilated by a collision with an unstable cycle forming a semistable limit cycle with multiplicity[22] two at $\mathcal{E}_{c_2}$.

We shall now discuss the physical implications. As the bias $\mathcal{E}_0$ is increased, the steady state $(sf)$ becomes unstable at $\mathcal{E}_{H_1}$ (5), and finite-amplitude current and voltage oscillations set on, corresponding to the stable limit cycle $slc$ in the space of the dynamic variables $(n_1, \mathcal{E}_\parallel, \Phi_B)^T$. The drifting carriers cycle periodically between the $GaAs$ and the $Al_xGa_{1-x}As$ layer, with a concomitant variation of the drift field $\mathcal{E}_\parallel$ and the potential barrier $\Phi_B$. Note that this occurs in the *positive* differential conductivity regime. Upon further increase of $\mathcal{E}_0$ the oscillations cease at $\mathcal{E}_{c_2}$ (9), and the steady state $(sf)$ on the other branch takes over (Fig.5, top). When $\mathcal{E}_0$ is decreased, this steady state persists until $\mathcal{E}_{H_2}$ (6) $< \mathcal{E}_{c_2}$ where

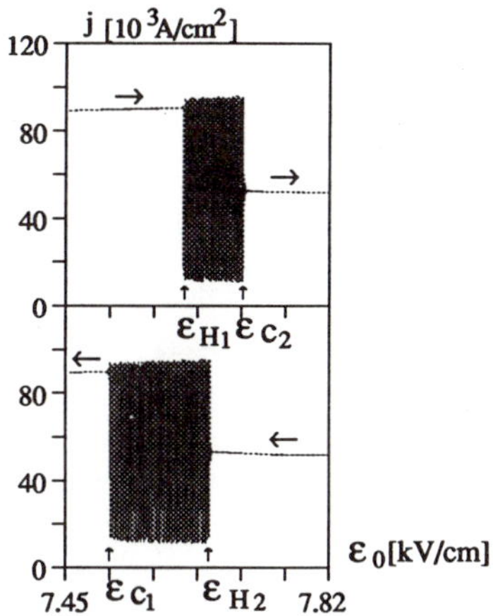

Figure 5. Hysteresis of periodic (shaded) and stationary states. The time-dependent current density is plotted versus the applied field $\mathcal{E}_0$ for slowly increasing (top) and slowly decreasing (bottom) $\mathcal{E}_0$. The oscillation frequency is $f = 103\,GHz$. (Parameters as in Table 1.)

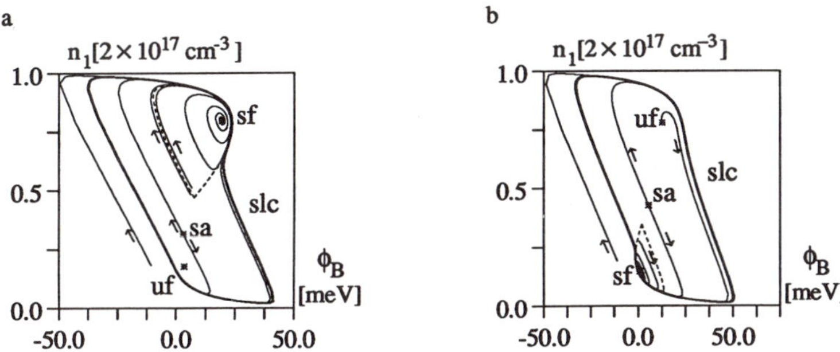

Figure 6. Coexistence of two stable attractors ($sf$ and $slc$). Phase portraits of $n_1$ versus $\Phi_B$ for (a) $\mathcal{E}_{ho_2} < \mathcal{E}_0 < \mathcal{E}_{H_1}$ and (b) $\mathcal{E}_{H_2} < \mathcal{E}_0 < \mathcal{E}_{ho_3}$. (Parameters as in Table 1 with (a) $\mathcal{E}_0 = 7.497\,kV/cm$ (b) $\mathcal{E}_0 = 7.7006\,kV/cm$). The fixed points are marked by *.

finite-amplitude oscillations set on, ceasing at $\mathcal{E}_{c_1}$ (2) $< \mathcal{E}_{H_1}$, and thus exhibiting dynamic hysteresis (Fig.5, bottom).

Fig.6 shows the coexistence of two attractors (a stable limit cycle $slc$ and a stable focus $sf$) in the hysteretic regimes $\mathcal{E}_{ho_2} < \mathcal{E}_0 < \mathcal{E}_{H_1}$ and $\mathcal{E}_{H_2} < \mathcal{E}_0 < \mathcal{E}_{ho_3}$. The two corresponding basins of attraction are separated by an unstable limit cycle (broken line). The motion along the stable limit cycle (bold line) is clockwise.

For smaller load resistance $R_L$ the path condensation of limit cycles [(2),(9) in Fig.4] can occur beyond the multistationary regime. The generated semistable limit cycle decomposes into a stable and an unstable limit cycle, the latter shrinks upon variation of the control parameter and vanishes via a subcritical Hopf bifurcation. Thus there are no homoclinic bifurcations in this scenario.

If $R_L$ is so small that $\sigma_L + \sigma_{diff} > 0$ throughout, where $\sigma_{diff} = dj/d\mathcal{E}_\parallel$ is the differential

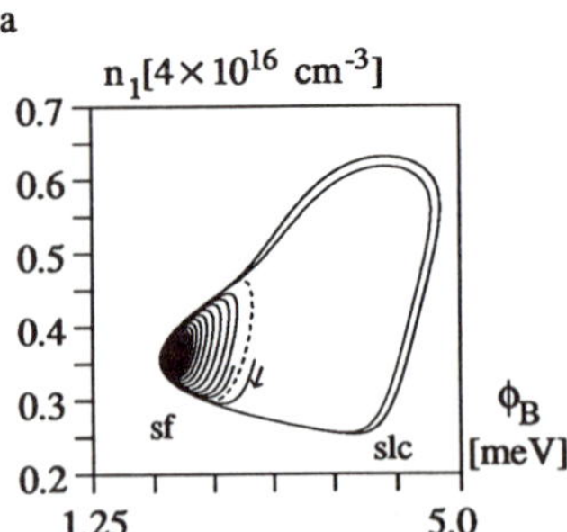
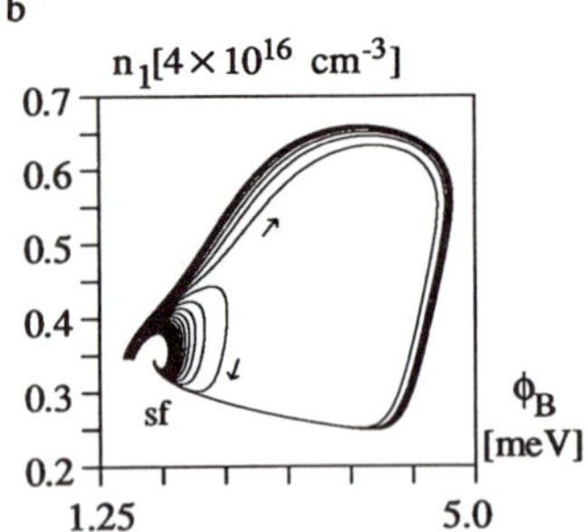


Figure 7. (a) Phase portrait for coexistence of two stable attractors (stable focus and limit cycle) for $\mathcal{E}_0 = 5.43\ kV/cm$. (b) Trajectories starting near by a stable focus and making a large excursion in the phase space for $\mathcal{E}_0 = 5.44\ kV/cm$. (Parameters as in Table 1., except for $N_D = 10^{16}\ cm^{-3}$, $L_2 = 450$ Å, $\mu_1 = 10000\ cm^2/Vs$, $\tau_E = 2.0 \times 10^{-12}\ s$).

conductance, multistationarity (SNDC) does not occur, but bistability between limit cycles and the stationary state is still possible.

## EXCITABILITY, CANARDS AND PHANTOM DUCKS

Interesting dynamical behavior occurs when the dielectric relaxation time of the electric field is large compared to the time constant of electron transfer out of the $GaAs$ channel. Our physical model is an example for this. In contrast to the above simulations, we have neglected the thermionic current contribution $\sim \dot{n}_1$ in (3). The dynamical system then behaves like a 'slow-fast' system. This means that there are time periods during which the trajectory moves slowly, but these time intervals are separated by small periods in which the solution jumps from one value to another. In such systems not only bifurcations (changes of the qualitative nature of the phase flow) but also sudden quantitative changes of the phase portrait in exponentially small ranges of the control parameter can occur. For example in Fig.7a a small unstable limit cycle has been generated in a subcritical Hopf bifurcation, and two stable attractors (a stable fixed point and a stable limit cycle) coexist at the same value of the control parameter. Upon increasing the bias voltage the large stable limit cycle shrinks slowly and the unstable cycle grows gradually. In an extremly narrow range of the control parameter, a relative change of the control parameter by $10^{-8}$ causes the unstable limit cycle to grow dramatically such that the two limit cycles suddenly come very close to each other and annihilate in a cyclic fold (condensation of paths). It is as if the existence of solutions with a intermediate amplitude would be a *canard*[19,23] (duck). Beyond the cyclic fold the dynamical system has only one attractor (stable focus), but in the phase portrait there are trajectories that almost hit the stable fixed points, and then make a large excursion before they eventually reach the attractor (see Fig.7b). This means that a small perturbation in the 'right' direction of the system residing at the fixed point will cause a large excursion in the phase space before returning to the stable equilibrium. Such behavior has recently been described for the van der Pol equation[24] driven by a constant force, and the trajectories which make a large excursion have been called *Phantom Ducks*, because they represent in some sense the ghost of the trajectories approaching the stable periodic attractor *before* the cyclic fold bifurcation (i.e., the *canards*).

The *Phantom Ducks* are associated with *excitability* of the dynamic system. This means that the system is capable of amplifying a small input signal (i.e., an initial fluctuation from the fixed point) into a much larger output signal (i.e.,a large excursion in phase space), if a threshold value of the initial fluctuation is exceeded. This is a general feature of 'slow-fast' systems.

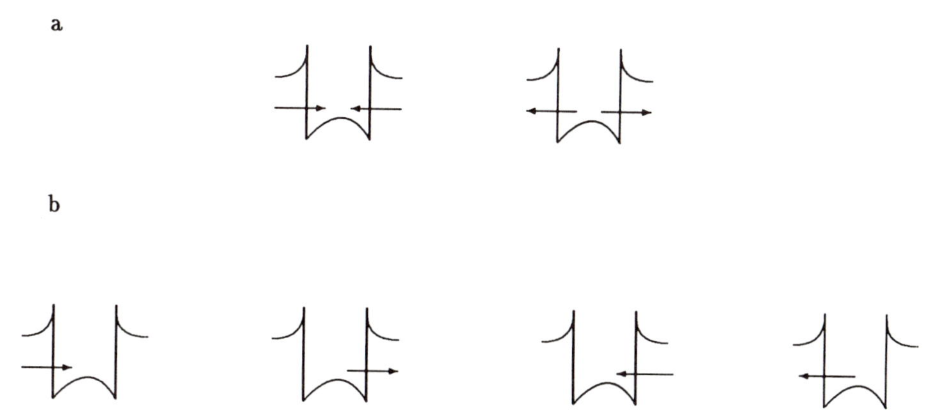

Figure 8. Modes of oscillation in a double-barrier-heterostructure: (a) symmetric periodic attractor (SA), (b) non-symmetric periodic attractor (NA). The arrows show schematically the electron transfer during an oscillatory cycle.

## TRANSIENT CHAOS

Qualitatively new dynamic behavior occurs if a double barrier heterostructure is considered, consisting of an $Al_xGa_{1-x}As - GaAs - Al_xGa_{1-x}As$ layer sequence of widths $L_2'$, $2L_1$, $L_2$, respectively. The model takes into account the two-dimensional density of states of the electrons as well as the possibility of thermionic emission and non-resonant tunneling currents between the layers. Denoting the corresponding dynamic variables in the top $Al_xGa_{1-x}As$-layer of widths $L_2'$ by $n_2'$, $\Phi_B'$ we obtain a dynamic system for the 5 variables $(n_2, n_2', \Phi_B, \Phi_B', \mathcal{E}_{\parallel})$. For $2L_1 = 65$ Å and $L_2 = L_2' = 250$ Å and $\mathcal{E}_0 > 4.25\ kV/cm$ there exists a symmetric periodic attractor with $n_2 = n_2'$, $\Phi_B = \Phi_B'$ as well as a periodic attractor off the symmetry hypersurface. These two different modes of oscillation are sketched in Fig.8. The 'symmetric attractor' (SA) attracts points mainly on, or in the neighborhood, of the symmetric hypersurface $n_2 = n_2'$, $\Phi_B = \Phi_B'$. The 'non-symmetric attractor' (NA) is attracting for stongly non-symmetric initial conditions. If one starts in the control parameter region $5.125\ kV/cm < \mathcal{E}_0 < 5.25\ kV/cm$ very closely to the symmetric hypersurface, a complicated dynamic behavior can be observed. The system performs period-two oscillations on an additional weakly symmetric attractor. Upon increasing the control parameter the attractor 'smears out' and becomes chaotic. This weakly asymmetric chaotic attractor (CA) is eventually transformed into a chaotic repeller (CR). The chaotic repeller is associated with transient chaos[25] of variable length before the 'non-symmetric' periodic attractor (NA) is suddenly approached. A typical example is shown in Fig.9.

## CONCLUSION

In conclusion, we have predicted complex bifurcation scenarios of periodic and fixed points for parallel transport in a modulation doped heterostructure. The oscillation mechanism is based upon the delayed feedback between the thermionic emission current across the layer interface and the resulting space charge dynamics of the potential barrier. A previous 5-variable model[12,13] has been reduced to a nonlinear dynamic system with the averaged carrier density in the $GaAs$ layer and the space charge potential barrier in the $Al_xGa_{1-x}As$ layer as the essential variables. We have investigated the influence of the circuit conditions on the onset of the oscillations and have demonstrated the possibility of oscillatory instabilities

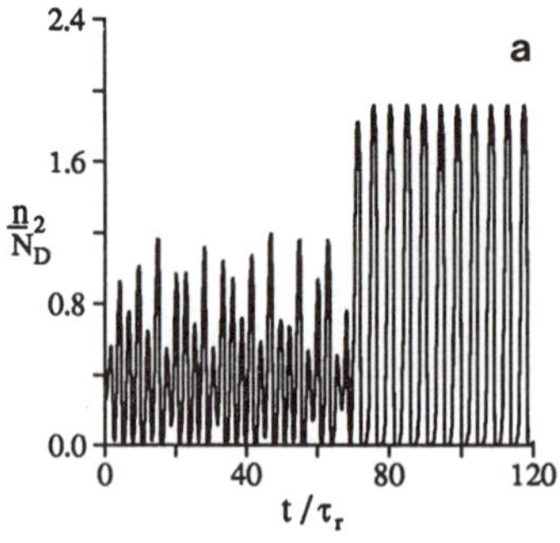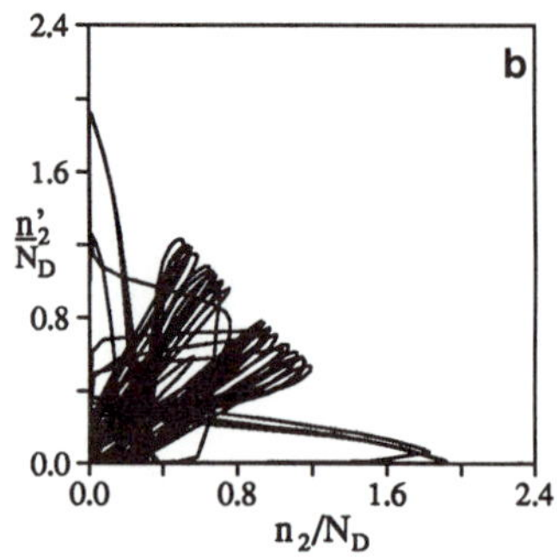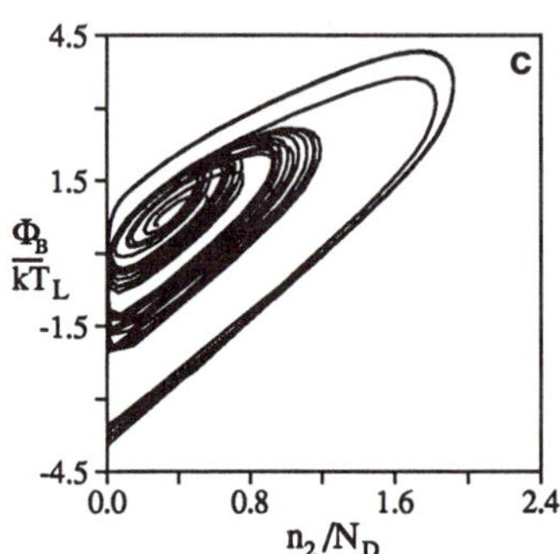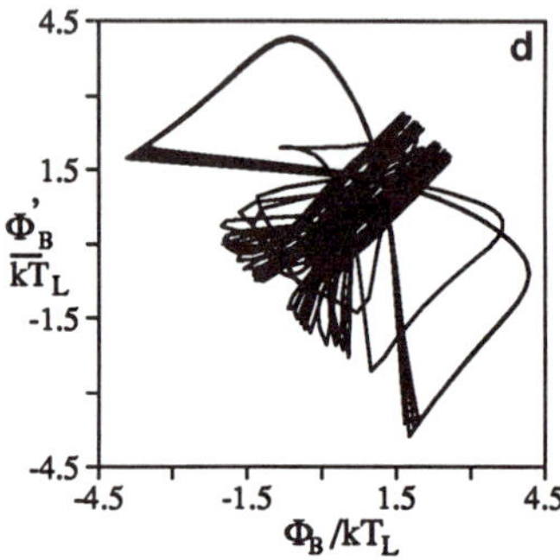

Figure 9. Transient chaos in a double-barrier heterostructure. (a) Carrier density $n_2$ versus $t$, (b) $n'_2$ versus $n_2$, (c) $\Phi_B$ versus $n_2$, (d) $\Phi'_B$ versus $\Phi_B$, for slightly asymmetric initial conditions $n_2(0) = n'_2(0) = 0.3\,N_D$, $\Phi_B(0) = 0.61\,k_B T_L$, $\Phi'_B(0) = 0.6\,k_B T_L$, $\mathcal{E}_\parallel(0) = 0$ and $\mathcal{E}_0 = 5.23\,kV/cm$. Parameters as in Table 1. ($\tau_r = 1.33 \times 10^{-12}\,s$)

for sufficiently large load resistance in regimes of positive differential conductivity via global bifurcations of a limit cycle and subcritical Hopf bifurcations. Bistability and hysteretic switching between periodic attractors and fixed points are found. Furthermore, the system shows 'canards', excitability and 'phantom ducks'. In a double-barrier heterostructure, transient chaos associated with a chaotic repeller is found.

## ACKNOWLEDGEMENTS

Helpful discussions with M. Asche, J. Kolodzey, M. Mocker, and P. Vogl are acknowledged.

## REFERENCES

1. E. Schöll, Nonequilibrium Phase Transitions in Semiconductors (Springer, Berlin 1987)
2. Z.S. Gribnikov, Sov. Phys. Semicond. 6 1204 (1973)
2. F. Pacha, F. Paschke, Electron. Commun. 32, 235 (1978)
4. K. Hess, H. Morkoc, H. Shichijo, and B.G. Streetman, Appl. Phys. Lett. 35, 469 (1979)
5. H. Shichijo, K. Hess, and B.G. Streetman, Sol. State Electron. 23, 817 (1980)
6. R. Sakamoto, K. Akai, and M. Inoue, IEEE Trans. Electron Devices 36, 2344 (1989)

7. K. Aoki, K. Yamamoto, N. Mugibayashi, and E. Schöll, Sol. State Electron. $\underline{32}$, 1149 (1989)

8. A. Kastalsky, M. Milshtein, L.G. Shantharama, J. Harbison, and L. Florez, Sol. State Electronics $\underline{32}$, 1841 (1989)

9. P.D. Coleman, J. Freeman, H. Morkoc, K. Hess, B.G. Streetman, and M. Keever, Appl. Phys. Lett. $\underline{40}$, 493 (1982)

10. A.J. Vickers, A. Straw, and J.S. Roberts, Semicond. Sci. Technol. $\underline{4}$, 743 (1989)

11. P. Hendriks, E.A.E. Zwaal, J.G.A. Dubois, F.A.P. Blom, and J.H. Wolter, J. Appl. Phys. $\underline{69}$, 302 (1991)

12. E. Schöll and K. Aoki, Appl. Phys. Lett. $\underline{58}$, 1277 (1991)

13. E. Schöll and K. Aoki, Proc. 20th Int. Conf. Physics of Semiconductors, ed. E.H. Anastassakis and J.D. Joannopoulos (World Scientific Singapore 1990) p. 1125

14. G. Hüpper and E. Schöll, Phys. Rev. Lett. $\underline{66}$, 2372 (1991)

15. A. Wacker and E. Schöll, to be published in Appl. Phys. Lett. $\underline{59}$ (1991)

16. R. Döttling and E. Schöll, to be published

17. K. Hess, Advanced Theory of Semiconductor Devices (Prentice Hall, New Jersey 1988)

18. D. Reznik, unpublished

19. J.M.T. Thompson and H.B. Stewart, Nonlinear Dynamics and Chaos (Springer, Berlin 1982)

20. J. Guckenheimer and P. Holmes, Nonlinear Oscillations, Dynamical Systems, and Bifurcations of Vector Fields, Applied Mathematical Sciences 42 (Springer, New York 1983)

21. S.N. Chow and J.K. Hale, Methods of Bifurcation Theory (Springer, Berlin 1982)

22. A.A. Andronov, E.A. Leontovich, I.I. Gordon and A.G. Maier, Vol. 2, Theory of Bifurcations of Dynamic Systems on a Plane (Isreal Program for Scientific Translations, Jerusalem 1971)

23. M. Diener, Math. Intell. $\underline{6}$, 38 (1984)

24. B. Braaksma, to appear in J. of Dyn. Systems and Diff. Eqs., 1991

25. T. Tél, in: Directions in Chaos, Vol.3, ed. by Hao Bai-lin (World Scientific, Singapore 1990), pp. 149-211

# NUMERICAL STUDY OF BIFURCATIONS IN THE BCS GAP EQUATION


T. Pavlopoulos, P.L. Christiansen, M.P. Soerensen, N. Lazarides, and P. Spathis[†]

Laboratory of Applied Mathematical Physics
The Technical University of Denmark, Bldg. 303
DK-2800 Lyngby, Denmark



## ABSTRACT

The classical BCS strategy is employed to study layered high $T_c$ superconductors. Starting from a tight binding description and a phenomenological pairing interaction between electrons an anisotropic BCS gap equation is derived. It will be shown that asymmetric pairing interaction within the layers leads to perturbed pitchfork bifurcations separating smoothly different phases of the gap parameter. If the hopping probabilities in the x and y direction differ, local minima of the Gibbs' free energy can turn into global minima and vice versa, leading to phase transitions which do not result from bifurcations.


## INTRODUCTION

According to the BCS (Bardeen, Cooper and Schrieffer) theory attractive interaction between electrons gives rise to formation of electron pairs with opposite momenta and spins characterizing the superconducting state[1]. The pair formation is described by the effective Hamiltonian[2-4]

---


†: Max-Planck-Institute fur Plasmaphysik, D-8046 Garching bei Munchen, Germany



*Chaotic Dynamics: Theory and Practice*
Edited by T. Bountis, Plenum Press, New York, 1992


"

$$H = \sum_{k,\sigma} \epsilon_k \alpha^+_{k\sigma} \alpha_{k\sigma} - \sum_{k,k'} V_{k,k'} \alpha^+_{k\uparrow} \alpha^+_{-k\downarrow} \alpha_{-k'\downarrow} \alpha_{k'\uparrow} \tag{1}$$

where $\epsilon_k$ is the quasiparticle energy of the normal state electrons measured relative to the Fermi energy which we shall denote by $\mu$ and $V_{k,k'}$ is the pairing interaction potential. The Fermi operators $\alpha^+_{k\sigma}$ and $\alpha_{k\sigma}$ create and annihilate an electron with wavevector $\mathbf{k}$ and spin $\sigma$, respectively. The $\mathbf{k}$ dependent gap parameter $\Delta_k$ is given by

$$\Delta_k = \frac{1}{N} \sum_{k'} V_{k,k'} F_{k'} \quad , \qquad F_k = \frac{\Delta_k}{2E_k} \tanh(\frac{1}{2}\beta E_k) \tag{2}$$

and it measures the energy gap between the superconducting ground state and the quasiparticle excitation state at the Fermi level. Here, $E_k = (\epsilon_k^2 + \Delta_k^2)^{1/2}$ is the quasiparticle excitation energy and N is the number of lattice sites. The temperature T enters through $\beta = 1/(k_B T)$ where $k_B$ is Boltzmann's constant. The BCS gap equation is derived by minimizing the expression for the Gibbs' free energy

$$G(\Delta_k) = \frac{1}{N} \sum_k (\epsilon_k - E_k + 2\Delta_k F_k - \frac{2}{\beta} \ln(1 + e^{-\beta E_k}))$$

$$- \frac{1}{N^2} \sum_{k,k'} V_{k,k'} F_k F_{k'} \tag{3}$$

In fact, Eq. (2) describes not only the minima of expression (3) but the extrema in general. Thus the BCS gap equation accepts a variety of solutions: stable solutions corresponding to local minima of (3) and unstable solutions corresponding to local maxima or saddle points of (3). The physical relevant solution is the one corresponding to the global minimum.

THE TIGHT BINDING BCS MODEL

We consider here a particularly simple model of a layered high $T_c$ superconductor consisting of generally rectangular sheets stacked as illustrated in Fig. 1. The size of the unit cell is given by the lattice constants $a_x$, $a_y$, and $a_z$. Only nearest intralayer neighbour interaction is taken into account. The pairing interaction strengths are denoted by $g_x$ and $g_y$ for the x and y direction respectively. The hopping probabilities are denoted by $t_1$, $t_2$ for hopping between nearest intralayer neighbours (x and y

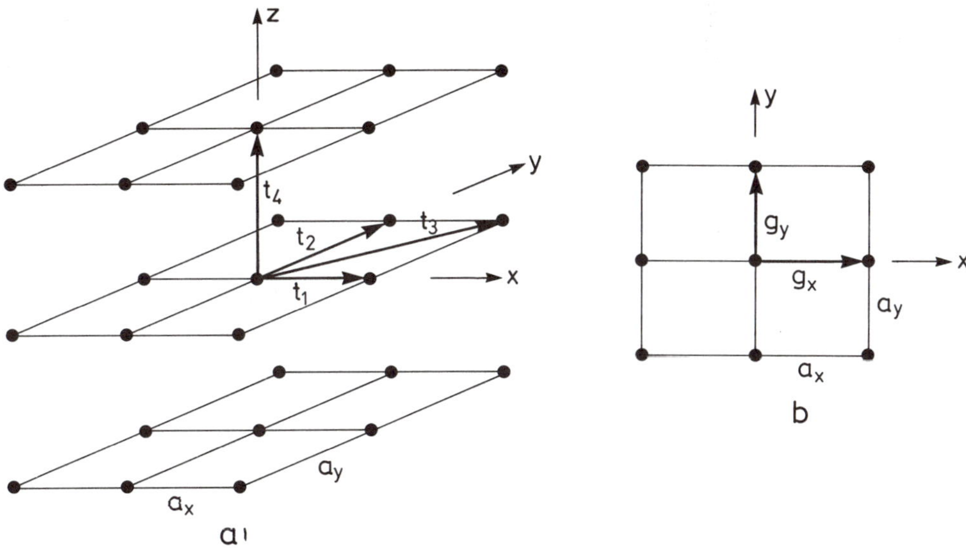


Fig. 1. (a): The hopping probabilities between intralayer nearest and next nearest neighbours and interlayer nearest neighbours and (b): the pairing interaction strengths between intralayer nearest neighbours.

direction respectively), $t_3$ for hopping between next nearest intralayer neighbours and $t_4$ for hopping between nearest interlayer neighbours. For this model we find the dispersion relation[2-4]

$$\epsilon_k = A[-2(\cos(k_x a_x) + A_y\cos(k_y a_y)) + 4B\cos(k_x a_x)\cos(k_y a_y) - 2C\cos(k_z a_z) - \mu] ,\qquad(4)$$

where $A = -2t_1$, $A_y = -2t_2/A$, $B = -t_3/(2t_1)$, $C = t_4/(2t_1)$, and $A\mu$ is the chemical potential. The wavevector is $\mathbf{k} = (n_x\pi/(N_x a_x), n_y\pi/(N_y a_y), n_z\pi/(N_z a_z))$ where $-N_q < n_q$

$\leq N_q$, q=x, y, z, and $N_q$ is the number of lattice sites in the q direction. Assuming that pairing interaction only takes place between electrons within the CuO planes and only between nearest neighbours the pairing potential becomes[4]

$$V_{kk'} = 2g_x\cos((k_x-k_x')a_x) + 2g_y\cos((k_y-k_y')a_y) \tag{5}$$

From expressions (4) and (5) we conclude that $\Delta_k$ is a symmetric and periodic function of $k$. Assuming a solution ansatz of the form

$$\Delta_k = 2\Delta_x \cos(k_xa_x) + 2\Delta_y \cos(k_ya_y) \tag{6}$$

the BCS gap equation reduces to a two dimensional nonlinear algebraic system

$$f_1(\Delta_x,\Delta_y) = \Delta_x - \frac{1}{N} g_x\sum_k \cos(k_xa_x)F_k = 0 \tag{7a}$$

$$f_2(\Delta_x,\Delta_y) = \Delta_y - \frac{1}{N} g_y \sum_k \cos(k_ya_y)F_k = 0 \quad . \tag{7b}$$

In the normal conducting state Eqs. (7a-b) possess the solution $(\Delta_x, \Delta_y) = (0,0)$. In the superconducting state multiple solutions of Eqs. (7a-b) exist corresponding to stationary points of the Gibbs' free energy. At fixed $\mu$ the solution corresponding to the lowest Gibbs' energy is to be chosen as the physical one which is always stable. The other solution points can be local minima (stable) or maxima (unstable) and hyperbolic points (unstable). The stability of a given solution point is determined by the eigenvalues of the Jacobian

$$\mathbf{J} = \begin{bmatrix} \dfrac{\partial f_1}{\partial \Delta_x} & \dfrac{\partial f_1}{\partial \Delta_y} \\[2mm] \dfrac{\partial f_2}{\partial \Delta_x} & \dfrac{\partial f_2}{\partial \Delta_y} \end{bmatrix} \quad . \tag{8}$$

If both eigenvalues are negative the point is stable. Hyperbolic points are characterized by having a negative and a positive eigenvalue. A necessary condition for the occurrence of bifurcations is that one eigenvalue is equal to zero. This condition may be fulfilled as the chemical potential $\mu$ and the absolute temperature T are varied. In the following we shall investigate pitchfork bifurcations emerging as T and $\mu$ are varied and in particular study bifurcation scenarios appearing as a result of perturbations. The perturbations of interest are asymmetric pairing interactions ($g_x \neq g_y$) and asymmetric hopping probabilities ($A_y \neq 1$).

## NUMERICAL RESULTS

Eqs. (7a-b) were studied numerically setting A and $k_B$ equal to unity which is equivalent to measuring the temperature in units of $A/k_B$ and the energies in units of A. A typical value for A is A=0.05 eV leading to temperatures measured in units of 580 K and typical values for the pairing strengths are $g_x=g_y=0.06$ eV (see ref. 4). In units of A we then have $g_x=g_y=1.20$. The values of B and C in Eq. (4) were chosen to be B=0.45 and C=0.1.

In the normal conducting state the gap parameter is zero. As the temperature is lowered the transition to the superconducting state occurs characterized by a non-zero gap. In terms of Eqs. (7a-b) this transition results from a pitchfork bifurcation. In the superconducting state the non vanishing solutions of the gap equation (7) are classified as s-wave solutions when $\Delta_x = \Delta_y$, d-wave solutions when $\Delta_x = -\Delta_y$, while all other cases are called mixed s- and d-wave solutions. A quick overview of all the solutions to Eq. (7) and their stability is best achieved by showing contour plots of the Gibbs free energy. In Fig. 2 we show a contour plot of the Gibbs energy in the case of symmetric pairing interactions and symmetric hopping probabilities in the x- and y-directions ($g_x = g_y$ and $A_y = 1$). Fig. 2 reveals four minima of the same depth meaning that four degenerate physical solutions exist to the gap equation. In addition four hyperbolic unstable points are observed together with a local maximum at the center. The dependence of these solutions on the absolute temperature T is shown in Fig. 3. As T is lowered from T=0.2 the first pitchfork bifurcation taking place at $T=T_c=0.17$ indicates the transition from the normal to a s-wave superconducting state. The second pitchfork bifurcation at T=0.12 signals a transition from the s-wave state to a mixed s- and d-wave state.

Variation of the Fermi energy gives rise to a rich structure of bifurcation phenomena. The Figs. 4 to 6 show schematically bifurcation scenarios emerging from

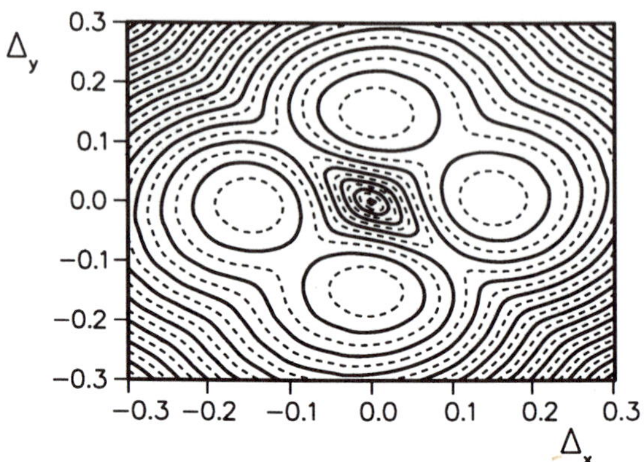

Fig. 2. Contour plots of the Gibbs free energy at $\mu = -2.0$, $T = 0$, $g_x = g_y = 1.2$, and $A_y = 1$.

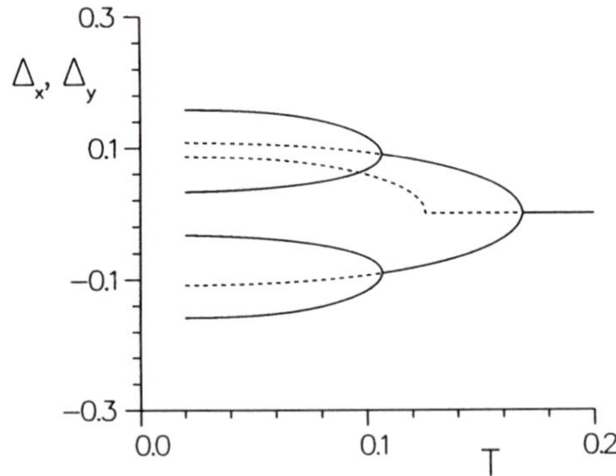

Fig. 3. Bifurcation diagram showing $\Delta_x$ and $\Delta_y$ vs. the temperature T at $\mu = -1.9$ in the symmetric case $g_x = g_y = 1.2$, and $t_1 = t_2$. Full curves mark stable solutions and the dashed curves mark unstable solutions.

varying $\mu$ in the following three cases: 1) $g_x = g_y$ and $A_y = 1$. 2) $g_x \neq g_y$ and $A_y = 1$. 3) $g_x = g_y$ and $A_y \neq 1$. The larger circles show the positions of global minima of the Gibbs energy and the smaller  The crosses are  circles show local minima.  hyperbolic points and the squares indicate local maxima. Fig. 4 provides a perspective view of the bifurcation scenario in the case 1 with symmetric pairing interactions and symmetric hopping probabilities. The first bifurcation in Fig. 4 signals the transition from the normal to the superconducting state and the existence of s-wave solutions. A second pitchfork bifurcation of the stable solutions produces mixed wave  solutions  and  by

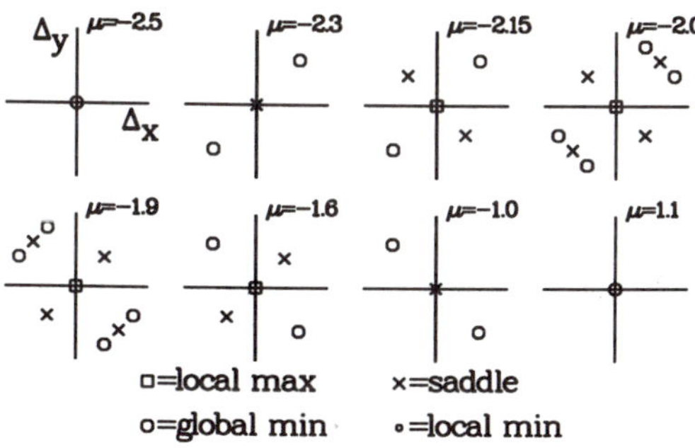


Fig. 4.  Schematic diagram of the extrema of the Gibbs  free energy for $A_y = 1$, $g_x = g_y = 1.2$ as $\mu$ increases.

increasing $\mu$ further the mixed wave solution disappears through another pitchfork bifurcation leaving behind d-wave solutions. Note that each "snapshot" in fig. 4 contains local minima of the same depth, representing thus the stable, physically relevant solutions. When the pairing interaction strenghts in the x- and y-direction differ slightly, the pitchfork bifurcations of the symmetric case described above are transformed into perturbed pitchfork bifurcations. After the second  bifurcation  the emerging  stable solutions correspond to minima of different depths (Fig. 5). The physically relevant solution is the one corresponding to the deepest minimum. Note that the initially physical solutions remain such through the whole sequence as $\mu$ increases. It should also be mentioned that now the symmetry is broken i.e. pure s-wave and pure d-wave solutions do not exist anymore. If the hopping probabilities in the x- and y-direction differ slightly then again perturbed pitchfork bifurcations emerge and s- and d-wave solutions cease to exist. However, now the physically relevant solution does not remain

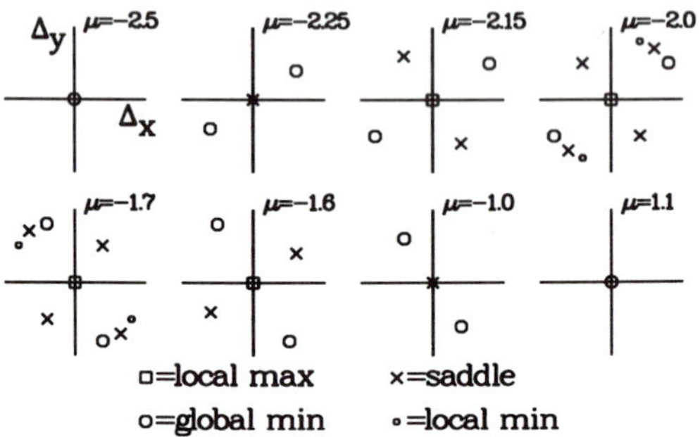

Fig. 5.  Schematic diagram of the extrema of the Gibbs free energy for $A_y = 1$, $g_x = 1.20$ and $g_y = 1.18$ as $\mu$ increases.

Fig. 6.  Schematic diagram of the extrema of the Gibbs free energy for $A_y = 0.9$, $g_x = g_y = 1.2$ as $\mu$ increases.

the same as $\mu$ increases. As is illustrated in Fig. 4 local and global minima swap i.e. the initially deepest minimum becomes shallow and finally disappears through a saddle node bifurcation. Thus, the final physical solution did not emerge from a bifurcation but from the gradual deepening of the initially shallow minimum and the simultaneous shallowing of the initially deep minimum. Hence, the competition between the two stable solutions led to a phase transition that did not result from a bifurcation as in the two first scenarios.

## SUMMARY

Applying the BCS strategy in modelling layered high $T_c$ superconductors it has been shown that the solutions of the associated BCS gap equation experience pitchfork and perturbed pitchfork bifurcations as the absolute temperature and the chemical potential are varying. The bifurcation points separate different superconducting phases. When the hopping probabilities in the x and y direction differ, phase transitions occur which do not result from bifurcations. When the temperature is used as a bifurcation parameter then again pitchfork (symmetric case) and perturbed pitchfork bifurcations (all nonsymmetric cases) emerge. The pitchfork bifurcations manifest themselves as discontinuities in the specific heat as a function of the temperature[5,6]. When the symmetry is broken, and the abrupt pitchfork bifurcations are replaced by saddle node bifurcations the discontinuities are removed.

## ACKNOWLEDGEMENTS

This work is supported by the Danish National Science Research Council (Grant Nos. 11-7075 and 11-9288) and the EC Science programme (Contract SCI /0229-C(AM)).

## REFERENCES

1.    A.J. Leggett, Rev. Mod. Phys., 47(2):331 (1975).

2.    T. Schneider and M.P. Soerensen, Z. Phys. B - Condensed Matter 81: 3 (1990).

3.    H. De Raedt, T. Schneider and M.P. Soerensen, Z. Phys. B - Condensed Matter 79:327 (1990).

4.    P.N. Spathis, M.P. Soerensen and N. Lazarides, to appear in Phys. Rev. B.

5.      H. Mori, Jour. Phys. Soc. Jap. 58:1394 (1989).

6.      T.C. Choy, M.P. Das and Hongzing He, Phase Transitions, 20:1 (1990).

# DYNAMICS OF FIBRE LASER WITH NONLINEAR EXTERNAL CAVITY


M.P. Sørensen, K.A. Shore[1], T. Geisler[2], P.L. Christiansen, J. Mark[3], and J. Mørk[3]

Laboratory of Applied Mathematical Physics
The Technical University of Denmark, Bldg. 303
DK-2800 Lyngby, Denmark


## ABSTRACT


Solid state lasers coupled to non-linear auxiliary cavities have been demonstrated to exhibit optical pulse compression using the Additive Pulse Mode-locking (APM) technique. The possibility of utilizing rare earth doped fibre lasers instead of solid state lasers for generating short optical pulses is investigated numerically in an APM configuration. It is demonstrated that pulse compression takes place and that soliton formation plays an important role in the mechanism for pulse shortening. Break down of an initial pulse into multiple compressed pulses has been observed in the numerical simulations.



1) Bath University, School of Electronic and Electrical Engineering, Bath, BA2 7AY, UK.

2) Danish Institute of Fundamental Metrology, Building 307, Lundtoftevej 100, DK-2800 Lyngby, Denmark.

3) TFL, Telecommunications Research Laboratory, Lyngsø Alle 2, DK-2970, Hørsholm, Denmark.



*Chaotic Dynamics: Theory and Practice*
Edited by T. Bountis, Plenum Press, New York, 1992

## INTRODUCTION

Rare earth doped fibre laser devices have been the subject of a rapid development from laboratory experiment to commercial exploitation. In optical communications systems fibre lasers are attractive amplification devices because of their direct compatibility with the components of optical systems. This also holds for application of the fibre laser in an Additive Pulse Mode-locking (APM) set up for generating short pulses with duration of the order of 100 femtoseconds. In an APM configuration the active fibre laser is coupled to a nonlinear optical fibre through a mirror. Pulse compression is then obtained by the interferometric addition of a pulse emitted from the fibre laser and a chirped pulse from the fibre cavity. Strictly the pulse travelling in the fibre laser will also be chirped due to nonlinearity. The pulse narrowing can be viewed as a result of constructive interference near the peak of the pulse and destructive interference occuring at the wings of the pulse[1,2,4].

Analytical approaches to describe the pulse compression in an APM configuration has been utilized in a number of publications[2,4]. However, analytical work has so far only been able to demonstrate pulse width reduction in each round trip in the APM set up. Here the aim is to study the details of the transients and the final steady state pulse propagation emerging from a given initial pulse in a full numerical simulation of the APM configuration.

## APM MODELLING TASK

In Fig. 1 a simple APM configuration set up is displayed. The left hand box represents the active nonlinear fibre laser which is coupled to a passive nonlinear fibre at the right hand side. The coupling takes place at the center mirror which forms a common mirror for the two cavities. The pulse propagation through the active fibre laser and the passive fibre cavity is governed by[3,4]

$$\frac{\partial A}{\partial Z} + \beta_1 \frac{\partial A}{\partial T} + \frac{i}{2}\beta_2 \frac{\partial^2 A}{\partial T^2} - i\gamma|A|^2 A = GA \quad . \tag{1}$$

$A = A(Z,T)$ is the electrical field amplitude measured in units of $\sqrt{W}$ (W = Watt) at the

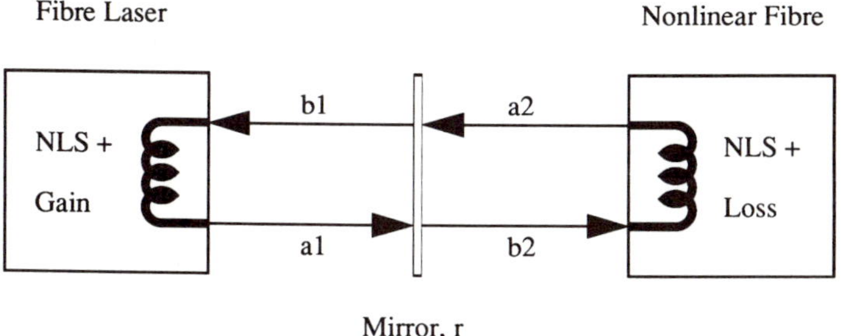


Fig. 1   Schematic diagram of the all-fibre pulse mode locking configuration.

cross sectional position Z and at time T in the fibre. $\beta_1$ is the inverse group velocity which equals $n_g/c$ where $n_g$ is the refractive index and c is the velocity of light in vacuum. The parameter $\beta_2$ is related to the dispersion parameter D through $\beta_2 = -\lambda_0^2 D/(2\pi c)$. $\lambda_0$ is the wavelength of the carrier wave. The parameter $\gamma$ denotes the nonlinearity of the fibre. Let $n_2$ be the nonlinear index coefficient and let $A_{eff}$ be the effective core area then $\gamma = n_2\omega_0/(cA_{eff})$ where $\omega_0$ is the frequency of the carrier wave. The G on the right hand side of Eq. (1) is a differential operator given by[2,5]

$$ G = -i\frac{\alpha}{2} + i\frac{G_0}{1 + P/P_{sat}} \left(1 + \frac{1}{\omega_b^2}\frac{\partial^2}{\partial t^2}\right) \quad . \tag{2} $$

$\alpha$ represents the linear damping in the fibre laser as well as in the passive fibre cavity. The second term in Eq. (2) models the bandwidth limited gain in the laser fibre. This term is zero in the passive fibre cavity. $G_0$ and $\omega_b$ are the gain and the bandwidth, respectively, of the fibre laser. P is the total pulse power defined by

$$ P = F_{rep} \int_{-\infty}^{\infty} |A(Z,T)|^2 dT \quad . \tag{3} $$

$F_{rep}$ is the repetition frequency of the pulse and $P_{sat}$ is the saturation power of the fibre laser. Using the variable transformations

$$t = k_t(T - \beta_1 Z) \ , \quad z = k_z Z \ , \quad A(Z,T) = k u(z,t) \ , \tag{4}$$

with the parameter definitions

$$k_z = -\beta_2 k_t^2 / \epsilon \ , \quad k^2 = k_z / \gamma \ , \tag{5}$$

$$\gamma = n_2 \omega_0 / (cA_{eff}) \ , \quad \Gamma = \alpha / (2k_z) \ ,$$

the dynamical equations (1) and (2) transform into their normalized form

$$i\frac{\partial u}{\partial z} + \frac{1}{2}\epsilon\frac{\partial^2 u}{\partial t^2} + |u|^2 u - i\Gamma u = igu \ , \tag{6}$$

where g is the differential operator

$$g = \frac{ig_0}{1 + P/P_{sat}} \left(1 + \frac{k_t^2}{\omega_b^2}\frac{\partial^2}{\partial t^2}\right) \ , \tag{7}$$

where the normalized unsaturated gain is given by $g_0 = G_0 / k_z$. The parameter $\epsilon$ is the normalized dispersion $\beta_2$. For $\epsilon = +1$ the fibre has anomalous dispersion and for $\epsilon = -1$ we are in the normal dispersion regime. Keeping $k_z$ and $k_t$ constant, increasing (decreasing) $\epsilon$ corresponds to increasing (decreasing) the dispersion of the fibre. In the APM configuration in Fig. 1 a laser pulse travels through the fibre laser and toward the partially transparent mirror. At the mirror, part of the pulse is transmitted into the passive nonlinear fibre, and part of the pulse is reflected back into the fibre laser. The two cavity lengths are adjusted so that the pulse from the fibre laser and the pulse from the passive fibre meet at the common mirror. After n round trips of the pulses in the APM system the amplitude of the outcoming pulse from the fibre laser is $a_{1,n}$ and the amplitude of the outcoming pulse from the passive fibre is $a_{2,n}$. The reflected/transmitted pulse amplitudes $b_{1,n}$ and $b_{2,n}$ are then given by the relations

$$b_{1,n} = ra_{1,n} + ta_{2,n} \tag{8}$$

$$b_{2,n} = ta_{1,n} - ra_{2,n}$$

and

$$a_{1,n} = \eta \hat{P}(\eta b_{1,n-1}; \epsilon, g_0)$$

$$a_{2,n} = \eta e^{-i\psi} \hat{P}(\eta b_{2,n-1}; \epsilon, 0)$$

(9)

where $\hat{P} = \hat{P}(u(z=0), \epsilon, g_0)$ is the operator accounting for propagation through the fibre section with gain $g_0$ and normalized dispersion $\epsilon$. r is the reflection coefficient and t = $(1-r^2)^{1/2}$ is the transmission coefficient of the mirror. Fine tuning of the passive cavity length is accounted for by multiplying the outgoing pulse ($a_{2,n}$) from the passive fibre with the factor $\exp(-i\psi)$, with $\psi$ being the phase shift[1]. Furthermore, there are coupling losses when the pulse enters and leaves a fibre. This coupling loss we denote $\eta$ and it is assumed to be the same for both fibres. In the numerical simulations we have used the parameter values in table 1 both for the fibre laser and for the passive fibre cavity.

TABLE 1. List of the parameter values used in the numerical calculations. The unsaturated gain $g_0$ for the fibre laser has been varied.

| | | |
|---|---|---|
| $n_g = 1.45$ | $n_2 = 3.2 \; 10^{-20} \; m^2/W$ | $c = 3.0 \; 10^8 \; m/s$ |
| $\beta_1 = 4.83 \; 10^{-9} \; s/m$ | $\beta_2 = -3.44 \; (ps)^2/km$ | $D = 2.7 \; ps/(km \; nm)$ |
| $\Gamma = 2.09 \; 10^{-3}$ | $A_{eff} = 39 \; (\mu m)^2$ | $k_t = 2.5 \; 10^{12} \; Hz$ |
| $k_z = 2.15 \; 10^{-2} \; m^{-1}$ | $k = 2.88 \; \sqrt{W}$ | $\epsilon = 1.0$ |
| $\eta = 0.8$ | $r = 0.9$ | $\psi = -2.36$ |
| $P_{sat} = 2 \; mW$ | Fibre length = 2.05 m | $F_{rep} = 1.0 \; 10^8 \; Hz$ |
| $\alpha = 9.0 \; 10^{-5} \; m^{-1} = 0.39 \; dB/km$ | | |

## NUMERICAL RESULTS

In the full numerical simulation of the APM configuration it is possible to obtain detailed information about the transient dynamics occuring during pulse compression and to obtain the final steady state pulse. So far analytical studies have only provided results showing the possibility of pulse compression from one roundtrip to the next. The transient dynamical behaviour is best illustrated by depicting the Full Width Half Maximum (FWHM) as function of the number of round trips, n. In Fig. 2a we show a 3-dimensional plot of the compression of the initial sech pulse

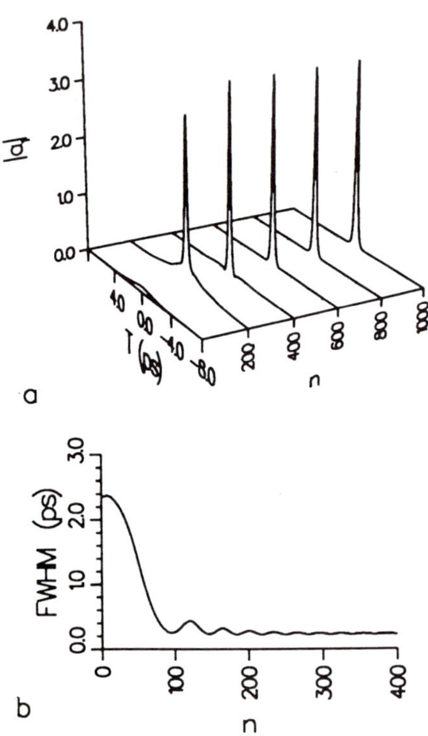



Fig. 2. (a) The normalized amplitude $a_{1,n}$ as function of T and n with the initial conditions given by (10) ($B_1 = 0.1$ and $B_2 = 0.3$). In the fibre laser $g_0 = 20.0$. The other parameter values are stated in table 1. (b) FWHM as function of the number of round trips, n.

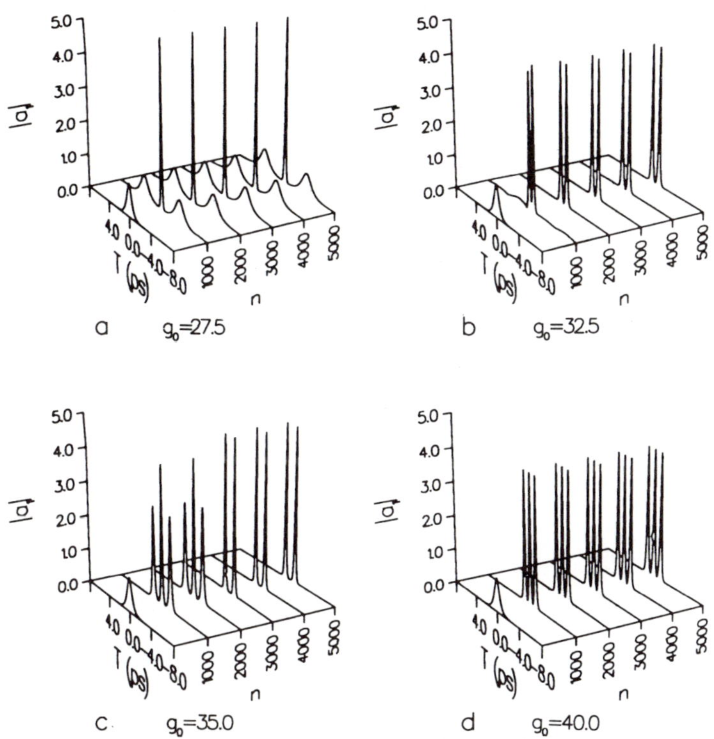


Fig. 3   Break up of the initial single pulse (10) ($B_1 = 1.0$ and $B_2 = 1.0$) into multiple pulses for high values of the gain in the fibre laser. (a) $g_0 = 27.5$, (b) 32.5, (c) 35.0, and (d) 40.0. Other parameters as in Fig. 2.

$$a_{1,0}(T) = B_1 sech(B_2 T), \quad a_{2,0}=0, \quad b_{1,0} = 0, \quad b_{2,0} = 0 \quad , \tag{10}$$

travelling toward the mirror from the fibre laser. Fig. 2b shows the FWHM as function of n. The final pulse width is 234 femtoseconds which in this case corresponds to a reduction by a factor of 10 of the initial width. Other choices for the initial pulse lead to the same modelocked state, but with wider pulses requiring more roundtrips to reach steady state. The initial pulses are believed to arise due to noise fluctuations, but we do not address this question further. As $g_0$ is increased the original single pulse becomes unstable and breaks up into various types of multiple travelling pulses as shown in Fig. 3. In the Figs. 3b and 3d two and three bunched pulse propagation are observed, respectively. However more complex types of multiple pulse propagations can appear as is evident from the Figs. 3a and 3c. The amplitude of the pulses are so large that we are in the soliton regime in both fibres. As the initial single pulse gains more and more energy in the fibre laser, this energy may become too large for sustaining just a single soliton pulse which then breaks down into multiple solitons. Due to the loss and energy input terms in the nonlinear Schrödinger Eq. (1) the solitons are strictly not pure solitons but merely solitary waves. Break down into solitons and/or solitary waves from arbitrary initial conditions is a well known phenomenon[3].

## CONCLUSION

A numerical model of a simple APM configuration has been implemented in order to simulate pulse formation. The simulations provide detailed information about the pulse compression including the transients and the final steady state pulse propagation. In the simulations the pulse amplitudes are so large that the pulses must be interpreted as quasi solitons and that the solitonic nature may play an important role in the formation of femtosecond pulses beside the usual interferometric mechanism beeing proposed so far. Furthermore, instability of the single pulse propagation for high gain values in the fibre laser leads to break up into multiple solitary waves. This supports the importance of solitonic effects in the interpretation of the mode of operation of the APM laser configuration utilized for generation of femtosecond pulses. Also higher order terms like self frequency shifts may be of importance for very narrow (100-200 fs) pulses. However, preliminary calculations show in accordance with Ref. 5 that stable compressed pulses can be obtained also in this case.

## ACKNOWLEDGEMENTS

The authors acknowledge Prof. E.P. Ippen, MIT, for useful discussions. One author (KAS) acknowledges financial support from the Danish Research Academy

(Forskerakademiet) under grant No. 5900103 and from TFL and expresses his appreciation for all the hospitality and assistance received during stays at TFL and at the MIDIT Centre of the Technical University of Denmark.

## REFERENCES


1.  J. Mark, L.Y. Liu, K.L. Hall, H.A. Haus, and E.P. Ippen, Opt. Lett. 14:48 (1989).

2.  E.P. Ippen, H.A. Haus, and L.Y. Liu, J. Opt. Soc. America, B6:1736, (1989).

3.  G.P. Agrawal, IEEE Photonics Techn. Letts., 2:875, (1990).

4.  T. Brabec, F. Krausz, Ch. Spielmann, E. Wintner, and M. Budil, J. Opt. Soc. Am. B8, 1818, 1991.

5.  M. Nakazawa, H. Kubota, K. Kurokawa, and E. Yamada, J. Opt. Soc. Am. B8, 1811, 1991.

# SYMBOLIC DESCRIPTION OF IRREGULAR BEHAVIOUR IN A LASER MODEL


C. Jung

Fachbereich Physik
Universität Bremen
2800 Bremen
Germany


and


F. Hollinger

Festkörper-Laser-Institut
Strasse des 17. Juni 137
1000 Berlin
Germany


## INTRODUCTION

When we encounter complicated motion in some dynamical system, then we are interested in finding a method to give a quantitative measure of the degree of chaoticity. One set of suitable quantities are the generalized entropies $K_q$ of order $q$ [1]. In general, it takes some numerical effort, to extract $K_q$ from a given time series. The computation becomes simple in those cases, in which we can describe the dynamics by an appropriate symbolic description [2]. Then, the question arises, how to find an appropriate symbolic description. So far we do not yet have a systematic method to solve this task. Therefore it is always interesting to find particular examples for which a symbolic description drops out in a quite natural way. The purpose of this paper is to show, how a surprisingly simple symbolic description emerges in some parameter range for the irregular dynamics of a laser model, which we have presented previously [3].

## THE MODEL

Our model describes an unidirectional ring resonator which is stabilized to one longitudinal mode. Complicated behaviour is created by the nonlinear interaction between the various transversal modes ( for a review about laser chaos caused by transversal effects see [4] ). This set up allows to restrict our considerations to a surface $S$ inside the resonator perpendicular to the direction of propagation of the light field. On $S$ there exists a set of eigenfunctions $f_j$ of the resonator with eigenvalues $\gamma_j$. $\mid \gamma_j \mid^2$ gives the relative part of the energy in mode $j$


*Chaotic Dynamics: Theory and Practice*
Edited by T. Bountis, Plenum Press, New York, 1992

which returns to $S$ after one roundtrip around the resonator. $\varphi_j$, the phase of $\gamma_j$, gives the phase shift of the complex electric field $E_j$ of mode $j$ during one round trip. Without loss of generality we can always assume that $\varphi_1 = 0$.

The electric field $E$ on $S$ and $N$, the projection of the inversion density on $S$, are decomposed into the eigenfunctions $f_j$ according to

$$E = \sum_j E_j f_j \tag{1}$$

$$N = \sum_j N_j f_j \tag{2}$$

The time is taken to be discrete where the unit of time is the round trip time around the ring resonator. The dynamics of the system is given by the polynomial map :

$$E_k(n+1) = \gamma_k E_k(n) + \frac{\sigma l}{2} \sum_{j,m} \gamma_m U_{k,j,m} N_j(n) E_m(n) \tag{3}$$

$$N_k(n+1) = (1-A)N_k(n) + wN_0 s_k - \frac{\epsilon \sigma L}{2\hbar\omega} \sum_{j,m,r} V_{k,j,m,r} N_r(n) E_j(n) E_m^*(n) \tag{4}$$

$\sigma$ is the cross section for induced emission, $l$ is the length of the amplifier, $w$ is the pump rate, $A$ is the decay rate of the inversion, $L$ is the legnth of the resonator, $\hbar\omega$ is the photon energy, $\epsilon$ is the dielectric constant, $N_0$ is the density of active atoms. $s, U, V$ are overlaps of the mode functions ( for more details see [3] ). In the following we restrict our considerations to the cases, where two modes are above threshold. Then 5 dimensions of the phase space are important, as their coordinates we use : $N_1, N_2, | E_1 |^2, | E_2 |^2$ and $\chi$, the relative phase between $E_1$ and $E_2$. As shown in [3] the most interesting parameter to vary is $\varphi_2$. In the case of small $\varphi_2$ we find permanent cooperative frequency locking and the energy output becomes constant. For large $\varphi_2$ we find multiperiodic behaviour where the two modes run almost independently of each other. In between there is a small range of $\varphi_2$ where we find complicated behaviour. For $| \gamma_1 | = 0.8$ , $| \gamma_2 | = 0.6$ and $\varphi_2 = 0.14\pi$ the total output energy as a function of time has been shown in Fig.5 in [3]. Also shown there is the Fourier transform of the power, clearly indicating the presence of genuine chaos. Let us use these parameter values to demonstrate our ideas on a symbolic description in the next sections.


CONSTRUCTION OF THE SYMBOLIC SEQUENCE

In the chaotic parameter region the model produces irregular pulses of light and regular relaxation oscillations in the time intervals in between these spikes. In Fig.1 we present the 5 important phase space coordinates as function of time, displaying a time interval containing 2 consecutive irregular outbursts. Part (a) shows $| E_1 |^2$ and part (b) shows $| E_2 |^2$. During the irregular outbursts the powers are maximal and show fast fluctuations. In between, in the laminar phases, the powers are smoother and show regular damped modulation, which are relaxation oscillations. Part (c) shows $\chi$, indicating that in the laminar phases the two modes are phase locked and accordingly the slow oscillations of $| E_j |^2$ are relaxation oscillations of the phase locked state. The clue, why the locked state is not permanently stable, is given by the behaviour of $N_1$ and $N_2$ shown in parts (d) and (e) of Fig.1. During the chaotic outbursts the inversion is depleted to a value at which the phase locked state exists. However, during the time intervals of phase locked laminar behaviour the inversion grows up to a level at which the phase locked state becomes unstable. Then the accumulated inversion triggers the next irregular outburst, during which the excess inversion is used up. Because of its inertia the system always overshoots and undershoots its equilibrium level of inversion and never comes to a rest. In addition, the exact amount of energy which is thrown out during the irregular outbursts is never the same and the system never comes back to exactly the same point in phase space. We find chaotic behaviour.

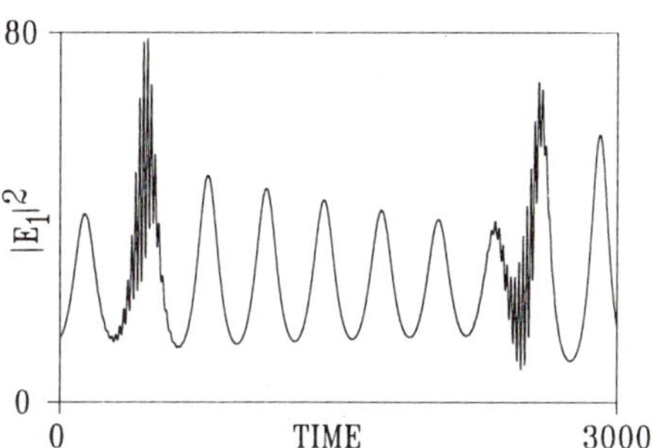

Fig.1a.  Plot of $|E_1|^2$ as function of time

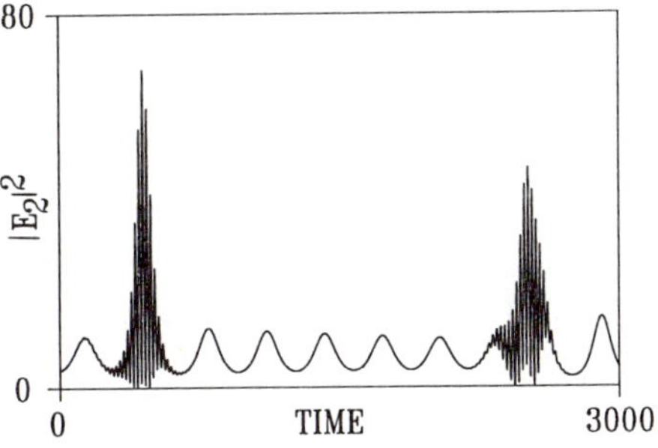

Fig.1b.  Plot of $|E_2|^2$ as function of time

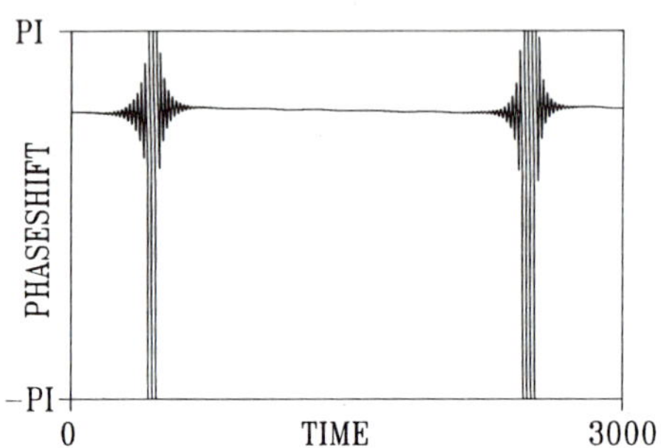

**Fig.1c.** Plot of $\chi$ as function of time

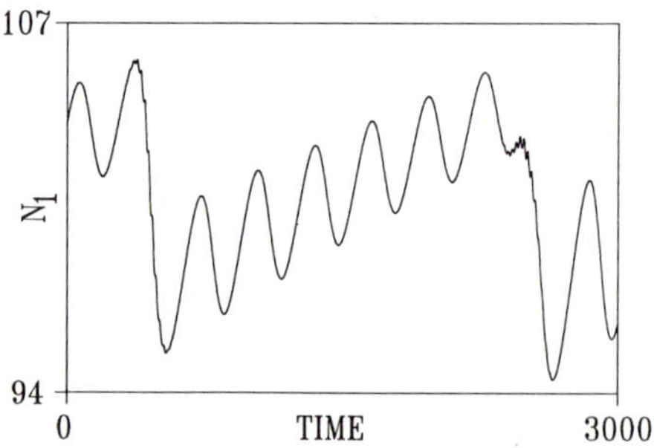

**Fig.1d.** Plot of $N_1$ as function of time

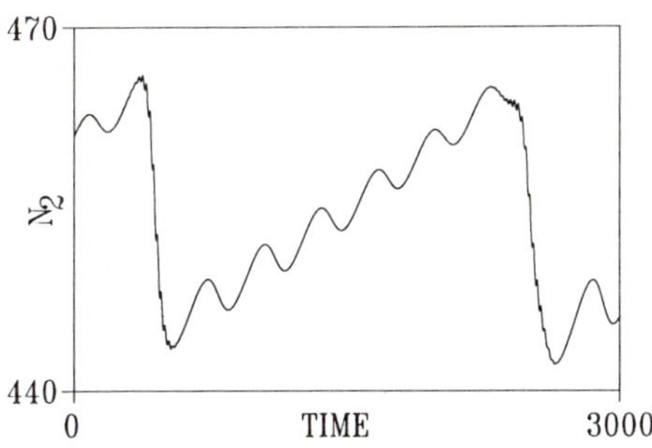



Fig.1e. Plot of $N_2$ as function of time

The most important observation for a quantitative evaluation of the amount of chaos comes from Fig.1c : In the laminar intervals the angle $\chi$ always comes back to the same value. Therefore, the total phase shift during the irregular outburst must be an integer multiple of $2\pi$. In this way to each irregular outburst there is assigned an interger number very naturally. And to the complete time series of the dynamics there is assigned a symbolic sequence of integers. For the example shown in Fig.1c we find a symbol value 3 for the left outburst and a symbol value 4 for the right one.

The question arises, whether this symbolic description is a symbolic dynamics in the mathematical sense. At the moment we do not have a proof for this conjecture, because we are not sure, that two different trajectories in phase space always lead to a different symbolic sequence in the long run. However, it is plausible to assume the validity of this conjecture. If this assumption would be wrong, then our computation of the entropies in the next section would only provide lower bounds to the correct values.

## EVALUATION OF THE SYMBOLIC SEQUENCE

Next we compute the entropies of the symbolic sequence : We take a long sequence of symbols created by the dynamics for the same parameter values as before. We cut the sequence into blocks of length $J$. Only the symbol values 2,3,4,5 occur with appreciable probability, therefore we disregard the very few blocks containing another symbol value. Because we take $M = 4$ values of the symbols into account, we have to deal with a total number of $L(J) = M^J$ different possible blocks $B_i^{(J)}$ of length $J$. We count the relative probabilities $P_i^{(J)}, i = 1, \cdots, L(J)$ to find the blocks $B_i^{(J)}$.

In Fig.2 we show $\ln P_i^{(J)}$ for the case of $J = 5$. In this histogram the blocks are ordered in lexicographical order along the horizontal axis. This plot is a multifractal structure, whose dimensions $D_q$ of order $q$ are related to the entropies $K_q$ of order $q$ of the symbolic sequence by [5] :

$$K_q = \ln(M) \cdot D_q \tag{5}$$

$D_q$ is computed from the probabilities $P_i^{(J)}$ as

$$D_q = -\ln[\sum_{i=1}^{L(J)} (P_i^{(J)})^q]/[(q-1)J \ln M] \tag{6}$$

in the limit of large $J$. In our case it was sufficient to take $J = 7$. Fig.3 gives a plot of $(q-1)D_q$ as function of q. The value for the fractal dimension $D_0$ very close to 1 indicates,

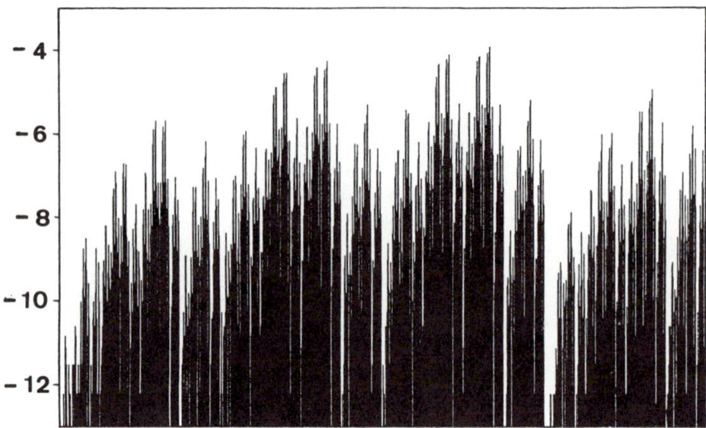

Fig.2. Histogram of the logarithms of the probabilities of symbol blocks of length $J = 5$ in $M = 4$ symbol values. The 1024 different blocks are arranged lexicographically.

that nearly all blocks in the symbols 2,3,4,5 have a probability different from zero, even though the probabilities can be extremely small. Correspondingly, the topological entropy $K_0$ is close to $\ln 4$. From the plot we read off the value $D_1 \approx 0.8$ for the information dimension leading to a value $K_1 \approx 1.11$ for the metric entropy. The asymptotic slope of 0.6 for $q \to \infty$ gives the value of the largest scaling factor of the symbolic dynamics. Because 4 is the most common symbol for the present parameter values, this means : If there is a long sequence of only 4's, then the probability is approximately $M^{-0.6} \approx 0.44$ that the next symbol is another 4.

What error do we make by disregarding the very few blocks containing a symbol value of 1 or 6 ? If we would take into account these additional symbol values, we would have to take $6^J$ different blocks of length $J$. Then all dimensions $D_q$ would be multiplied by a factor of approximately $\ln 4/\ln 6$ compared to the previous values, because all additional new blocks would either be completely absent in the symbolic sequence or would have an extremely small occupation probability. On the other hand, the connection between $K_q$ annd $D_q$ would be $K_q = \ln(6) \cdot D_q$, so that the factor $\ln 6$ would cancel out and $K_q$ would have approximately the same value as before. In this sense it is acceptable for the computation of the entropies, to disregard symbol values which occur with very small probability only.

In our system it is not useful to consider negative values of $q$ because of the following reason : If we take longer blocks, then there are some symbolic sequences ( e.g. the sequence consisting of 2's only ) whose probability is different from zero but converges to zero faster than exponentially with growing block length. And the scaling factors in the symbolic sequence do not possess a lower limit larger than zero. In the terminology of Ref.[6] our system shows a phase transition at $q = 0$ and formally all $D_q$ and $K_q$ for negative values of $q$ are $-\infty$.

Fig.4 shows the relative probability to find a particular time distance between adjacent chaotic outbursts. We observe : For smaller values of the time distance there are strong tendencies for this time to be an integer multiple of the cycle time of the beat oscillation between the two modes. For longer times this strong correlation fades out and the triggering of a new chaotic outburst is no longer restricted to a particular phase value of the beat motion. There still remain slower modulations related to the relaxation oscillation.

## DISCUSSION

The system switches between laminar phases of regular relaxation oscillations and turbulent

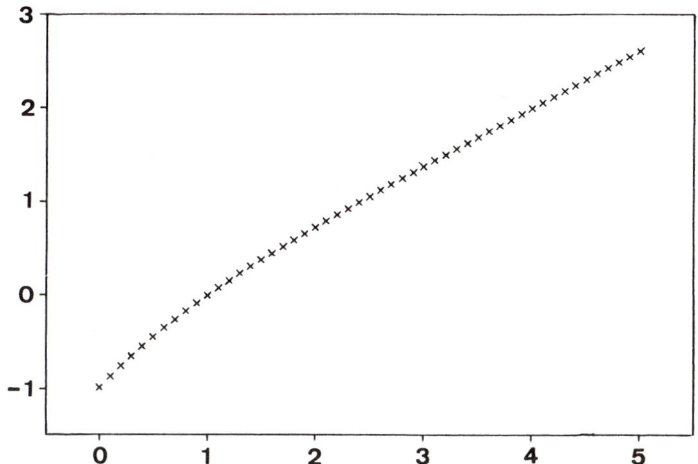

Fig.3.  Plot of $(q - 1)D_q$ versus $q$ for the histogram of Fig.2.

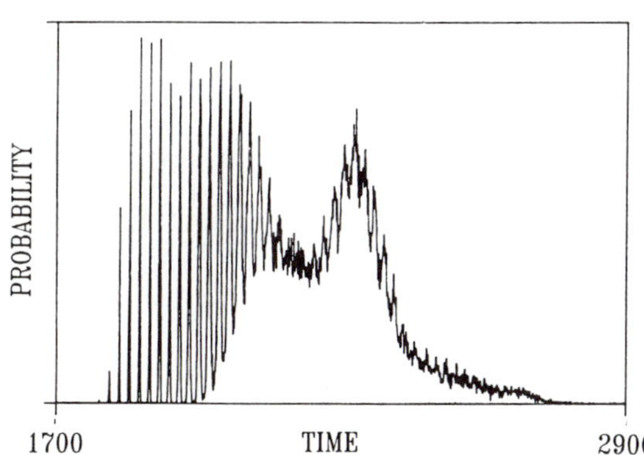



Fig.4. Relative probabilities for different values of the time distances between adjacent chaotic outbursts.

phases of irregular outbursts. Naturally, the question arises, whether this is a case of intermittency. To test for this we can check, whether the behaviour fits into one of the 3 well known classes of intermittency as described e.g. in [7].

One criterion is the distribution of the time lengths of the laminar phases. A comparison between Fig.4 in this paper and the figures in [7] indicates, that it does not fit into one of the usual classes of intermittency. Another criterion is the type of one-dimensional map to which the system can be reduced. For our system we have tried out many possibilities to construct a 1-D map. Unfortunately, all efforts were complete failures. In addition, for intermittency we expect, that the curve of Fig.3 tends towards slope zero for large values of $q$. On the other hand, an argument in favour of intermittency is the variation of the behaviour under variation of the parameter $\varphi_2$ : There is a critical value $\varphi_{c-}$ such that for $\varphi_2 < \varphi_{c-}$ the phase locked state is asymptotically stable. For $\varphi_2 > \varphi_{c-}$ the average time distance between adjacent irregular outbursts goes approximately like $T \propto (\varphi_2 - \varphi_{c-})^{-1/2}$. In total we conclude, that the behaviour of our system is not intermittency of one of the well known classes, even though the switching between laminar and turbulent phases looks like a case of intermittency. The reason for the unusual behaviour could be, that 5 dimensions in phase space are important in this system and that it is therefore impossible, to reduce the dynamics to a 1-D map, which is essential for the usual classification of intermittency.

The central idea of this paper is the construction of a symbolic sequence for a system where two oscillators switch between locked and unlocked states. Because the relative phase always returns to the same value in each time interval of phase locking, the total phase shift during any time interval of irregular behaviour can only be an integer multiple of $2\pi$, leading to an assignment of an integer to each irregular outburst in a natural way and thereby to a symbolic sequence for the whole trajectory of the system. Even when this construction does not provide a symbolic dynamics in the mathematical sense, this symbolic sequence is useful to establish lower bounds for the entropies $K_q$. We expect these lower bounds to be not very much smaller than the exact values, because in the worst case the symbolic sequence does not describe correctly the laminar phases, which do not contribute essentially to the entropies anyway, because of their regular character. Our values of the entropies are computed relative to the symbolic sequence and therefore they correspond to time units being the average time distance between adjacent chaotic outbursts. To get the entropies related to physical time we have to divide our values by the average time belonging to one step in the symbolic sequence.

An alternative explanation of the whole szenario is the following : By the nonlinear coupling of the modes with the amplifier the original modes $f_k$ of the empty resonator are no longer relevant. Some new modes $\tilde{f}_k$ have been created, which are mixtures of the old modes. For our choice of parameters the new mode $\tilde{f}_1$ is always above threshold, whereas the new mode $\tilde{f}_2$ switches on ( during the chaotic outbursts ) and off ( during the laminar time intervals ) irregularly. This permanent irregular relaxation behaviour of the second new mode causes the irregular time dependence of the total output power. Vice versa the old modes are linear combinations of the new modes. While both new modes are above threshold during the chaotic outbursts, the energy density of the old modes shows fast interference oscillations at the beat frequency between the two new modes ( see Figs.1a,b ).

The relative probabilities to find the various symbol values and blocks are very sensitive to the parameter value of $\varphi_2$. The general tendency is like this : For increasing $\varphi_2$ the higher symbol values become more likely until at $\varphi_2 > \varphi_{c+} \approx 0.148\pi$ the system switches from chaotic to multiperiodic and the new mode $\tilde{f}_2$ is always above threshold in the long time behaviour. For decreasing values of $\varphi_2$ the lower symbol values take over until at $\varphi_2 < \varphi_{c-} \approx 0.138\pi$ the system switches to a long time behaviour with constant output power. Then the new mode $\tilde{f}_2$ is always below threshold and in the picture of the new modes we find single mode operation with only the mode $\tilde{f}_1$ running. For $\varphi_2 < \varphi_{c-}$ but very close to $\varphi_{c-}$ we find long chaotic transients. This indicates the presence of a chaotic saddle in phase space which turns into a chaotic attractor at $\varphi_2 = \varphi_{c-}$. The exact values of $\varphi_{c-}$ and $\varphi_{c+}$ depend on the other parameter values, especially on $\mid \gamma_1 \mid$, $\mid \gamma_2 \mid$.

The phenomenon described in this article is not restricted to the case of two modes. Our numerical model has always four modes taken into account and so far only two of them have been above threshold. We have made a few runs with three modes above threshold and discovered similar phenomena for suitable parameter values : The system always comes back to a laminar phase, in which all three modes are phase locked. After a while one mode decouples from the other two, a chaotic outburst occurs and the relative phase moves. Soon the coupling is restored at the same phase value as before and the total phase shift during the outburst has been an integer multiple of $2\pi$. Each of the three modes is able to decouple from the other two which remain locked among themselves. When three modes are above threshold, then sometimes it also happens that all three modes decouple simultaneously from each other during a chaotic outburst.

## REFERENCES


1. P. Grassberger and I. Procaccia, Phys. Rev. A28, 2591 (1983)
2. A. Csordas and P. Szepfalusy, Phys. Rev. A38, 2582 (1988)
3. F. Hollinger, C. Jung and H. Weber, J. Opt. Soc. Am. B7, 1013 (1990)
4. N. B. Abraham, J. Opt. Soc. Am. B7, 951 (1990)
5. J. D. Farmer, Z. Naturforsch. 37a, 1304 (1982)
6. A. Csordas and P. Szepfalusy, Phys. Rev. A39, 4767 (1989)
7. H. G. Schuster, " Deterministic Chaos ", Physik Verlag, Weinheim (1984)

# CHAOTIC DYNAMICS IN PRACTICE: AN ELECTRONIC CIRCUIT FOR A NONLINEAR OSCILLATOR WITH DAMPING, FORCING, AND POSSIBLE ESCAPE TO INFINITY


E. del Río, A. Rodríguez-Lozano and M.G. Velarde

Facultad de Ciencias, U.N.E.D. Apartado 60.041
Madrid 28.080, Spain


## 1. INTRODUCTION

Nonlinear dynamical systems were already studied by H.Poincaré and subsequently by other authors(see Ref 1 for a comprehensive review ). However, it was not until 1963 with the discovery made by E. N. Lorenz of "deterministic chaos" in the **long time evolution** of a model for Bénard convection that the field received a high momentum. Today Lorenz's model is a paradigm in the study of nonlinear phenomena[2,3]. Moreover, recently, it was shown how for suitable parameter values this model and further generalizations of it can be transformed to a Duffing oscillator[4] or combination of oscillators[5]. Duffing's as well as other nonlinear oscillators, e. g., van der Pol's ,etc. have received a great deal of attention[6] for their regular and chaotic behaviour, which can be explored analytically, numerically with the digital computer, or analogically with a suitable electronic circuit.

In the present work we describe aspects of the dynamics of a nonlinear oscillator proposed about a century ago (1885) by Helmholtz[7] to account for combinational tones that must arise whenever the vibrations are so large that the square of the displacements has a sensible influence on the motions. According to Helmholtz[7] if we assume that in the vibrations of the tympanic membrane and its appendage, the square of the displacement has an effect on the vibrations, the magnitude of the force called into action namely $(aX + bX^2)$ implies that when $X$ changes its sign, the force changes not merely in sign, but also its absolute value. This assumption can hold only for an elastic body which is **unsymmetrically** related to positive and negative displacements. Now among the vibrating parts of the human ear, the drumskin is especially distinguished by its want of symmetry, because it is forcibly bent inwards to a considerable extent by the handle of the hammer. He ventured therefore to conjecture that this peculiar form of the tympanic membrane conditions the generator of combinational modes. Then the model oscillator he

proposed was one describing the motion of a single heavy point
under the influence of an **unsymmetrical** force early mentioned
together with an external forcing of sinusoidal character loosely
accounting for the effect of sound waves in space. In simple
terms he wrote

$$\ddot{X}+g\dot{X}+X+X^2=Asin(\omega t) \tag{1}$$

where X denotes the vibratory displacement, with a dot
accounting for a time derivative, g is a damping factor and A and
ω are the forcing amplitude and frequency, respectively. Further
details about the model and its possible relevance to physiology
can be found in Helmholtz's monograph[7]. Eq.(1) is the standard
form of the equation scaled with its "natural" units. However,
later on we shall see that (1) is not the most suitable form for
experiment, i.e., for an electronic analog circuit.

Recently, Thompson[8,9] has advocated the use of the archetypal
eq.(1) to describe ship stability to waves in windy situations
and its potential and eventual capsize. Indeed, the Helmholtz-
Thompson driven-oscillator has an underlying asymmetric potential
energy with a minimum corresponding to the upright state and a
saddle point beyond which the system can escape to a capsized
equilibrium state, thus reducing the problem of capsize to an
example of escape from a potential well.

It is interesting to note that the escape of a dynamical
system from a potential well is a recurrent theme in physics and
engineering and under periodic forcing it is known that the
escape will often be triggered by chaotic motions[8,10].

## 2. THE ELECTRONIC CIRCUIT

For the Helmholtz equation (1) we have constructed a model
analog circuit. Fig 1 is a sketch of the model. Let us summarize
its different parts, its mode of operation and its technical
performance.

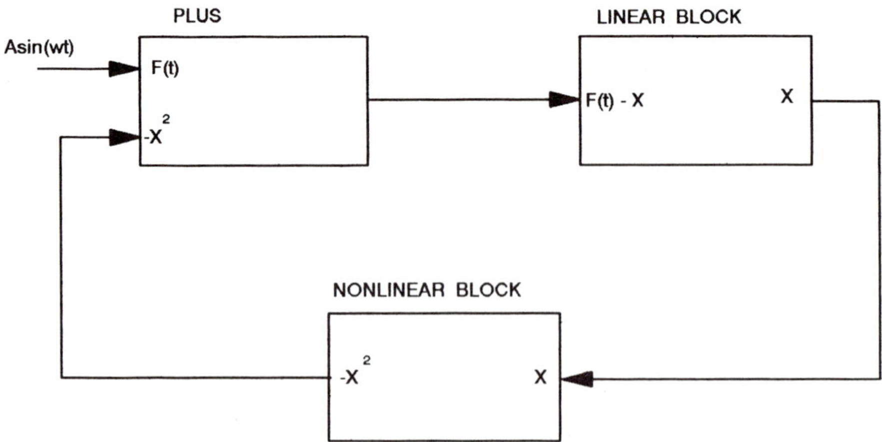


Fig.1    Block-diagram of Helmholtz's analog circuit.

## i) Linear part of the circuit

The usual approach to devise the linear part is by means of integrations[11,12]. However, such circuits act as integrators under the assumption of virtual ground[13] thus leading to large errors at low frequencies. They are not suitable to accurately account for nonlinear equations. Low frequency mode are highly important when dealing with Feigenbaun cascades[14] or intermittencies with 1/f noise[15]. Another problem with the circuit with integrators as in Fig 2 is that the output tends to wander off, even with the input is grounded, due to op-amp offsets and bias current because there is no feedback at dc, and of course the dc is very important in nonlinear circuits. Because this residual drift it may be necessary to put a large resistor across the capacitor but in this case even though we may use the virtual ground approximation we never have an optimum match between the integrator and the analytical problem. Fig 2 shows the ratio $V^{exp}_{out}/V^{th}_{out}$ between experiment and theory for a standard integrator and the electronic solution of the problem in our circuit.In both cases we have used as op-amp the op-07 and data have been adjusted by least squares fit. As in our circuit we do not obtain dX/dt in a straight way Fig 2 provides and upper bound to the error in the whole linear parts of the circuit.

The above given considerations led us to the choice sketched in Fig. 3. We have the <u>input</u> at resistance $R_2$ whereas the <u>output</u> is taken at the non-inverting entry of the operational amplifier (OP in what follows). The equation related to the circuit is:

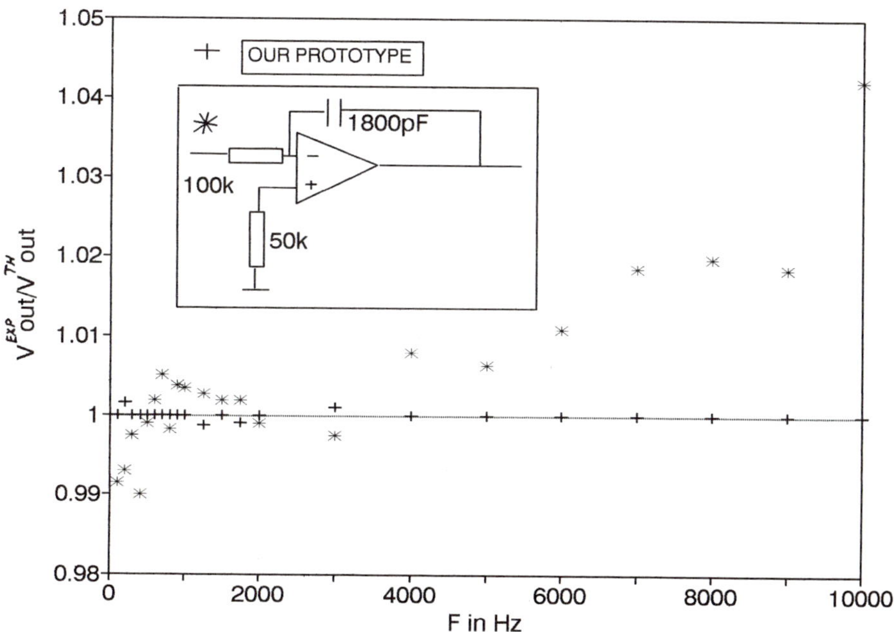


Fig. 2. Comperation between the standard integrator (*) and our
solution (+). Super and subscripts denote experimental
and theoretical outputs, respectively. The inset refers
to the standard integrator found in the literature.

$$X_{in}=L\ddot{X}_{out}+R\dot{X}_{out}+\frac{X_{out}}{C} \tag{2}$$

with

$$L=C_1 C_2 R_1 R_2 \tag{3}$$

$$R=(R_1+R_2)\,C_1+C_2 R_2\left(1-G+\frac{R_1}{R_{in}}\right) \tag{4}$$

$$C=\frac{R_{in}}{R_1+R_2+R_{in}} \tag{5}$$

were G is the gain for the block formed by IC1 and M1.
Eqs.(2)-(5)  deserve some pertinent comments:

i.a) We have no problem in controlling the damping factor g
(Adjustable by M1). Obviously , according to the result obtained
for R, Eq.(4) which is our g, we may  continuously proceed from
negative  to  positive  values  of  g.  This  is  particularly
interesting  for at g = 0 and A = 0 the equation is reduced to
a **hamiltonian integrable** system and this permits to  cross-check
the circuit's performance.

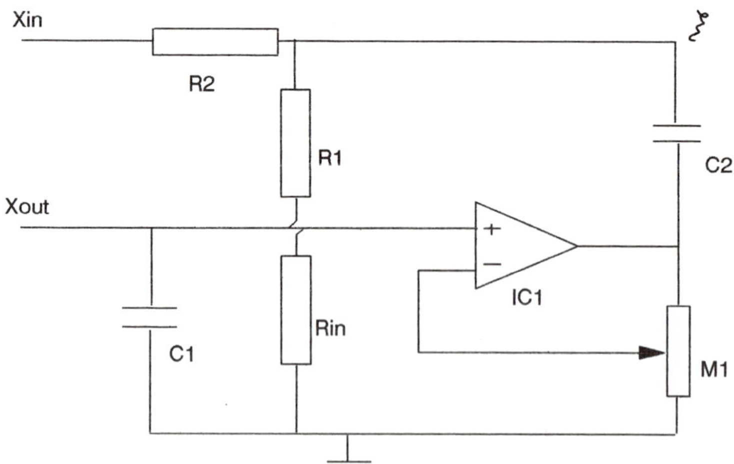


Fig. 3.    Linear Block of the circuit.

i.b) The accuracy of the circuit's performance rests on the accuracy of its components and does not depend on any theoretical approximations which are left out here. If we adjust it in a way that the OP works with feedback near unity, the high gains obtained so far with OP in an open cycle, ensure a fairly linear amplifier in a wide frequency range. Note that we shall neglect frequencies whose contribution to the solution is about the system's noise level.

## ii) **Obtaining dX/dt**

A shortcoming of our circuit is that we must obtain dX/dt in an indirect way. In fact we take

$$\dot{X}_{out} = \frac{1}{R_1 C} \, [\xi - (1 + \frac{R_1}{R_{in}}) \, X_{out}] \tag{6}$$

Thus using OPs as addition elements we obtain dX/dt, and so the full phase space of the dynamics. Note that the way we obtain dX/dt, the output error of dX/dt does not affect the accuracy in X.

## iii) **Initial condition control**

At t=0 we must specify **X** and **dX/dt**. Then the circuit carries the  time evolution on its own in an independent way. To fix **X** it suffices to fix the <u>input</u> potential at the non-inverting entry in Fig. 3. However, as indicated above we cannot directly choose **dX/dt** as we ought to act on the voltage at $\xi$ to obtain **dX/dt**. Once we have given such a voltage, the **dX/dt** device provides the **dX/dt** value corresponding to that voltage, as the circuit works in c.c. too. Thus we  incorporate two variable voltage power supplies, one for **X** and another for $\xi$ whose control is done from outside.

## iv) **Further details about the circuit**

The circuit has been designed to operate with a ±12V, with error below 0.001V. It is remarkable that the accuracy that IC1 (we use the OP77) has to compute X depends on the dielectric loss and charge retention of $C_1$ and $C_2$. Thus we are led to the use of capacitors capable of supporting up to 200V with all voltages in the circuit being of much lower value. Please note that although the capacitor $C_1$ looks like in a non-inverting sample-hold circuit, here plays a rather different role (see Sect. 2.i).

## v) **A test of the circuit's overall operation**

In Fig. 4  we depict the potential together with data E-E$_{kin}$ for g=0, A=0 and for each one of these two cases (always g=0, A=0): the inset of the separatrix and the outset.


## 3. MELNIKOV'S CRITERION AND ITS EXPERIMENTAL TEST

Melnikov[16]  provided a function M(A,g,$\omega$,t) which is a measure of the separation between the stable, Ws, and unstable, Wu, manifolds and if M=0 it implies the existence of a structure like a Smale horseshoe associated to some iteration  of the Poincaré map. This is  paradigmatic to "topological" chaos.

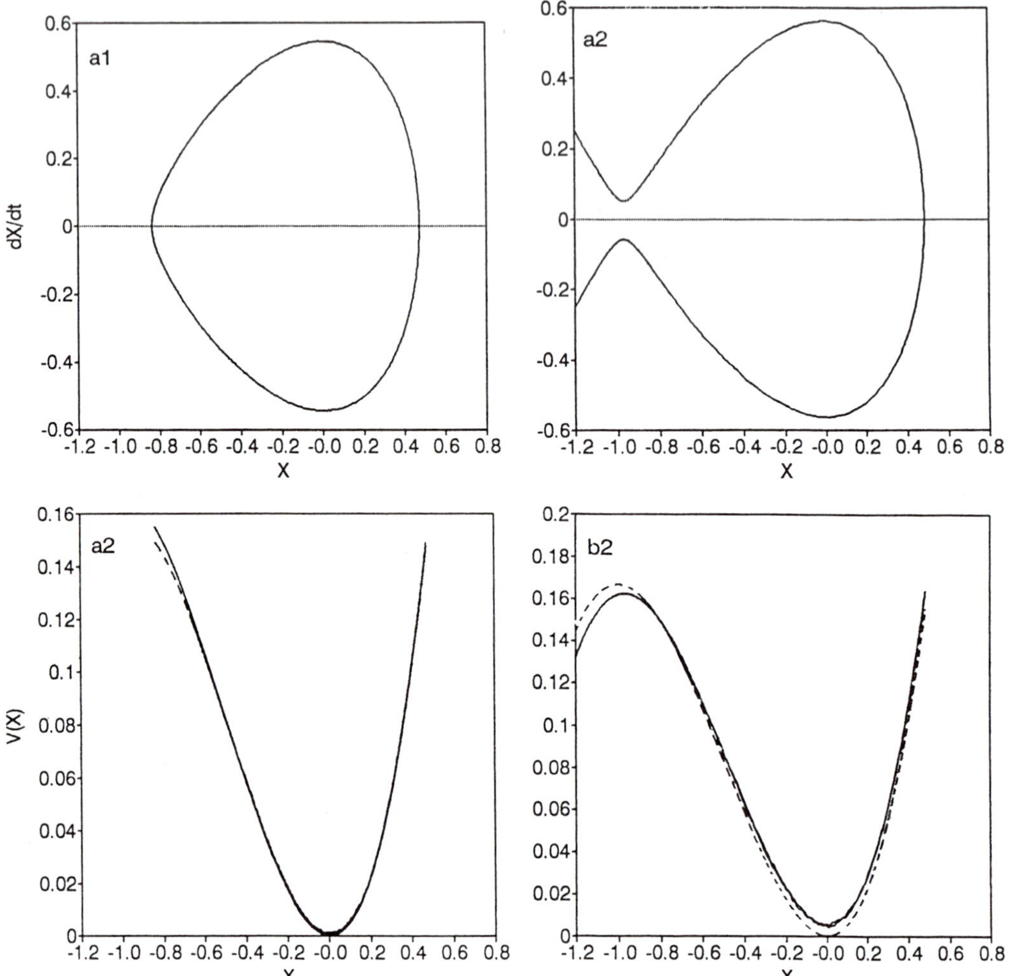

Fig. 4.  Phase space trajectories (a) and potential curves (b)
inset (a1,b1) and outset (a2,b2) the homoclinic
trajectory. Experimental point are on solid lines and the
theoretical values are on the broken lines.

For our nonlinear Helmholtz-Thompson oscillator equation we have indeed a perturbed hamiltonian system for which the Melnikov function is

$$M(A,g,\omega,t_0)=\frac{6\pi\omega^2 A\cos(\omega t_0)}{\sinh(\pi\omega)}-\frac{6}{5}g \qquad (7)$$

Thus when M=0 we have

$$\left|\frac{A}{g}\right|=RM(\omega)=\frac{\sinh(\pi\omega)}{5\pi\omega^2} \qquad (8)$$

which defines Melnikov's ratio[16] . Then for $|A/g|$ greater than this Melnikov's ratio we have Smale's horseshoes. Note, however, that in accordance with Melnikov's analysis the nearer $\alpha$ ($\alpha^2=g^2+A^2$) is to zero the better  the fit of theory to experiment is to be expected. It should also be noted that the existence of Smale's  horseshoes does not imply the existence of an attractor.

Fig. 5 provides a comparison of our experimental values with Melnikov's ratio. Different values of g have been used. The initial conditions correspond to the hyperbolic point. Each experimental point in Fig. 5 has been obtained by holding fixed $\omega$ and g  and letting A vary until we find a **seemingly chaotic signal** or the orbit escapes to the infinity. There are three distinct regions in (Fig. 5). (i) Region A from $\omega=0$ to $\omega=0.45$. (ii) Region B from $\omega=0.45$ to $\omega=0.8$, and (iii) Region C for $\omega>0.8$.

In  region A we clearly see how much our experimental results depart from Melnikov's ratio.

In region B we notice that the Melnikov's ratio is perfectly matched. As all experimental points obtained are very near to the system's escape, this is an indication that Smale's horseshoes come quite near to the escape point too.

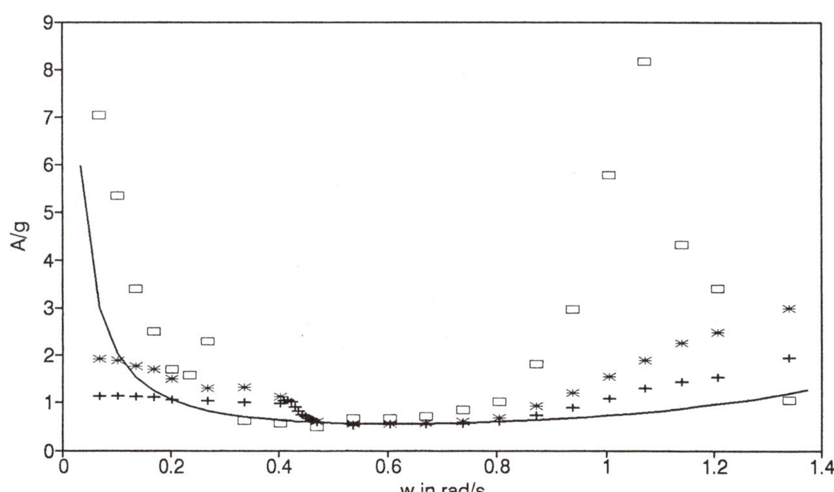



Fig.5 Melnikov's ratio for three different values of g, the damping factor. Solid line denotes theoretical value; +(g=0.228), ✳ (g=0.138), and ▢ (g=0.0189) denote experimental data.

Region C can be analyzed into two subregions. Subregion $C_1$ covers the range $\omega=0.8$ to $\omega=1$. Here the experimental points drastically depart from Melnikov's ratio and we have detected a seemingly "chaotic" signal. When crossing the Melnikov curve, a situation obtained by increasing A, we do not see "chaos". However, "chaos" is soon found when A/g is above the Melnikov's ratio.

Further increasing the frequency we detect chaos nearer and nearer the escape point. As a matter of fact, in subregion $C_2$, $\omega>1$, the "chaotic" behaviour disappears then leading to a period doubling for $\omega<1.2$ and single periodic orbit for $\omega>1.2$.

One expects that when $\alpha$ goes to zero, i. e., when g goes to zero the agreement is better between theory and experiment. It just happens the opposite as we can see in  Fig. 5.


## 4. DETAILED ANALYSIS OF BEST FIT TO MELNIKOV'S RATIO. AN ESTIMATE OF ERRORS

According to Melnikov's theory[16,17], for a given $\alpha$ there is a Melnikov's function $M_\alpha(A,g,\omega)$ whose limiting value as $\alpha$ goes to zero is a known value $M(A,g,\omega)$. Thus we expect that the lower the value of $\alpha$ is, the smaller the discrepancy is between our data and Melnikov's ratio. We found just the opposite. There appears the need of a discussion of noise and other experimental errors, and their evolution when $\alpha$ varies.

First we see from Fig. 5 that the discrepancy does not depend on $\omega$ and so the error does not change with the clock but rather with the natural time in the system as scaled with the forcing period. For  this reason an error estimate must be stroboscopically done using the Poincaré map. We know that a Smale horseshoe is associated with some iteration of the map. Let N be the Nth iterate of Poincaré's map. Our **conjecture** is that the experimental errors scale with $N(\alpha)$, i. e., the evolution of the experimental errors with $\alpha$ matches the evolution of $\alpha$ with N. Then we have[16]

$$N(\alpha) = L + a \cdot \ln \left( \frac{b}{\alpha \cdot supM(t_0)} \right) \tag{9}$$

where L, a and b are constants and $supM(t_0)$ is the  supreme value of M for all $t_0$. When $\alpha$ decreases, $N(\alpha)$ increases, which supports our conjecture.

Our assumption is that

$$\frac{A'}{g'} - RM_\alpha(\omega) = kN(\alpha) \tag{10}$$

To a first approximation

$$\frac{A'}{g'} - RM(\omega) - c\alpha = kN(\alpha) \tag{11}$$

with higher order terms in $\alpha$ neglected. c is  assumed  small [17]. Consider now two measurements of the ratio A/g for a given $\omega$, corresponding to two different values of $\alpha$. Using Eq.(11) and (9)

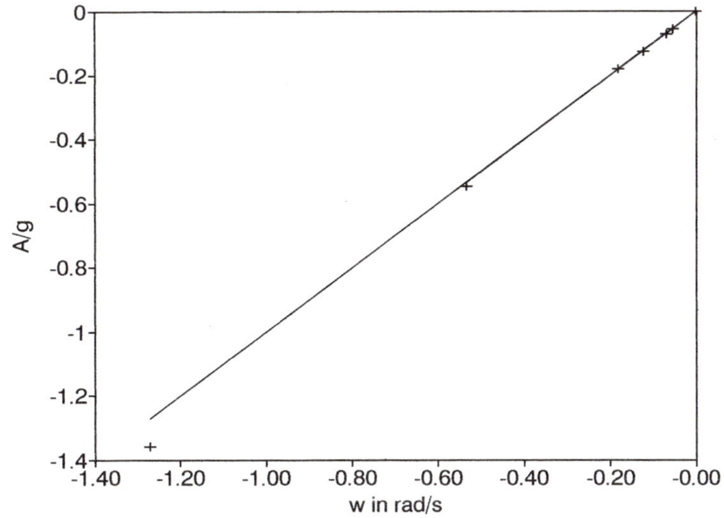


Fig.6        A test of errors of A/g (see main text).

for two diferent measurements and subtracting one from the other
we have (j=2,3)

$$\frac{(A_j/g_j)-(A_1/g_1)-c(\alpha_j-\alpha_1)}{\ln(\alpha_j/\alpha_1)}=-ak \qquad (12)$$

Then if our initial assumption is valid, a plot of (12) with j=2
versus (12) with j=3 for values of ω in the region B must show
the bisector line of the third quadrant. It remains to identify
c. We choose the value of c that provides the least square fit
to the bisector line c=0.0089. Thus $c(\alpha2-\alpha1)<10^{-3}$ which is at
least two orders of magnitude smaller than the difference
(A2/g2)-(A1/g1). This clearly points out that c does not really
matter for the final results. The excellent agreement of the
results with our assumptions is depicted in Fig. 6.


## 5. DIGITAL COMPUTER CROSS-CHECK OF THE RESULTS

In order to cross-check the accuracy and usefulness of our
analog circuit of the Helmholtz-Thompson nonlinear oscillator we
have numerically explored with the digital computer the different
regions discussed above. We have used a fourth order Runge-Kutta
method. Fig. 7 illustrates the excellent agreement obtained
between the digital computer results and the analog circuit data.
Fig. 8 is a cross-check of our assumption concerning the errors
in region B.
We believe having shown in this paper that our analog
circuit is a highly reliable instrument to account for both
qualitative and quantitative features of the Helmholtz-Thompson
oscillator, and thus for its eventual applications.

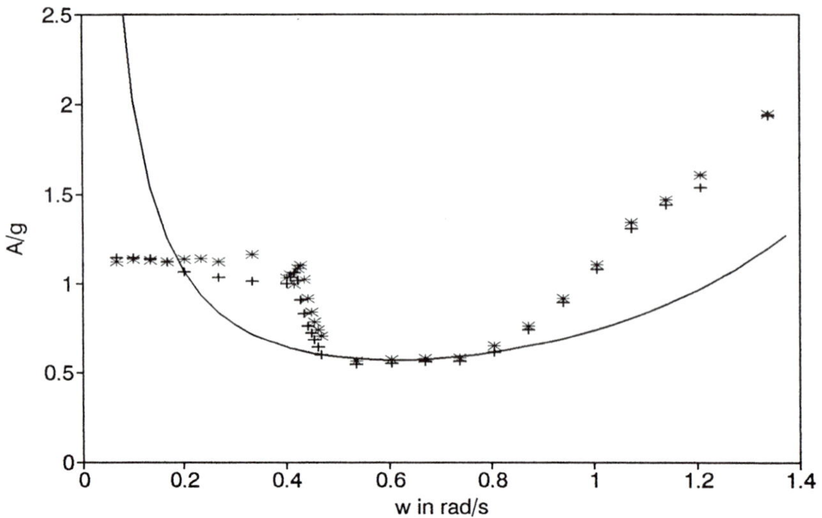

Fig.7 Melnikov's ratio. Comparison of theory, analog and digital
      results for g=0.228.

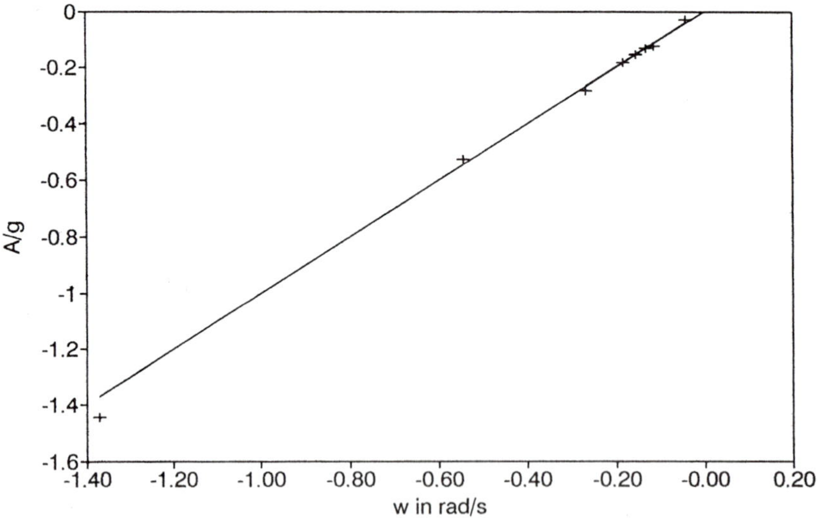

Fig.8 A test of errors in obtaining Melnikov's function (see main
      text).

**ACKNOWLEDGMENTS**

This research has been sponsored by CICYT GRANT PB86-651. The authors have benefited from fruitful discussions with Proff. E. Holzhauer, R. Mackay, M.I. Rabinovich, E. Räuchle, M.A.F. Sanjuan, M. Savage and J. Williams.

**References**

[1] J. GUCKENHEIMER,and P. HOLMES, Nonlinear oscillations, dynamical systems and bifurcations of vector fields.(Springer-Verlag. Berlin,1983).

[2] E. N. LORENZ J. Atm. Sci. **20** 130 (1963).

[3] H. HAKEN, Synergetics, 3rd edition, (Springer-Verlag, Berlin 1984).

[4] G. P. FLESSAS, Phys. Lett. **108A** 4 (1985).

[5] M. A. F. SANJUAN,J. L. VALERO, and M. J. VELARDE, Il Nuovo Cimento-D (1991 to appear).

[6] J. M. T. THOMPSON and H. B. STEWART, Nonlinear dynamics and chaos (Wiley, Chichester, 1986)

[7] H.L.F. HELMHOLTZ On the sensation of tone as a physiological basis for the theory of music.(Dover reprints, N.Y.1954).

[8] J. M. T. THOMPSON, R. C. T. RAINEY and M. S. SOUMAN Phil. Trans. R. Soc. london **A332** 149 (1990).

[9] For a succinct account see also the section on Nonlinear Dynamics and the Capsize of Boats in "The remarkable World of Nonlinear Systems", Science and Engineering Research Council SERC, London 1989.

[10] J. M. T. THOMPSON, F. R. S. and M. S. SOLIMAN, Proc. R. Soc. Lond. A **428** 1-13 (1990)

[11] G. CICOGNA and L. FRONZONI Phys. Rev. A **42** 4 (1990).

[12] G. QIN,D. GONG, R. LI and X. WEN Phys. Lett. A **141** 8,9 (1989)

[13] P.HOROWITZ and W. HILL. The art of electronics.Cambridge univ. press - 2nd Ed (1989).

[14] M. J. FEIGENBAUM. Physica 7D 16 (1983)

[15] Y. POMEAU and P. MANNEVILLE.Comm. Math. Phys. **74**, 189 (1980)

[16] For a comprehensive and critical account see P. J. HOLMES, and J. E. MARSDEN,Commun.Math.Phys. **82**, 523 (1982).

[17] S. CHOW, J. HALE, and J. MALLET-PARET, J.Diff.Eq. **37**, 351 (1980)

# DYNAMIC ANALYSIS OF ENCEPHALIC ACTIVITY

George Zouridakis, Henrik Nyberg,
and Ben H. Jansen

University of Houston, Department of Electrical Engineering
Houston, TX 77204-4793, USA

## ABSTRACT

The application of nonlinear dynamical analysis techniques to electroencephalograms (EEGs) and evoked potentials (EPs) is described. The first method involves the extraction of features from state-space trajectories reconstructed from EEG segments recorded immediately prior to the presentation of a light flash. The feature vectors were used in a clustering procedure to produce classes of similar prestimulus EEG segments. Averaged evoked potentials were computed for each class and distinct differences between EPs were observed. In a second experiment, a neurophysiologically-based model of EEG generation was tested to determine its response to impulse-like inputs. It was found that this model produced EP-like activity. This suggests that the generators of the EP are located in the cortex.

## INTRODUCTION

Electroencephalograms (EEGs) are recordings of the minute (generally less than $300\mu V$) electrical potentials produced by the brain. Soon after the "discovery" of the human EEG by Berger [1], attempts were made at developing objective, quantitative methods to aid in the interpretation of the rather complex, almost noise-like looking signals. Until very recently, the basic approach to EEG analysis involved the assumption that the EEG is *random*. Consequently, statistical pattern recognition techniques, segmentation procedures, syntactic methods, knowledge-based approaches, and even artificial neural network methods have been developed with different levels of success. A fundamentally different approach to computerized EEG analysis, however, is making its way into the laboratories. The basic idea, inspired by recent advances in the area of nonlinear dynamics, and especially the theory of chaos, is to view an EEG as the output of a *deterministic* system of relatively simple complexity, but containing *nonlinearities*. This suggests that studying the geometrical dynamics of EEGs, and the development of

*Chaotic Dynamics: Theory and Practice*
Edited by T. Bountis, Plenum Press, New York, 1992

neurophysiologically realistic models of EEG generation may produce more successful automated analysis techniques than the classical, stochastic methods.

An approach to EEG interpretation on the basis of nonlinear dynamics is justified, because the basic building blocks of the brain are neurons, which are highly nonlinear elements. Although the exact origin of the scalp-recorded EEG remains unknown, the consensus is that it results mainly from temporal and spatial summation of the relatively slow postsynaptic potentials on cortical pyramidal cells. Hence, the EEG appears to be produced by a system containing many nonlinear, interconnected elements. The vast number of neurons in the brain (about $10^{11}$) and the even larger number of interconnections ($10^{14}$) make it likely that the EEG can display chaotic behavior. This has inspired a number of researchers to compute the correlation dimension of the "EEG attractor". Most of the efforts have been in the determination of the (correlation) dimension of EEG tracings recorded from subjects in different neurophysiological states. Babloyantz and co-workers have studied the attractor dimensions for an extensive range of EEG states, including wakefulness, different sleep stages, and pathologies such as epilepsy and Creutzfeld-Jakob disease [2, 3, 4]. These workers have reported that the correlation dimension increased with the level of mental activity. Subjects in sleep stage 4 displayed a dimension of around 4, EEG from alert but resting subjects was found to have a dimension of about 6, and a dimension of close to 10 was observed for EEG recorded from mentally active subjects. Creutzfeld-Jakob disease reportedly has a dimension slightly less than sleep stage 4, and an attractor with a dimension close to 2 has been found during petit-mal epileptic seizures. Furthermore, evidence is mounting that the Lyapunov exponent is positive for a variety of EEG patterns (see for example [5]). These factors combined point to a possibly deterministic, chaotic nature of the EEG.

One should be aware that the techniques to determine the fractal dimension or (largest) Lyapunov exponent require that the system which produces the signal under observation is in its *asymptotic* state, i.e., has passed its transient behavior. Furthermore, the system is assumed to be *time invariant*, i.e., the system characteristics do not change over the duration of the observation. Determining if the conditions of asymptotic behavior and time invariance hold is virtually impossible in the case of EEG signals. First, the brain continuously receives sensory input relayed through the brainstem and the thalamus to the cortex. These inputs may be viewed as (small?) perturbations, possibly causing in the state-space trajectory to jump suddenly to other orbits. In other words, an asymptotic state may never be reached. Secondly, assessing time invariance of the system is complicated by the fact that changes in EEG characteristics are not necessarily due to changes in the underlying system (i.e., the neuronal circuitry). In contrast to linear systems, where changes in system output indicate a change in system parameters, nonlinear systems can produce nonstationary signals without any change in system parameters or inputs. Therefore, one should be careful to assume, for example, that the fast, irregular beta activity seen in the EEG of a mentally active person represents a different brain state than the regular, almost periodic alpha rhythm observed when no mental processing takes places. The aforegoing implies that the EEG dimension values published in the literature are of questionable significance, because EEG segments were selected on the basis of *signal* characteristics rather than on *system* and/or *input* characteristics. Consequently, one may have been studying only part of the "EEG attractor".

Calculating the correlation dimension or any other characteristic of the EEG attractor may not be the most important contribution of chaos theory to EEG analysis. More relevant is the introduction of a *deterministic* viewpoint to EEG generation, rather than the *stochastic* viewpoint held virtually since the discovery of the EEG. Taking this *deter-*

*ministic* viewpoint to EEG generation has implications for quantitative and visual EEG interpretation. First, a number of new analysis tools, taken from the area of nonlinear dynamics, become available to investigate the *dynamic* characteristics of the EEG (e.g., state-space trajectory reconstruction, first-return maps, bifurcation diagrams). These representations can be used to deduce the complexity of the underlying system (i.e., the brain), and may lead to useful, predictive models of EEG generation. It is this aspect of chaos theory that we are pursuing in our laboratory. Specifically, we are studying the responses of the brain to distinct sensory stimuli (in our case, light flashes). These so-called visual evoked potentials (VEPs) may be viewed as the impulse response of the visual system. Given the nonlinear nature of the brain, VEPs may vary in shape and amplitude, depending on the stimulus characteristics and the initial conditions (i.e., the psychophysiological state of the subject under study). A complicating factor is that the response to a single stimulus is much smaller than the 'spontaneous' EEG activity. Averaging techniques, utilizing the time-locked nature of the response to the stimulus, are used to improve the signal-to-noise ratio. An extensive study on normal volunteers provided us with insight in the relationship between prestimulus EEG characteristics (i.e., the EEG immediately prior to stimulus presentation) and the subsequent shape of the VEP [6, 7, 8]. The basic approach taken involved "selective" averaging, i.e., prestimulus EEG segments were clustered on the basis of similarity in power spectral content, and average VEPs were computed for each cluster. We are currently in the process of repeating this study, using clustering of state-space trajectories. The techniques developed and some preliminary results are presented in the next section. A second line of inquiry involves the development of a neurophysiologically-realistic model of EP generation, that can be used to explain some of the observations seen in actual VEP recordings. This is the topic of the third section. A discussion concludes this chapter.

## TRAJECTORY-BASED CLUSTERING

It has been suggested that a reconstructed state-space trajectory (by way of time delay embedding [9]) reveals more of the dynamics of the underlying system than the original time series [10]. Based on this premise, a quantitative analysis of a signal in the state-space domain may be a more plausible way of characterizing the dynamics of a system, than an analysis of the original time series. We have explored this concept using clustering analysis. In this procedure, a state-space trajectory is constructed from a time series interval, and segments having "similar" state-space trajectories are lumped together. Features quantifying the geometry and topology of trajectories have been defined, and the similarity between two trajectories is measured in terms of the Euclidean distance between the feature vectors.

Two different approaches to feature extraction may be defined. In a *local* feature extraction scheme the state-space is divided into subspaces, and feature extraction is done locally in each subspace. A comparison between trajectories is made between corresponding subspaces, and a measure of similarity is based on the cumulative action of all subspaces (this approach has been successfully used in a speech recognition task [11]). In a *global* feature extraction scheme, the feature extractor operates on the whole trajectory and each trajectory is attributed with a number derived from its geometry and topology (all "conventional" chaotic measures are extracted in this fashion). A measure of similarity between two trajectories is here defined as the difference in this "global" number.

In the present work, we have set out to cluster selected EEG segments (in our case, the 1-second intervals prior to the presentation of a flash of light). Similar trajectories have been grouped together in a hierarchical clustering procedure. Several measures of similarity have been defined, both on the basis of a local feature extraction scheme and a global feature extraction scheme. The clustering algorithm has been evaluated on simulated data, both on EEG-like data generated by an AR-model, and deterministic data generated by Duffing's oscillator, and proved to work reliably.

Preliminary experiments (with a measure of similarity based on the presence or absence of a trajectory in a subspace) on actual EEG data from one subject resulted in three clusters: a "waxing-waning alpha activity" cluster; a "steady alpha activity" cluster; and a "drowsiness" cluster. Averaged evoked potentials were computed for the poststimulus EEG segments associated with each of the three clusters and the results are shown in Figure 1. As one can see, substantial differences between the two alpha EPs and the drowsiness EP are found. More surprising is the fact that there are differences between the steady alpha and the waxing-waning alpha EP, especially during the first 100 msec after the presentation of the stimulus.

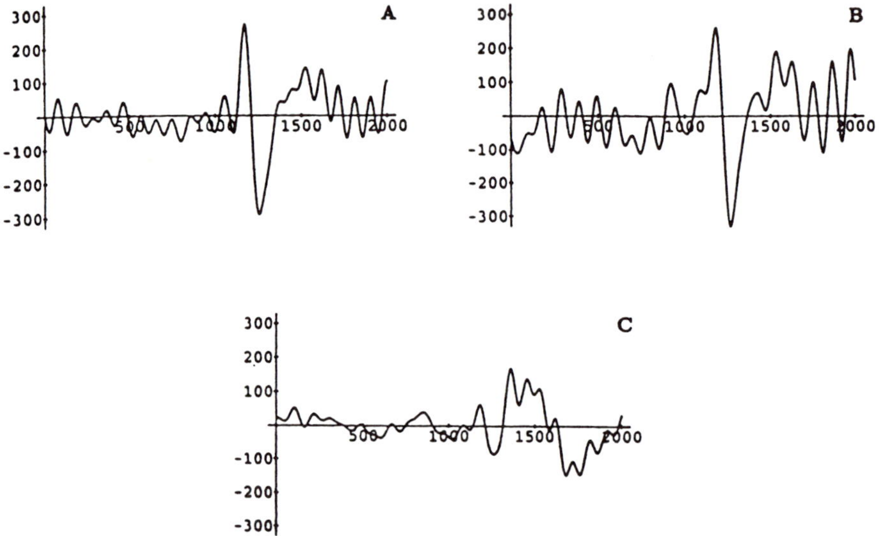

Figure 1. *Averaged evoked potentials (including the 1-second prestimulus segment and the 1-second poststimulus segment) for the "waxing-waning alpha" (A), "steady alpha" (B) and the "drowsiness" (C) cluster. Stimuli were presented at the 1000 msec tick mark. Amplitudes are in arbitrary units.*

## NEUROPHYSIOLOGICAL MODELING

As stated before, better use of chaos theory in EEG analysis can be made by developing realistic models for EEG generation. Model outputs can be compared to actually observed data through dimension calculations, analysis of state-space trajectories, etc.

Freeman and co-workers have made substantial progress along aforementioned lines in the area of modeling the olfactory system [12, 13]. Freeman reports that this model, when presented with an arbitrarily small initiating pulse at the receptor input, generates continuing activity that has the statistical properties of the background EEG of resting animals. The signals produced by the model appear to be chaotic on the basis of state-space plots. Chaotic activity persists even when an input is given that simulates sustained receptor input during inhalation, even though the model output under these conditions is a near-periodic oscillation.

Starting from a model originally developed for alpha rhythm generation [14], we have designed a computer model that can produce "spontaneous" EEG and evoked potentials [15]. This lumped parameter model represents cortical columns and is based on two interacting populations of neurons: one consisting of main cells and the other of local interneurons (see Figure 2). The population of main cells is characterized by two linear transfer functions representing the excitatory and inhibitory postsynaptic potentials (EPSPs and IPSPs, respectively) and a static, nonlinear element which relates the average level of membrane potential to the pulse density of action potentials fired by the neurons. The population of interneurons is similarly described by analogous linear and nonlinear functions, albeit that only one linear transfer function is used per class of interneurons to take into account the fact that interneurons receive excitatory input only. Finally, interconnectivity constants representing the average number of synaptic contacts from main cells to interneurons ($c_1$ and $c_3$), and from interneurons to main cells for both excitatory ($c_4$) and inhibitory ($c_2$) branches are used. The model receives nonspecific input from other cortical structures (such as the association areas), and specific afferent activity relayed through the thalamus.

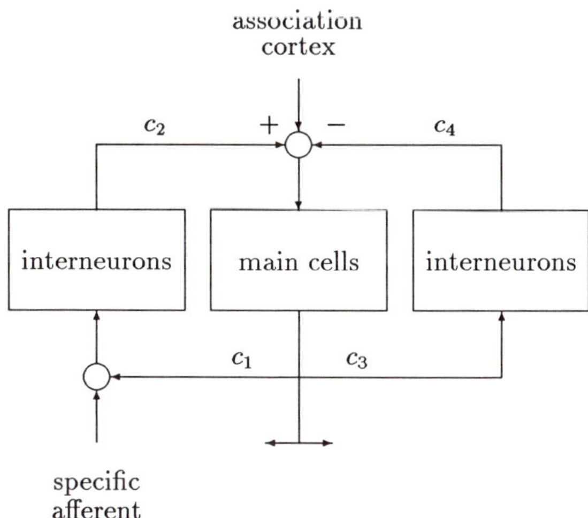


Figure 2. *The EEG model.*

A linear analysis of this model revealed its preference for producing alpha-like rhythms when stimulated with random noise [16]. Experiments conducted by the authors showed that the same model is capable of much more complex behavior when it is allowed to operate in a nonlinear mode [15, 17]. For instance, a version of the model containing an inhibitory loop only for which the connection strengths were gradually increased, produced rhythmic outputs varying in frequency from 12.5 cycles per second to 6.5 cycles per second in response to random stimulation. Evoked potential-like activity was generated by the model when a pulse-like wave was presented to the interneurons receiving specific afferent input. An example is shown in Figure 3.

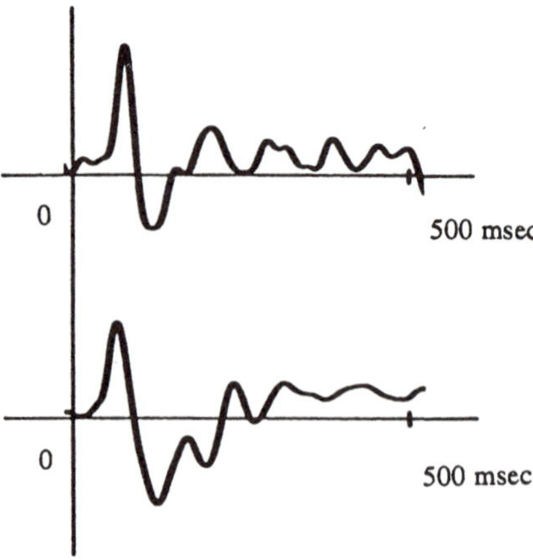

Figure 3. **Top**: *Evoked potential generated with the EEG model.* **Bottom**: *Visual (flash) evoked potential measured from human.*

## CONCLUDING REMARKS

The aforegoing two sections illustrate how concepts, borrowed from the area of non-linear dynamics, can be used in unraveling the functioning of the human brain. Taking a deterministic, but nonlinear viewpoint to EEG generation forces a reevaluation of many EEG findings reported in the literature. For example, the waxing-waning of the alpha activity points to a nonlinear oscillator mechanism responsible for alpha activity generation. The occurrence of intermittent, single slowwaves (K-complexes) during sleep stage 2, and the trains, and ultimately almost continuous slow-wave activity (delta activity) during the deeper sleep stages 3 and 4, could mark a path away from chaos. The inconsistent relationship between intensity of the sensory stimulus and the amplitude of the resulting EP is indicative of the nonlinearity of the sensory cortex involved. We are convinced that a rigorous application of nonlinear dynamic analysis techniques to EEG activity will result in a better understanding of the mechanisms involved in the generation of the EEG, and, consequently, better diagnostic techniques.

## REFERENCES


[1] Berger, H.: Über das Elektrenkephalogramm des Menschen, *Archiv. Psychiatrie,* 87: 527-570, 1929.

[2] Babloyantz, A., Nicolis, C. and Salazar, J.M.: Evidence of chaotic dynamics of brain activity during the sleep cycle, *Physics Letters A,* 111: 152-156, 1985.

[3] Babloyantz, A. and Destexhe, A.: Low-dimensional chaos in an instance of epilepsy, *Proc. Natl. Acad. Sci. USA,* 83: 3513-3517, 1986.

[4] Babloyantz, A. and Destexhe, A.: The Creutzfeld-Jakob disease in the hierarchy of chaotic attractors. In: *From Chemical to Biological Organization,* Markus, M., Müller, S. and Nicolis, G. (Eds.), Springer-Verlag, 1988.

[5] Frank, G.W., Lookman, T., Nerenberg, M.A.H., Essex, C., Lemieux, J. and Blume, W.: Chaotic time series analyses of epileptic seizures, *Physica D,* 46: 427-438, 1990.

[6] Brandt, M.E., Jansen, B.H. and Carbonari, J.P: Pre-stimulus spectral EEG patterns and the visual evoked response, *Electroencephal. clin. Neurophysiol.*, 80:16-20, 1991.

[7] Jansen, B.H., and Brandt, M.E.: The effect of the phase of prestimulus alpha activity on the averaged visual evoked response, *Electroencephal. clin. Neurophysiol.*, 80:241-250, 1991.

[8] Brandt, M.E., and Jansen, B.H.: The Relationship between Prestimulus Alpha Amplitude and Visual Evoked Potential Amplitude, *International Journal of Neuroscience*, in press, 1991.

[9] Packard, N.H., Crutchfield, J.P., Farmer, J.D. and Shaw, R.S.: Geometry from a time series. *Physical Review Letters*, 45: 712-715, 1980.

[10] Breeden, J.L. and Packard, N.H.: *Technical Report CCSR-91-3*. Center for Complex Systems Research, University of Illinois, Urbana-Champaign, 1991.

[11] Gong Y. and Haton, J.P.: Text-independent speaker recognition by trajectory space comparison. *ICASSP-90*, Albuquerque, April 1990.

[12] Freeman, W.J.: Simulation of chaotic EEG patterns with a dynamical model of the olfactory system, *Biol. Cybern.*, 56: 139-150, 1987.

[13] Skarda, A. and Freeman, W.J.: How brains make chaos in order to make sense of the world, *Behav. Brain Sci.*, 10: 161-195, 1987.

[14] Lopes da Silva, F.H., Hoeks, A., Smits, A. and Zetterberg, L.H.: Model of brain rhythm activity, *Kybernetik*, 15: 27-37, 1974.

[15] Zouridakis, G.: *Simulation of EEG and VEP Activity*, M.Sc.-thesis, University of Houston, Houston, 1990.

[16] van Rotterdam, A., Lopes da Silva, F.H., van den Ende, J., Viergever, M.A. and Hermans, A.J.: A model of the spatial-temporal characteristics of the alpha rhythm, *Bull. Math. Biol.*, 44: 283-305, 1982.

[17] Jansen, B.H.: Quantitative analysis of electroencephalograms: is there chaos in the future?, *Int. J. Biomed. Comput.*, 27: 95-123, 1991.

# CHAOTIC RESPONSE OF A PERIODICALLY FORCED SYSTEM OF TWO
# COMPETING MICROBIAL SPECIES

Petros Lenas and Stavros Pavlou

Institute of Chemical Engineering and High Temperature Chemical
Processes, and Department of Chemical Engineering, University of
Patras, GR-26110 Patras, Greece

## INTRODUCTION

Competition is one of the most common interactions between two microbial species
inhabiting the same environment. It occurs when the microorganisms compete with each
other for nutrients and other resources of the environment. Microbial competition is an
interaction common in natural ecosystems, but it occurs also in several industrial processes
utilizing mixed cultures, such as wastewater treatment.

The simplest case of microbial competition is the so-called *pure and simple competition*
(Fredrickson and Stephanopoulos, 1981), an idealized situation where the two species
compete for a single nutrient and where there is no other interaction besides competition
between the two species. This situation can be realized in the laboratory in a chemostat, i.e.,
a well-stirred vessel in which the two microbial species grow together and which is fed with
medium containing the nutrient competed for by the two species. The dynamics of the
autonomous system of pure and simple competition in a chemostat has been studied by
several workers theoretically (Powell, 1958; Aris and Humphrey, 1977; Hsu et al., 1977;
Hsu, 1978; Butler and Wolkowicz, 1985) and experimentally (Meers, 1971; Jost et al.,
1973). These studies show that coexistence of the two species in the chemostat can be
obtained only in cases where the specific growth rate curves of the two species cross and for
discrete values of the chemostat dilution rate. However, in practice it is not feasible to
operate the system at exactly those values of the dilution rate because of random fluctuations
in the flowrate (Stephanopoulos et al., 1979a). Thus, although the autonomous system of
microbial competition in a chemostat predicts coexistence of the two species at specific
values of the operating parameter (dilution rate), in practice the coexistence state is not
obtainable in a chemostat operating at steady state conditions. Periodic variation of one of
the system's operating parameters has been suggested as a means of obtaining coexistence
of the two species in the chemostat. Indeed, it has been shown by several workers that
periodic variation of the chemostat dilution rate (Stephanopoulos et al., 1979b; Butler et al.,
1985; Matsubara et al., 1986) or of the nutrient concentration in the feed (Hsu, 1980; Smith,
1981; Hale and Somolinos; 1983) allows coexistence of the two microbial species in a state
of limit cycle oscillations for some range of values of the operating parameters and thus in
this way the coexistence state is obtainable in practice.

In this work it is shown that by periodic variation of the chemostat dilution rate it is
possible to obtain coexistence of the two species not only in a state of limit cycle oscillations
but also in some cases in a quasiperiodic or in a chaotic state. Quasiperiodicity results by
Hopf bifurcation from a periodic state and chaos results from a sequence of successive
period doublings.

*Chaotic Dynamics: Theory and Practice*
Edited by T. Bountis, Plenum Press, New York, 1992

MODEL EQUATIONS

The equations describing the autonomous system of two competing microbial species in a chemostat can be written in dimensionless form as

$$\frac{dx_1}{dt} = - u\, x_1 + f_1(z)\, x_1 \tag{1a}$$

$$\frac{dx_2}{dt} = - u\, x_2 + f_2(z)\, x_2 \tag{1b}$$

$$\frac{dz}{dt} = u\, (z_F - z) - f_1(z)\, x_1 - f_2(z)\, x_2 \tag{1c}$$

where $x_1$, $x_2$ and $z$ are the dimensionless concentrations of the two microbial species and of the nutrient, respectively, $z_F$ is the dimensionless concentration of the nutrient in the feed, $u$ is the dimensionless dilution rate of the chemostat and $f_1(z)$ and $f_2(z)$ are the dimensionless specific growth rate functions for the two microbial species. The expressions for the specific growth rates are

$$f_1(z) = \frac{z}{1 + z + \gamma_1\, z^2} \tag{2a}$$

$$f_2(z) = \frac{\alpha\, z}{\beta + z + \gamma_2\, z^2} \tag{2b}$$

where $\alpha$, $\beta$, $\gamma_1$ and $\gamma_2$ are constants. By using arguments similar to those of Aris and Humphrey (1977) we can reduce the system of equations (1) to a system of two differential equations by use of the algebraic equation

$$x_1 + x_2 + z = z_F \tag{3}$$

to eliminate one of the state variables:

$$\frac{dx_1}{dt} = - u\, x_1 + f_1(z_F - x_1 - x_2)\, x_1 \tag{4a}$$

$$\frac{dx_2}{dt} = - u\, x_2 + f_2(z_F - x_1 - x_2)\, x_2 \tag{4b}$$

The two systems of differential equations (1) and (4) have asymptotically similar dynamic behavior, since all the solution trajectories of system (1) are attracted to the plane in the $(x_1, x_2, z)$ space described by eq. (3). Thus one may use the system of equations (4) to study the dynamic behavior of the system of equations (1).

For a coexistence state of the autonomous system to be possible, there must be a steady state solution for the system of equations (1) such that $x_1 > 0$ and $x_2 > 0$. As can be seen from eq. (1), such a solution is possible only if there exists a solution $z$ for the following equations:

$$u = f_1(z) = f_2(z) \tag{5}$$

Apparently, eq. (5) does not have a solution for all $u$, but for specific values $u_c$ such that

$$u_c = f_1(z_c) = f_2(z_c) \tag{6}$$

where $z_c$ is a root of the equation $f_1(z) = f_2(z)$, i.e., a $z$ value corresponding to a point of intersection of the two curves defined by $f_1(z)$ and $f_2(z)$. Therefore, a necessary condition for the coexistence state to be obtainable in the autonomous system is that the two specific growth rate curves cross. Since these two curves may have at most two points of intersection, there can be up to two values of $u$ for which the coexistence state is possible. In such a case the two species coexist in a steady state of indeterminate values of the state variables (the concentrations of the two species). The coexistence steady state is represented by a straight line in the state space. Any point on this line is a steady state of the system. The linearized system about the coexistence steady state has one zero eigenvalue corresponding to perturbations along the straight line (Aris and Humphrey, 1977). Thus the stability of any point on the coexistence steady state line is determined by the sign of the other eigenvalue. The sign of the second eigenvalue depends on the slopes of the specific growth rate curves at the point of their intersection. This is illustrated in Fig. 1. If both slopes at the point of intersection are positive, the second eigenvalue is negative at every point of the steady state line and the trajectories in the phase plane are attracted to the straight line (Fig. 1a). The steady state in this case is characterized as semistable (Aris and Humphrey, 1977). If both slopes at the point of intersection are negative, the second eigenvalue is positive at every point of the steady state line and the trajectories in the phase plane move away from the straight line (Fig. 1b). The steady state in this case is characterized as semiunstable. An interesting case arises when the slopes of the specific growth rate curves at the point of intersection have opposite signs. In this case the second eigenvalue is positive on a segment of the straight and negative on the rest, so part of the line attracts and part of it repels the trajectories (Fig. 1c). The coexistence steady state line is characterized as partly semistable and partly semiunstable. The intermediate point on the line between the semistable and the semiunstable segments has a double zero eigenvalue.

As shown by Powell (1988), the coexistence steady state of the autonomous system is a structurally unstable state. Thus, if we add periodic forcing to the system we get a qualitatively different picture. We are going to use the chemostat dilution rate as the forcing variable, $u = u_o + a \cos(\omega t)$. In this case, the reduction to a system of two differential equations by use of eq. (5) still holds (Stephanopoulos et al., 1979b; Butler et al., 1985; Matsubara et al., 1986):

$$\frac{dx_1}{dt} = - [u_o + a \cos(\omega t)] x_1 + f_1(z_F - x_1 - x_2) x_1 \tag{7a}$$

$$\frac{dx_2}{dt} = - [u_o + a \cos(\omega t)] x_2 + f_2(z_F - x_1 - x_2) x_2 \tag{7b}$$

The other operating parameter $z_F$ could also be used as the forcing variable, but in that case an equation different from eq. (3) should be used for the reduction of the system (Hale and Somolinos, 1983).


RESULTS

The periodically forced system is analyzed for different values of the parameters $\alpha$, $\beta$, $\gamma_1$ and $\gamma_2$ of the specific growth rate expressions resulting to different types of intersection of the specific growth rate curves. We begin the analysis by computing the limit cycles of the forced system as fixed or periodic points of the stroboscopic map (which is a special case of the Poincaré map for periodically forced systems), using a shooting algorithm and full Newton-Raphson iteration. The stability of the limit cycles is determined by computing the characteristic (or Floquet) multipliers as the eigenvalues of the monodromy matrix resulting from integration over one period of the variational equations around the limit cycle with initial conditions the unit matrix. By continuation of the fixed and periodic points with respect to one of the operating parameters of the system we compute a one-parameter bifurcation diagram and we determine the bifurcation points, i.e., points where one or both of the characteristic multipliers crosses the unit circle in the complex plane. For the continuation we used AUTO, a continuation/bifurcation algorithm written by Doedel (1986). Then by continuation of the bifurcation points with respect to two operating parameters we determine the regions in the operating parameter space where different qualitative behavior of the system is observed. AUTO does two-parameter continuation of certain codimension-1 bifurcations of maps, like saddle-node bifurcations, period doubling bifurcations and Hopf bifurcations. However, this is not the case with transcritical bifurcation points. AUTO can locate these points in the one-parameter bifurcation diagram, but cannot do two-parameter continuation of them. Thus we have developed an algorithm to trace the transcritical bifurcation curves in the operating parameter space. Description of this algorithm as well as details of the numerical methods used are given elsewhere (Lenas and Pavlou, 1991). In this way we determine the boundary of the region in which coexistence of the two microbial species is observed. From the four operating parameters of the forced system, i.e., the average dilution rate, $u_o$, the nutrient concentration in the feed, $z_F$, the forcing amplitude, a, and the forcing frequency, $\omega$, we chose $u_o$ and $z_F$ as continuation parameters.

The analysis shows that the coexistence state is realized in some range of the operating parameters and not for discrete values as in the case of the autonomous system. Thus, from

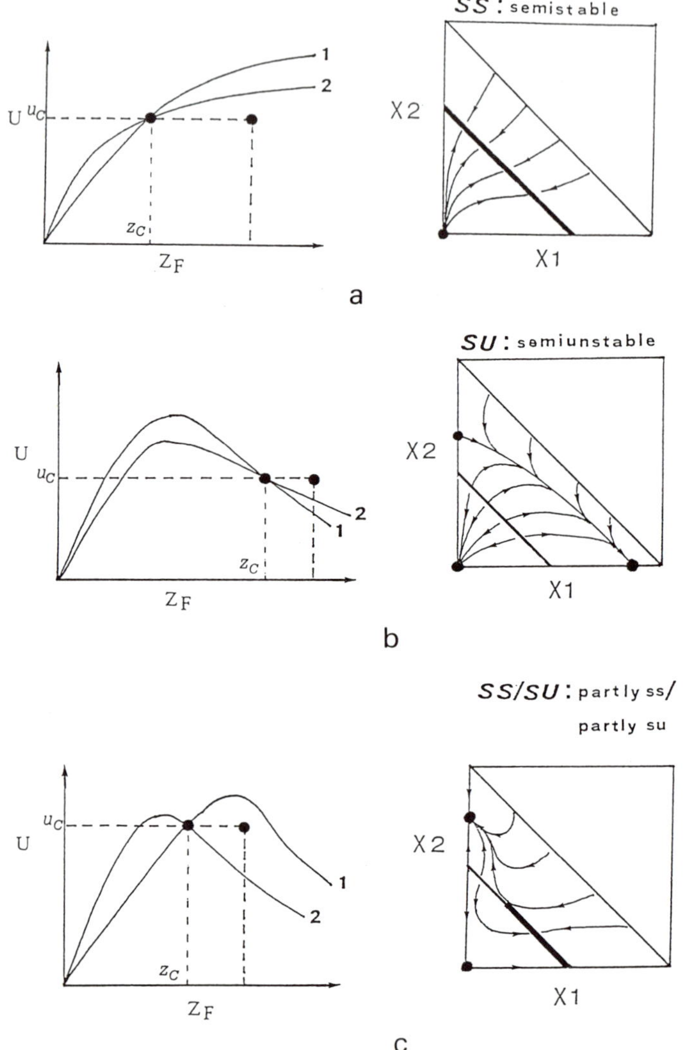



Fig. 1. Coexistence steady state line in the phase plane of the autonomous system; (a) semistable; (b) semiunstable; (c) partly semistable and partly semiunstable.

a practical point of view, one may operate the chemostat in the coexistence region without worrying for the effect of small random fluctuations. The characteristics of the coexistence state of the forced system depend on the type of the perturbed coexistence steady state of the autonomous system. A semistable steady state line of the autonomous system (Fig. 1a) results in a stable coexistence limit cycle of the forced system with period the same as the forcing period $T=2\pi/\omega$. A semiunstable steady state line of the autonomous system (Fig. 1b) results in an unstable (saddle-type) coexistence limit cycle of the forced system with period again the same as the forcing period T. However, when the coexistence steady state of the autonomous system is partly semistable and partly semiunstable (Fig. 1c), Hopf bifurcations and period doublings may be observed in the forced system resulting in quasiperiodicity and chaos. These bifurcations are probably the result of perturbing the point on the steady state line with a double zero eigenvalue. An example of such a case is shown in Fig. 2, where there exist two coexistence regions. These regions begin from the points of intersections of the two curves which are described by the equations $u_o = f_1(z_F)$ and $u_o = f_2(z_F)$. The coexistence state in the lower region results from a semiunstable steady state line of the autonomous system and therefore it is in the form of a saddle-type limit cycle. The coexistence state in the upper region results from a partly semistable and partly semiunstable steady state line and in some part of the region it is a stable limit cycle with period equal to the forcing period T. However, there is also a sequence of period doubling curves defining regions where there is a stable coexistence limit cycle of period 2T, 4T, 8T, etc. (Fig. 2b) until a region of chaotic behavior is defined. A one-parameter bifurcation diagram for fixed value of $z_F$ is shown in Fig. 2c, where the successive period doublings are observed with each limit cycle losing stability and giving rise to a limit cycle of double period.

Another case is shown in Fig. 3, where the two specific growth rate curves have one point of intersection with slopes of opposite signs. Therefore, the coexistence steady state line of the autonomous system is partly semistable and partly semiunstable. Again there is a sequence of period doubling curves leading to a region of chaotic behavior. In this case we were able to determine also within the chaotic region a narrow region where there is a stable limit cycle of period 3T.This is shown in detail in Fig. 3b. The limit cycle of period 3T gives rise by period doubling to a limit cycle of period 6T and this to a limit cycle of period 12T and so on up to the chaotic region. This is illustrated in the one-parameter bifurcation diagram of Fig. 4, where the three branches of the period 3T state are shown together with the states of period 6T and 12T arising from the middle branch. From Fig. 3b we can see also that there are regions of multistability, where we have two stable limit cycles of period T and 2T or of period T and 4T. This is the result of subcritical period doubling of the period T limit cycle as illustrated in the one-parameter bifurcation diagram of Fig. 5. The chaotic attractor of the system is depicted in Fig. 6 together with its stroboscopic map. We can see that the stroboscopic map of the attractor is one dimensional implying that there exists a strongly attracting direction.

A third case is shown in Fig. 7 where the crossings of the two specific growth rate curves are the same as in the case of Fig. 2. In the lower coexistence region there is a saddle-type limit cycle of period T. However, as can be seen in more detail in Fig. 7b, in the upper region, besides the sequence of period doubling curves there is a Hopf bifurcation curve below which the limit cycle loses stability and an invariant circle in the stroboscopic plane bifurcates from it. The continuous trajectory is quasiperiodic ($T^2$ torus) with two incommensurate frequencies (their ratio is irrational), one of them being the frequency of forcing. An example of such an invariant circle is shown in Fig. 8. As shown in Fig. 7b, the higher the value of $z_F$, the higher the period of the limit cycle which will undergo Hopf bifurcation. This is better illustrated in the one-parameter bifurcation diagrams shown in Fig. 9. Thus, if the $z_F$ value is high enough, there will be a sequence of period doublings of unstable limit cycles.


CONCLUSIONS

A mathematical model describing a system of two microbial species competing for a single nutrient in a chemostat predicts that the two species can coexist only at discrete values of the system's operating parameters. In such a case the two species coexist in a steady state

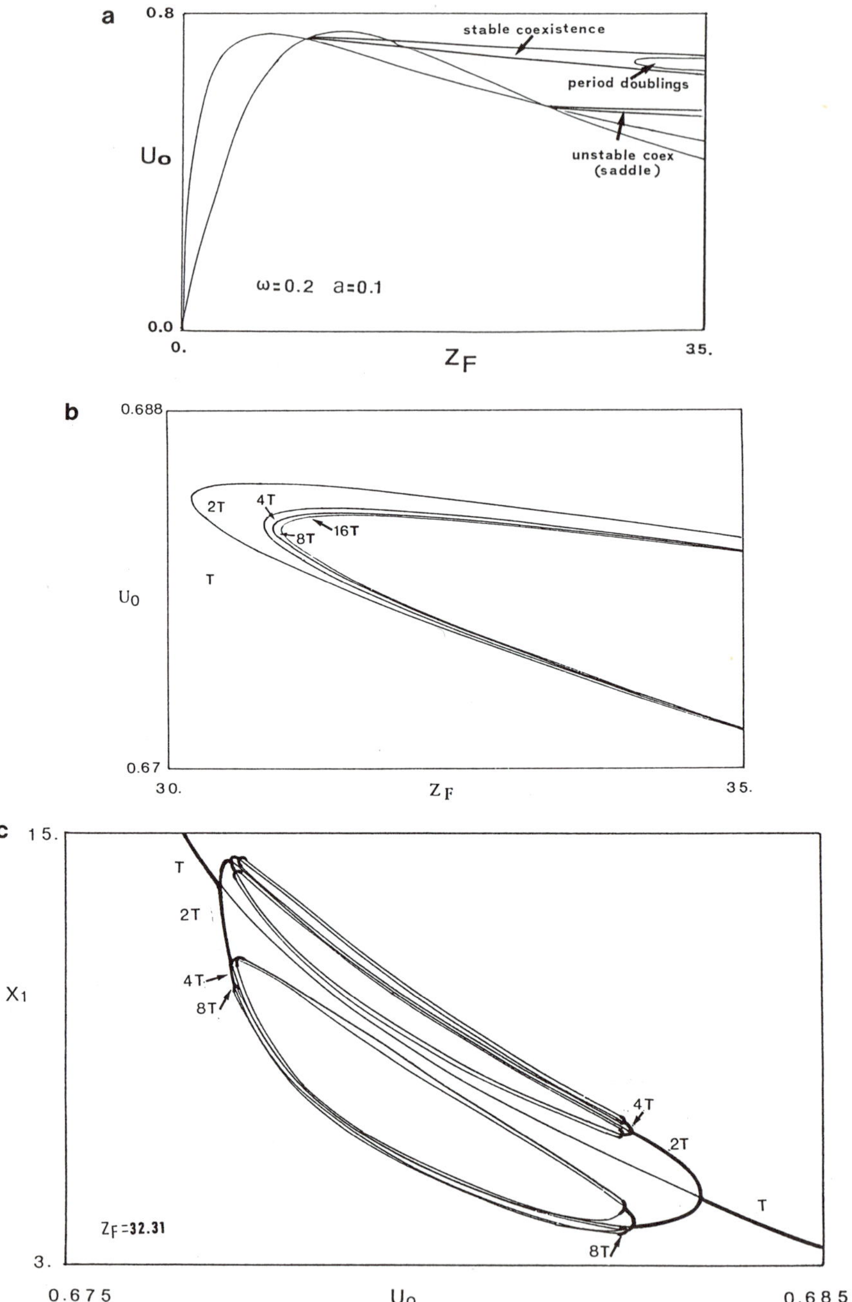


Fig. 2. (a) Coexistence regions in the operating parameter space of the forced system; (b) period-doubling curves in the upper coexistence region; (c) one-parameter bifurcation diagram for fixed $z_F$ (———— stable, ——— saddle). Parameter values of specific growth rate expressions: $\alpha = 5$, $\beta = 30$, $\gamma_1 = 0.03$, $\gamma_2 = 0.27$.

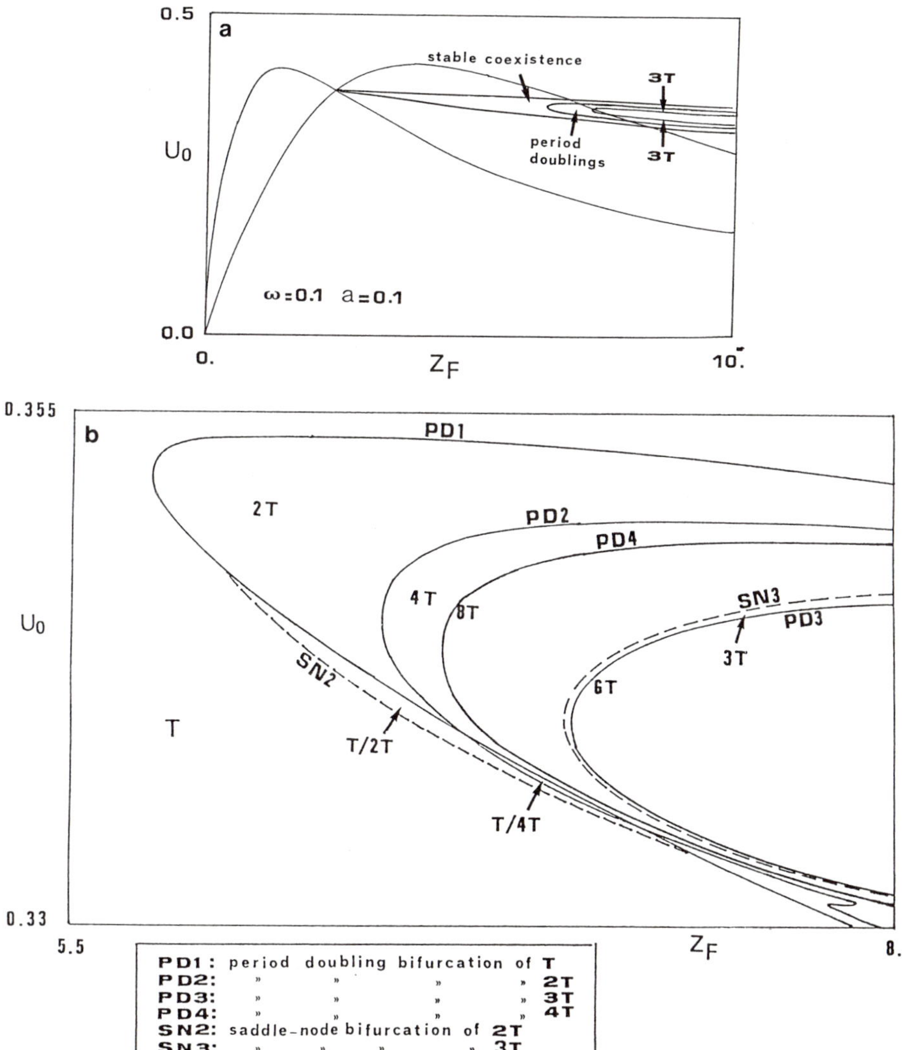


Fig. 3. (a) Coexistence region in the operating parameter space of the forced system; (b) period-doubling and saddle-node bifurcation curves in the coexistence region. Parameter values of specific growth rate expressions: $\alpha = 7$, $\beta = 30$, $\gamma_1 = 0.5$, $\gamma_2 = 2$.

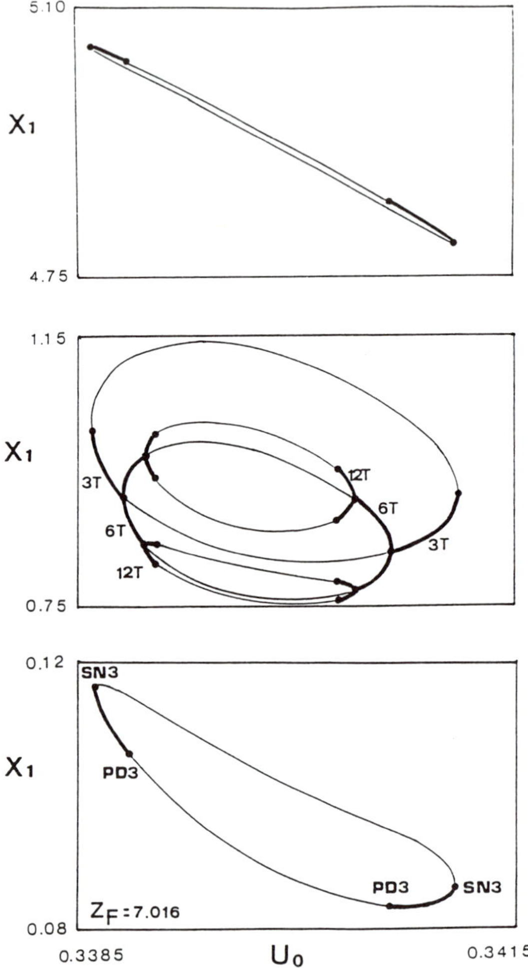



Fig. 4. One-parameter bifurcation diagram showing the period 3T state of the case of Fig. 3 and the period doublings arising from the middle branch (————— stable, ————— saddle).

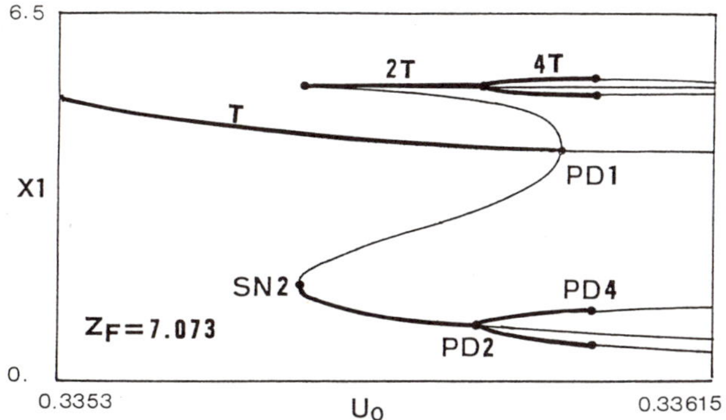



Fig. 5.   One-parameter bifurcation diagram showing subcritical period doubling resulting in multistability in the case of Fig. 3 (———— stable, ———— saddle).

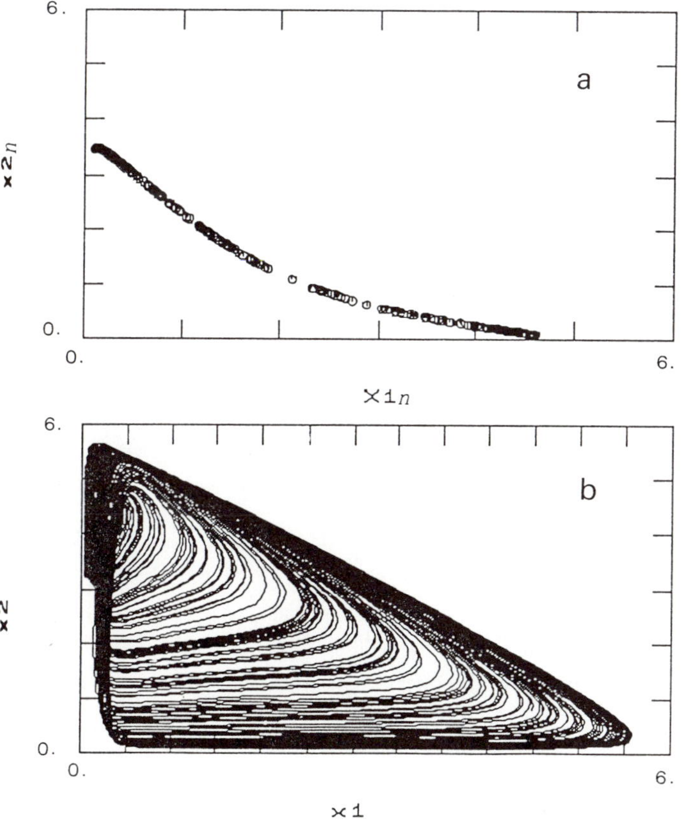



Fig. 6.   Chaotic attractor of the case of Fig. 3. (a) Stroboscopic map; (b) phase plane representation. Operating parameter values: $u_o = 0.343$, $z_F = 7.016$.

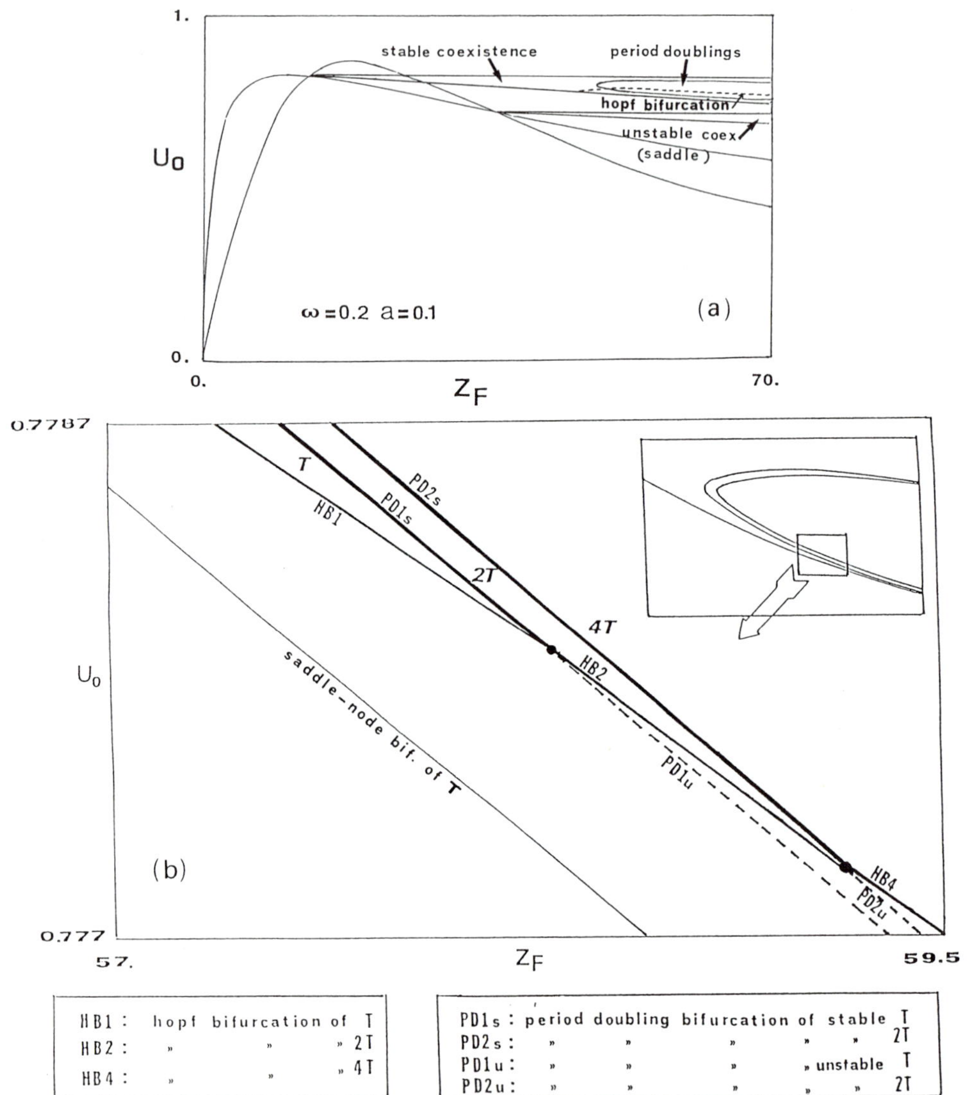


| HB1 : | hopf bifurcation of | T |
|---|---|---|
| HB2 : | " " | " 2T |
| HB4 : | " " | " 4T |

| PD1s : | period doubling bifurcation of stable | | T |
|---|---|---|---|
| PD2s : | " " | " " | " " 2T |
| PD1u : | " " | " " | " unstable T |
| PD2u : | " " | " " | " " 2T |

Fig. 7.  (a) Coexistence regions in the operating parameter space of the forced system; (b) period-doubling and Hopf bifurcation curves in the upper coexistence region. Parameter values of specific growth rate expressions: $\alpha = 10.5$, $\beta = 100$, $\gamma_1 = 0.01$, $\gamma_2 = 0.305$.

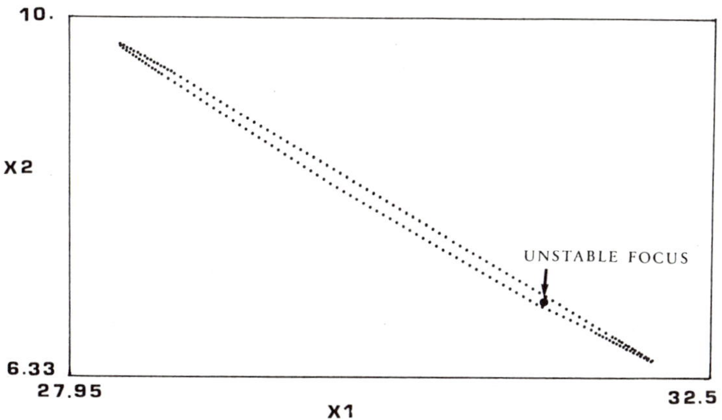



Fig. 8. Invariant circle (stroboscopic map of $T^2$ torus) of the case of Fig. 7. Operating parameter values: $u_o = 0.78$, $z_F = 56.3$.

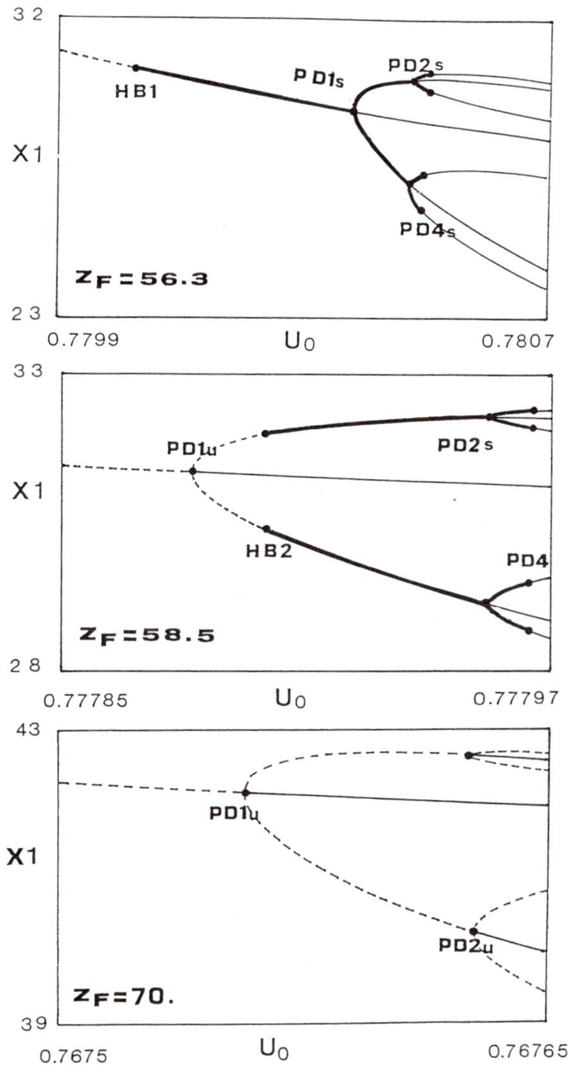



Fig. 9. One-parameter bifurcation diagrams for different values of $z_F$ of the case of Fig. 7 (————— stable, ————— saddle, --------- unstable).

of indeterminate values of the state variables (the concentrations of the two species). The coexistence steady state is represented by a straight line in the state space. Any point on this line is a steady state of the system. The linearized system about the coexistence steady state has one zero eigenvalue corresponding to the direction of the straight line. Periodic variation of one of the system's operating parameters results in coexistence of the two microbial species in a state of limit cycle oscillations for a wide range of operating conditions. However, depending on the form and relative position of the specific growth rate curves of the two microbial species, it is possible to have Hopf bifurcation of the coexistence limit cycle leading to quasiperiodic behavior or successive period doublings leading to chaotic behavior, although the autonomous system has only real eigenvalues about each steady state and does not possess any natural frequency to interact with the externally imposed one. Quasiperiodicity and chaos probably result from perturbation of a steady state of the autonomous system with a double zero eigenvalue.

From a practical point of view periodic variation of one of the operating parameters of the system allows coexistence of the two microbial species for some range of values of the chemostat operating parameters and not for discrete values of them as in the case of the autonomous system. Thus operation of the chemostat is not affected by small random fluctuations in the operating parameters. However, the coexistence state is not always in the form of limit cycle oscillations, but quasiperiodic or chaotic behavior may also occur making control of the system difficult.


## REFERENCES

Aris, R., and Humphrey, A. E., 1977, Dynamics of a chemostat in which two organisms compete for a common substrate, *Biotechnol. Bioeng.*, **19**:1375.

Butler, G. J., Hsu, S. B., and Waltman, P., 1985, A mathematical model for the chemostat with periodic washout rate, *SIAM J. Appl. Math.*, **45**:435.

Butler, G. J., and Wolkowicz, G. S. K., 1985, A mathematical model of the chemostat with a general class of functions describing nutrient uptake, *SIAM J. Appl. Math.*, **45**:138.

Doedel, E. J., 1981, AUTO: a program for the bifurcation analysis of autonomous systems, *Cong. Num.*, **30**:265; also, Doedel, E. J., 1986, *AUTO86 User Manual*, Pasadena.

Fredrickson, A. G., and Stephanopoulos, G., 1981, Microbial competition, *Science*, **213**:972.

Hale, J. K., and Somolinos, A. S., 1983, Competition for fluctuating nutrient, *J. Math. Biol.*, **18**:255.

Hsu, S. B., 1978, Limiting behavior for competing species, *SIAM J. Appl. Math.*, **34**:760.

Hsu, S. B., 1980, A competition model for a seasonally fluctuating nutrient, *J. Math. Biol.*, **9**:115.

Hsu, S. B., Hubbell, S., and Waltman, P., 1977, A mathematical theory for single-nutrient competition in continuous cultures of micro-organisms, *SIAM J. Appl. Math.*, **32**:366.

Jost, J. L., Drake, J. F., Fredrickson, A. G., and Tsuchiya, H. M., 1973, Interactions of *Tetrahymena pyriformis, Escherichia coli, Azotobacter vinelandii*, and glucose in a minimal medium, *J. Bacteriol.*, **113**:834.

Lenas, P., and Pavlou, S., 1991, Periodic, quasiperiodic and chaotic coexistence of two competing microbial populations in a periodically operated chemostat, *Math. Biosci.*, submitted.

Matsubara, M., Watanabe, N., and Hasegawa, S., 1986, Bifurcations in a bang-bang controlled mixed culture system, *Chem. Eng. Sci.*, **41**:523.

Meers, J. L., 1971, Effect of dilution rate on the outcome of chemostat mixed culture experiments, *J. Gen. Microbiol.*, **67**:359.

Powell, E. O., 1958, Criteria for the growth of contaminants and mutants in continuous culture, *J. Gen. Microbiol.*, **18**:259.

Powell, G. E., 1988, Structural instability of the theory of simple competition, *J. Theor. Biol.*, **132**:421.

Smith, H., 1981, Competitive coexistence in an oscillating chemostat, *SIAM J. Appl. Math.*, **40**:498.

Stephanopoulos, G., Aris, R., and Fredrickson, A. G., 1979a, A stochastic analysis of the growth of competing microbial populations in a continuous biochemical reactor, *Math. Biosci.*, **45**:99.
Stephanopoulos, G., Fredrickson, A. G., and Aris, R., 1979b, The growth of competing microbial populations in a CSTR with periodically varying inputs, *AIChE J.*, **25**:863.

# IRREGULAR BURSTING IN MODEL NEURONES

Julie Hyde

Departments of Applied Mathematical Studies and Physiology
University of Leeds
Leeds LS2 9JT   United Kingdom

Electrophysiological recordings from isolated neurons or from neurons *in situ*, form continuous time series, where activity can be either a regular, patterned or irregular discharge of action potentials. Models, often systems of coupled ordinary differential equations, can be constructed that accurately simulate observed behaviours in neurons.

Here, I examine one mathematical model - the Hindmarch-Rose equations and consider how our knowledge and prediction of the behaviour of the system can be interpreted physiologically.

## 1. Introduction

Early electrophysiological studies on excitation were on isolated nerve trunks, whose axons have a stable resting state and which generate a propagating, all-or-none action potential in response to a supra-threshold stimulus [1]. This gave rise to the idea of the threshold behaviour of neurons. However, chronic recordings from neurons *in situ* in unanaesthetised animals, or from isolated neuronal somata show that activity- a regular or patterned discharge of action potentials - is common, and that a stable resting potential is just one possible type of behaviour, others being small amplitude oscillations, a regular repetitive discharge of action potentials (beating), or repetitive patterned bursts of action potentials which is irregular. The Hindmarch-Rose (HR) equations [3] are a simple excitation system similar to the better known FitzHugh-Nagumo equations [3] and adequately simulate the behaviours of molluscan neurones from electro-physiological experiments. The problem is to understand the structure of the model and its behaviour in parameter space, and to relate this to physiological interpretation.

## 2. The Hindmarch-Rose equations

$$dx/dt = y - ax^3 + bx^2 - z + I$$

$$dy/dt = c - dx^2 - y \tag{1}$$

$$dz/dt = r(s(x - x_1) - z)$$

where the parameters $a, b, c, d, s$ and $x_1$ are real and take fixed values of 1.0, 3.0, 1.0, 5.0 4.0 and -1.6 respectively throughout the following

*Chaotic Dynamics: Theory and Practice*
Edited by T. Bountis, Plenum Press, New York, 1992

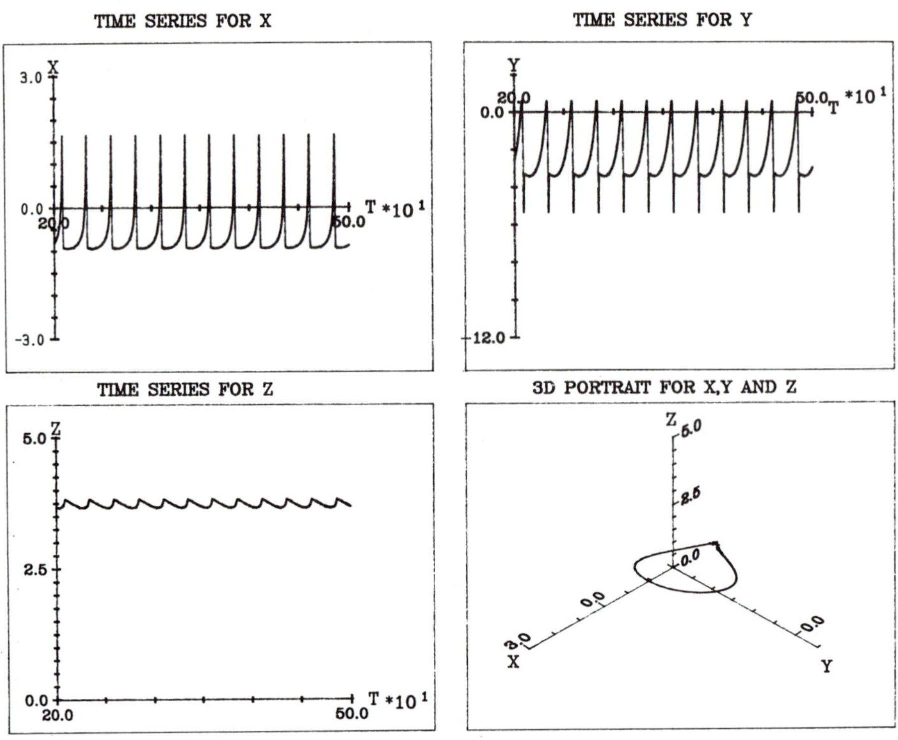

Figs 1a-1d. *Simple periodic behaviour. r = 0.01, I = 3.7*

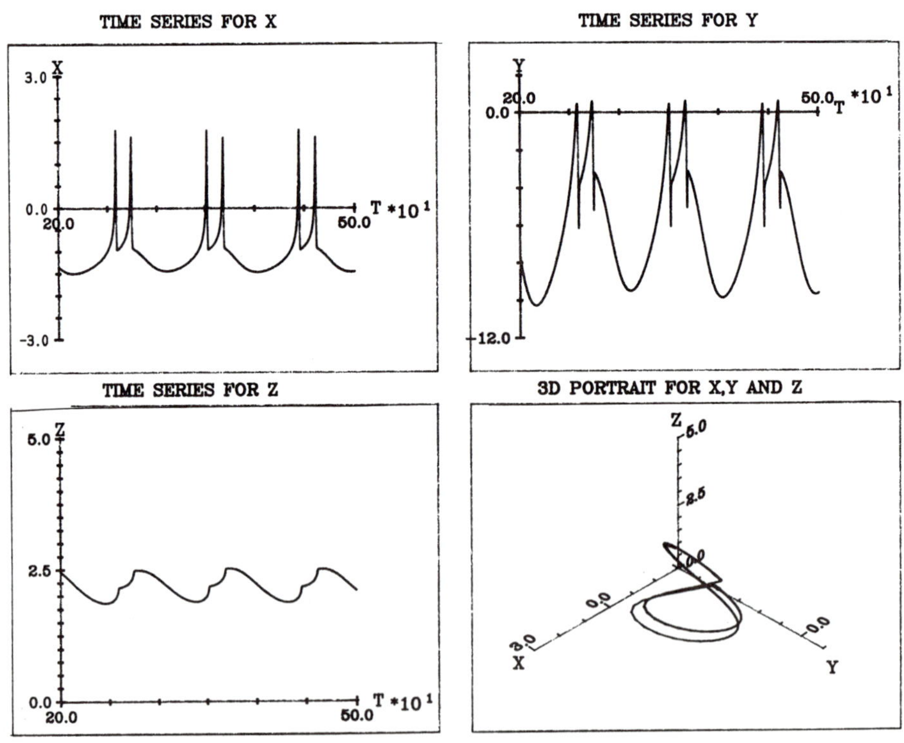

Figs 2a-2d. *Period two solution. r = 0.01, I = 2.3*

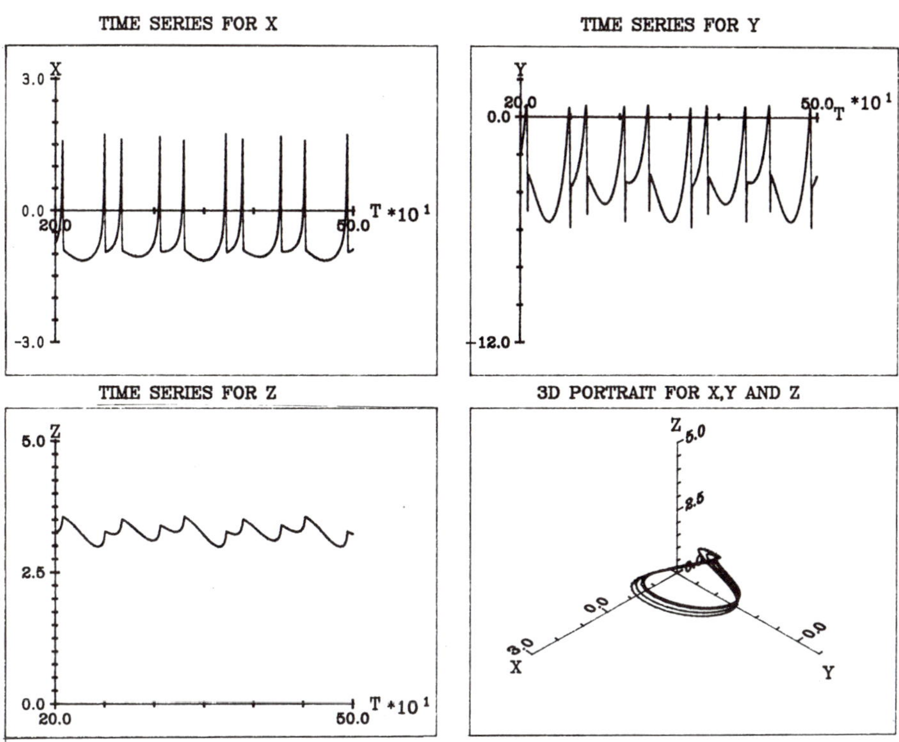

Figs 3a-3d. *Period four solution. r = 0.015, I = 3.3*

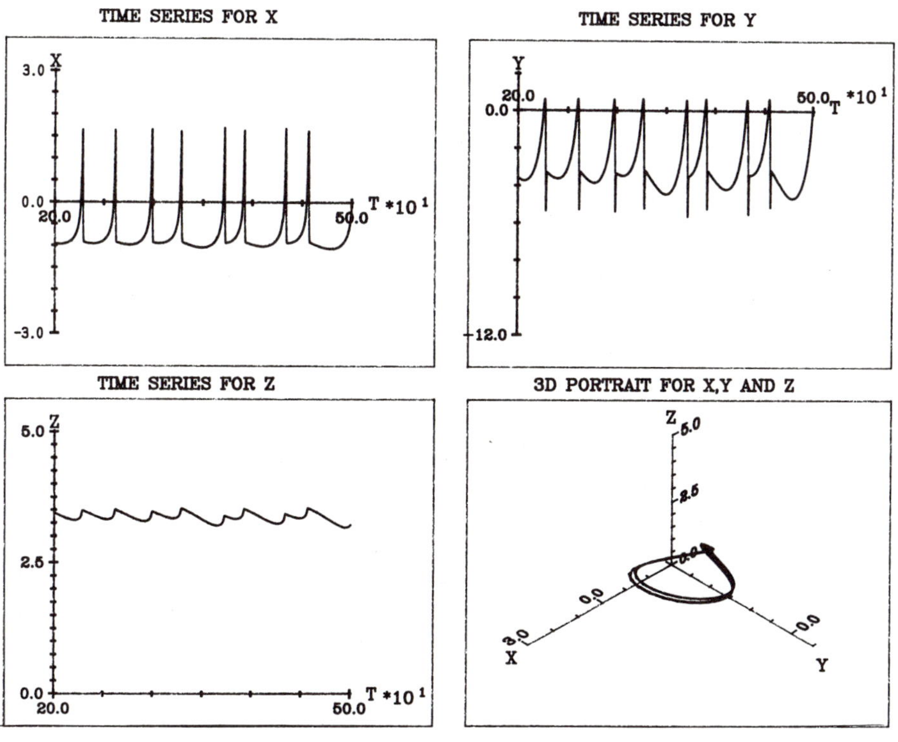

Figs 4a-4d. *Chaotic behaviour. r = 0.01, I = 3.36*

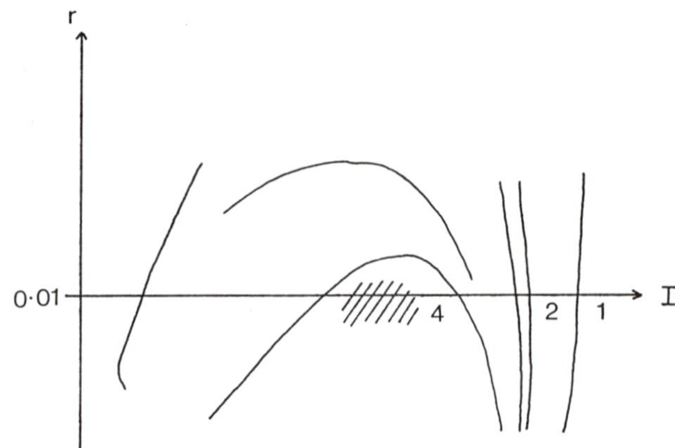

Fig 5. *One section of the bifurcation diagram for the Hindmarch-Rose equations*

examples. It is for different values of r and I that we observe different patternings.

The variables x,y and z do not have any direct physiological meaning but they can be thought of as being related to membrane potential, a recovery variable and an adaption variable respectively.The parameter I can be thought of as being analogous to a maintained injected current.

r and I may be treated as bifurcation parameters, and the behaviour examined in the (r,I) plane. For a small varying r and different values of I the time series solutions are shown with their corresponding three dimensional state space diagram. Fig 1d is a simple closed loop indicating that the solution is simple and periodic whereas in fig 2d, the trajectory winds round twice before repeating itself indicating a period two solution. In fig 3d the state space diagram shows a period four solution. Fig 4d shows a strange attractor and therefore a chaotic solution. These different behaviours have arisen via bifurcations between periodic solutions where a stable periodic orbit loses its stability as a Floquet multiplier crosses the unit circle in the complex plane [4]. The patterns of bifurcations may be understood by approximating Poincaré sections by a piecewise continuous interval map [5].

The different behaviours illustrated in figs 1-4 could be produced in the same cell by changing the intensity of a maintained injected current, or could be observed in different cells whose membranes had different leakage conductances. Changes analogous to the r parameter could be produced by $K^+$ channel blockers, or by different $K^+$ selective channel densities.

## References

[1] A.V. Holden. The mathematics of excitation. In *Biomathematics in 1980*. L.M. Ricciardi & A.Scott (eds) North-Holland, Amsterdam (1982)
[2] J.L. Hindmarch & R.M. Rose. *A model of neuronal bursting using three coupled first order differential equations*. Proc. Roy. Soc. London. **B221** 87,(1984)
[3] R. FitzHugh. Mathematical models of excitation and propagation in nerve. In *Biological engineering*, H.P. Schwann (Ed). McGraw-Hill: New York (1967)
[4] A.V. Holden, J. Hyde & M.A. Muhamad. *Periodicity, bursting and chaos in neural activity*. 9th Summer workshop on mathematical physics. July 17-20 (1990). Arnold-Sommerfeld institute for mathematical Physics. H.D. Doebner & F. Pasemann (Eds)
[5] A.V. Holden & F. Yinshui. *From simple to complex oscillatory behaviour via intermittent chaos in the Hindmarch-Rose model*. CNLS preprint (1991)

# SPECTROSCOPY AND DYNAMICS OF VIBRATIONALLY EXCITED

# MOLECULES: A PHASE SPACE STRUCTURE ANALYSIS

Stavros C. Farantos

Department of Chemistry, University of Crete, and
Institute of Electronic Structure and Laser, FORTH
Iraklion, Crete 711 10, Greece

## ABSTRACT

Methods for assigning and interpreting the dynamical behaviour of highly vibrationally excited polyatomic molecules, based on a phase space structure analysis, are
presented. Specific examples are shown for the bound motions of hydrogen cyanide in
the ground electronic state, and for the unbound states of ozone (resonances) in the
electronically excited state occupied during the photodissociation of the molecule.

## 1. INTRODUCTION

Spectroscopy provides the means to deduce the dynamics of an excited molecule.
For years now chemists and physicists study the vibrational spectra of polyatomic
molecules using well described techniques [1]. The first and basic step in solving the
molecular Schrödinger equation, is the Born-Oppenheimer approximation, which adiabatically separates the electronic from nuclear motion. The result of this approximation
is the construction of the potential energy surface, i.e. the potential which describes
the motion of nuclei at a specific electronic state of the molecule [2].

For a given potential energy surface (pes) the nuclear hamiltonian is usually expressed in normal coordinates, based on the assumption that a set of harmonic oscillators, which is equal to the number of the internal degrees of freedom, gives a good
zero order hamiltonian. The solutions of Schrödinger equation for the full hamiltonian,
which includes anharmonic terms as well, are then computed with perturbation theory.

This strategy assumes that a Taylor expansion of the potential energy surface
around the equilibrium geometry of the molecule is adequate. The assignment of
the spectrum is accomplished when we label each peak of the spectrum with a set of
quantum numbers; one quantum number for each degree of freedom. Since the zero
order hamiltonian in the normal coordinate formulation describes decoupled harmonic

oscillators, the dynamics of the molecule can be easily extracted. Sometimes complications arise because of the resonances between two vibrational levels of equal energy. The known methodology can also treat these cases.

Spectra assigned with a complete set of quantum numbers cover the low energy regime, and could be named *regular*. The theory of the regular molecular spectroscopy is the content of the three volumes of Herzberg's books [3].

In the last years the experimental methods have been advanced and low and high resolution spectra at highly excited vibrational levels have been produced. Such techniques are: (i) Multiphoton absorption spectroscopy, (ii) dispersion fluorescence spectroscopy, (iii) stimulated emission pumping spectroscopy, and (iv) electron photodetachment spectroscopy. In methods (ii), and (iii) an excited electronic state is used as an intermediate step, from which the molecule is deexcited to high vibrational states of the ground electronic state. In the electron photodetachment spectroscopy the anion of the molecule is used as an intermediate step, which is ionized to produce the vibrationally excited neutral molecule.

For high energies a normal coordinate representation of the nuclear hamiltonian as well as perturbation theory are not valid and the exact solution of the Schrödinger equation is required. The latter is achieved by variational calculations using large basis sets. Still accurate calculations can be done only for triatomic molecules and using appropriate coordinate systems for the representation of the hamiltonian and the construction of the basis set [4].

In such large scale quantum calculations the extraction of dynamics and the visualization of molecular motions are not easy. Fortunantly, for the motions of nuclei in the molecule the semiclassical assumption provides a good approximation, and therefore, classical mechanics is expected to give a good description of the dynamics.

The classical dynamics of a bound multidimensional system has been studied extensively the last thirty years [5]. The generic picture emerged from these studies for a system with large coupling and anharmonic terms in the potential, is a phase space prevailed by chaotic trajectories, and which has small islands of stability embedded in it. Even in the absence of regular regions there are unstable periodic orbits with small Lyapunov exponents, which differentiate the region around them from other regions of phase space. Now the question comes, what is the quantum mechanical analogue of the above classical picture?

This question has led researchers in investigating the problem of *what is quantum chaos*, and admittedly there is no satisfactory answer up to now. However, the research has produced some important results [6, 7].

Quantum mechanics may remain regular and localized at energies where classical mechanics has strongly exponentially divergent trajectories. It was also found, that at high energies there are localized eigenfunctions in configuration space, even if delocalized (*irregular*) states exist at lower energies (as irregular we define those states for which there are not $N$ good quantum numbers for their assignment). The localization occurs in regions of phase spase where there are classical resonances [8] or around unstable periodic orbits [9].

> Thus, we reach at the important conclusion that, *the classical phase space structure reveals topological characteristics of the quantum wavefunctions.*

Furthermore, the localized eigenfunctions are responsible for the high intensity peaks in a spectrum (higher than the peaks due to transitions to delocalized states), and the regularities which are observed in low resolution spectra [10]. Gutzwiller [11] was the first to produce formulae for the semiclassical calculations of the density of states by using families of periodic classical orbits of the dynamical system.

Experimental support for the theoretical results has come from the photoionization of hydrogen atom in strong magnetic fields [12, 13], as well as the stimulated emission pumping spectra of acetylene [14, 15, 16], $HCN$ [17], $Na_3$ [18, 19], $HCP$ [20], and $SO_2$ [21]. Another well studied case is the photodissociation spectrum of $H_3^+$ [22].

The advances in nonlinear dynamics open new prospects for analysing the spectroscopy, and producing information about the dynamical behaviour of vibrationally excited molecules. New questions arise such as:

1. What causes the localization of the wavefunctions in specific regions of phase space?

2. Is there a systematic way to study the phase space structure of a molecule?

3. How can we use the classical mechanical results of a nonlinear analysis to assist the quantum mechanical calculations for producing and assigning spectra?

In this article we present methods to answer the last two questions. The first question is more fundamental and although some theories have emerged, none of them seems to be satisfactory at present [8, 9, 23, 24].

## 2. PHASE SPACE STRUCTURE ANALYSIS

We assume that the molecular pes of a specific electronic state is known. In the classical mechanical approximation we are interested in investigating the *phase space structure* for a range of energies. The total energy of the system is the varying parameter in a nonlinear mechanical search.

With the term "phase space structure" we mean to search for regions of phase space which contain trajectories of similar type. For example a coarse graining characterization of phase space is the detection of the regions with quasiperiodic and chaotic trajectories. A more detailed study will reveal which resonance dominates in certain region of phase space. Of course in classical mechanics there is structure at every scale. However, in a molecular phase space search we are not interested in investigating volumes less than $h^N$, i.e. the phase space volume which corresponds to a quantum state [25]. $N$ denotes the number of degrees of freedom, and $h$ is Planck's constant.

A systematic search of the phase space structure is obtained by first locating the following geometrical objects with increasing dimensionality, $d$.

1. Stationary (or equilibrium or critical) points, $d = 0$.

2. Periodic orbits, $d = 1$.

3. Reduced dimension tori, $d < N$, and

4. KAM tori semiclassically quantizable, $d = N$.

The next step is to investigate the behaviour of the trajectories in the neighborhood of the above mentioned classical objects, and perhaps to study stable and unstable manifolds [8, 23, 24, 26, 27]. This requires a linear stability analysis.

Let $H$ be the hamiltonian of the system, and $(q_i^0, p_i^0, i = 1, ..., N)$ an initial point in phase space. The time evolution of this state is described by Hamilton's equations;

$$\frac{dq_i^0(t)}{dt} = \frac{\partial H}{\partial p_i^0}$$

$$\frac{dp_i^0(t)}{dt} = -\frac{\partial H}{\partial q_i^0} \quad i = 1, ..., N. \tag{1}$$

The stationary points are defined by

$$dq_i^0/dt = dp_i^0/dt = 0 \quad i = 1, ..., N. \tag{2}$$

For a hamiltonian of the form,

$$H = \sum_{i=1}^{N} \frac{p_i^{02}}{2m_i} + V(q_1^0, ..., q_N^0), \tag{3}$$

the stationary points are the roots of equations,

$$p_i^0 = 0, \quad \frac{\partial V}{\partial q_i^0} = 0, \quad i = 1, ..., N, \tag{4}$$

i.e. they coincide with the extrema of the pes. Molecular pess have usually more than one minima and saddle points.

The theorem of Weinstein-Moser [28, 29] predicts the appearance of at least $N$ families of periodic orbits from each potential minimum. These periodic orbits describe the basic motions of the system and they are called *principal*.

A systematic search for periodic orbits is achieved by constructing a bifurcation diagram. By increasing the energy of the system we follow the evolution of the principal periodic orbits and we find the energies at which bifurcations occur. The bifurcating families are also followed with the energy and new bifurcations are found. Numerical algorithms have been developed for finding critical points, bifurcation points, and periodic orbits; efficient continuation schemes for following the evolution of these objects with energy have been proposed [30]. Since all the algorithms are based on the solution of nonlinear equations, practically there are numerical difficulties in the location of periodic orbits, especially those which emerge from saddle stationary points and they are not associated with the principal families or bifurcations of them.

To find the stability properties of a trajectory $(q_i^0(t), p_i^0(t), i = 1, ..., N)$, we study the difference of equations of motion;

$$\frac{d}{dt}[q_i(t) - q_i^0(t)] = \frac{\partial H}{\partial p_i} - \frac{\partial H}{\partial p_i^0}$$

$$\frac{d}{dt}[p_i(t) - p_i^0(t)] = -\frac{\partial H}{\partial q_i} + \frac{\partial H}{\partial q_i^0}$$

$$i = 1, ..., N, \tag{5}$$

where $(q_i(t), p_i(t))$ is a nearby trajectory.

Linearizing the rhs of the above equations we have,

$$\frac{d}{dt}[q_i(t) - q_i^0(t)] = \sum_{j=1}^{N} \left[ \frac{\partial^2 H}{\partial q_j^0 \partial p_i^0}(q_j - q_j^0) + \frac{\partial^2 H}{\partial p_j^0 \partial p_i^0}(p_j - p_j^0) \right]$$

$$\frac{d}{dt}[p_i(t) - p_i^0(t)] = -\sum_{j=1}^{N} \left[ \frac{\partial^2 H}{\partial q_j^0 \partial q_i^0}(q_j - q_j^0) + \frac{\partial^2 H}{\partial p_j^0 \partial q_i^0}(p_j - p_j^0) \right]$$

$$i = 1, ..., N. \tag{6}$$

These are the *variational equations* and can be written in the general form,

$$\frac{d\mathbf{x}(t)}{dt} = A(t)\mathbf{x}(t). \tag{7}$$

The second derivatives of the hamiltonian are calculated at $(q_i^0(t), p_i^0(t) \quad i = 1, ..., N)$. It is well known that the solutions of the above linear differential equations are given in terms of $x_i(t) = c_i \exp(\lambda_i t)$, where $c_i$ are constants, and $\lambda_i, i = 1, ..., N$, the eigenvalues, of the matrix:

$$A = \begin{vmatrix} \dfrac{\partial^2 H}{\partial q_j^0 \partial p_i^0} & \dfrac{\partial^2 H}{\partial p_j^0 \partial p_i^0} \\ -\dfrac{\partial^2 H}{\partial q_j^0 \partial q_i^0} & -\dfrac{\partial^2 H}{\partial p_j^0 \partial q_i^0} \end{vmatrix}. \tag{8}$$

Therefore, we may conclude that the nearby trajectories diverge exponentially if the eigenvalues have a positive real part, and are stable if the eigenvalues are imaginary with norm equal to one.

Since both trajectories $(q_i^0(t), p_i^0(t))$ and $(q_i(t), p_i(t))$ are functions of time the matrix elements in eq. (6) are also implicit functions of time. On the other hand the coefficient matrix $A$, could also be considered as a function of the phase point $(q_i^0, p_i^0)$.

If $(q_i^0(t), p_i^0(t))$ is a periodic solution, then it can be shown that the variational equations (6) have also periodic solutions, and they can be expressed with the eigenvalues of the monodromy matrix, $M(t = T)$. The monodromy matrix is obtained by integrating the variational equations for a period $T$ and with the unit matrix as initial conditions [30]. Thus, the nearby trajectories are unstable in the direction of those eigenvectors of the monodromy matrix, which have eigenvalues with positive real part. In this way we can determine whether the periodic orbit is surrounded by tori, reduced dimension tori or complete chaotic trajectories.

The above strategy cannot be exhaustive and we cannot claim that we reveal every important structure of phase space. Furthermore the procedure is time consuming for realistic multidimensional systems.

There is an approximate method which relates the pes's domains with the unstable regions of phase space. Brumer and Duff [31, 32] and Toda [33] suggested the use of the variational equations for studying the stability properties of a system by examining the eigenvalues of $A$ as functions of the phase point $(q_i^0, p_i^0)$, and ignoring the dependence on time. Cerjan and Reinhardt [34] critisized this procedure and suggested some corrections, such as the finding of all critical points of the difference equations and then performing the linear stability analysis at each critical point. Several application [32], and particularly applications on realistic molecular systems [35], show that this method can associate regions of the potential energy surface with strong chaotic behaviour.

## 3. THEORETICAL LOW RESOLUTION SPECTRA

In this section we show how the phase space structure analysis assists to assign spectra and to extract the dynamics. At high energies or for polyatomic molecules, where the density of states is large, it is difficult to analyse spectra at the level of isolated states. Thus, with low resolution we mean, that we obtain spectra with resolution $\Delta E$, an energy range which could hide several eigenstates. Time dependent methods are the most suitable for the construction of such spectra, since we propagate the wavefunction for the time interval $\Delta t = h/\Delta E$. Of course for high densities, $\Delta E$ may be so small, that the numerical integration of equation of motions for the whole interval $\Delta t$ becomes prohibited.

### 3.1 Quantum calculations

We adopt a time dependent formalism as developed by Heller [10]. An initial excitation is assumed to be induced by a dipole moment operator $\hat{\mu}$. Then, the initial wavefunction is $|\phi(0)> = \hat{\mu}|\chi_i>$. $|\chi_i>$ denotes the eigenstate of another electronic state in SEP spectroscopy. The time evolution of the initial wavefunction is given by solving the time dependent Schrödinger equation;

$$i\hbar \frac{\partial |\phi(t)>}{\partial t} = \hat{H}|\phi(t)> . \tag{9}$$

Several schemes have been proposed for the propagation of the wave packet $|\phi>$ in time, [36]: (i) Second order difference method, (ii) the split operator algorithm, (ii) representation of the propagator operator in Chebyshev series, and (iv) the Lanczos method.

The second order difference methode (SOD) is based on a Taylor expansion of the time propagator,

$$\hat{U} = \exp(-i\Delta t\hat{H}/\hbar) \approx 1 - i\Delta t\hat{H}/\hbar - \frac{1}{2}\Delta t^2 \hat{H}^2/\hbar^2, \tag{10}$$

and on symmetrizing appropriately in order to make the algorithm stable:

$$|\phi_{n+1}(\mathbf{q})> - |\phi_{n-1}(\mathbf{q})> = (\hat{U} - \hat{U}^+)|\phi_n(\mathbf{q})> . \tag{11}$$

This procedure leads to the result,

$$|\phi_{n+1}> = |\phi_{n-1}> - \frac{2i}{\hbar}\Delta t\hat{H}|\phi_n> . \tag{12}$$

$\Delta t$ is the time elapsed between the $n$ and $n+1$ iterations.

The split operator method (SOP) proposed by Feit et al. [37] is based on the equation,

$$|\phi(\mathbf{q}, t_0 + \Delta t)> \approx \exp(i\Delta t\hbar\nabla^2/4\mu)\exp(-i\Delta tV/\hbar)\exp(i\Delta t\hbar\nabla^2/4\mu)|\phi(\mathbf{q}, t_0)> . \tag{13}$$

A description of the other methods and comparisons among them are given in ref. [36]. For the calculation of the kinetic part of the hamiltonian we discretize the configuration space, and we use fast Fourier transform techniques to find the action of the hamiltonian on the wavepacket.

The spectrum can be expressed by first evaluating the correlation function;

$$C(t) = <\phi(0)|\phi(t)>, \tag{14}$$

and then computing its Fourier transform,

$$I(E) = \frac{1}{2\pi\hbar}\int_{-\infty}^{\infty}\exp(iEt/\hbar)C(t)dt. \tag{15}$$

To see how the eigenvalues could be extracted from eq. (14), we expand $|\phi(t)>$ as a series of the eigenfunctions $|n>$ of the hamiltonian $\hat{H}$.

$$|\phi(t)> = \exp(-i\hat{H}t/\hbar)|\phi(0)> = \sum_n \exp(-iE_n t/\hbar)|n><n|\phi(0)> . \tag{16}$$

By introducing the overlap integral, $c_n = <n|\phi(0)>$, the spectrum becomes,

$$I(E) = \frac{1}{2\pi\hbar}\sum_n |c_n|^2 \int_{-\infty}^{\infty}\exp[-i(E_n - E)t/\hbar]dt =$$

$$\frac{1}{2\pi\hbar}\sum_n |c_n|^2 \lim_{T\to\infty}\int_{-T}^{T}\exp[-i(E_n - E)t/\hbar]dt =$$

$$\frac{1}{2\pi}\sum_n |c_n|^2 \lim_{T\to\infty}\frac{2\sin((E_n-E)T/\hbar)}{E_n-E}=\sum_n |c_n|^2\delta(E_n-E). \tag{17}$$

Thus, for infinite integration time we have a sum of delta functions located at the eigenvalues $E_n$. Finite integration in time will give a sum of broaden peaks centered at $E_n$.

We could also extract the eigenfunctions $|n>$ by computing the Fourier transform,

$$|n>=\frac{1}{c_n}\int_{-\infty}^{\infty}\exp(iE_nt/\hbar)|\phi(t)>dt. \tag{18}$$

It is clear, that in a time dependent calculation, we locate those eigenfunctions which overlap significantly with the initial wavefunction $|\phi(0)>$. Since we expect localization of the eigenfunctions around periodic orbits or in resonance zones, we simulate the spectrum by taking the appropriate initial wavepacket centered at the place of interest. It can be either an experimental spectrum or a theoretical one, which will reveal those states that are localized at particular regions of phase space.

Taylor and coworkers obtained low resolution spectra for the hydrogen atom in a magnetic field [38], and the $H_3^+$ [39], by distributing a basis set of Gaussian functions along the periodic orbit and its neighborhood. The diagonalization of the hamiltonian matrix in association with stabilization techniques [40] reveals the eigenfunctions localized in the region of periodic orbit.

## 3.2 Classical calculations

Although quantum mechanical calculations are always what we should ask, however they are not feasible at present for more than three degrees of freedom systems. It is useful then to find the classical analogue of the quantum spectrum. Thus, the correspondence between spectrum and phase space structure is straightforward.

The following formulation has been used in the past [41, 42]. By taking the square of the absolute value of the correlation function $C(t)$, (*the survival probability function*),

$$|C(t)|^2 = |<\phi(0)|\phi(t)>|^2 =<\phi(0)|\phi(t)><\phi(t)|\phi(0)>, \tag{19}$$

and using the indentity relation,

$$\hat{1}=\sum_n |n><n|, \tag{20}$$

we get,

$$|C(t)|^2 =<\phi(0)|\phi(t)><\phi(t)|\hat{1}|\phi(0)>=\sum_n [<\phi(0)|\phi(t)><n|\phi(0)><\phi(t)|n>]=$$

$$\sum_n [<n|\phi(0)><\phi(0)|\phi(t)><\phi(t)|n>]=tr[\hat{\rho}(0)\hat{\rho}(t)]. \tag{21}$$

$\hat{\rho}(0)$ is a density operator,

$$\hat{\rho}(0)=|\phi(0)><\phi(0)|$$

and

$$\hat{\rho}(t)=|\phi(t)><\phi(t)|=e^{-i\hat{H}t/\hbar}|\phi(0)><\phi(0)|e^{i\hat{H}t/\hbar}, \tag{22}$$

is the Heisenberg representation of $\hat{\rho}(0)$.

We can pass to the classical analogue by replacing the trace in eq. (21) with an integral over the phase space, and by replacing the density operators with classical distribution functions,

$$\Omega(t)=\int \rho(\mathbf{q}(0),\mathbf{p}(0))\rho(\mathbf{q}(t),\mathbf{p}(t))d\mathbf{q}d\mathbf{p}. \tag{23}$$

The classical initial distribution $\rho(0)$ is usually a Wigner or Husimi transform of the initial quantum wavefunction, $|\phi(0) >$, [43, 44]. The spectrum is then defined as the Fourier transform,

$$I_c(\omega) = \int e^{i\omega t}\Omega(t)dt. \tag{24}$$

The classical survival probability function, eq. (23), and its Fourier tranform, have succesfully been used and advanced by Gomez Llorente, Pollak, and Taylor the last years [45, 46].

## 4. APPLICATIONS

In this section we present results of the methodology described in the previous section from the applications into two typical triatomic molecules, $HCN$, and $O_3$. For $HCN$, we examine bound motions in the ground electronic state. The characteristic of this molecule is the extended regular behaviour and despite of the isomerization process, $HCN \rightarrow HNC$, which supports.

For $O_3$, we examine unbound motions in the excited electronic state which the molecule occupies during its photodissociation. The classical trajectories are strongly chaotic at the excitation energies, but the quantum dynamics shows localization.

### 4.1 Hydrogen cyanide

The pes of the ground electronic state has two minima, which correspond to the stable linear molecule, $HCN$, and to the metastable isomer, $HNC$, at the energy of 0.484 eV above the absolute minimum. The barrier of isomerization is 1.51 eV. The potential is shown in figure 1.

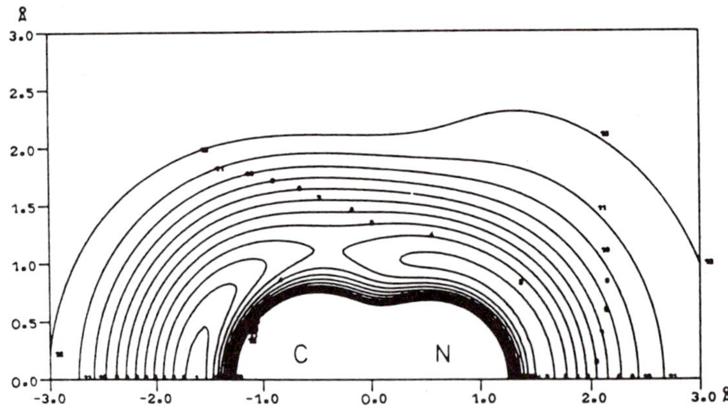

Figure 1.    Contours of the potential energy surface of $HCN$ for $H$ moving around $CN$ with the $C - N$ bond length fixed at 1.16 Å. Contour 1 = -13.60 eV and contour intervals = 0.50 eV [2].

In figure 2 we show diagrams of characteristics for the families of periodic orbits which we have located. Those families which emanate from the absolute minimum are denoted as $M1A$, $M1B$, etc. Families emanating from the second minimum are denoted as $M2$, whereas the periodic orbits which appear above the barrier of isomerization are denoted with the prefix $I$. Solid lines correspond to stable periodic orbits, dotted lines to single unstable (the monodromy matrix has one pair of real eigenvalues), and the squares to double unstable (two pairs of real eigenvalues) periodic orbits. Crosses are for complex unstable periodic orbits, i.e. the monodromy matrix gives one quadruplet of complex eigenvalues.

From these diagrams we can draw some important conclusions about the dynamical behaviour of the molecule. The motions which describe stretches of $CH$, and $CN$ (these correspond to the periodic orbits $M1A$ and $M1B$ respectively) remain stable at energies even above the first dissociation limit. This regularity has been confirmed in quantum mechanical calculations as well [47]. Trajectories associated with excitation of the bending mode (periodic orbits of $M1C$ type) are also regular up to 0.75 eV above the potential barrier of isomerization. States related with the isomerization process are localized along the $I$ type periodic orbits, which we have named *rotating periodic orbits*.

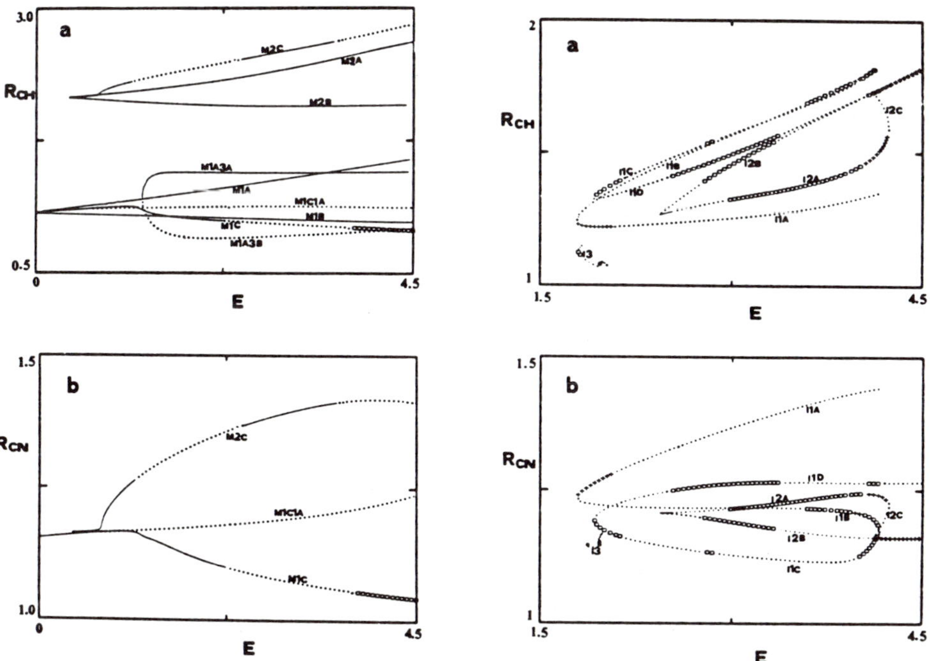

Figure 2. Diagrams of characteristics for $HCN$. $R_{CH}$ and $R_{CN}$ are bond lengths. Solid lines correspond to stable periodic orbits, dotted lines to single unstable, and squares to double unstable periodic orbits, and crosses for complex unstable periodic orbits. $M1$ denotes families of periodic orbits which originate from the absolute minimum, $HCN$, $M2$ denotes families which start from the metastable $HNC$ minimum, and $I$ those families which originate above the barrier of isomerization. For details see ref. [47].

These conclusions are confirmed not only from the quantum mechanical calculations [47, 48], but also from recent SEP spectroscopic experiments [17]. In figure 3 we compare the classical survival probability function with its quantum mechanical analogue. In figure 4 we compare the theoretical classical mechanical spectrum with the experimental dispersed fluorescence spectrum [49].

The agreement between classical and quantum correlation functions demonstrates that localized quantum eigenstates should exist at the same regions of phase space where the periodic orbits and tori have been found. Therefore, the assignment of the spectrum can be obtaind by using these classical objects [50, 51, 52].

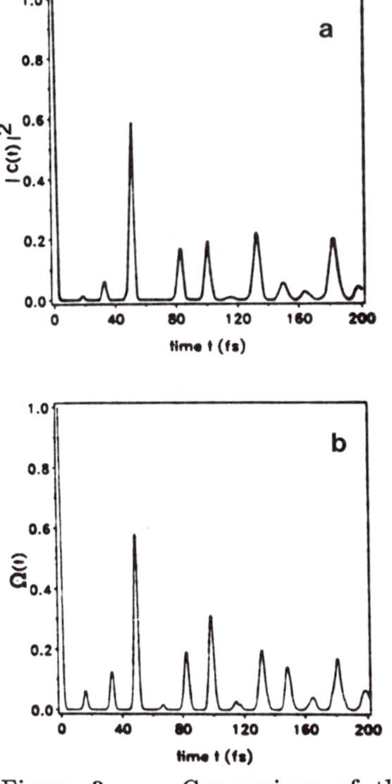

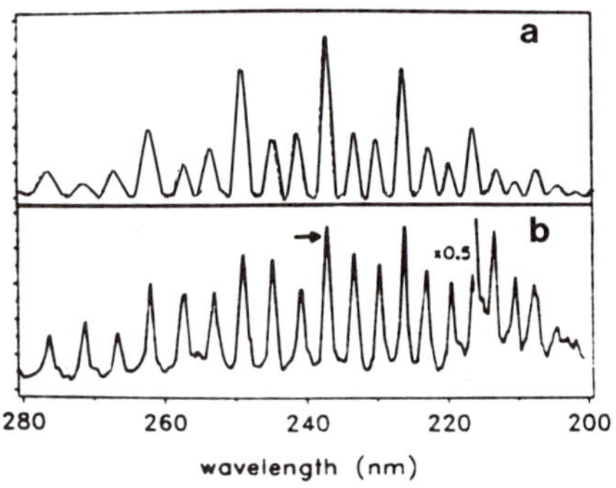

Figure 3. Comparison of the quantum survival probability function (a) with its classical analogue (b).

Figure 4. Comparison of the spectrum obtained from the classical survival probability function (a), and the experimental fluorescence spectrum [49].

## 4.2 Ozone

Contrary to $HCN$, ozone has been found to be highly unstable [53]. Figure 5 shows the pes of the $^1B_2$ electronic state, from which the molecule dissociates. $R_1$, and $R_2$ are the two bond lengths and the angle between them is held fixed at $115.75^0$. Large values of $R_1$ or $R_2$ denote the dissociation channels. In the figure the Wigner transform of the initial wavefunction, the ground vibrational electronic state, is also shown. This is a Gaussian function centered on the ridge of the potential which separates two symmetric minima.

The absorption spectrum $D^1B_2 \leftarrow X^1A_1$, shows a broad peak, the Hartley band [54]. The main interest in this spectrum is a weak structure on the top of Hartley's band, which consists of small peaks separated by approximately 250 $cm^{-1}$. Johnson and Kinsey [53] produced the survival probability function from the experimental spectrum, and carried out a periodic orbit search. Farantos and Taylor [55] also carried out 3D classical calculations, but found no oscillatory structure in the survival probability function. At the same time Le Quéré and Leforestier [56] published a 3D quantum mechanical correlation function with rich oscillatory structure. No interpretation was given for the dynamics of the molecule.

310

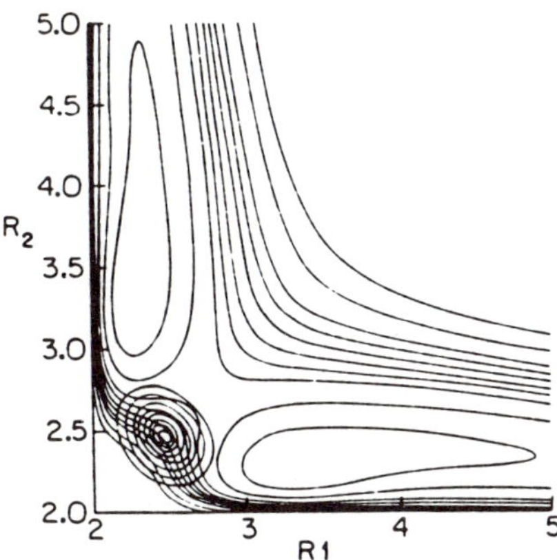

Figure 5.    Contours of the ozone pes of Sheppard and Walker [57]. $R_1, R_2$ are the bond lengths and the angle between them is kept constant at $\phi = 115.75^0$. Distances in $a_0$ and the energy of the contours are -4.6, -4.2, -3.8, -3.4, -3.2, -3.0, -2.8, -2.4, and -2.0 eV. The Wigner transform of the initial wavefunction is also shown.

In order to elucidate these discrepancies we have performed quntum mechanical calculations by solving the time dependent Schrödinger equation and restricting the problem into two dimensions. The angle between $R_1$, and $R_2$ was kept fixed. Figure 6 shows the 2D quantum survival probability function, which contrary to the classical one, has indeed very rich structure. By plotting snapshots of the evolving wavefunction it is not difficult to understand the origin of the oscillations in the survival probability function.

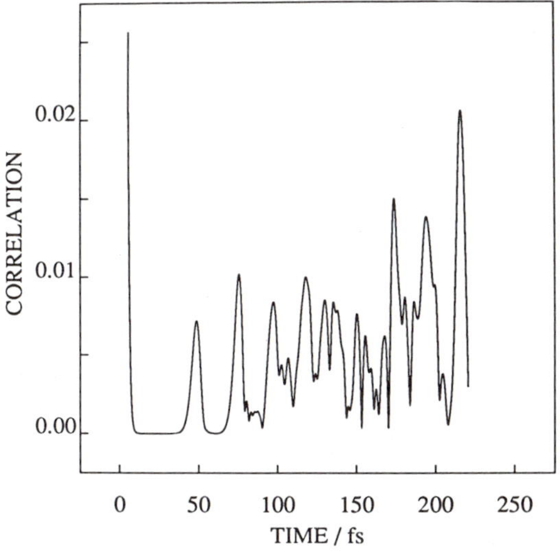

Figure 6.    The absolute value of the 2D quantum correlation function.

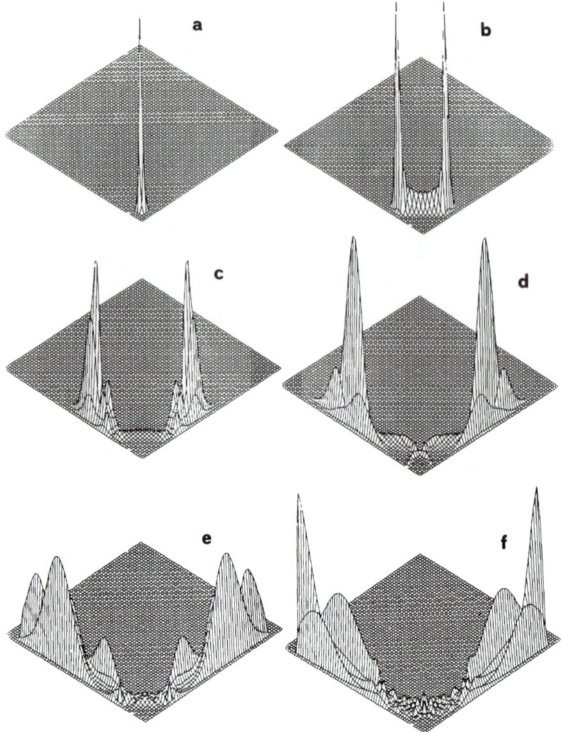

Figure 7.    Snapshots of the time evolution of the wavepacket. (a) 0 fs,
(b) 16.22 fs, (c) 32.54 fs, (d) 48.81 fs, (e) 65.07 fs, and (f) 81.34 fs.

In figure 7 we can see that while the bulk of the wavepacket moves towards the
exit channels (the molecule dissociates), part of the wavepacket returns to the initial
interaction region, and that results in the peaks of function $C(t)$.

Still it is difficult to understand the dynamics of the system from these plots. Therefore, we try to locate the periodic orbits which have the same periods as the recurrence
times extracted from the correlation function (figure 8). The orbit 8a corresponds to
the symmetric stretch and has a period of 47 fs. This is the first recurrence time,
and it is in good agreement with the first peak in the quantum survival probability
function. Figure 8b is an asymmetric stable periodic orbit which does not overlap
with the initial wavefunction and therefore it does not influence the evolution of the
wavepacket. However, we believe that it can explain recurrence times found in the
experimental correlation function, which are not reproduced with the present potential
[55]. Because of the symmetry of periodic orbits we should expect the recurence times
in the correlation function to coincide with the half and full period of the periodic
orbits, and this is what happens [58].

In conclusion we can state, that for ozone although classical mechanics fails to
reproduce the quantum mechanical survival probability function, and this is because
of the high instability of the system, however periodic orbits can be used to explain
the quantum localization, and to elucidate the detail dynamics of the system.

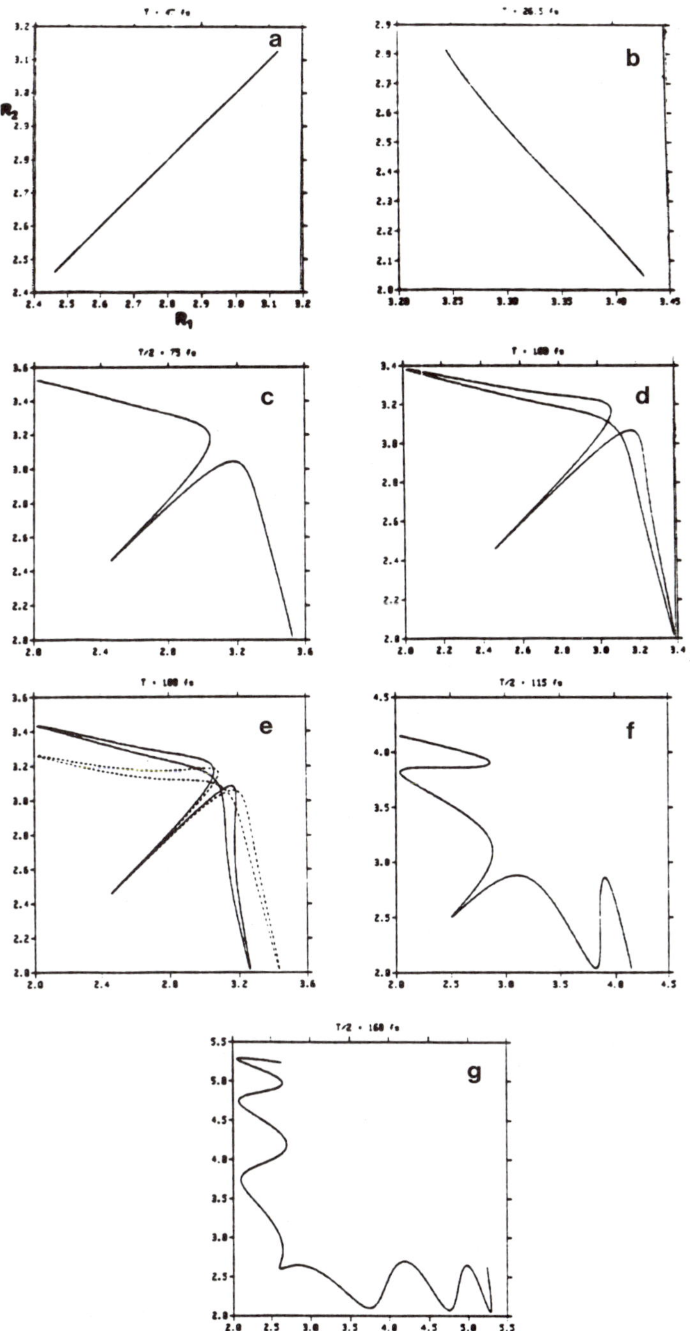

Figure 8.    Located periodic orbits in the 2D surface of ozone.  The periods are (a) 47 fs, (b) 26.5 fs, (c) 150 fs, (d) 108 fs, (e) 216 fs, (f) 230 fs, and (g) 320 fs.

## 5. CONCLUSIONS

The research of the last years on the comparison of classical and quantum calculations on real molecules has shown the importance of knowing the phase space structure for understanding, and thus, assigning the vibrational spectra of highly excited molecules. Unstable periodic orbits are the stationary classical objects which seem to be the most important for assigning the spectra. It is found that the eigenfunctions of the system are localized (scared [9]) around particular periodic orbits. However, in highly unstable regions it may happen one eigenfunction to be influenced by several periodic orbits. In this case we associate the peaks of the time correlation function with the periods of the periodic orbits.

We foresee that in the future phase space structure analysis will become a basic procedure for analysing the spectra of highy excited molecules.

## ACKNOWLEDGMENTS

I am grateful to Prof. H. S. Taylor for stimulating discussions.

# References


[1] E. B. Wilson, J. C. Decius, and P. C. Cross. *Molecular Vibrations*. McGraw-Hill, New York, 1955.

[2] J. N. Murrell, S. Carter, S. C. Farantos, P. Huxley, and A. J. C. Varandas. *Molecular Potential Energy Functions*. Wiley, 1984.

[3] G. Herzberg. *Vol. I, II, III*. Van Nostrand Reinhold Company, New York, 1950.

[4] R. E. Wyatt. In J. O. Hirschfelder, R. E. Wyatt, and R. D. Coallson, editors, *Lasers, Molecules, and Methods*, page 231. Wiley, N.Y., 1989.

[5] A. J. Lichtenberg, and M. A. Liberman. *Regular and Stochastic Motion*, volume 38. Applied Math. Sciences, Springer - Verlag, 1983.

[6] B. V. Chirikov. *Chaos*, 1:95, 1991.

[7] S. C. Farantos, and J. Tennyson. In R. Lefebvre and S. Mukamel, editors, *Stochasticity and Intramolecular Redistribution of Energy*, page 15, by D. Reidel, Dordrecht, 1987.

[8] M. J. Davis. *J. Phys. Chem.*, 92:3124, 1988.

[9] E. J. Heller . *Phys. Rev. Letters*, 53:1515, 1984.

[10] E. J. Heller. *Acc. Chem. Res.*, 14:368, 1981.

[11] M. C. Gutzwiller. *Chaos in Classical and Quantum Mechanics*, volume 1. Springer-Verlag, 1990.

[12] A. Holle, G. Wiebusch, J. Main, B. Hager, H. Rottke, and K. H. Welge. *Phys. Rev. Letters*, 56:2594, 1986.

[13] D. Wintgen and H. Friedrich . *Phys. Rev.*, A36:131, 1987.

[14] E. Abramson, R. W. Field, D. Imre, K. K. Innes, and J. L. Kinsey. *J. Chem. Phys.*, 83:453, 1985.

[15] J. P. Pique, Y. Chen, R.E. Field, and J. L. Kinsey. *Phys. Rev. Letters*, 58:4715, 1987.

[16] J. P. Pique, Y. M. Engel, R. D. Levine Y. Chen, R.E. Field, and J. L. Kinsey. *J. Chem. Phys.*, 88:5972, 1988.

[17] X. Yang, C. A. Ragashi, and A. M. Wodtke. *J. Opt. Soc. Am. B*, 7:1835, 1990.

[18] M. Broyer, G. Delacretaz, G. -Q Ni, R. L. Whetten, J. -P. Wolf and L. Wöste. *Phys. Rev. Letters*, 62:2100, 1989.

[19] M. Broyer, G. Delacretaz, G. -Q Ni, R. L. Whetten, J. -P. Wolf and L. Wöste. *J. Chem. Phys.*, 90:4620, 1989.

[20] Y. -T. Cheng, D. M. Watt, R. W. Field, and K. K. Lehmann. *J. Chem. Phys.*, 93:2149, 1991.

[21] K. Yamanouchi, S. Takeuchi, and S. Tsuchiya. *J. Chem. Phys.*, 92:4044, 1990.

[22] A. Carrington, and R. A. Kennedy . *J. Chem. Phys.*, 81:91, 1984.

[23] M. Founargiotakis, S. C. Farantos, G. Contopoulos, and C. Polymilis. *J. Chem. Phys.*, 91:1389, 1989.

[24] S. C. Farantos, M. Founargiotakis, and C. Polymilis. *Chem. Phys.*, 135:347, 1989.

[25] L. D. Landau, and E. M. Lifshitz. *Statistical Physics*, volume 5. Pergamon Press, New York, 1970.

[26] S. Wiggins. *Physica D*, 44:471, 1990.

[27] R. E. Gillilan, and G. S. Ezra. *J. Chem. Phys.*, 94:2648, 1991.

[28] A. Weinstein. *Inv. Math.*, 20:47, 1973.

[29] J. Moser. *Commun. Pure Appl. Math.*, 29:727, 1976.

[30] R. Seydel. *From Equilibrium to Chaos: Practical bifurcation and stability analysis.* Elsevier, 1988.

[31] P. Brumer, and J. W. Duff. *J. Chem. Phys.*, 65:3566, 1976.

[32] J. W. Duff, and P. Brumer. *J. Chem. Phys.*, 67:4898, 1977.

[33] M. Toda. *Phys. Letters*, A48:335, 1974.

[34] C. Cerjan, and W. P. Reinhardt. *J. Chem. Phys.*, 71:1819, 1979.

[35] S. C. Farantos. *Chem. Phys.*, 71:157, 1982.

[36] C. Leforestier, R. Bisseling, C. Cerjan, M. D. Feit, R. Friesner, A. Guldberg, A. Hammerich, G. Jolicard, W. Karrlein, H. D. Meyer, N. Lipkin, O. Roncero, and R. Kosloff. *J. Comput. Phys.*, 94:59, 1991.

[37] M. D. Feit, J. A. Fleck, and A. Steiger. *J. Comput. Phys.*, 47:412, 1982.

[38] H. S. Taylor and J. Zakrzewski. *Phys. Rev. A*, 38:3732, 1988.

[39] J. M. Gomez Llorente, J. Zakrzewski, H. S. Taylor, and K. C. Kulander. *J. Chem. Phys.*, 90:1505, 1989.

[40] A. U. Hazi, H. S. Taylor. *Phys. Rev. A*, 1:1109, 1970.

[41] M. Baranger. *Phys. Rev.*, 111:481, 1958.

[42] E. J. Heller, and M. J. Davis. *J. Phys. Chem.*, 84:1999, 1980.

[43] Y. Weissman, and J. Jortner. *J. Chem. Phys.*, 77:1486, 1982.

[44] S. W. McDonald. *Phys. Rep.*, 158:337, 1988.

[45] J. M. Gomez Llorente, and H. S. Taylor . *J. Chem. Phys.*, 91:953, 1989.

[46] J. M. Gomez Llorente, and E. Pollak. *J. Chem. Phys.*, 90:5406, 1989.

[47] S. C. Farantos, and M. Founargiotakis. *Chem. Phys.*, 142:345, 1990.

[48] J.A. Bentley, J.-P. Brunet, R. E. Wyatt, R. A. Friesner, and C. Leforestier. *Chem. Phys. Letters*, 161:393, 1989.

[49] A. P. Baronavski. *Chem. Phys. Letters*, 61:532, 1979.

[50] S. C. Farantos, J. M. Gomez Llorente, O. Hahn, and H. S. Taylor. *Chem. Phys. Letters*, 166:71, 1990.

[51] S. C. Farantos, J. M. Gomez Llorente, O. Hahn, and H. S. Taylor. *J. Chem. Phys.*, 93:76, 1990.

[52] S. C. Farantos, J. M. Gomez Llorente, O. Hahn, and H. S. Taylor. *J. Chem. Phys.*, 94:2376, 1991.

[53] B. R. Johnson, and J. L. Kinsey. *J. Chem. Phys.*, 91:7638, 1989.

[54] D. E. Freeman, K. Yoshino, J. R. Esmond, and W. H. Parkinson. *Planet. Space Sci.*, 32:239, 1984.

[55] S. C. Farantos, and H. S. Taylor. *J. Chem. Phys.*, 94:4887, 1991.

[56] F. Le Quéré, and C. Leforestier. *J. Chem. Phys.*, 94:1118, 1991.

[57] M. G. Sheppard, and R. B. Walker. *J. Chem. Phys.*, 78:7191, 1983.

[58] S. C. Farantos. *Chem. Phys., in press*, 1991.

INHIBITION OF CHAOTIC BEHAVIOUR IN COUPLED

MODELS OF ATMOSPHERIC DYNAMICS AND CLIMATE EVOLUTION

John Brindley and Tomasz Kapitaniak

Department of Applied Mathematical Studies
and Center for Nonlinear Studies
University of Leeds
Leeds LS2 9JT, U.K.

## INTRODUCTION

The difficulty of carrying out long-term predictions of atmospheric
dynamics and the evolution of climate is a problem of obvious concern.
Nowadays there is increasing awareness that deterministic chaos might
provide a natural prototype for complexity of atmospheric and climatic
dynamics.

The occurrence of multiple equilibria and of chaotic behaviour in nonlinear oscillators subjected to periodic forcing is widespread and well-
-known. An example is the Duffing equation, which we may write as

$$\ddot{x} + k\dot{x} + x - x^3 = B\cos\Omega t \tag{1}$$

arising almost ubiquitously in models of mechanical oscillations, which has
been extensively studied [1,2] ( for a useful summary see [3]). The parameter
space of (1) is divided with great complexity into regions of different
qualitative behaviour, and the space of initial states is divided with
similar complexity into the basins of attraction of competing attractors,
which may be steady, periodic or chaotic. Solutions in chaotic regimes are
often characterised by behaviour which may show some regularity over short
timescales with occasional substantial deviations, as the trajectory spends
time slow-moving near one unstable equilibrium between sporadic excursions
to the neighbourhood of another.

Aperiodicity is the first apparent characteristic of the behaviour of
many geophysical fluid dynamic systems, but atmospheric and oceanic flows
often exhibit substantial coherent features, localised in either or both of
space and time, which occur sporadically and unpredictably but with a
certain statistical regularity. Such features are exemplified by blocking
patterns in the mid-latitude atmosphere, or by persistent anomalies (of
which El Nino is the most spectacular) in sea surface temperature and
associated rainfall. Figure 1 demonstrates the variability of rainfall at
Nauru Island in the West Pacific. The time series shown is characteristic as
usually very little, if any, precipitation falls in this region of the central
Pacific, and only during El Nino do large amounts of precipitation occur.

*Chaotic Dynamics: Theory and Practice*
Edited by T. Bountis, Plenum Press, New York, 1992

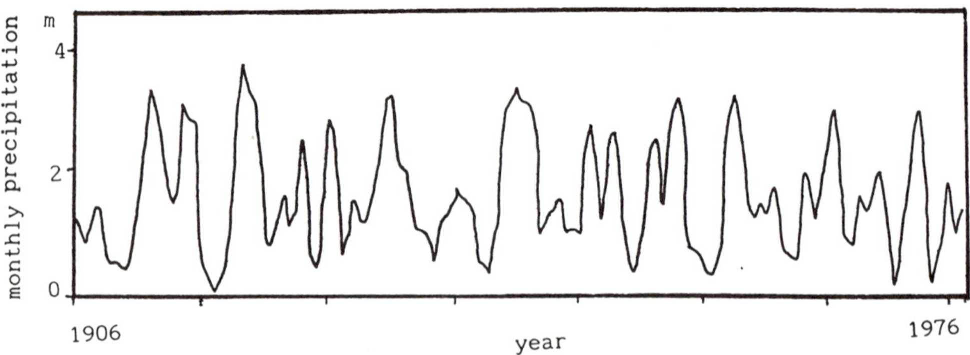

Figure 1. Mean monthly precipitation at Nauru Island.

On a different timescale, fluctuations in climate display similar characteristics, distinguished by quasi-regular "anomalies" on timescales of decades to hundreds of thousands of years. As an example one can take the ice core and deep ocean core data of quaternary glaciations which show a clear out average periodicity as shown in Figure 2 where near-periodicity of roughly 100000 dominates records of ice volume, with smaller peaks near 40000 and 20000 years.

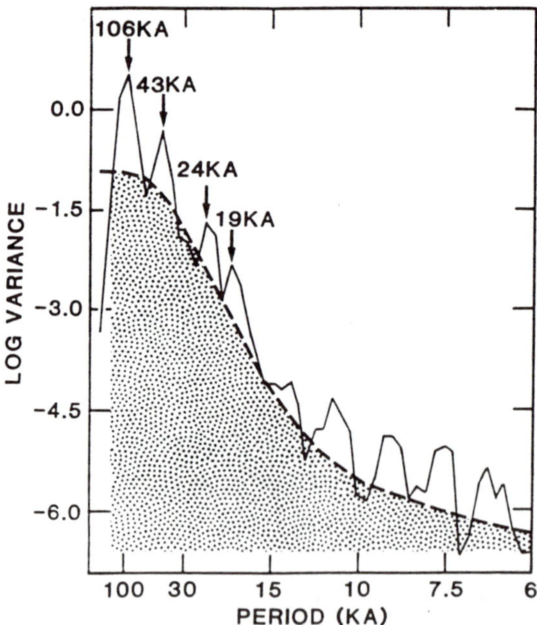


Figure 2.   Power spectrum of the ice volume based on deep-sea cores (after Imbrie and Imbrie[16]); KA=1000 years; broken line indicates spectrum of red-noise-like background.

318

The existence and persistence of such features is reminiscent of the
behaviour described by a solution trajectory of a nonlinear oscillator
wandering near to one unstable equilibrium and making occasional excursions
to the neighbourhood of a second unstable equilibrium.  This similarity in
behaviour has stimulated the development of simple conceptual models,
usually taking the form of forced or coupled nonlinear oscillators,
representing to a greater or less extent the assumed essential physics. The
common feature of such models is  the assumption that ocean-atmosphere
coupling (perhaps enriched by variation of ice extent) is a crucial physical
ingredient, and that the annual cycle of solar heating, modified as
necessary to include effects of orbital variations, is an important driving
mechanism. An excellent overview of coupled ocean-atmosphere models of El
Nino and the so called Southern Oscillation has appeared recently [4] whilst
similar reviews of climatic models are contained, for example, in [5,6].

In this paper we examine 2 models, each consisting of coupled nonlinear oscillators, which represent coupled ocean-atmosphere dynamics in a
simple way on appropriate timescales.  In the first, based on Saltzman's
model [7] of coupling between sea-ice extent and ocean temperature, we have
added to the existing astronomical forcing an extra periodic forcing
compatible with the earth's orbital variations. We have also considered an
extra forcing in the form of band limited white noise. In the second, based
on a model by Vallis [9] coupling local temperatures in the East and West
Pacific with a current strength generated in part by these temperature
differences, we have added to the annual forcing a second stochastic forcing
compatible with the 20-60 day period of forced atmospheric Kelvin modes. In
each case, numerical experiments have indicated the replacement of chaotic
behaviour in the singly periodically forced problem by  noisy periodicity in
the "quasi-periodically" forced problem. The mean period of response is
surprisingly different from either of the forcing periods, and is in the
second case qualitatively similar to that of the El Nino events. Finally, we
report that numerical experiments on coupled Saltzman oscillators, representing the North and South hemispheres, again show noisy periodicity in regimes
where each oscillator, existing uncoupled, shows chaotic behaviour.


THE SALZMAN OSCILLATOR


Long-term climatic variation, characterised most dramatically by the
sequence of ice ages during the last million or so years, has been widely
attributed to self-sustained oscillations arising from coupling between sea
ice  extent  and mean ocean temperature.  A simple model proposed by
Saltzman [7] has  received much attention, and Nicolis [9,10] has shown that
the  addition  of  periodic forcing to Saltzman's model, intended to
simulate variations in radiative  heat input associated with variations in
the earth's orbit (on timescales of $O(10^5)$ years), can produce a chaotic
response.

Following Nicolis, we can write the system in the succinct form

$$\dot{x} = y, \quad \dot{y} = \gamma_2 x - x^3 + \mu(\gamma_1 y - x^2 y + q\sin\Omega t) \qquad (2)$$

where x and $\mu x + y$ are respectively the scaled deviations of the size of
sea ice extent and of mean ocean temperature from reference states. For
varying $\gamma_1$, $\gamma_2$, $\mu$, q and $\Omega$, the response of this oscillator of course
demonstrates the usual range of qualitative behaviour. For some values of
the parameters of climatic significance, the behaviour is unmistakably
chaotic, and Figure 3(a) comprises a Poincaré map for such a set of values.

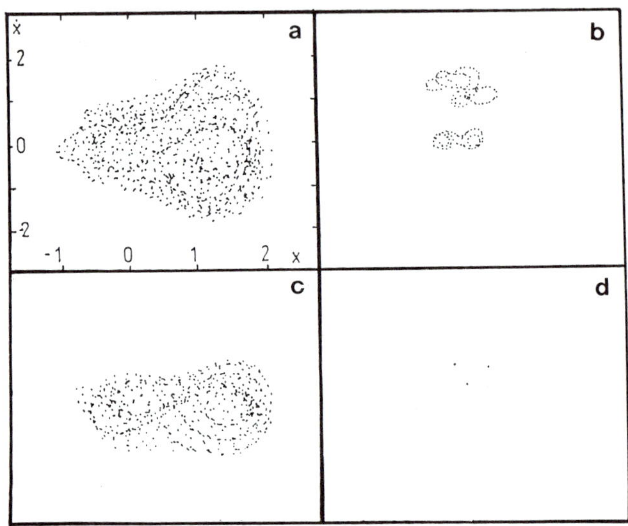

Figure 3   (a)  Poincare map (interval $2\pi/\Omega$ ) for equation (2).   Following
Nicolis [9] we have chosen $\gamma_1$ = 1.01, $\gamma_2$ = 1, $\mu$ = 0.1, q = 0.74,
$\Omega$ = 0.3.
(b)  Poincare map for equations (2) with an additional term
asin$\omega$t, with a = 0.05, $\omega$ = 3 x $10^4$; other parameter values as
in Figure 3(a).
(c)  Mean Poincare map for equations (2). Here $\omega_1 \in$ [0.2, 0.4] and
other parameter values are as for Figure 3(b).
(d)  Mean Poincare map for equations (2) with larger value of a
(a=0.25).

The addition to the forcing in (2) of a term $a\sin\omega t$, choosing $a = 0.05$ and $\omega = 30,000$, to represent the weak annual forcing, has a dramatic effect on the Poincaré map, now reproduced as Figure 3(b). The chaotic map is replaced by a map which corresponds to motion on a torus, indicating a quasi-periodic response by the oscillator. The qualitative character of the behaviour is highly sensitive to the value of a, and in Figure 4(a) we demonstrate this sensitivity over a range of values of a.

A second modification to (2) comprises the addition of band limited white noise to the basic single frequency forcing. Thus, by considering the equation

$$\ddot{x} - \mu(\gamma_1 - x^2)\dot{x} - \gamma_2 x + x^3 = \mu q\sin\Omega t + a \sum_{i=1}^{N} \sin\omega_i t \tag{3}$$

and taking $\omega_i \in [0.2, 0.4]$, we have, for the same values of parameters as in Figure 3(b), a mean Poincaré map as indicated in Figure 3(c). Note that we have defined the mean Poincaré map to be the set $M \subset R^2$:

$$\langle M(x(t_o))\rangle = \left\{\langle x(t)\rangle, \langle \dot{x}(t)\rangle \;\middle|\; t = 2\pi k/\Omega \;;\; k = 1, 2, \ldots\right\}$$

where $x(t)$ are the realisations of the solutions of (3) for the initial condition $x(t_o)$, and $\langle . \rangle$ indicates an ensemble average obtained from a number of realizations of $\omega_i$ [11]. When a is increased, we find that the mean Poincare map takes the form of Figure 3(d) corresponding to noisy periodicity; chaotic behaviour has been replaced by regular stochasticity.

As a final experiment with the Saltzman model we examine a pair of weakly Coupled Saltzman oscillators (representing two hemispheric regions coupled by weak energy transfer across the Equator). The equations are

$$\begin{array}{r}
\ddot{x} - \mu(\gamma_1 - x^2)\dot{x} - \gamma_2 x + x^3 = \mu q\sin\Omega t + ky \\[1em]
\ddot{y} - \mu(\gamma_1 - y^2)\dot{y} - \gamma_2 y + y^3 = \mu q\sin\Omega t + kx
\end{array} \right\} \tag{4}$$

where k is constant.

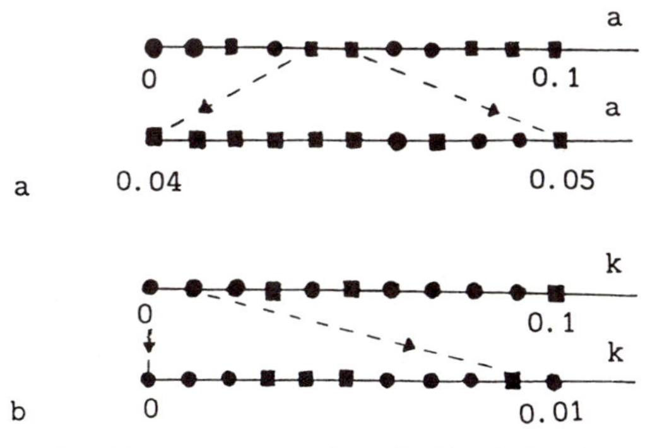


$\bullet$ – chaotic,  $\blacksquare$ – quasiperiodic behaviour

Figure 4    (a) Qualitative change in behaviour of solutions to (4) as a
varies; other parameter values as in Figure 1(b).
(b) Variation of qualitative behaviour of solutions to (4) as k
varies; values of $\gamma_1$, $\gamma_2$, $\mu$, $\Omega$ as in earlier cases.

Choosing again the same parameter values as  above, so that each decoupled
oscillator behaves chaotically for k = 0, we find that, for k > 0, periodic
oscillations are possible. Dependence on k is very sensitive, as indicated in
Figure 4(b).

In summary, we see that the chaotic behaviour of equation (2) is readily
destroyed by the addition of further weak forcing or coupling, and that the
dependence on "perturbation strength" is extremely sensitive.


EL NINO - SOUTHERN OSCILLATION


The El Nino Southern Oscillation (ENSO) is the predominant interannual
variability of the tropical ocean-atmosphere system. It results in conside-
able fluctuations in rainfall, sea surface temperature (SST) and the intensity
of the trade winds over the Pacific Ocean. Its range of variation and
essential unpredictability impose considerable stress on ecological systems,
and its effects on global weather and climate are accepted but poorly
understood. Many "simple" models have been designed to represent ocean-
-atmosphere interactions which may be involved in ENSO dynamics, and a recent
very comprehensive survey[4] admirably summarises present thinking. We consider
here a very simple two-point model of ENSO proposed by Vallis[8] which is based
essentially on the East-West temperature advection equation and the equation
of continuity. The model comprises three ordinary nonlinear equations
describing the evolution of ocean current and of ocean mixed-layer
temperatures at Eastern and Western stations in the Pacific, which may be
written

$$\left.\begin{aligned}
\frac{du}{dt} &= B(T_E - T_W)/2\Delta x - C(u - u^\bullet) \\[2mm]
\frac{dT_W}{dt} &= u(\overline{T} - T_E)/2\Delta x - A(T_W - T^\bullet) \\[2mm]
\frac{dT_E}{dt} &= u(T_W - \overline{T})/2\Delta x - A(T_E - T^\bullet)
\end{aligned}\right\} \qquad (5)$$

where $\overline{T}$ is the constant temperature of deep ocean, A, B and C are constants,
the term  $B(T_E - T_W)/2\Delta x + Cu^\bullet$ represents wind-produced stress, $-Cu$
represents mechanical damping, $T^\bullet$ is the temperature to which the ocean
would relax in the absence of motion. The independent variable t is
considered to vary on timescales of the ENSO phenomena, ie, 2-4 years.

In our numerical experiments we have used the Vallis values, namely
$A = 1$ year$^{-1}$, $C = 0.25$ month$^{-1}$, $B = 2m^2$ sec$^{-2}$ C$^{-1}$, $u^\bullet = -0.45m$ sec$^{-1}$,
$T^\bullet = 12°C$ and $\overline{T} = 0°$.  These values characterise the background state and
may have an important impact on the dynamics.

Vallis considered both the "free" behaviour of this system, and the
effect on it of an annual forcing, and concluded that behaviour reminiscent
of El Nino was possible.  Our conjecture here is that the effects of the
so-called Madden-Julian Oscillations (MJO) of the atmosphere, associated
with propagating and reflecting Kelvin waves, are also important in ENSO
dynamics, as has been suggested by recent observational evidence[12]. For a
detailed discussion see[13].

We thus introduce a forcing in the ocean current equation consisting of

(1)   a component of constant amplitude frequency to represent the annual
      forcing
(2)   a term intended to represent the MJO effects in the form of an
      atmospheric wind stress.
We have chosen various forms for this forcing varying from pure white noise
to sharply peaked distributions, with randomly chosen frequency (in the 40-
-60 day period range) and phase.
Some results are shown for the case in which we have taken

$$\dot{u} = -0.45 + 0.1\,\cos\Omega t + \sum_{i=1}^{100} \delta_i \cos(\omega_i t + \phi_i)$$

where $\delta_i$, $\omega_i$, $\varphi_i$ are random variables such that

$$\delta_i \in (0.05,\ 0.3)\ ,\quad 2\pi/\omega_i \in (20,\ 60)\ \text{days}$$
$$\phi_i \in (0,\ 2\pi)\ ,\ \text{and}\quad 2\pi/\Omega = 365\ \text{days}.$$

Other parameters have been set at values found by Vallis to give a chaotic
response. Figure 5 shows the variation of u,$\Delta$t over a period of 100 years
after 1000 years of integration.
Roughly periodic large anomalies are very prominent in these plots. Gross
features are summarised in the spectrum (Figure 5), which shows a strong
noisy peak corresponding to a mean period of 3.5-4 years. All these features.

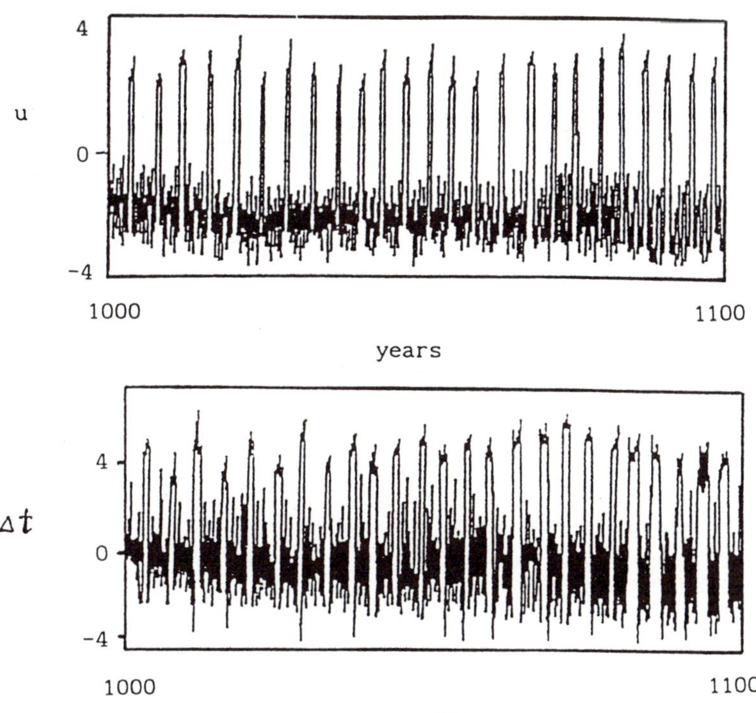


Figure 5    Variation of u and $\Delta$t over a period of 100 years as a result of
            integrating equations (4) for a 1000; values of parameters as
            specified in text; $\Delta t = T_E - T_W$.

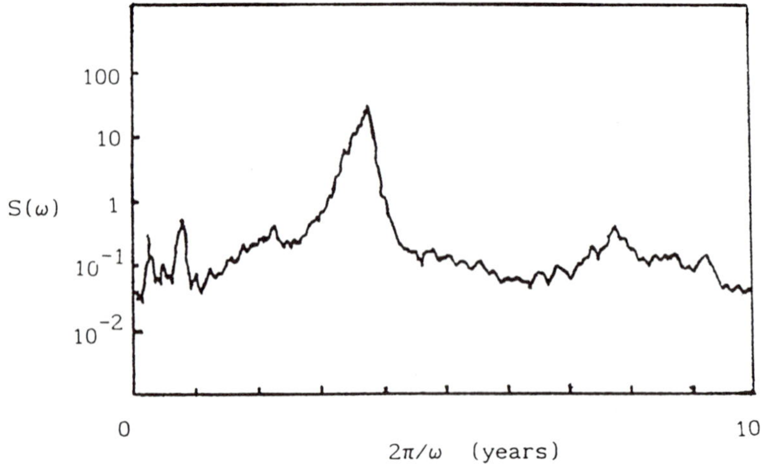


Figure 6    Spectral density distributions for Δt over the 100 years of
            Figure 5.

are reminiscent of El Nino and are discussed fully elsewhere [13]; the main
point we wish to stress here is the appearance of a noisy periodicity in
this "quasi-periodically" forced system at parameter values where a simple
forcing gives chaotic behaviour.

Some support for the description of the real El Nino dynamics as noisy
periodicity rather than chaos may be adduced by analysis of the example of
rainfall data given in Figure 1. Briefly, any dissipative system such as
this ocean-atmosphere system, has the property that its phase space volume
contracts as the system approaches its asymptotic state (attractor). If we
consider the attractor as a set of points embeddded in the phase space then by
a well-known procedure[17] we can assign to it a number called itsdimension. The
dimension turns out to be a lower bound to the number of independent variables
necessary to describe the attractor. If this dimension is non-integer (as in
the case of chaotic systems) then the bound is the next higher integer.

Following Grassberger-Procaccia[17] we constructed an m-component vector
from the time series p(t) shown in Figure 2 as

$$P_i(t_i) = \{p_1(t_1+\tau), p_2(t_1+2\tau), \ldots, p_m(t_1+m\tau)\} \ ,$$

where $\tau$ is an appropriate time delay and calculated the correlation integral
defined for N vectors distributed in an m-dimensional embedded space as a
function of distance r:

$$C(r,m) = \lim_{N\to\infty} N^{-2} \sum_{i=1}^{N} \sum_{j=1}^{N} \Theta(r-|P_i-P_j|)$$

where $\Theta$ is the Heaviside step function. If the number of points is large
enough, as assumed above, this distribution will obey a power-law scaling
with r for small r: $C(r,m) \sim r^{\nu}$, where $\nu$ is the correlation dimension.

Correlation dimension can be used to quantify presence of chaos in our
system. In the case of chaos, as we increase embedding dimension m, the
correlation dimension is seen to converge to a fixed value. If the system
is random the correlation dimension is seen to increase monotonically with m

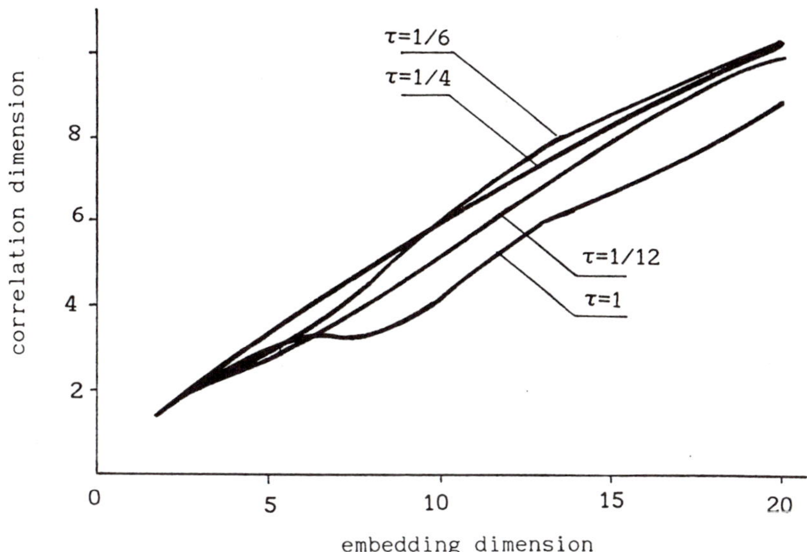


Figure 7.    Correlation dimension $\nu$ versus embedding dimension m for a time
series of monthly mean percipitation of Nauru (see Figure 2).


Figure 7 displays the results of a calculation from the time series of
rainfall at Nauru shown in Figure 1. There is no evidence of a low-
dimensional attractor such as is characteristic for chaotic phenomena and
the suggestion is that ENSO events cannot be adequately modelled by a low-
dimensional  dynamical system.



## REFERENCES


[1]   Y.Ueda,   Steady motions exhibited by Duffing's equation; a picture
      book of regular and chaotic motions, in: "New Approaches to Nonlinear
      Problems in Mechanics", P J Holmes ed., 311-322, SIAM, Philadelphia
      (1980).
[2]   P.J. Holmes, A nonlinear oscillator with a strange attractor,Phil Trans
      R. Soc Lond A 292: 419 (1979).
[3]   T. Kapitaniak, Chaotic Oscillations in Mechanical Systems, Manchester
      University Press, Manchester (1991).
[4]   J. P. McCreary and D.L. Anderson, An overview of coupled ocean-
      atmosphere models of El Nino and the southern oscillation, J Geophys
      Res. 96 Supp, 3125 (1991).
[5]   H. Le Treut and M. Ghil, Orbital forcings, climatic interaction and
      glaciation cycles, J Geophys Res. 88:5167 (1983).

[6]   A. Wiin-Nielsen, On simple climatic models with periodic and stochastic forcing, Geophys, Astrophys, Fluid Dyn $\underline{28}$:1 (1984).

[7]   B. Saltzman, A. Sutera and A. Evenson, Structural stochastic stability of a simple auto-oscillatory climate feedback system, J Atmos Sci $\underline{38}$: 484 (1981).

[8]   G.K. Vallis, Conceptual models of El Nino and the southern oscillation J Geophys Res $\underline{93}$:13979 (1988).

[9]   C. Nicolis, Self oscillations and predictability in climate dynamics – periodic forcing and phase locking, Tellus $\underline{36A}$:217 (1984).

[10]  C. Nicolis, Long-term climatic variability and chaotic dynamics, Tellus $\underline{40A}$:50 (1987).

[11]  T. Kapitaniak, Chaos in Systems with Noise, World Scientific, Singapore (1990).

[12]  K.M. Weickmann, El Nino/Southern oscillation and Madden-Julian (30--60 Day) oscillations during 1981-82, J Geophys Res. $\underline{96}$ Supp 3187 (1991).

[13]  A. Barcilon, J. Brindley and X-S. Chen X-S, A conceptual coupled ocean-atmosphere model, in preparation (1991).

[14]  J. Brindley and T. Kapitaniak, Analytic predictors for strange non-chaotic attractors, Phys Lett $\underline{A155}$:161 (1991).

[15]  J. Brindley, T. Kapitaniak and M.S. El Naschie, Analytical conditions for strange chaotic and nonchaotic attractors of the quasiperiodically forced van der Pol equation, Physica D , in press (1991).

[16]  J.Imbrie and K.P. Imbrie, Ice Ages: Solving the mystery. Enslow Publ., Short Hills, N.J (1979).

[17]  P. Grassberger and I. Procaccia Measuring the strangeness of strange attractors, Physica, $\underline{9D}$:189 (1983).

EVIDENCE FOR CHAOTIC DYNAMICS IN THE

OUTER SOLAR PLASMA AND THE EARTH MAGNETOSPHERE



G.P. Pavlos[1], A.G. Rigas[1], D. Dialetis[2], E.T. Sarris[1,2],
L.P. Karakatsanis[1] and A.A. Tsonis[3]

[1]Department of Electrical Engineering
 Demokritos University of Thrace, Xanthi, Greece
[2]National Observatory of Athens
 Institute of Ionosphere and Space Physics, Athens, Greece
[3]University of Wisconsin, Milwaukee, Wisconsin


ABSTRACT


In this paper we present a new approach of space plasmas dynamics
based on methods of chaos theory which are applied to the time series
sunspot index, solar wind temperature and B-field, as well as to plasma
sheet B-field and AE index measurements of the earth magnetosphere.
Moreover we develope an interesting set of criteria against the colored
noise pseudo-chaos. Finally we provide evidence for the existence of
chaotic attractor dynamics for the outer solar plasma (including solar
wind) and for the earth magnetospheric system with fractal dimensions
~4.5 and ~3.5 respectively.



1.INTRODUCTION

The outer solar plasma including the convection zone, the solar
corona and the solar wind is a "turbulent" dissipative dynamical system
with complicated temporal evolution remaining far away of thermodynamical
equilibrium. Prigogine [1] has supported the hypothesis that physical
systems far from thermodynamic equilibrium can exhibit a rich variety of
self-organizing behavior. Some years ago one of the authors applied these
concepts to space plasma [2]. Especially, he proposed the non-local
chaotic dynamics of strange attractors as a complementary paradigm to
local processes for the explanation of magnetospheric substorms. The
energy flux produced in the core of the sun by thermonuclear reaction is
radiatively transfered outwards. The resulting total light flux from the
core amounts to ~$4 \times 10^{26}$ Joule/sec. The outward energy flow causes
temperature gradient of the outer solar plasma according to the relation

$$\frac{dT}{dr} = (1 - \frac{1}{\gamma}) \frac{T}{P} \frac{dP}{dr}$$

where P is pressure in the solar wind. Here we assume that the perfect
gas law is a natural approximation. Parker [3] showed that the solar
corona plasma is subjected to supersonic expansion according to the
equation

$$\frac{1}{u}\frac{du}{dr}(u^2 - c^2) = \frac{2c^2}{r} - \frac{GM_{sun}}{r^2}$$

where $c = (RT/\mu)^{1/2}$ is the isothermal speed of sound. The solution of this equation gives a critical radius $r(c) = GM_{sum}/2c^2$ where the coronal expansion starts. This outward flow of coronal plasma is known as solar wind and spacecraft observations at 1AU showed that the mean solar wind velocity is ~400Km/sec. However it has been shown that often the solar wind velocity increases drastically to high values (~700Km/sec). Also it has been shown that these high speed solar wind streams have their sources in solar corona regions with open magnetic lines. From this point of view the solar wind dynamics is connected with the solar magnetic activity and sunspots dynamics. It is supposed that the solar magnetic fields are generated beneath the base of the convective zone, where toroidal fields are created from poloidal fields by differential rotation [4]. An important manifestation of solar magnetic activity is the flare events with explosive release of magnetic energy ~$10^{25}$J for an explosive event. These events occur in active regions with enhanced and complex magnetic field causing strong disturbance of the solar wind. The interaction of the supersonic magnetized solar wind plasma with the earth magnetic field lead to the formation of a complex magnetospheric plasma subsystem of the outer solar corona. The magnetospheric plasma dynamics is also turbulent and irregular especially as it is appeared at substorm events.

Chaotic dynamics has been already suggested for the description of solar magnetic activity [5] as well as for the magnetospheric substorm dynamics [6,7,8]. Burlaga [9] has studied the multifractal structure of the solar wind magnetic field by analyzing large scale fluctuations. In the following we present briefly some new results of the chaotic analysis applied to experimental time series of the space plasma.

## 2. EXPERIMENTAL CHAOTIC ANALYSIS OF SPACE PLASMAS

According to chaotic dynamics an erratic signal can be generated by a nonlinear deterministic system with few degrees of freedom in the absence of external noise. Figure 1 shows five time series corresponding to: a) monthly values of sunspot index for the period 1750-1985 b) solar wind temperature (Tsw) and solar wind magnetic field (Bsw) with 1 hour time resolution measured in situ in the interplanetary space and c) 15 sec time resolution magnetic field measurements in situ in the magnetotail plasma sheet of the earth magnetosphere (Bps) and auroral electrojet index (AE index) with 1 min time resolution. The first time series (sunspot index) shows the subconvective zone of solar plasma dynamics which causes the solar magnetic activity. The next two time series Tsw, Bsw show the expanding outer solar corona dynamics and finally the last two time series show the earth magnetospheric dynamics during substorm events. The above time series are clearly erratic and aperiodic signals as their power spectra can reveal. It has been found that the corresponding power spectra approximately follow the power law of the form $P(f) \sim f^{-a}$ where the exponents $a$ take values in the range $1.1 < a < 3.2$. In particular we have found $a \cong 3.2$ for the sunspot index, $a \cong 1.1$ for the Tsw time series, $a \cong 1.8$ for the Bsw time series, $a \cong 1.3$ for the Bps time series and $a \cong 1.9$ for the AE index. The broadband character of the power spectra reveals quasiperiodic ergodicity with infinite rationally independent frequencies $\omega_1, \omega_2, \ldots$, or low dimensional chaotic ergodicity. These power spectra are the Fourier transforms of the autocorrelation

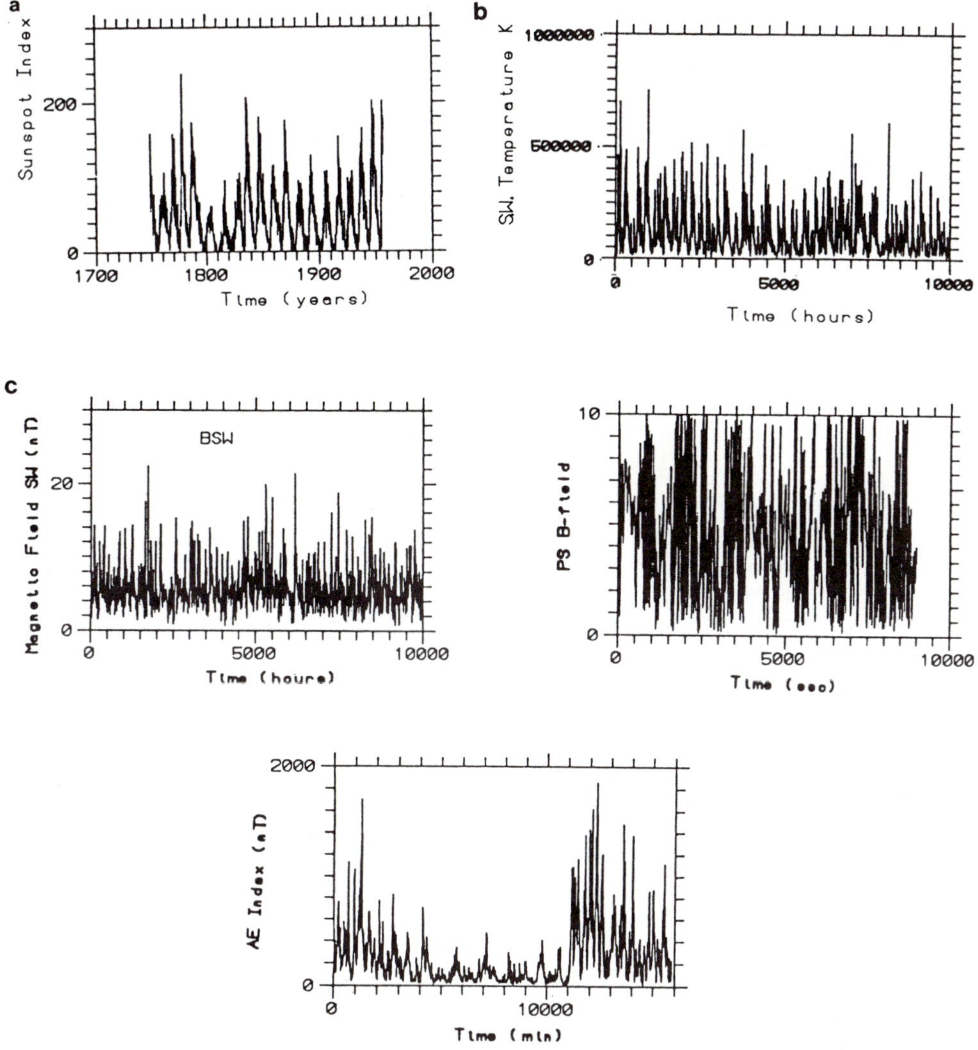


Fig.1. a) Daily values of Sunspot index for the period 1750-1985.  b) Solar wind temperature (Tsw) and magnetic field (Bsw) data with 1 hour time resolution. c) Magnetotail plasma sheet magnetic field (Bps) data with 1.2 sec time resolution and AE index time series with 1 min time resolution.

functions $C(\tau)$, that is

$$P(f) \simeq \int_{-\infty}^{\infty} C(\tau)\ e^{-i2\pi f\tau} d\tau$$

It follows from this relation that, when the power spectrum obeys the law $P(f) \sim f^{-\alpha}$, the autocorrelation function $C(\tau)$ decays as the lag time $\tau$ increases. Indeed the computation of $C(\tau)$ for the above time series reveals an abrupt exponential decay of the autocorrelation function except in the case of sunspot index time series. The sunspot index autocorrelation function decays as a damped oscillation. If we suppose that the experimental signals are caused by finite-dimensional nonlinear dynamical systems with evolution process in the form of differential equation

$$d\vec{x}(t)/dt = \vec{g}(\vec{x}(t),\lambda),$$

then the decaying of the autocorrelation function described above reveals a mixing evolutionary process, on an attracting set $\Lambda$, known as chaotic or strange attractor.

Since colored noises also reveal decaying autocorrelation function, the spectral analysis of time series is not sufficient for the experimental verification of chaotic dynamics. For this reason we study the experimental time series by using algorithms implied by chaos theory. Dissipative systems possess the important property of the contraction of volumes in the state space according to the relation

$$\frac{dV(t)}{dt} = \int_{A(t)} \mathrm{div}\ [\vec{g}(\vec{x}(t)]\ d\vec{x}$$

where A is a bounded set in the state space $R^n$ and V is its volume. For flows with divg<0 asymptotic motions occur on attracting sets (attractors) with zero volume. In the case of strange attracting sets (strange attractors) two kind of characteristics are important and experimentally verifiable: a) They have complex geometric structures with non integer dimensions which do not disappear upon magnification. That is, they are fractal sets. b) Nearby orbits on the attractor reveal local exponential expansion according to the sensitive dependence on initial conditions (S.I.C). On a strange attractor the expansion of a set A in some direction is possible even if its volume vanishes asymptotically. For signals with broadband spectrum caused by rationally independent quasiperiodicity neither of these conditions is satisfied. The second characterisric of a strange attractor (S.I.C.), which is related to the exponential divergence of nearby orbits, causes the broadband power spectrum character and the abrupt decay of the autocorrelation function of the observables.

For the experimental verification of these characterestics we use the observed time series $X(t_i)$, i=1,2,..,N in order to reconstruct the dynamics of the system in an embedding pseudo-phase space topologically equivalent to the real state space of the system, according to Takens embedding theorem [10]. This theorem states loosely that if $X(t_i)$ is a discrete-time scalar cross-section from a deterministic time chaotic process with an D-dimensional attracting fractal set (strange attractor), then there exist an m $\leq$2D+1 and a smooth function such that

$$X(t_n) \simeq g(\ X\ (t_{n-m}),\ \ldots,X\ (t_{n-1})\ )$$

Thus, given an observable time series $X(t_i)$, we can generate the equivalent state vector $\vec{X}(t_i)$ as follows

$$\vec{X}(t_i) \equiv \{\ X(t_i),\ X(t_{i+\tau}),\ldots,X(t_{i+(m-1)\tau})\}.$$

The parameter m is the embedding dimension and $\tau$ is an appropriate delay time. If the observed time series $X(t_i)$ is caused by the evolution of the system on a strange attractor structure, then its fractal dimension D can be estimated according to Grassberger and Procaccia [11] by the relation

$$D_{cor} \simeq \lim_{r\to 0,\ m\to\infty} \frac{d\ln\ C(r;m)}{d\ln(r)}$$

C(r;m) denotes the Grassberger - Procaccia correlation integral

$$C(r;m) = \lim_{m \to \infty} \frac{1}{m^2} \sum_{i,j=1}^{m} \theta(r - \|\vec{X}(t_i) - \vec{X}(t_j)\|),$$

where $\theta(a)$ is the Heaviside function and $\|\vec{X}(t_i) - \vec{X}(t_j)\|$ denotes the distance between states $\vec{X}(t_i)$, $\vec{X}(t_j)$ in the m-dimensional reconstructed pseudo phase-space. The reconstructed equivalent phase space can also be used in order to verify the second characteristic of the strange attractor (sensitivity to initial conditions) by the calculation of the largest exponent (L.L.E) according to the relation

$$L_{max} = \frac{1}{t_m - t_0} \sum_{i=1}^{m} \log_2 \left[ L'(t_i)/L(t_{i-1}) \right]$$

where $L(t)$ is the displacement vector of two nearby trajectories in the (2D+1) - dimensional reconstructed phase [12]. The sensitive to initial conditions evolution in a strange attractor implies that at least one of the Lyapunov exponents of the dynamical system must be positive.

Fig.2 shows the slopes $dlnC(r;m)/dlnr$ of the correlation integral $C(r;m)$ corresponding to the sunspot index and the magnetospheric plasma sheet magnetic field (Bps) time series. We observe that for low values of distance r a scaling $C(r;m) \sim r^{dm}$ occurs, while the scaling exponent $d_m$ saturates at the values ~4.5 and ~3.5 respectively for embedding dimensions $m \geq 5$. Figs. 3(a,b,c,d,e) show the dependence of the scaling exponent $d_m$ as a function of the embedding dimension m for the sunspot index, solar wind (Tsw, Bsw) and the magnetospheric (Bps, AE index) time series. Clearly the saturation values for the solar wind time series were found to be $D_{cor} \simeq 4.8$ for the Tsw time series and $D_{cor} \simeq 4.5$ for the SW B-field (Bsw). For the magnetospheric time series we have found $D_{cor} \simeq$ 3.6 for Bps and $D_{cor} \simeq 3.5$ for the AE index. A correct estimation of time series for the embedding dimension of the system dynamics in a reconstructed phase space depends greatly on the amount of data (time series length). Smith[13] has supported that the minimum $N_{min}$ number of

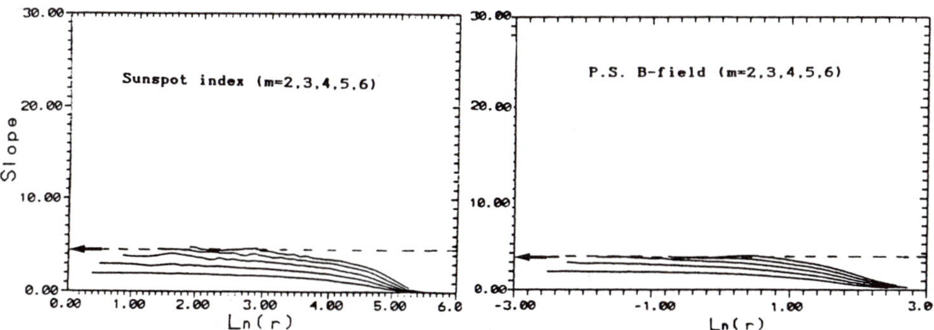


Fig.2.  Slopes  $dlnC(r;m)/dlnr$  (m=2,3,4,5,6) corresponding to the correlation integrals $C(r;m)$ calculated for the Sunspot index and the magnetotail plasma sheet magnetic field time series.

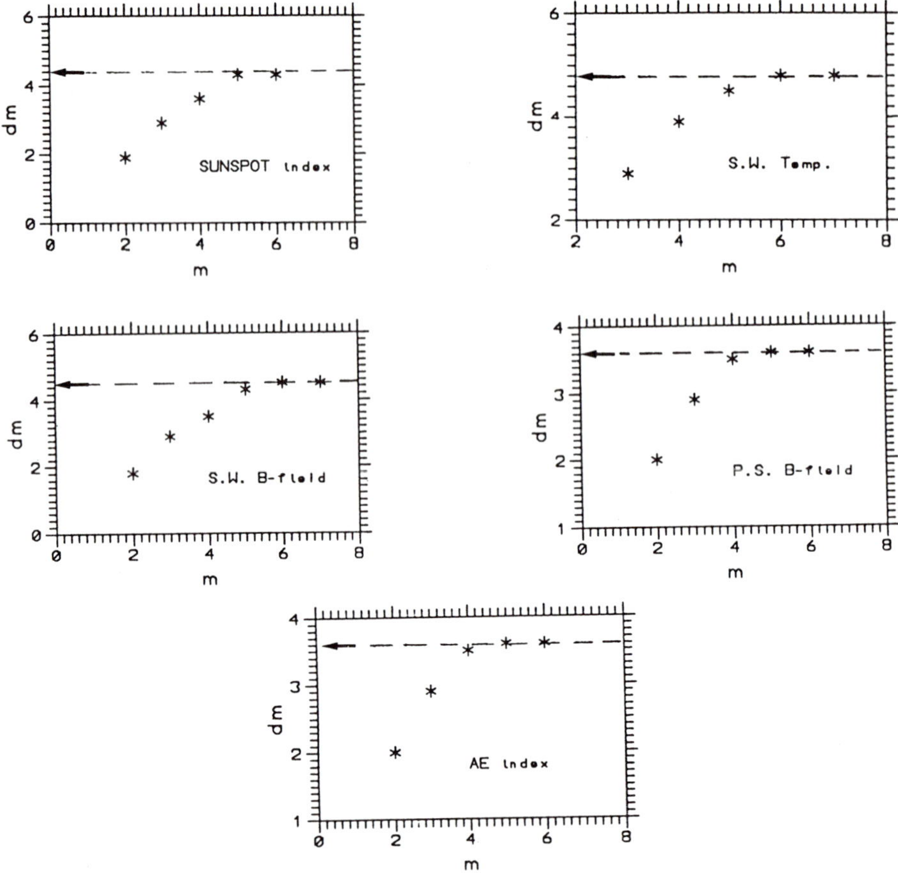

Fig.3. The scaling exponents d$_m$ as a function of the embedding dimension
m for Sunspot index, Tsw, Bsw, Bps and AE index.

independent data points required for the existence of a scaling region
with a meaningful plateau (d$_m$ = dln C(r;m)/dlnr =constant) must be of the
order N$_{min}$ $\simeq$ 40$^{Dcor}$. Recently it has been estimated that in cases where an
attractor of dimension D$_{cor}$ exists, the required minimum amount N$_{min}$ of
data is given by N$_{min}$ $\simeq$ 10$^{2+0.4Dcor}$ [14]. In our cases the last
criterion was satisfied. We have also taken into account the dependence
of the calculated value D$_{cor}$ with the amount of data points. The values
of D$_{cor}$ given here did not seem to change when the number of points N was
increased further.

Table 1 describes the calculation of the L.L.E. L$_{max}$ obtained by the
Wolf algorithm. Clearly for all cases the values of L$_{max}$ are positive.
These characteristics provide evidence to us in order to suppose that the
observed time series correspond to strange attractor structures referred
to the solar magnetic activity and to the magnetospheric activity.
According to this supposition the sunspot index and solar wind time

Table 1. Calculated values of the largest Lyapunov exponents corresponding to the observed space plasma time series.

| Space plasma signals | Sunspot index | Solar wind | | Magnetosphere | |
|---|---|---|---|---|---|
| | | $T_{sw}$ | $B_{sw}$ | $B_{ps}$ | AE index |
| L.L.E | Lmax≃1.5 bit/month | Lmax≃ 2 bit/h | Lmax≃ 1.5 bit/h | Lmax≃ 0.5 bit/s | Lmax≃ 2 bit/min |

series reveal the evolution in an attracting set ( solar strange attractor) with fractal dimension D≈4.5, while the AE index and the plasma sheet B-field (Bps) time series show the evolution of the magnetospheric dynamics in a magnetospheric strange attractor with fractal dimension D≈3.5. However, in order to make sure that this supposition is correct, we have to defend the interpretation of the results of the chaotic analysis described above, as well as the spectrum characteristics of the space plasma time series, against the stochastic or random process (colored or white noises), which under certain conditions can mimic deterministic chaos and appear as pseudo - chaotic low - dimensional attractors. This problem is analysed in the next section.

3.CHAOS AND NOISE

In Osborne and Provenzale [15] has been shown that stochastic systems with power-low spectra (colored noises) can reveal pseudo -chaotic profiles with finite low-dimensional fractal dimensions. This means that the experimental tracing of chaos is controversial. In general the standard expansion of a real random function is given by the relation

$$X(t) = \int_0^\infty \cos\omega t\, df_1(\omega) + \int_0^\infty \sin\omega t\, df_2(\omega),$$

where $f_1(\omega)$, $f_2(\omega)$ are appropriate real random functions when the time parameter becomes discrete the above relation can be expressed in the form of the standard Fourier series

$$X(t_n)=\sum_{k=1}^{M/2} C_k\cos(\omega_k t_n+\varphi_k), t_n=n\Delta t; n=1,2,\ldots,M$$

where $\varphi_k$ are random phases, and $C_k$ are related to the colored noise power spectrum $P(\omega_k)$ by the expression $C_k = 2\pi P(\omega_k)/M\Delta t$. Such a random signal can deceive or fool the Grassberger - Procaccia algorithm when the power spectrum is of the form $P(\omega)=c\omega^{-\alpha}$ showing low fractal dimension. For $1<\alpha<3$ the fractal dimension D of the above colored noises is given by D=2/α-1. Because of this the Grassberger and Procaccia algorithm cannot distinguish between deterministic chaotic dynamics and stochastic fractal system, where small space scales are related to small time scales. The above stochastic signals (colored noises) are known as self-affine fractal Brownian motions (fBm) and satisfy the self-affinity relation

$$X(t_i + b\tau) - X(t_i) \overset{d}{=} b^H [X(t_i + \tau) - X(t_i)]$$

This relation holds independent of time, H is known as the scaling exponent while " $\overset{d}{=}$ " mean equality in the sense of distributions. In general a stochastic process $X(t)$ with spectral density proportional to $\omega^{-\alpha}$ corresponds to self-affine fBm with $H=(\alpha-1)/2$ and fractal dimension $D \approx 1/H$. It can be proved for fBm that the average increments $[X(t_i) - X(t_0)]_{av}$ are zero, while the variance of increments diverges with time according to the relation

$$Var[X(t_i) - X(t_0)] \sim (t_i - t_0)^{2H}$$

where
$$Var[X(t_i) - X(t_0)] = av\{[X(t_i) - X(t_0)]^2\}.$$

This means that fBm with pseudo-chaotic profile, although they may have low fractal dimension, they are not stationary signals. This is in contrast to the dynamical evolution of a system in a strange attractor structure. Phenomenologically the dynamics in a strange attractor is a probabilistic process which can be described by an invariable probability density $\rho(\vec{x})$ given as the time average of Dirac $\delta$-distributions along a trajectory $\vec{x}(t)$ in the phase space

$$\rho(\vec{x}) = \lim_{T \to \infty} \frac{1}{T} \int_0^T \delta(\vec{x} - \vec{x}(t)) dt$$

Since the strange attractor dynamics is ergodic, the following relation must hold

$$\int_{phase\ space} g(\vec{x}) \rho(\vec{x}) d\vec{x} = \lim_{T \to \infty} \frac{1}{T} \int_0^T g(\vec{x}(t)) dt$$

Usually we can have available experimental measurements only for one coordinate of state space (say $x_1$). This means that we can calculate the projection $\rho(x_1) dx_1$ according to the relation

$$\rho(x_1) dx_1 = \int \rho(\vec{x}) dx_2 dx_3 \ldots dx_n$$

Stationarity for a certain range of time can be tested by calculating the invariable measure $\rho(x_1)$ for the entire time range and for the first half of the same time range [16]. Moreover, the autocorrelation function $C(\tau)$ of fBm decays slowly according to the relation

$$C(\tau) \simeq \int_{\omega_{min}}^{\omega_{max}} \omega^{-\alpha} e^{i\omega\tau} d\omega \quad ,$$

where the limits of integration are the bandwidth of the components of the measurement process. If $X(t_i)$ is a deterministic signal caused by a strange attractor process in a m-dimensional state-space, then as we have described before there must exist a smooth function g such that

$$X(t_n) = g(X(t_{n-1}), \ldots, X(t_{n-m}))$$

However $X(t_n)$ may not have been observed exactly. Instead we may observe $Y(t)=X(t)+W(t)$, where $W(t)$ is an external noise and $X(t)$ is the trend of the observed signal $Y(t)$ corresponding to the law dimensional attractor. $W(t)$ is the stochastic component possibly corresponding to a high dimensional attractor. By applying a moving average smoothing we reduce the stochastic component $W(t)$, while by differencing the signal we remove the trend $X(t)$. These operations produce new signals according to the relations

$$S(t_i) = \sum_{j=-p}^{p} W_j \, Y(t_{i+j}) \qquad\qquad \nabla Y(t_i)=Y(t_i)-Y(t_{i-1})$$

where p is a positive integer and $W_j$ are positive weights with $\Sigma W_j =1$ and $W_j = W_{-j}$.

These characteristics of random signals permit us to develope some helpful criteria against the pseudo-chaotic profile of the experimental time series. If we suppose now that the space plasma chaotic characteristics described previously are the profile of a Brownian colored noise pseudo-chaos, then the observed space plasma signals must be non-stationary. Moreover chaotic characteristics of the space plasma signals as fractal correlation dimension $(D_{cor})$, Largest Lyapunov exponent $(L_{max})$, power spectrum exponents $a = dlnP(\omega)/dln\omega$, and signal decorrelation times corresponding to the first zero of the autocorrelation function $C(\tau)$ must be similar to the corresponding characteristics for colored noises.

Table 2 summarizes the chaotic characteristics of the observed space plasma signals. According to the relation $D=2/a-1$ the correlation dimensions of colored noises with power spectra exponents $a$ equal to the power spectra exponents of the space plasma time series were found to be rather different than the correlation dimension values ~4.5 and ~3.5 estimated for solar and magnetospheric dynamics. In parallel to the space plasma time series we have obtained the chaotic characteristics of colored noises with fractal dimensions ~4.5 and ~3.5. In this case, if we suppose that the observed space plasma chaos is a colored noise pseudo-chaos, it would be natural to have similar values for the L.L.Es and decorrelation times corresponding to space plasma signals and colored noises which have the same fractal dimension as the space plasma signals. As we can see in Table 2 the L.L.Es corresponding to the space plasma dynamics, were found one or two orders higher than the L.L.E corresponding to random signals with fractal dimension ~4.5 and ~3.5. Moreover, in the case of random signals the corresponding decorrelation times were almost two orders higher than the decorrelation times of space plasma time series. In contrast to the slow decay of the noise autocorrelation the abrupt decay of the space plasma autocorrelation functions is in agreement with the S.I.C.,which must be true as the system dynamics evolves in a fractal sub-manifold of phase space.

The calculation of the two probability densities $\rho(x_1)$, $\rho'(x_1)$ for all the space plasma signals, which correspond to the entire time range $(\rho(x_1))$ and to the first half of the same time range $(\rho'(x_1))$, revealed that the processes are stationary. The $X^2$-test showed that the coincidence between $\rho(x_1)$ and $\rho'(x_1)$ was very good (over 90%). However, in the case of colored noises with the same correlation dimensions as the space time series the coincidence was very bad (below 10%). The smoothing of the space plasma time series by using a moving average procedure showed that the fractal dimensions of the smoothed signals are remarkably lower than the correlation dimensions estimated by using the

Table 2. Summarized results of the comparison between chaos and noise for space plasmas dynamics.

| Time Series | Sunspot index | Solar wind $T_{sw}$ | Solar wind $B_{sw}$ | Noise $D_{cor} \cong 4.5$ | Magnetosphere $B_{pl.sh.}$ | Magnetosphere AE index | Noise $D_{corr} \cong 3.5$ |
|---|---|---|---|---|---|---|---|
| Pow. Spec. exp. ($\alpha$) | ~3.0 | ~1.8 | ~1.1 | ~1.4 | ~1.2 | ~1.9 | ~1.6 |
| Fractal dim. $D_{cor}$ | 4.8 | 4.5 | 4.5 | 4.5 | 3.5 | 3.5 | 3.5 |
| Fract. Dim. for noise with simil. power spec. $D = 2/\alpha - 1$ | 0.9 | 2.5 | 2.0 | 4.5 | 8 | 2.2 | 3.5 |
| Deccorel. time ($\tau^d$) | 40 months | 40 Hours | 80 Hours | 600 | 10 sec | 60 min | 1500 |
| L.L.E. $L_{max}$ | 0.15 $\frac{bit}{month}$ | 0.15 $\frac{bit}{hour}$ | 0.2 $\frac{bit}{hour}$ | 0.01 | 0.5 $\frac{bit}{sec}$ | 0.22 $\frac{bit}{min}$ | 0.02 |
| Stationarity $x^2$ - test significance | 96% | 93% | 95% | 5% | 94% | 96% | 5% |
| $D_{cor}$ after moving average filtering | 4.8 | 4.3 | 4.3 | — | 3.2 | 2.8 | — |

original unsmoothed signals. In contrast the differencing of the original time series gave signals which do not reveal any saturation of the scaling exponent $d_m$ at least for embedding dimensions m lower than ~ 10. These two results indicate that the original time series contain a stochastic high-dimensional component W(t) which, however, does not cover the low-dimensional deterministic component X(t) of the original signal Y(t).

Finally we can use the Fourier presentation of the observed signals given by relation

$$X(t_i) = \sum_k C_k \cos(\omega_k t_i + \varphi_k),$$

in order to distinguish between the chaotic and the pseudo-chaotic processes, as the randomization of phases $\varphi_K$ transforms the original time series to a colored noise with similar power spectrum. In the case of a pseudo-chaotic colored noise signal $X(t_i)$ the phase randomization does not alter the fractal correlation dimension. Conversely, for a deterministic chaotic signal $X(t_i)$ the randomization of $\varphi_k$ alters drastically the correlation dimension. As we can see in Fig.4 the phase randomization of space plasma signals does not produce low-dimensional saturation of the scaling exponent $d_m$. This result as well as the previous comparison of the observed space plasma time series with colored noises strongly support the hypothsesis for the existence of strange attractor structures corresponding to the dynamics of solar magnetic activity and the solar wind streams ejection, as well as of the magnetospheric substorm activity.

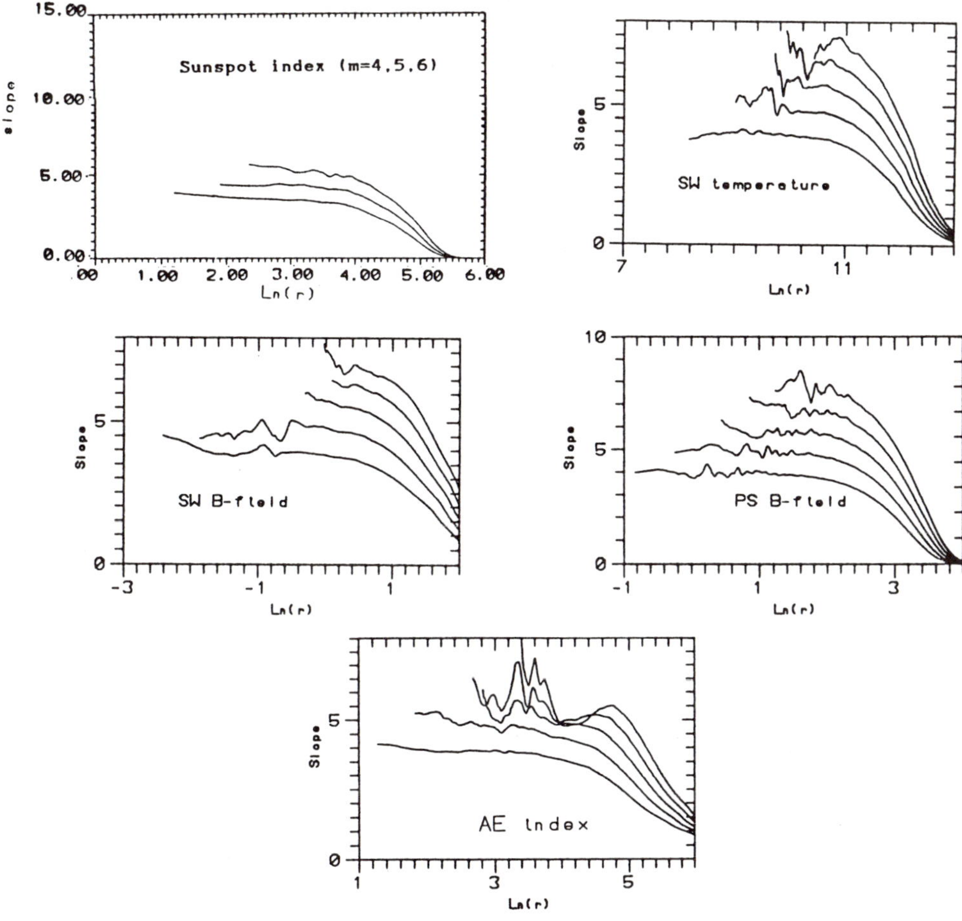


Fig.4.  Slopes  dlnC(r;m)/dlnr  corresponding  to  time  series  with randomized phases for Sunspot index, Tsw, Bsw, Bps and AE index.


## 4.DISCUSSION

Although different ways of interpreting the chaotic results discussed in this work are not completely excluded,  we believe however that we have provided strong evidence about the existense of low-dimensional chaos in the dynamics of outer solar plasma (including the solar wind) and the magnetospheric plasma. It is important to emphasize here that the supposition of deterministic chaos is not without physical meaning for space plasmas as they are  non-linear dissipative systems. For such systems some kind of non-local self-organizing holistic process may be present implying the reduction of the infinite degrees of freedom in a few macroscopic degrees of freedom which are sufficient for the macroscopic description of the system dynamics. Possibly this reduction to low dimensionality could be explained by the slaving principle [17]. According to the Whitney's embedding theorem [18] the fractal dimension $D_{cor} \simeq 4.5$ corresponding to the dynamics of the outer solar plasma reveals a strange attractor embedded in a m-dimensional phase space where $m \leq 2D_{cor}+1$, i.e. $m \leq 10$. Similarly the magnetospheric dynamics must be described in a phase space with dimension m≤8, since the fractal dimension of the magnetospheric strange attractor was founded to be ~ 3.5.

The first natural paradigm of low-dimensional chaotic dynamics corresponding to a system of infinite degrees of freedom was the study of fluids by Lorenz [19]. Lorenz used free boundary conditions to the Navier - Stokes system including continuity of mass and heat conduction equations , and retained the lowest order terms in the Fourier expansion in order to obtain the 3-dimensional flow model (Lorenz model) :

$$dX/dt = -\sigma X + \sigma Y, \quad dy/dt = -XZ + \tau X - Y, \quad dZ/dt = XY - bZ$$

For $\sigma=10$ , $b=8/3$ and $\tau>24.7$ this model bifurcates to a strange attractor dynamics with fractal dimension $D\approx2.05$. The Lorenz strange attractor is the prototype for the explanation of the hydrodynamical chaos as, for example it can be observed experimentally in Taylor-Couette flow, or in the Rayleigh-Benard convection. Both these systems can be used as explanatory paradigms of the space plasma chaos, since the Taylor-Couette angular velocities $W_i$ (inner) and $W_0$ (outer) correspond to the solar differential rotation together with convection drive, the solar dynamo and the solar magnetic activity. Moreover the temperature gradient $\nabla T$ in the Rayleigh-Benard system contained between parallel plates heated from below corresponds to the solar chromosphere and corona heat conduction, according to the following equation

$$\frac{d}{dr}(r^2 T^{5/2} \frac{dT}{dr}) = 0$$

described in [21]. Similar analogies can be found for the magnetospheric plasma. These analogies together with the solar and magnetospheric low-dimensional strange attractor dynamics supported by our previous chaotic analysis applied to space plasma signals [8], reveal the necessity for the extension of the Lorenz model to magnetized fluids. According to such a generalised model the solar and magnetospheric fractal dimensions (~ 4.5 and ~3.5),which are higher than the fractal dimension ~2.06 of the Lorenz attractor, suggest that an extension of the Lorenz model is necessary in order to include the magnetic field motion. Some more results in this direction will appear in detail in the near future [21].



REFERENCES

1. Prigogine, I., From Being to Becoming Time and Complexity in the physical science, W.H. Freeman and Company (1979).
2. Pavlos, G.P., Magnetospheric dynamics, Proc. Symposium on Solar and Space Physics, National Observatory of Athens, Greece (1988).
3. Parker, E.N., Dynamics of the interplanetary gas and magnetic fields, Astrophys. J. 128: 664 (1958).
4. Weiss, N.O., In Accretion discs and magnetic fields in Astrophysics, G. Belvedere, ed., Kluwer, Dordrecht (1989).
5. Weiss, N.O., Periodicity and aperiodicity in solar magnetic activity Phil. Trans. R. Soc. Lond. A 330: 617 (1990).
6. Baker, D.N., Klimas, A.J., McPherron, R.L., and Buchner J., The evolution from weak to strong geomagnetic activity: an interpretation in terms of deterministic chaos, Geophys. Res. Lett. 17: 41 (1990).
7. Vassiliadis,D.V.,Sharma,A.S.,Eastman,T.E.,  and  Papadopoulos,  K.,

Low-dimensional chaos in magnetospheric activity from AE time series, Geophys. Res. lett. 17: 841 (1990).

8. Pavlos, G.P., Kyriakou, Rigas, A.G., Liatsis, P.J., Trochoutsos, P.C., and Tsonis, A.A., Evidence for strange attractor structure in space plasmas, accepted for publication to Ann. Geophys. (1991).

9. Burlaga, L.F., Multifractal structure of the interplanetary magnetic field, Geophys. Res. Lett. 18: 69 (1991).

10. Takens, F., Detecting strange attractors in turbulence, Lecture notes in Mathematics, Vol.898, pp.366-381, Springer Verlag, Berlin (1981).

11. Grassberger P.,and I.Procaccia, Measuring the strengness of strange attractors, Physica D 9 : 189 (1983).

12. Wolf,A.,Swift,J.B.,Swinney,H.L.,and Vastano J., Determining Lyapunov exponents from time series,Physica D 16: 285 (1985).

13. Smith, A., Intrinsic limits on dimensional calculations, Phys. Lett. A 133 : 283 (1988).

14. Nerenberg, M.A.,and Essex C.,Correlation dimension and systematic geometric effect, Phys.Review A 42 : 7065 (1990).

15. Osborne, A.R., and Provenzale, A., Finite correlation dimension for stochastic systems with power-law spectra, Physica D 35: 357 (1989).

16. Isliker H., and Kurths, J., On the interpretation of correlation dimensions found in Astronomical time series, Private communication at Institute fur Astronomie, ETH Zentrum CH-8092 Zurich Switzerland.

17. Haken, H., Information and self organization, a macroscopic approach to complex systems, Springer series in Synergetic, Berlin Heidelberg (1988).

18. Whitney, H., Differentiable Manifolds, Ann. Math. 37 : 645 (1936).

19. Lorenz, E.N., Deterministic Nonperioddoc flow, Journal of the Atm. Scien. 20 : 130 (1963).

20. Stix, M., The Sun, Astronomy and Astrophysics Library, Springer-Verlag, Berlin Heidelberg (1989).

21. Pavlos G.P., Rigas A.G., Karakatsanis L.P., Kyriakou G.A., and Dialetis D., Experimental and theoretical aspects of the solar and magnetospheric chaotic activity, Submitted to Solar Physics.

# TIME SERIES ANALYSIS OF MAGNETOSPHERIC ACTIVITY

# USING NONLINEAR DYNAMICAL METHODS


D. Vassiliadis, A.S. Sharma, and K. Papadopoulos

Department of Physics and Astronomy
University of Maryland
College Park, MD 20742, USA


## ABSTRACT


Several attempts have been made to describe and predict the dynamical behavior of the
Earth's magnetosphere as it is reflected in geomagnetic time series. Early results indicated
that the dimension is low so that it might be possible to model the complex interactions
giving rise to the irregular time series as a dynamical system with intrinsically unstable
dynamics. However, it was recently pointed out that the low dimension may have been due
to the long autocorrelation time scale of the system. A subsequent study of different time
intervals and algorithmic and physical parameters gave further evidence that, indeed, the
convergence of the correlation dimension was due to time correlations rather than global
state space structure. Further examination of other diagnostics (exponents, forecasting)
in this light shows that, although at first sight they pass several tests and conform to the
interpretation of a low-dimensional chaotic system, several of their features are inconsistent
with the assumptions of the analysis. The above difficulties inherent to analysis of a variety
of real-world time series are pointed out and briefly discussed.


## INTRODUCTION

The *magnetosphere* is the outermost layer of the Earth's atmosphere, an environment
of tenuous plasma and strong terrestrial magnetic fields [1]. Both its shape and dynamics
are influenced by its surroundings, the *solar wind*, i.e. the particles and radiation that are
continuously emitted outward from the Sun at speeds of the order of 100 km/sec, and the
*ionosphere*, the immediately lower atmospheric layer with which it exchanges particles and
energy.

In particular the solar wind pressure deforms the magnetosphere into a extended
magnetic cavity whose nightside is much longer than the compressed dayside and has
a roughly conical shape. The magnetic fields of the Earth and the solar wind frequently
merge at the boundary and in this process 1% of the solar wind particle flux is funneled to
the Earth. This flux is strong enough to load the nightside of the magnetosphere with energy
which is dissipated when it exceeds a threshold value. The dissipation is usually rapid and

follows the pattern of the solar wind activity and flux injection. A number of accompanying phenomena are observed at every such unloading throughout the magnetosphere and in particular in the nightside [1,2]. Part of the accumulated energy is ejected antisunward in the form of a long, fast drop of plasma (a *plasmoid*) embedded in a closed magnetic field. Closer to the Earth *magnetic substorms* occur on time scales of 20 min to 1h, during which strong currents flow downwards to the ionosphere where they are suddenly dissipated with part of the energy converted in the familiar *auroral lights*.

While extended observations of isolated events and continuous monitoring of the magnetospheric activity has been going on for the last three decades, space physics has been focussing mostly on local plasma physical processes. The global models that have been introduced cannot adequately explain large-scale phenomena or forecast the occurrence of substorms and plasmoid releases. At the same time several features of the magnetospheric system, such as the constancy of response patterns at the solar wind perturbations, the presence of macroscopic structures (plasmoids), and spectral analysis indicate that the system is strongly dissipative and nonlinear, and cannot be described by traditional linear spectral methods. Therefore new methods for phenomenological modeling and forecasting which have recently been put forward in the dynamical systems theory were applied to the time series data. In these analyses conventional time series were used that monitor the global activity. Such time series are the AL and AE geomagnetic indices which measure the strength of the ionospheric currents in terms of the magnetic field fluctuations they cause on the surface of the Earth. The indices measure the maximum simultaneous fluctuations over several stations of the same latitude.

Although initial results were encouraging and indicated that the system was low-dimensional and chaotic, it was subsequently shown that those findings were due to features of the system unrelated to the assumptions of nonlinear dynamical systems theory. In general, it is realized that several nonlinear dynamical techniques may give spurious indications for low dimensionality or chaos when applied to real-world time series. In what follows some of these pitfalls are presented together with tests or techniques on how to avoid them. Most of the emphasis is on the correlation dimension followed by additional comments on Lyapunov exponent calculations and forecasting.

CORRELATION DIMENSION

A calculation of the correlation dimension [3] from the time series of a deterministic system, is a quantitative feature of its state space structure and promises to give an estimate for the number of degrees of freedom for a model that reproduces the observed behavior. Such calculations make several assumptions for the observed time series that are not satisfied at all times in real systems. Therefore the result of the calculation should be subjected to several tests before it is related to the properties of the system.

Following the standard application of the method the data are first embedded in an m-dimensional space, usually with the method of delays [4]. The delay must be long enough so that successive components of the same vector are maximally independent, i.e. all autocorrelations between them due to dynamical effects are removed. The only correlations that should remain are of geometric origin, due to the underlying state space structure (e.g. that of an attractor). Thus the time scale for the delay is usually of the order of the autocorrelation time, which, according to a variety of definitions, is the time when the *autocorrelation function* crosses zero, when its first minimum occurs, or when it drops below 1/e of its initial value. Alternatively, the *mutual information* [5] can be used instead of the autocorrelation function as a nonlinear estimate of dynamical correlations. In the case of the geomagnetic indices the delay was taken as the autocorrelation time (>2h for AE, 3h for AL), which is long compared to the sampling time. The sampling time scale was 1 min for AE data from January 1984, and 2.5 min for AE and AL in 1967. For these time series smaller and larger delays were also tried, but results did not change beyond the statistical error.

From these data the correlation sum ("integral") was calculated. This quantity measures the spatial correlation, or degree of relative closeness, between all pairs of m-D(imensional) points created from the time series and for self-similar sets of points it shows a power law scaling with the relative distance r:

$$C(r; m) = \frac{1}{N^2} \sum_{i=1}^{N} \sum_{j\neq i}^{N} \Theta\left(r - \|x_i - x_j\|_m\right) \underset{r\to 0}{\to} Ar^{\nu}$$

where N is the number of points in the series, A is a constant and $\Theta$ is the Heaviside or step function. The second sum in the equation counts all the neighbors whose distance is smaller than r, while the first one averages that distance over all points in state space. The distance r can be calculated using different norms [6]; generally the Euclidean norm is used.

In the limit as r goes to zero the limit of the above exponent defines the correlation dimension:

$$\nu = \lim_{r\to 0} \frac{\ln C}{\ln r}$$

In experimental or observational time series the resolution sets a lower limit to the smallest r that can be obtained. The AE and AL indices are measured in terms of positive integers (units: nT=nanoTesla) so that the smallest r is unity. The highest values are of the order of a few thousand nT. As m was increased the correlation dimension was seen to saturate at noninteger values below 4 [7].

An important condition on the time series refers to the *stationarity* of the signal over the length of the time series. Stationarity is defined as the constancy of the average and standard deviation throughout this length. An operational definition is a low level of power at the low-frequency end of the spectrum [8]. Stationarity is required so that the time series is long enough to cover the attractor, or other underlying structure in the state space. The geomagnetic indices are *not* stationary all the time, so the intervals analysed were chosen so as to maintain this condition. More specifically the correlation dimension was observed to roughly remain the same for a variety of stationary intervals (N = 5–10k) of different levels of activity (the activity being measured as the second statistical moment). After that the number of points was increased. The longer time series (20–30k) were not everywhere stationary, yet they yielded similar correlation dimension estimates. This was seen as evidence that the dimension is fairly constant and seems to characterize the system as a whole over a wide range of activity. The magnitude of the dimension also satisfied the criterion: $\nu < \log_{10}(N)$, referring to the scaling of the dimension with the number of points in the time series.

The test of *surrogate data* [8] was also applied to the time series. The time series was Fourier transformed, its phases randomized and inverse-Fourier transformed. The procedure gives a time series that preserves the spectral characteristics and autocorrelation, but any state space structure is destroyed. For a deterministic system the dimension of the randomized time series should be much higher than that of the original. The AL and AE index time series passed this test, in other words calculation of the correlation dimension for the respective randomized time series gave a much higher or nonconvergent estimate for the dimension.

Recently, however, it was pointed out [9] that the low correlation dimension may have been the consequence of the relatively *high autocorrelation time* (compared to the sampling time) of the time series, in spite of all of the tests mentioned above. Following [9] this group repeated the calculations, but now neglecting the pairs of those points that are too close in time [8–10]. These pairs are correlated dynamically, as mentioned above, and their inclusion is an additional, overweighted contribution of the correlation integral at the small scales. This is the reason that finite time series from colored noise seem to have a low dimension [11], which is in contrast with their large number of degrees of freedom. The time scale below which all pairs of points are excluded is the Theiler parameter w, which should be comparable to the autocorrelation time or some similar estimate of the dynamical correlation. In contrast, the value of w in the "normal" correlation dimension algorithm [3] is the sampling time (minimum time resolution).

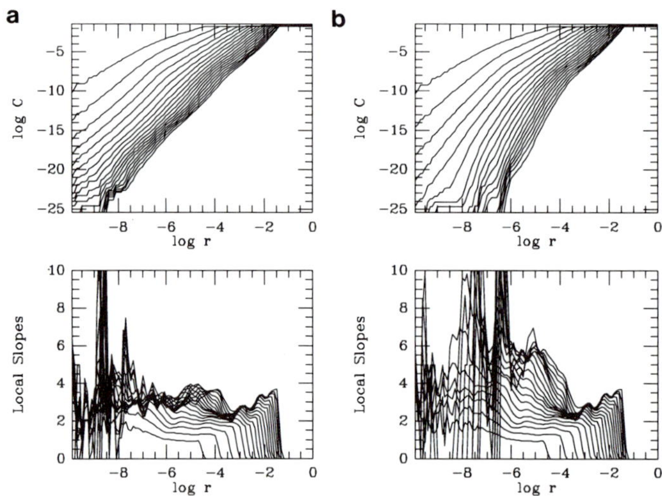

*Figure 1. The correlation integral C(r) and its local slope for the AL index. N=11k, m=2–20. a) w=1, b) w=10.*

For the geomagnetic index time series it is seen that when increasing w up to a few sampling times (w=5–20 min), the correlation dimension as obtained from the linear range of log C(log r) does not change. However, when the Theiler parameter becomes comparable to the autocorrelation time (w=1–3h) the linear range disappears completely and the correlation dimension does not converge (Fig. 1). The number of pairs not counted is still much smaller (w/N<1%) than the total number of pairs contributing to the C(r) integral. The Theiler test loses its validity when w/N becomes of the order of a few percent or larger. Alternatives to the Theiler test would be to either increase the number of points, N, or decrease the sampling rate, while keeping w=1 in both cases. However, the first alternative is costly computationally (for a scaling of the proper N in the case of colored noise see [8]), while the second one may excessively thin out a short time series.

Two more methods, namely *Poincare sections* and search for *unstable periodic orbits*, were used to look for structure of the state space. Both methods gave results that indicated the presence of structures and periods of closed orbits. However, these results were not stationary, i.e. they were significantly different for different successive segments of the geomagnetic time series. Furthermore using a Theiler-like test, the structures observed on Poincare sections of finite thickness were found to be due to successive points which were highly correlated.


OTHER DIAGNOSTICS: FIRST LYAPOUNOV EXPONENT AND PREDICTION

In this part a brief overview of other diagnostics for nonlinear structure and chaos is presented. It has been seen that for real-world time series their calculation may be problematic, while they pose minor problems for time series from known low-dimensional dynamical systems.

The Lyapounov exponents of a chaotic system are important both for its classification as well as for study of its dynamical characteristics. The existence of positive exponents characterizes the system as chaotic, however their calculation from time series is not easy. There is a growing literature on algorithms that calculate the Lyapunov spectrum (all the exponents) by state space reconstruction as outlined above [12]. However, up to now these algorithms lead to spurious exponents coming from the extra embedding dimensions, although the first article in [12] has tried to address this problem. An alternative method [13] calculates only the highest exponent; if it is positive, the system is chaotic. For random systems that have no underlying state space structure the method should in principle give zero, since such systems are not characterized by exponential divergence of initial conditions. However, its application on the AL geomagnetic index [14] gave a finite positive exponent with a time scale of 10 min (4 sampling times) for all intervals examined, close to the physical time scale of interest, but smaller than the autocorrelation time. The test that eventually eliminated the exponent was *variation of the algorithmic parameters* (SCALMX, SCALMN, EVOLV [13]) during which the value of the exponent varied as well. However, it should be noted that during variation of other parameters (time delay, number of points) the exponent was approximately constant.

Prediction algorithms based on nonlinear dynamical schemes [15] can also be used as diagnostics for the chaotic nature of the system. Application of a low-dimensional local linear prediction scheme showed that the logarithm of the average *normalized prediction error*

$$\langle E^2 \rangle = \frac{\left\langle \left( y_{\text{pred}} - y_{\text{obs}} \right)^2 \right\rangle}{\sigma_{y_{\text{pred}}} \sigma_{y_{\text{obs}}}}$$

(where the normalization involves the standard deviations of both the predicted and the observed time series) increased slowly with time and reached unity after a long time (1h) comparable to the substorm time scale. This meant that the model could predict to a good accuracy up to the physical time scales of interest. For some cases the time of good forecasting agreed with the time scale obtained earlier from the Lyapounov exponent algorithm (10 min).

However, the time scale of prediction is still short compared to the autocorrelation time, since there is no underlying geometry of the state space that can aid the model in making longer predictions. Under these conditions the local linear model becomes equivalent to an *autoregressive/moving average* model which can statistically predict a stochastic time series. This was a first indication of the nature of prediction of the series.

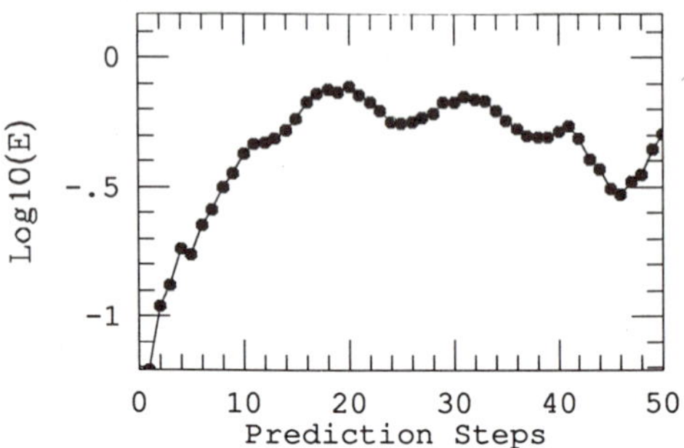

*Figure 2. Evolution of the normalized prediction error (see text) with prediction time for the AL index time series. A training set of 20k is embedded in a 4-dimensional space with the delay equal to the autocorrelation time (70 steps). 1 time step = 2.5 min.*

A further test to distinguish between deterministic versus stochastic prediction is the behavior of the prediction error with embedding dimension and delay. For a nonchaotic time series the error grows with the embedding dimension and decreases for smaller delays. In contrast a chaotic time series the error has a minimum for the correct embedding dimension and delay chosen as mentioned above. The AL time series gave best results for a 4–dimensional model (agreeing with the spurious estimate of the correlation dimension!); however the error increased with the delay indicating a nonchaotic system. Aside from the prediction error, other estimators (prediction-observation correlations, $\chi^2$ of the fitted model) as well as a case-by-case inspection of time series intervals also showed that their prediction was stochastic (the algorithm did well in a statistical sense only) and not deterministic. As with other diagnostics, nonlinear dynamical prediction of the time series gave positive results for some tests, but did not pass others.

## CONCLUDING REMARKS

In this paper several conventional methods of time series analysis using nonlinear dynamics have been presented. These methods were applied on geomagnetic time series in order to describe the complex behavior of the Earth's magnetospheric system. Almost all of these methods (with the exception of Poincare sections and unstable periodic orbits as well as the singular spectrum analysis [16] which was not described here) gave results that were initially consistent with the picture of the magnetosphere as a low-dimensional chaotic system. Each one of these methods (correlation dimension, highest Lyapunov exponent, forecasting) even passed several tests devised to distinguish between chaotic and nonchaotic irregular time series. However, there are tests that eventually showed that the complexity of the time series is not due to the global features of the state space structure, but to local, nonpermanent characteristics caused by the high autocorrelation time of the time series. Reducing this time scale or properly weighting its effects may eliminate spurious contributions to the diagnostics. Alternatively, systematic examination of the result with the variation of algorithmic parameters is another litmus test for their validity. In general these diagnostics should be seen as serving to reject a null hypothesis [8], namely that the system is low-dimensional and chaotic; their passing one or more tests does not guarantee that the hypothesis is true. Finally for the magnetospheric system these tests proved to be inadequate in showing if there is a chaotic or regular response to the solar wind. It seems that the solar wind input needs to be included in the state space and that different, recently described methods for input-output systems [17] will be necessary.


<u>References</u>


1.  S.-I. Akasofu (ed.), *Dynamics of the Magnetosphere*, D. Reidel Publishing Company, Dordrecht, Holland, 1980; A.T. Lui (ed.), *Magnetotail Physics*, The John's Hopkins University Press, Baltimore, 1987
2.  Y. Kamide and J. A. Slavin (eds.), *Solar Wind- Magnetosphere Coupling*, Terra Sci. Pub., Tokyo, 1986
3.  P. Grassberger and I. Procaccia, *Physica 9D*, 189 (1983)
4.  F. Takens, in: A. Dold and B. Eckmann (eds.), *Dynamical Systems and Turbulence, Warwick 1980*, Lecture Notes in Mathematics 898, Springer Verlag, Berlin, 1981
5.  A. M. Fraser, Ph.D. thesis, University of Texas at Austin, May 1988
6.  N. Gershenfeld, this volume
7.  The dimension was calculated to be $\nu=3.6$ by: D. Vassiliadis, A. S. Sharma, T. Eastman, and K. Papadopoulos, *Geophys. Res. Lett. 17*, 1841 (1990) for a variety of intervals of AE and AL with different activity levels; $\nu=4$ by: D. A. Roberts, *J. Geoph. Res. 96*, 16031 (1991) again for various intervals of AL; $\nu=2.5$ by L.-H. Shan, C. K. Goertz, and R. Smith, *Geophys. Res. Lett. 18*, 147 (1991) for a single interval; and $3<\nu<4$ by G. Pavlos, private communication. All these low-dimensional results are due to high autocorrelations as indicated by the Theiler test.
8.  J. Theiler, *Phys. Lett. A 155*, 480 (1991)
9.  D. Prichard and C. Price, preprint (submitted to *Geophys. Res. Lett.*)
10. J. Theiler, *Phys. Rev. A 34*, 2427 (1986)
11. A. R. Osborne and A. Provenzale, *Physica D35*, 357 (1989)
12. H. Abarbanel, *Phys. Rev. Lett. 65*, 1523 (1991); X. Zeng, R. Eykholt, and R. A. Pielke, *Phys. Rev. Lett. 66*, 3229 (1991); G. Mayer-Kress (Ed.), *Dimensions and entropies in chaotic systems*, Springer Verlag, Berlin, 1986.
13. A. Wolf J. B. Swift, H. L. Swinney, and J. A. Vastano, *Physica 16D*, 285 (1985)
14. D. Vassiliadis, S. Sharma, and K. Papadopoulos, *Geophys. Res. Lett. 18*, 1643 (1991)
15. J. D. Farmer and J. J. Sidorowitch, in: Y. C. Lee (ed.), *Evolution, Ecology and Cognition*, World Sci., Singapore, 1989; also references in that article
16. D. S. Broomhead, R. Jones, and G. P. King, *J. Phys. A 20*, L563 (1987)
17. M. Casdagli,to appear in: *Nonlinear Prediction and Modeling*, Addison-Wesley, 1991

SPECTRAL PROPERTIES OF TRAJECTORIES IN NEAR INTEGRABLE

HAMILTONIAN SYSTEMS

G. Voyatzis and S. Ichtiaroglou

Department of Physics
University of Thessaloniki
54006 Greece


Spectral properties have often been used in the study of trajectories of dynamical systems. It is well known that the spectrum of a quasiperiodic motion is characterized by a few distinct major peaks, located at positions which are integer combinations of the fundamental frequencies, instead of a continuous distribution of peaks which appears in the case of chaotic motion [1-4], in which case the appearance of $1/f$ noise has been shown in Hamiltonian systems of two degrees of freedom [5,6]. Spectral properties of trajectories associated with the deformation of tori in near integrable Hamiltonian systems are studied in [7]. The purpose of this paper is to apply this method on a system of three degrees of freedom in order to study the transition from regular motion to chaos. The method is based on the spectral analysis of the integrals of motion of an integrable Hamiltonian along trajectories of the non-integrable perturbed system.

We consider a perturbed Hamiltonian of $n$ degrees of freedom

$$H = H_o ( I_j ) + \varepsilon H_1 ( I_j , \varphi_j )  \tag{1}$$

where $I_j$ $(j=1..n)$ are integrals of $H_o$ in involution and smooth functions of the action variables, $\varphi_j$ are angle coordinates on the torus and $\varepsilon$ a small parameter. In the perturbed system $I_j$ vary slowly and in general their variation is $O(\varepsilon)$ for time intervals of order $1$ [8]. If the Hamiltonian satisfies special conditions of non-degeneracy and steepness, the variation of $I_j$ remains of order $\varepsilon$ for large time intervals [9].

We suppose that $H_1$ is mod$2\pi$ with respect to $\varphi$. From the equations of motion we obtain

$$I_j = I_{jo} + \varepsilon \sum_m \frac{m_j h_m}{m\omega} \exp(im\omega t) + O(\varepsilon^2) \tag{2}$$

where $m=(m_1,\ldots,m_n)\in\mathbf{Z}^n$ and $h_m$ the Fourier coefficients of $H_1$. The appearance of small denominators makes the convergence of the above expression doubtful, while the same problem emerges also for the power spectrum of $I_j$ which has the form [7]

$$P_j(\Omega) = \sum_m \left|\varepsilon \frac{m_j h_m}{m\omega}\right|^2 \delta(m\omega - \Omega) + O(\varepsilon^4) \tag{3}$$

The denominators $m\omega$ may take arbitrarily small values, so one might have thought that significant amplitudes will appear at the low band of the spectrum. For a regular trajectory which lies on an invariant torus the following relations hold [10]

$$|m\omega| \geq k|m|^{-(n+1)} \qquad\qquad |h_m| \leq M e^{|m|\rho} \tag{4}$$

where $k,M,\rho$ are positive numbers and $|H_1|<M$. By taking into account (4), it can be shown that the terms $h_m/m\omega$ decrease in geometric progression as $|m|\to\infty$ [10]. The same behaviour is also true for the terms $m_j h_m/m\omega$ since $m_j$ increases linearly. So for a trajectory on an irrational torus, $P_j(\Omega)$ must converge rapidly to zero as $\Omega\to 0$. Let $\Gamma_\omega=\{\Omega/\Omega=m\omega,\ m\in\mathbf{Z}^n\}\subset\mathbf{R}$. We define the function $A(\Omega)$ which is related to $P_j(\Omega)$, as follows

$$A(\Omega) = \begin{cases} |\varepsilon m_j h_m / \Omega|^2 & \Omega\in\Gamma_\omega \\ 0 & \Omega\notin\Gamma_\omega \end{cases} \tag{5}$$

It can be shown that [7]

$$A(\Omega) \leq \left(\varepsilon M \frac{\exp(-a\Omega^{-1/p})}{\Omega}\right)^2 \tag{6}$$

where $p=n+1$ and $a>0$. The above estimation shows an exponential convergence of the power spectrum as $\Omega\to 0$.

So for a quasiperiodic motion, the low band of the power spectrum of $I_j$ does not show significant amplitudes. In the case of a trajectory which evolves in a region of broken tori, the estimation (4a) does not hold. The small denominators destroy the convergence of (2) and the evolution of $I_j$ is dominated by the superposition of very slow oscillations in a random fashion. The low band of the power spectrum is filled by a continuous distribution of significant peaks. Since the integrals $I_j$ are

constants of motion in the unperturbed system, their power
spectra contain information about the deformation of tori under
the perturbation exclusively. Tori near a significant resonance
are strongly deformed and the power spectra consist of a set of
major peaks in contrast to robust invariant tori which exhibit a
neat spectrum with insignificant peaks in the low band which
converges to zero as $\Omega \to 0$. Numerical results show that the
transition to chaos is reflected on the power spectrum by the
generation of new peaks in the low band which increase their
amplitudes as the perturbation increases. When a stochastic layer
is formed the low band is abruptly enriched by a continuous
distribution of peaks .

### *Example*

In [7] spectral analysis has been applied to the planar
Hamiltonian

$$H = \frac{1}{2} ( p_x^2 + p_y^2 ) + \frac{1}{4} ( x + y )^2 + \frac{1}{4} ( x - ky )^4 \qquad (7)$$

which is integrable for *k=1* with a second invariant

$$I_1 = ( p_x + p_y )^2 + ( x + y )^2 \qquad (8)$$

regarding $\varepsilon=1-k$ as the perturbation parameter. For $\varepsilon=0$ the
existence of two stable periodic orbits at *x=0*, $p_x=\pm 0.5$ can be
shown, surrounded by invariant tori. Near one of these periodic
orbits and for *E=0.25*, a torus which corresponds to *1:3* resonance
exists. By increasing $\varepsilon$, this resonant torus breaks up forming a
chain of islands and a thin chaotic zone along the homoclinic
connections of the hyperbolic fixed points on the Poincare
section.

Here we consider an extension of (7) in three degrees of
freedom. The Hamiltonian which is studied is

$$H = \frac{1}{2} ( p_x^2 + p_y^2 + p_z^2 ) + \frac{1}{4} ( x + y )^2 + \frac{A}{2} z^2 + \frac{1}{4} ( x - ky )^4 + \frac{1}{4} z^4 - \varepsilon xyz^2 \qquad (9)$$

and for *k=1*, $\varepsilon=0$, possesses as invariants $I_1$ and

$$I_2 = p_z^2 + A z^2 + \frac{1}{2} z^4 \qquad (10)$$

We restrict ourselves on the effect of the perturbation on the
region close to the *1:3* resonance of (7) by considering the power
spectrum of (8) and (10) along the trajectory with initial
conditions $x=y=p_z=0$, *z=0.1*, $p_x=0.46$, $p_y=0.537$, and for several
values of $\varepsilon$ in the domain *[0.05 , 0.3]*.

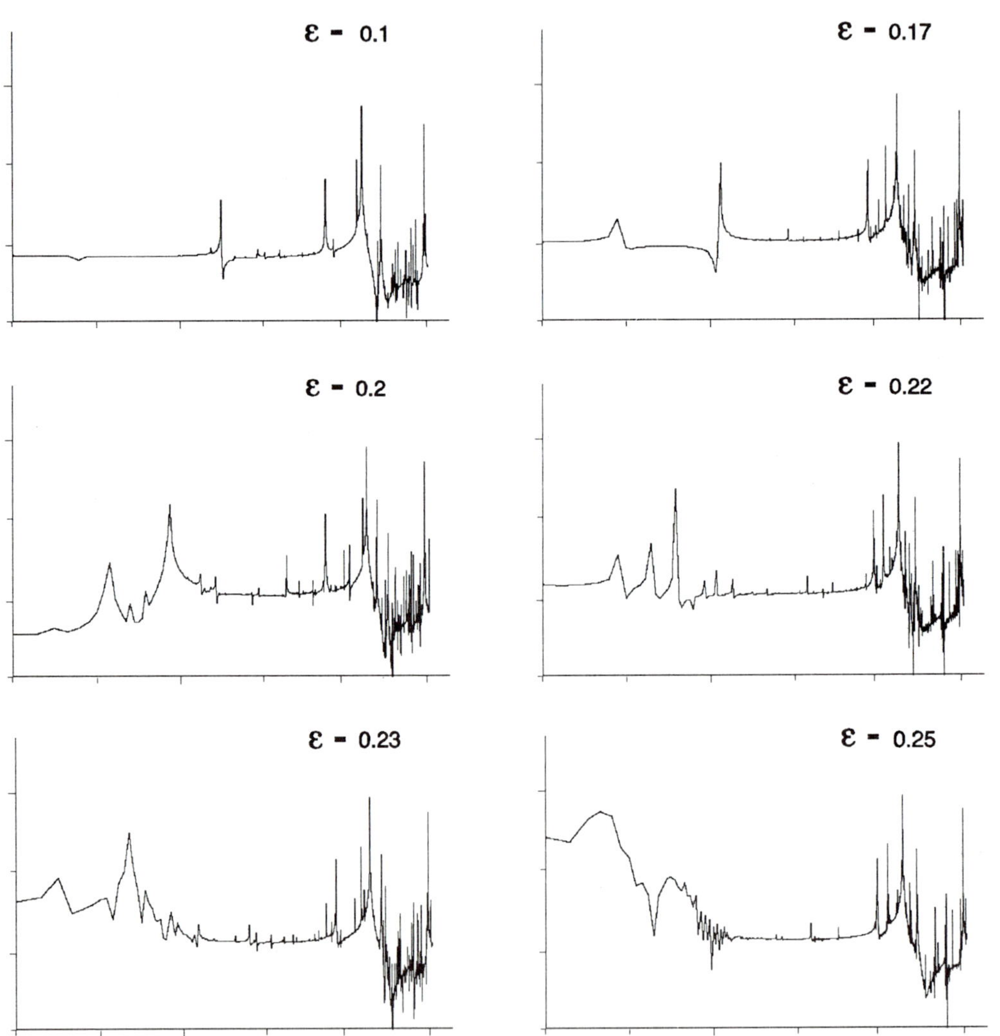



Fig. 1. Power spectra of $I_1$ for several values of $\varepsilon$ in log-log scale. The spectra suggest a transition to chaos for $\varepsilon > 0.22$.

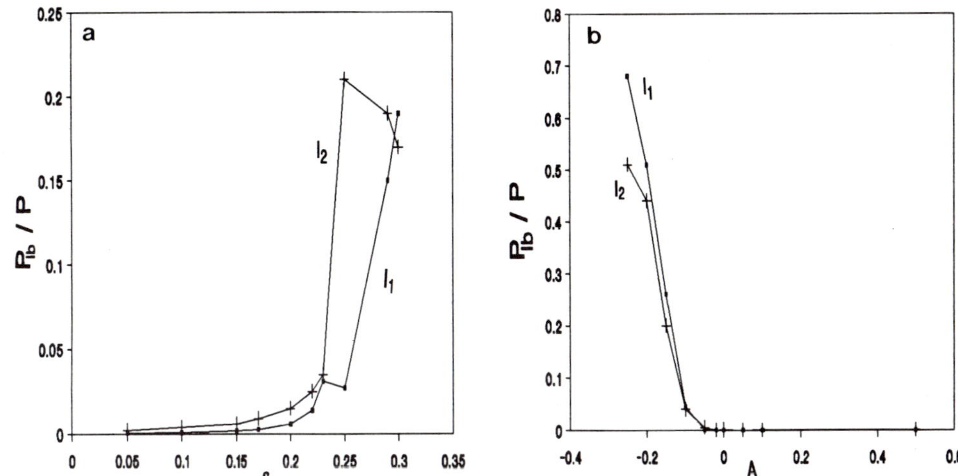

Fig. 2.    The ratio of the low band per total power. a) for
$A=1/2$, versus $\varepsilon$ b) for $\varepsilon=0.1$ versus $A$. An abrupt
increase occurs at the transition from regular to
chaotic motion.

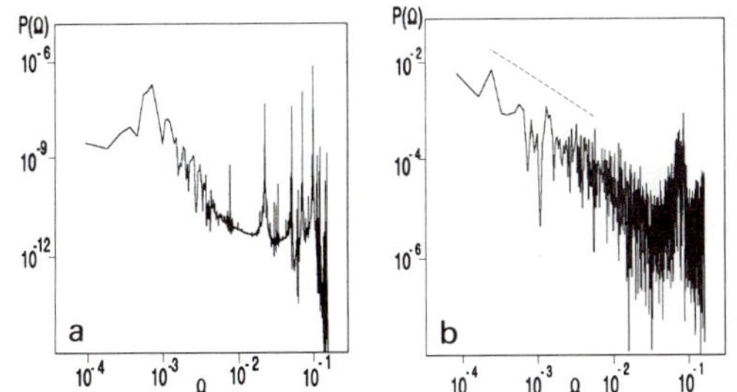

Fig. 3. Power spectra of two chaotic orbits for $A>0$ and $A<0$
respectively. An $1/\Omega$ divergence as $\Omega\to0$ is observed in the
second case.

In figure 1 power spectra of $I_1$ are shown. For $\varepsilon=0.1$, no peaks appear as $\Omega\to0$, suggesting that the trajectory is quasi-periodic. By increasing $\varepsilon$, we observe the generation of peaks in the low band. The trajectory seems to be regular until $\varepsilon=0.22$. For $\varepsilon=0.23$, significant peaks fill the low band suggesting the formation of a thin stochastic zone where the trajectory evolves. By increasing $\varepsilon$ even more, chaotic motion occupies a larger region in phase space and the divergence of the low band becomes more obvious (see also figure 3a for $\varepsilon=0.3$).

It is worth mentioning that the formation of a thin chaotic zone gives rise to a remarkable change in the low band, in contrast to the region of high frequencies, the form of which seems to be more or less unaffected. In figure 2a the variation of the ratio of the power of the low band per total spectral power with respect to $\varepsilon$ for both $I_1$ and $I_2$ is shown. We observe that for $\varepsilon>0.22$ where a transition from regular to chaotic motion occurs, an abrupt increase of this ratio appears. The same behaviour is observed also in figure 2b which corresponds to the spectra of $I_1,I_2$ on a trajectory close to a periodic orbit for various $A$. For $A>0$, $z$ exhibits oscillatory motion and the region around the periodic orbit consists of invariant tori. For $A<0$ the motion in $z$ becomes unstable and for $\varepsilon\neq0$ this instability leads to the generation of chaotic motion. Thus the qualitative features of the low band of the spectrum of $I_j$ seem to be highly indicative for the distinction between regular and chaotic trajectories and the transition from one kind of motion to the other.

The spectral behaviour of the low band for the chaotic motion after long time integration is different in the two cases examined. In the case of the chaotic motion near the $1{:}3$ resonance of the $2D$ system, $P(\Omega)$ seems to level off to a finite value at $\Omega=0$ (figure 3a) indicating a normal diffusion process characterized by a finite diffusion coefficient [6]. In the case of chaos for $A<0$, $1/f$ noise has been observed (figure 3b) indicating that this phenomenon may also be generic for Hamiltonian systems of more than two degrees of freedom.


REFERENCES


1.  G. E. Powell and I. C. Percival, "A spectral entropy method for distinguishing regular and irregular motion of Hamiltonian systems", *J.Phys.A: Math.Gen* **12** 2053 (1979)
2.  S. Kim, S. Ostlund and G. Yu, "Fourier analysis of multi-frequency dynamical systems", *Physica D* **31** 117 (1988)
3.  R. Roy, B. G. Sumpter, G.A. Pfeffer, S. K. Gray and D. W. Noid, "Novel methods for spectral analysis", *Phys.Rep.* **205** 110-150 (1991)

4.  R. Z. Sagdeev, D. A. Usicov and G. M. Zaslavsky  "Nonlinear Physics", *Switzerland:Hardwood Acad. publ.* (1988)

5.  T. Geisel, A. Zacherl and G. Radons, "Generic 1/f Noise in Hamiltonian Dynamics", *Phys.Rev.Lett.* **59** 2503 (1987)

6.  T. Geisel, A. Zacherl and G. Radons, "Chaotic diffusion and 1/f noise of particles in two-dimensional solids", *Z.Phys.B* **71** 117 (1988)

7.  G. Voyatzis and S. Ichtiaroglou, "On the spectral analysis of trajectories in near integrable Hamiltonian systems", submitted to *Nonlinearity* (1991)

8.  V. I. Arnold "Dynamical systems III", *Berlin:Springer* (1988)

9.  N. N. Nekhoroshev, "An exponential estimate of the time of stability of nearly-integrable Hamiltonian systems", *Russian Math. Surveys* **32:6** 1 (1977)

10. V. I. Arnold "Small denominators and problems of stability in Classical and Celestial Mechanics", *Russian Math. Surveys* **18:6**  85 (1963)

SYMMETRY BREAKING IN THE

PERIOD DOUBLING ROUTE TO CHAOS


J.P. van der Weele

Center for Theoretical Physics
University of Twente
P.O. Box 217, 7500 AE Enschede
The Netherlands


## 1. INTRODUCTION

The period doubling route to chaos is not quite as universal as one is often made to believe. In many practical situations the well known sequence $1 \to 2 \to 4 \to 8 \ldots \to$ chaos is interrupted or even completely broken off before chaos is reached [1-6]. The present paper is about one particular (rather mild) kind of interruption: a symmetry breaking bifurcation, already at the level of period 2. The period 2 solution, instead of period doubling to a period 4 solution, bifurcates into two non-symmetric solutions of period 2. Subsequently both of the two newly formed period 2 solutions resume the period doubling route as if nothing had happened. The complete scenario can thus be described by the following sequence: $1 \to 2 \to 2 \times (2 \to 4 \to 8 \to 16 \to \ldots \to$ chaos$)$.

As early as 1979 this sequence was found by R.M. May to be the typical route to chaos for a one-dimensional map with two critical points [7,8], but at the time the relevance to practical problems seemed to be marginal and even May himself thought that it was interesting mainly for its "mathematical intricacies". However, there is more to it than that. The main point of the present paper is that the symmetry breaking *can* be seen in real life systems and therefore the emphasis will lie on two examples: in section 2 we describe a ball bouncing on a vibrating table, and in section 3 a pendulum which is being driven up and down at its point of support. In doing so it will become clear that the symmetry breaking interruption is not restricted to one-dimensional systems. In section 4 we extend May's analysis to two dimensions and come up with a model map which accurately represents all of the previously observed behaviour.

## 2. FIRST EXAMPLE: THE BOUNCING BALL

Consider a small ball, of negligible mass, which is bouncing up and down on a vibrating table. This system, known as the Fermi-Ulam model, has been studied by a number of people, both theoretically [9,10,11] and experimentally [12-20]. The (vertical) motion of the table is given by $z(t) = z_0 \cos\omega t$, in which the amplitude $z_0$ is taken to be much smaller than the height of the jumps of the ball. The situation is sketched in figure 1.

*Chaotic Dynamics: Theory and Practice*
Edited by T. Bountis, Plenum Press, New York, 1992

Figure 1. A small ball bouncing up and down on a vibrating table.

Let $v_i$ be the (upward) velocity of the ball just after the i-th bounce, which takes place at time $t_i$. Neglecting air resistance and neglecting any variations in the height of the table, it takes the ball a time $2v_i/g$ to make its jump and return to the table. So we have:

$$t_{i+1} = t_i + \frac{2}{g} v_i . \tag{1}$$

Furthermore, assuming that the bounce is a perfectly elastic one, the velocity of the ball is augmented by twice the velocity of the table. That is,

$$v_{i+1} = v_i - 2z_0\omega \sin\omega t_{i+1} . \tag{2}$$

The above two equations fully describe the dynamics of the system, in the absence of dissipation. Eliminating the velocity via $v_i = g (t_{i+1}-t_i)/2$ the two equations can be combined into one, relating the times $t_i$, $t_{i+1}$ and $t_{i+2}$ of successive bounces:

$$t_{i+2} = 2t_{i+1} - \frac{4z_0\omega}{g} \sin\omega t_{i+1} - t_i . \tag{3}$$

Equivalently, with $x_i = \omega t_i$ and $y_i = \omega t_{i-1}$ (both modulo $\omega T(\omega) = 2\pi$, where $T(\omega)$ is the period of the table) this can be written in the form of a two-dimensional area preserving map:

$$\begin{cases} x_{i+1} = 2x_i - A \sin x_i - y_i , & (4a) \\ \\ y_{i+1} = x_i . & (4b) \end{cases}$$

Here

$$A = \frac{4z_0\omega^2}{g} \tag{5}$$

is the so-called chaos parameter $(A \geq 0)$. If we increase the value of A, by increasing the value of $\omega$ (*not* the value of $z_0$, since this must remain small) we witness a transition from order to chaos via period doubling bifurcations.

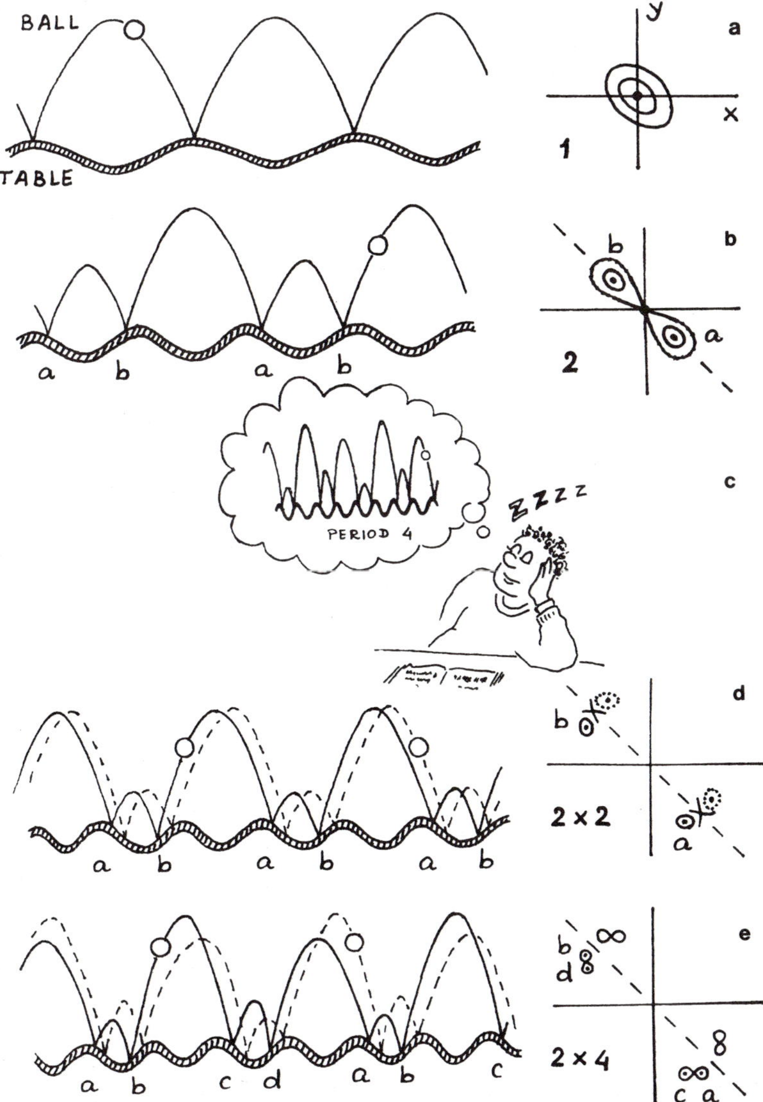



Figure 2. The route to chaos for the bouncing ball. As discussed in the text the successive stages are (a) a 1 cycle, (b) a (symmetric) 2 cycle, (d) two non-symmetric 2 cycles and (e) two non-symmetric 4 cycles. A symmetric 4 cycle, as depicted in figure (c), is *not* part of the route.

For $0 < A < 4$ the behaviour of the ball is quite simple. It just follows the rhythm of the table, jumping to the same height after each bounce, with a periodicity $T(\omega) = 2\pi/\omega$. This behaviour is sketched in figure 2a, along with the corresponding picture of the $(x,y)$ plane: the map (4a)-(4b) has a stable fixed point (1-cycle) in the origin $(x,y) = (0,0)$.

At $A = 4$ a period doubling bifurcation takes place. To keep up with the pace of the table the ball has to change its pattern of jumps. The simple one-step behaviour has become unstable. and in its place comes a stable two-step pattern, with period $2T(\omega)$. This is shown in figure 2b. The corresponding picture of the $(x,y)$ plane shows the same bifurcation in another way: the origin has become unstable and in doing so has given birth to a stable 2-cycle.

Now, an unsuspecting student of Universality in the Period Doubling Route to Chaos (depicted in figure 2c) might be lured into thinking that the next bifurcation would be a period doubling one to period $4T(\omega)$. But this is not the case. In fact, the next bifurcation is a symmetry breaking one. It takes place at $A = 6.283$. The period $2T(\omega)$ remains the same, but the original (symmetric) 2-cycle becomes unstable and gives birth to two new stable 2-cycles. These newly born 2-cycles are not symmetric with respect to the origin, as one can see in figure 2d.

This bifurcation (in which the system as a whole does not lose its symmetry, but the individual orbits do) may seem quite a surprising phenomenon when one first encounters it. Nevertheless, it is not at all a rare or pathological bifurcation. It is found in a number of systems, in various fields, and it is known under at least four different names, viz. symmetric saddle node bifurcation [21], Rimmer bifurcation [22], double point bifurcation [23] and symmetry breaking bifurcation [34].

The next bifurcation, which takes place at $A = 6.594$, is a period doubling one again. The two 2-cycles become unstable and both of them give birth to a stable 4-cycle. So the system has now two non-symmetric 4-cycles (the ball follows either one of them, depending on the initial conditions) as sketched in figure 2e. From that point on the system follows a normal period doubling cascade. That is, the two 4-cycles bifurcate into two 8-cycles, which in turn bifurcate into two 16-cycles, and so on until chaos is reached at about $A = 6.635$. The successive bifurcation values $A_k$ are given in table 1.

Included in the same table are the values of

$$\delta_k = \frac{A_k - A_{k-1}}{A_{k+1} - A_k} , \qquad (6)$$

that is, the ratio between the successive bifurcation intervals. For large k this ratio tends to the limiting value $\delta_k \rightarrow 8.72$, in agreement with the well known result for period doubling in Hamiltonian systems and area preserving maps [24,25]. True enough, in the usual definition of the approximants $\delta_k$ only period doubling bifurcations are counted but this does not affect the limiting behaviour, only the first approximants. Indeed, if we were to exclude the symmetry breaking bifurcation (at $A = 6.283$) the first approximant would not be 7.35 but $\delta_1 = (6.5938-4)/(6.6302-6.5938) = 71.3$. This anomalously large value only helps to illustrate how well embedded the symmetry breaking bifurcation is (also quantitatively) into this route to chaos.

Up to now, we have been talking about the conservative case, without

Table 1. The successive bifurcation values $A_k$ for the bouncing ball without dissipation, as given in eqns. (4a)-(4b). In the right hand column the corresponding values of $\delta_k$ (defined in eq. (6)) are given, and in the left hand column the nature of the successive bifurcations is indicated. For example, the symmetry breaking bifurcation is indicated by $2 \to 2\times2$, i.e. one stable 2-cycle is replaced by two stable 2-cycles.

|  | $A_k$ | $\delta_k$ |
|---|---|---|
| $1 \to 2$ | 4 |  |
| $2 \to 2\times2$ | 6.2832 | 7.35 |
| $2\times2 \to 2\times4$ | 6.5938 | 8.53 |
| $2\times4 \to 2\times8$ | 6.63019 | 8.71 |
| $2\times8 \to 2\times16$ | 6.634366 | 8.72 |
| $2\times16 \to 2\times32$ | 6.63484389 | 8.72 |
| $2\times32 \to 2\times64$ | 6.634898808 | .. |
| .... | .... | .. |

any dissipation. This is all very well, but a real bouncing ball will always be slightly dissipative, and so we'd better include some dissipation in our model. How does one do that? The generally adopted view [11-17,20] is that the dissipation is mainly due to the fact that the bounces are not really elastic. Let us say that at each bounce the ball loses a fraction $1-J^2$ of its kinetic energy $\frac{1}{2}$ m $(v-\dot{z})^2$ with respect to the moving table. In that case equation (2) takes the following form:

$$v_{i+1} = Jv_i - (1+J) z_0 \omega \sin \omega t_{i+1} , \qquad (7)$$

while equation (1) remains the same. This leads to a dissipative version of (4a)-(4b), in which J is the jacobian of the mapping $(0 \leq J < 1)$. For this dissipative map one finds the same route to chaos as in the conservative case (J=1), including the symmetry breaking bifurcation; only now (for $0 \leq J < 1$) the approximants $\delta_k$ tend to another limiting value, namely 4.67. This is the famous Feigenbaum value for period doubling in one dimensional maps with a quadratic maximum and for dissipative two-dimensional maps [26,27].

So one might expect the symmetry breaking bifurcation to have been detected in bouncing ball experiments. But this is not the case! In none of the experimental studies in which the period doubling route has been followed [12,13,17,18] is there any hint of a symmetry breaking bifurcation. Apparently, the generally accepted model described above is not entirely correct. But what is wrong with it? We shall come back to this question at the end of section 4.

## 3. SECOND EXAMPLE: THE DRIVEN PENDULUM

In this section we consider a pendulum, as sketched in figure 3, which is being moved up and down (at the point of support) with amplitude $a$ and period $\Omega$. The motion of the point of support is given by

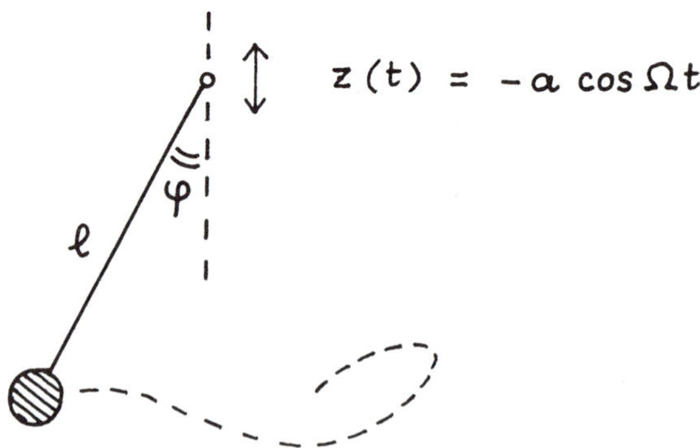


Figure 3. A pendulum (with an inflexible rod of negligible mass) which is being moved up and down at its point of support.

$z(t) = -a \cos\Omega t$. This system, just like the bouncing ball, is quite a popular one. It has gained attention from a number of people, theoretically [28-32] as well as experimentally [33-36].

The equation of motion reads, in the absence of dissipation,

$$\ddot{\varphi} + \frac{1}{\ell}\,(g + a\Omega^2 \cos\Omega t)\,\sin\varphi = 0 \tag{8}$$

where $\ddot{\varphi} + (g/\ell)\,\sin\varphi = 0$ is the well known equation for a non-damped pendulum. The driving force acts as a periodic perturbation on the gravitational force, in such a way that g is replaced by $g + a\Omega^2\cos\Omega t$ (a small insect sitting on the bob of the pendulum would feel itself grow heavier and lighter all the time). This way of driving a system, acting directly on one of the system's parameters instead of introducing a separate extra forcing term, is called parametric pumping.

The characteristic thing about this way of pumping is that the pendulum reacts to the driving force not when this has the eigenfrequency of the pendulum, but when it has about *twice* this frequency. In formula:

$$\Omega \approx 2\omega_0 \;, \tag{9}$$

where $\omega_0 = \sqrt{g/\ell}$ is the eigenfrequency of the pendulum. If $\Omega \approx 2\omega_0$ the simple downward equilibrium of the pendulum loses its stability and the pendulum starts to swing, in such a way that its motion resembles a Lissajous figure in the form of an infinity sign ($\infty$). This is an example of parametric resonance. This resonance does not require a lot of fine tuning. In fact for any choice of $\Omega$ and $a$ inside the hatched tongue of figure 4 one gets parametric resonance. Now, let us follow a vertical path through figure 4 (indicated by the dashed line), with $\Omega$ fixed at $\Omega = 2.24\ \omega_0$ and $a$ growing steadily. Thus $a$ (or equivalently $a/\ell$) plays the role of chaos parameter.

For $0 < (a/\ell) < 0.098$ the behaviour of the pendulum is very simple. It is just hanging downward, meekly following the up and down motion of the point of support. This situation is shown in figure 5a, along with

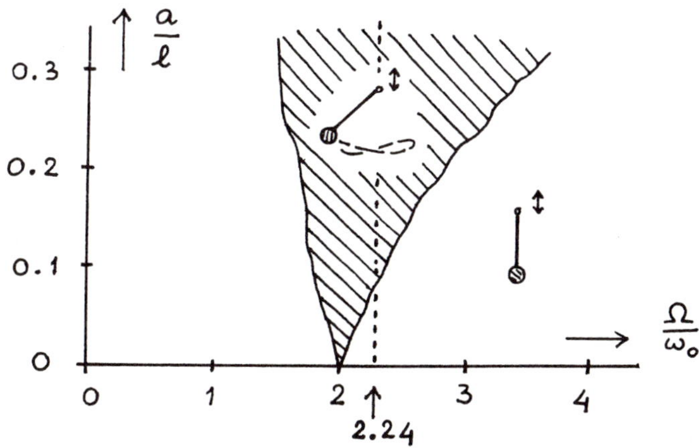


Figure 4. The $(\Omega, a)$ plane for the driven pendulum governed by eq. (8). Along the horizontal axis we have plotted $\Omega/\omega_0$, where $\omega_0$ is the eigenfrequency of the pendulum, and along the vertical axis we have plotted $a/\ell$.
For parameter values within the hatched area the downward equilibrium is unstable and the pendulum gets in a resonant motion. The dashed vertical line at $\Omega = 2.24\ \omega_0$ is the path along which we follow the route to chaos (see text).

the corresponding stroboscopic phase portrait. This portrait is obtained by looking at the pendulum only at discrete times $t_n = 2n\pi/\Omega$ with $n = 0,1,2,\ldots$ (when the point of support is at its lowest), as if we were watching it by the flickering light of a stroboscopic lamp synchronized with the driving force. At each flickering the values of $\varphi$ and $\dot{\varphi}$ are measured and the corresponding point in the $(\varphi, \dot{\varphi})$ plane is plotted. The points $\varphi(t_n)$, $\dot{\varphi}(t_n)$ at successive times $t_n$ constitute a two-dimensional map; for $0 < (a/\ell) < 0.098$ this map has a stable 1-cycle in the origin.

At $(a/\ell) = 0.098$ we enter the tongue of parametric resonance. At this point the origin of the stroboscopic $(\varphi, \dot{\varphi})$ plane becomes unstable and gives birth to a stable 2-cycle. This period doubling is depicted in figure 5b. The pendulum is swinging in (very flat) Lissajous fashion, symmetrically around the vertical line $\varphi = 0$.

Then, at $(a/\ell) = 0.345$, the orbit bifurcates again. But it does not double its period. This bifurcation is a symmetry breaking one, leaving the original 2-cycle unstable and giving birth to two new non-symmetric 2-cycles. In figure 5c these two new 2-cycles have been sketched.

The next bifurcation, at $(a/\ell) = 0.361$, is a period doubling one again. At that point two (non-symmetric) 4-cycles are born, which in turn bifurcate into two 8-cycles, etc. Just as in the case of the bouncing ball one can evaluate the approximants $\delta_k$ (the ratios between the successive bifurcation intervals) and again they go to the limiting value 8.72 for large k.

In an experimental situation, in the presence of dissipation, this limiting value will change to 4.67. But our main question is of course: will the symmetry breaking bifurcation survive ? The answer turns out to

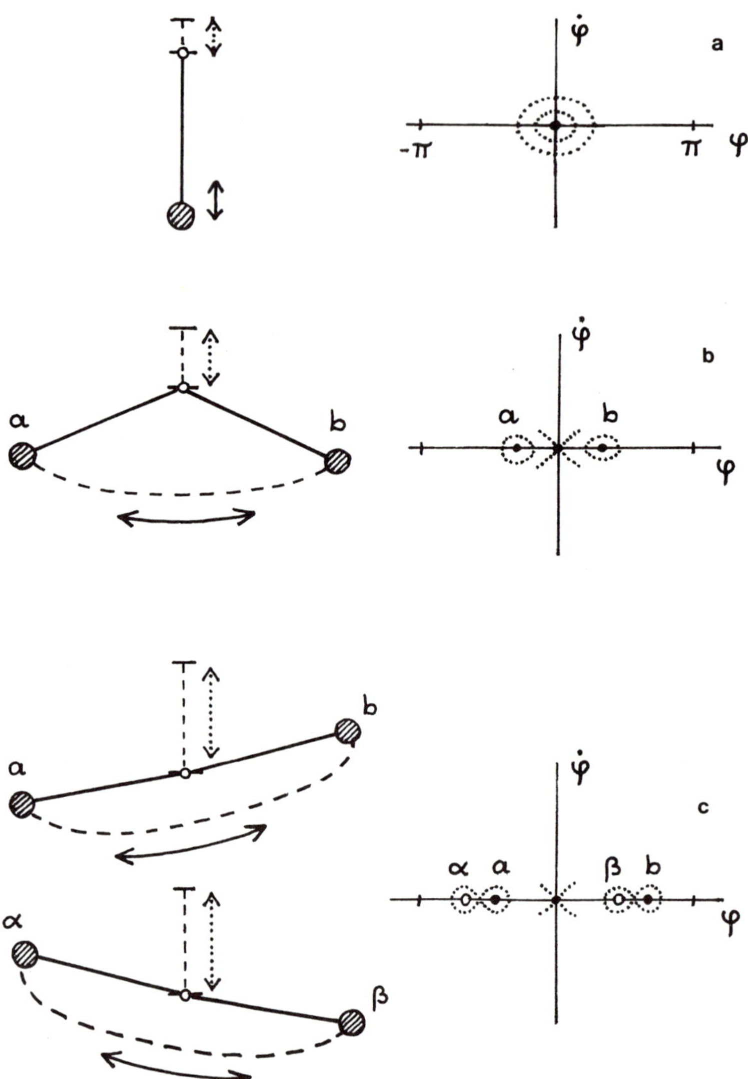

Figure 5. The first three stages in the route to chaos for the parametrically driven pendulum, at a fixed driving frequency $\Omega = 2.24\ \omega_0$. We see, for growing values of the driving amplitude, first (a) a 1-cycle, then (b) a (symmetric) 2-cycle, and then (c) two non-symmetric 2-cycles.

be yes. Koch, Leven, Pompe and Wilke [34,35] found the bifurcation in a real-life experiment. They also found (experimentally) that the dissipation for their pendulum could be written as $\lambda_0 \text{sign}(\dot{\varphi}) + \lambda_1 \dot{\varphi}$, with $\lambda_0$ and $\lambda_1$ both small, which means that the equation of motion takes the form:

$$\ddot{\varphi} + \lambda_0 \text{sign}(\dot{\varphi}) + \lambda_1 \dot{\varphi} + \frac{1}{\ell} (g + a\Omega^2 \cos\Omega t) \sin\varphi = 0 \qquad (10)$$

The term $\lambda_0 \text{sign}(\dot{\varphi})$ corresponds to "dry" friction at the point of support, and the term $\lambda_1 \dot{\varphi}$ represents the so-called viscous damping due to air resistance. With the help of a computer it is straightforward to check that equation (10) indeed shows the symmetry breaking bifurcation. So (fortunately) the difficulties of the bouncing ball do not exist here; the model just agrees with the experiment.

## 4. A MODEL TO DESCRIBE IT ALL

We have now seen two examples of period doubling cascades which are interrupted, at the period 2 level, by a symmetry breaking bifurcation. In this section we shall present a model which explains this behaviour.

First, let us recall that the sequence $1 \to 2 \to 2 \times (2 \to 4 \to 8 \ldots \to$ chaos) was found originally for one-dimensional maps with two critical points [7,8]. More specifically, May investigated the map $x_{i+1} = ax_i^3 + (1-a)x_i$ for $0 < a < 4$ and found that the $1 \to 2$ bifurcation takes place at $a = 2$, the $2 \to 2 \times 2$ bifurcation at $a = 3$, the $2 \times 2 \to 2 \times 4$ bifurcation at $a = 1 + \sqrt{5} = 3.236$, etc. The special feature of this map, which makes it different from the usual models for period doubling such as the logistic map, is the absence of a quadratic term. The first nonlinear term is a *cubic* one. Now, this same special feature is present in our two examples. The map for the bouncing ball does not contain a quadratic term:

$$\begin{cases} x_{i+1} = 2x_i - A \sin x_i - y_i & (4a) \\ \\ y_{i+1} = x_i & (4b) \end{cases}$$

and neither does the equation of motion for the driven pendulum

$$\ddot{\varphi} + \frac{1}{\ell} (g + a\Omega^2 \cos\Omega t) \sin \varphi = 0 \qquad (8)$$

The dominant nonlinear term is a cubic one, since $\sin\varphi = \varphi - \varphi^3/6 + \ldots$ . The dissipative versions which show symmetry breaking do not have any quadratic term either. All in all, it seems evident that the model map (also in two dimensions) should be a cubic one. So let us try the following form, being one of the simplest forms possible:

$$\begin{cases} x_{i+1} = 2Cx_i + 3x_i^3 - y_i & (11a) \\ \\ y_{i+1} = Jx_i & (11b) \end{cases}$$

It resembles the standard form of the Hénon map [37,27], the only difference being that the quadratic term $2x_i^2$ has been replaced by a cubic term $3x_i^3$. In this map C is the chaos parameter and J, a constant with a value between 0 and 1, is the jacobian. For $J = 1$ the map is area-preserving ($\delta_k \to 8.72$) and for $0 \le J < 1$ it is dissipative ($\delta_k \to 4.67$).

We put our model map to the test by *decreasing* the chaos parameter from C = 0 downward. In doing so we observe that the stable 1-cycle at the origin period-doubles into a 2-cycle at

$$C = - \frac{1}{2} (1 + J) ,$$

(12)

and the next bifurcation is seen to be a symmetry breaking one, just as in the two examples. This bifurcation takes place at

$$C = - (1 + J) .$$

(13)

The next bifurcation is a period doubling one again, in which the two (non-symmetric) 2-cycles bifurcate into two 4-cycles, and from this point on the period doubling cascade is not interrupted anymore. In figure 6 the successive approximants $\delta_k$ are shown for three different values of J, namely J = 1, J = 0.5 and J = 0. For J = 1 (the conservative case) the $\delta_k$ are seen to tend to the limiting value 8.72, and for J = 0.5 and J = 0 (i.e. in the presence of dissipation) they tend to 4.67. For J = 0.5, probably the most realistic value of the three from an experimental point of view, the $\delta_k$ show a nice crossover behaviour between a nearly conservative situation and the dissipative limit. This is all just as it should be, and the conclusion is clear: the model map (11a-11b) represents the interrupted period doubling cascade perfectly in every way.

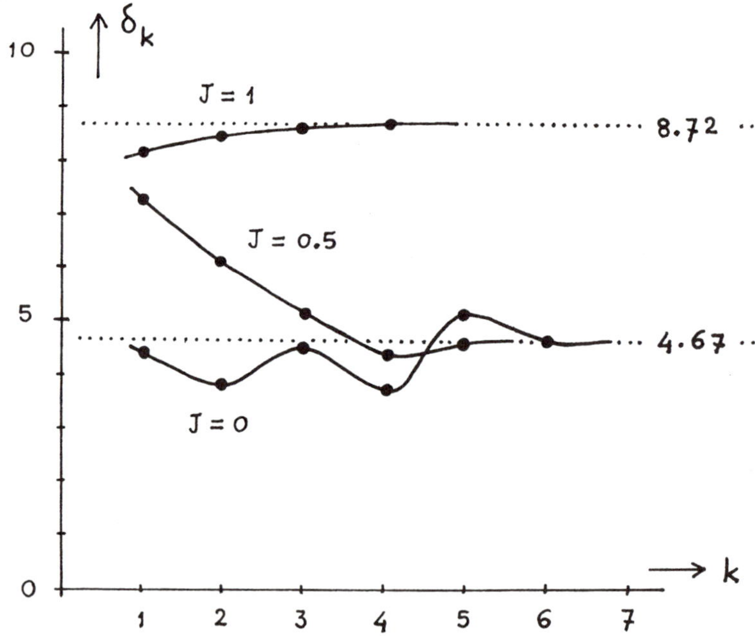


Figure 6. The approximants $\delta_k$ for the model map (11a)-(11b), for three different values of the jacobian J. In the evaluation of the approximants the symmetry breaking bifurcation has also been counted; otherwise the first approximant (for k = 1) would have been much larger.

Finally we can also understand the curious fact that in the bouncing ball experiments no symmetry breaking bifurcation has ever been observed. Apparently the experimental set-up introduces some small quadratic term which has been neglected in the theoretical model. There is plenty of room for that, since the model was full of idealizations. It may be a term due to the air resistance which we neglected, or it may be due to the variations in the height of the table (which we took to be negligible compared to the height of the ball jumps) or it may even be due to the mass of the ball (which we neglected in comparison with the mass of the table). In any case, if we add a small quadratic term to the map (4a)-(4b), as follows,

$$\begin{cases} x_{i+1} = 2x_i - A\,\sin x_i + 0.01\,x_i^2 - y_i & (14a) \\ \\ y_{i+1} = x_i & (14b) \end{cases}$$

the symmetry breaking bifurcation vanishes into smoke. The first period doubling $(1 \rightarrow 2)$ at $A = 4$ is followed by just another period doubling $(2 \rightarrow 4)$ at $A = 6.520$. So even the tiniest quadratic term is already sufficient to suppress the symmetry breaking bifurcation.

So we have come to the end of this paper. In conclusion one can say that whenever we come across a system without a quadratic term, the dominant nonlinear term being a *cubic* one, we should watch out for symmetry breaking in the period doubling cascade. We now know that such systems are not just mathematical curiosities. They really do occur in practice.


ACKNOWLEDGEMENT

It is a pleasure to thank the students G.J. Wiggers, M.E. Buitenhuis, I. Huisman, P. Krechting and P. Spelt for their enthousiastic help during the early stages of this work.

REFERENCES

1.  V. Franceschini, "Bifurcations of tori and phase locking in a dissipative system of differential equations", <u>Physica</u> 6D:285-304 (1983).
2.  E. Knobloch and N.O. Weiss, "Bifurcations in a model of magnetoconvection", <u>Physica</u> 9D:379-407 (1983)
3.  G.L. Oppo and A. Politi, "Collision of Feigenbaum cascades", <u>Phys. Rev.</u> A30:435-441 (1984).
4.  M. Bier and T.C. Bountis, "Remerging Feigenbaum trees in dynamical systems", <u>Phys. Lett.</u> 104A:239-244 (1984).
5.  T.P. Valkering and J.H.J. van Opheusden, Bouncing ball behaviour of a solitary wave, <u>in</u>: "Proc. of the conference on Nonlinear Dynamics", Bologna, Italy, G. Turchetti, ed., World Scientific, Singapore, 225-235 (1988).
6.  Th. Zeegers, "On the existence of infinite period doubling sequences in a class of 4D semi-symplectic mappings", <u>J. Phys. A: Math. Gen.</u> 24:2287-2314 (1991).
7.  R.M. May, "Bifurcations and dynamic complexity in ecological systems", <u>Annals of the New York Academy of Sciences</u> 316:517-529 (1979).

8.  R.M. May, "Nonlinear phenomena in ecology and epidemiology", in: Nonlinear Dynamics, edited by R.H.G. Helleman, _Annals of the New York Academy of Sciences_ 357:267–281 (1980).

9.  L.D. Pustyl'nikov, "Stable and oscillating motions in nonautonomous dynamical systems II", _Trans. Moscow Math Soc._ 2:1–101 (1978).

10. P.J. Holmes, "The dynamics of repeated impacts with a sinusoidally vibrating table", _J. Sound Vib._ 84:173–189 (1982).

11. R.M. Everson, "Chaotic dynamics of a bouncing ball", _Physica_ 19D:355–383 (1986).

12. P. Pieranski, "Jumping particle model. Period doubling cascade in an experimental system", _J. Physique_ 44:573–578 (1983).

13. P. Pieranski, Z.J. Kowalik and M. Franaszek, "Jumping particle model. A study of the phase space of a non-linear dynamical system below its transition to chaos", _J. Physique_ 46:681–686 (1985).

14. P. Pieranski and R. Bartolino, "Jumping particle model. Modulation modes and resonant response to a periodic perturbation, _J. Physique_ 46:687–690 (1985).

15. M. Franaszek and Z.J. Kowalik, "Measurements of the dimension of the strange attractor for the Fermi-Ulam problem", _Phys. Rev._ A33:3508–3510 (1986).

16. M. Franaszek and P. Pieranski, "Jumping particle model. Critical slowing down near the bifurcation points", _Can. J. Phys._ 63:488–493 (1985).

17. N.B. Tufillaro, T.M. Mello, Y.M. Choi and A.M. Alabano, "Period doubling boundaries of a bouncing ball", _J. Physique_ 47:1477–1482 (1986).

18. N.B. Tufillaro and A.M. Alabano, "Chaotic dynamics of a bouncing ball", _Am. J. Phys._ 54:939–944 (1986).

19. T.M. Mello and N.B. Tufillaro, "Strange attractors of a bouncing ball", _Am. J. Phys._ 55:316–320 (1987).

20. K. Wiesenfeld and N.B. Tufillaro, "Suppression of period doubling in the dynamics of a bouncing ball", _Physica_ 26D:321–335 (1987).

21. R. Abraham and J.E. Marsden, "Foundations of mechanics", 2nd edition Benjamin Cummings, Reading Mass., (1978); the "symmetric saddle-node bifurcation" is discussed on page 504.

22. R.S. MacKay, "The dominant symmetry of reversible maps", in his Ph.D. thesis, Princeton University (1982); here the bifurcation is named after R.J. Rimmer, _J. Diff. Eqns._ 29:329 (1978).

23. E.A. Jackson, "Perspectives of nonlinear dynamics", Cambridge University Press, Cambridge, vol. 1, (1989); the "double point bifurcation" is discussed in chapter 3.2.

24. T.C. Bountis, "Period doubling bifurcations and universality in conservative systems", _Physica_ 3D:577–589 (1981).

25. J.M. Greene, R.S. MacKay, F. Vivaldi and M.J. Feigenbaum, "Universal behaviour in families of area-preserving maps", _Physica_ 3D:468–486 (1981).

26. M.J. Feigenbaum, "Quantitative universality for a class of nonlinear transformations", _J. Stat. Phys._ 19:25–52 (1978).

27. J.P. van der Weele, H.W. Capel, T. Post and Ch.J. Calkoen, "Crossover from dissipative to conservative behaviour in period doubling systems", _Physica_ 137A:1–43 (1986).

28. L.D. Landau and E.M. Lifschitz, "Mechanics", volume 1 of Course of Theoretical Physics, Pergamon Press, Oxford, 1960; the parametrically driven pendulum is treated in the exercises 5.3c, 27.3 and 30.1.

29. J.B. MacLaughlin, "Period-doubling bifurcations and chaotic motion for a parametrically forced pendulum", _J. Stat. Phys._ 24:375–388 (1981).

30. R.W. Leven and B.P. Koch, "Chaotic behaviour of a parametrically excited pendulum", _Phys. Lett._ 86A:71–74 (1981).

31. B.P. Koch and R.W. Leven, "Subharmonic and homoclinic bifurcations in a parametrically forced pendulum", Physica 16D:1-13 (1985).
32. J.P. van der Weele and T.P. Valkering, Orde en chaos in de parametrisch aangedreven slinger (in Dutch), in: "Dynamische Systemen en Chaos", H.W. Broer and F. Verhulst, eds., Epsilon, Utrecht (1990), 232-255.
33. W. Case, "Parametric instability: an elementary demonstration and discussion", Am. J. Phys. 48:218-221 (1980).
34. B.P. Koch, R.W. Leven, B. Pompe and C. Wilke, "Experimental evidence for chaotic behaviour of a parametrically forced pendulum" Phys. Lett. 96A:219-224 (1983); an experimental picture of the "symmetry breaking bifurcation" is presented on page 222.
35. R.W. Leven, B. Pompe, C. Wilke and B.P. Koch, "Experiments on periodic and chaotic motions of a parametrically forced pendulum", Physica 16D:371-384 (1985).
36. W. van de Water and M. Hoppenbrouwers, Chaos and the experiment, in: "Proceedings of the Spring Meeting on Nonlinear Dynamics Twente", University of Twente, Enschede, March (1990) 58-79.
37. R.H.G. Helleman, with an appendix by R.S. MacKay, One mechanism for the onset of large-scale chaos in conservative and dissipative systems, in: "Long-time prediction in dynamics", C.W. Horton Jr., L.E. Reichl and A.G. Szebehely, eds., Wiley and Sons, New York, (1985) 95-126.

# DISSIPATIVE DETERMINISTIC CHAOTIC DYNAMICS  BETWEEN THE LYAPOUNOV TIME AND THE LONG TIME LIMIT: A PROBABILISTIC DESCRIPTION

Donal MacKernan

Faculté des Sciences
and Center for Nonlinear Phenomena and Complex Systems
Université Libre de Bruxelles, Campus Plaine C.P.231
Boulevard du Triomphe, B-1050 Bruxelles, Belgium

Realistic initial conditions for dynamical systems always have an inherent uncertainty due to limited experimental or numerical precision. For chaotic systems the uncertainty grows exponentially so that after a time of the order of the reciprocal of the largest Lyapounov exponent (the Lyapounov time) a point like evolutionary description fails. For times much greater than a characteristic long time limit, the state of the system,  to some extent , can be described in terms of long time average properties, which, if the system is ergodic, can be readily calculated numerically. However for intermediate times neither description is valid, indeed comparatively little is known. It may be more natural and efficacious to describe the point like initial condition as a narrow peaked probability distribution with width equal to the resolution limit, and then to evolve the distribution with a Liouville like operator. In this communication we shall  explore the possibility  of setting up a workable probabilistic description of one dimensional chaotic maps. In particular we shall discuss results  which we have obtained  for piecewise linear maps whose dynamics remains in a bounded region of the real line. Knowledge of the spectral properties of these systems can be helpful in predicting the time behavior of observables of the system for initial probability distributions very far from their associated asymptotic distributions[1] ( usually the equilibrium distribution).

The Frobenius Perron operator[1]

$$\rho_{n+1}(\vec{x}) = U[\,\rho_n(\vec{x})\,] = \int_{\Omega} \rho_n(\vec{y})\delta(\vec{x} - \vec{f}(\vec{y}))\,d\vec{y}$$

(1)

gives the exact evolution of probability densities for iterative chaotic maps. For one dimensional maps (1) takes the form

*Chaotic Dynamics: Theory and Practice*
Edited by T. Bountis, Plenum Press, New York, 1992

$$\rho_{n+1}(x) = U[\rho_n(x)] = \sum_{a=1}^{m} \rho_n(f^{-1}{}_a(x))\left|\frac{d}{dx}f^{-1}{}_a(x)\right|$$

$$(2)$$

where the $f^{-1}{}_a(x)$ are the local inverse branches of the iterative map over each of it's $m$ monotonic regions. Relating the density at time n to an initial density at time zero one obtains the formula

$$\rho_n(x) = \sum_{a_1,a_2,\dots,a_n} \rho_0(f^{-1}{}_{a_1} \circ f^{-1}{}_{a_2} \circ \dots \circ f^{-1}{}_{a_n}(x))\left|\frac{d}{dx}f^{-1}{}_{a_1} \circ f^{-1}{}_{a_2} \circ \dots \circ f^{-1}{}_{a_n}(x)\right|$$

$$(3)$$

The number of terms in the sum grows as $m^n$ ( ignoring the effects of possible pruning, that is unphysical and thus forbidden sequences of inverse branches, whose contributions to (2) and (3) are defined to be zero). Thus conventional analytical and numerical methods of solution of (3) for reasonably general initial densities for all but small times are possible only for a few simple problems[1].

Li[2] has proven for a large class of ergodic maps, that the invariant measure can be calculated via a finite approximation of the Perron-Frobenius operator, thus verifying the conjecture of Ulam. The approximation, essentially, consists of dividing the interval on which the dynamics is defined into n equal subintervals and then projecting the Perron Frobenius operator onto a set of n corresponding characteristic functions, thus obtaining a stochastic transition matrix. The approximation to the invariant density corresponds to the eigenvector of the transition matrix with unit eigenvalue. Unfortunately, the proof did not indicate how well such an approximation technique would work in describing the evolution( iteration by iteration) of reasonably general initial densities( for example initial densities which are piecewise constant over the subintervals defined above). Moreover, the choice of the above characteristic functions, as a basis for projection, is expected to be, for most cases, far from optimal, that is the number of basis functions required for a good approximation are likely to be very high.

Clearly, another approach towards solution of (4) is needed. We have recently developed a method of solving (4) which is somewhat similar to the Markov Coarse Graining approach[3,4,5,6,7]. The method consists of finding a finite basis of functions( which need not be complete) appropriate for each individual dynamical system which either exactly or approximatively span the functional space containing the evolving densities. Thus, the evolution of densities which initially are linear combinations of the basis functions are generated by a time independent transition matrix . In addition eigenvectors of the transition matrix can be used to define exact/approximate eigenfunctions of the Perron-Frobenius operator. The elements of a given eigenvector equal the expansion coefficients of a corresponding eigenfunction when expressed as a linear combination of functions of the basis. Constructing an individual basis for each dynamical system has at least three advantages over the use of a more standard basis, such as that proposed by Li: <u>first</u>, for many maps one is able to evolve densities exactly while with a standard basis approximations must be made; <u>second</u>, the basis size is considerably smaller even when approximations are required in both types of basis; <u>third</u>, eigenfunctions of the Perron-Frobenius operator are more easily calculated, even in the case of approximation. The distinguishing feature of this type of basis is that its elements possess many of the symmetries( in the general sense of the word) found in (3). It should be noted that in the results and arguments about to be presented we have assumed that the iterative maps in question are either everywhere expansive or are equivalent via smooth topological isomorphism to maps that have this property. Thus the possibility of singularities in (3) have been avoided so that we can treat the evolving distributions as ordinary functions.

In the following two examples a finite set of basis states exists such that densities which initially are linear combinations of elements of the basis remain so throughout their subsequent evolution( we shall say that the basis set is closed under the Perron-Frobenius operator). In terms of these states, the evolution operator is a finite dimensional stochastic matrix and the corresponding stochastic process is a Markovian.

1. Dynamical systems which have the property that initial local equilibrium densities remain in local equilibrium throughout their subsequent evolution. Local equilibrium here means that the evolving density can be expressed as a linear combination of characteristic functions( defined over the cells of an appropriate partition) times an invariant "equilibrium" density.

$$\rho_n(x) = \rho_{eq}(x)\Sigma_i A_i(n)\chi_i(x)$$

2. Piecewise linear maps whose non-differentiable points fall in a finite number of iterations onto a periodic orbit/orbits( this means that there exists a topological Markov partition on the region where the dynamics asymptotically lies), and dynamical systems which are equivalent to them with respect to smooth topological isomorphism. This class was alluded to by Grossman and Thomae[8] in work concerning the existence of unique invariant measures and stationary correlation functions.

Unfortunately, for typical iterative maps a _finite_ basis of functions that is closed under the Perron-Frobenius operator is not expected to exist( with the exception of the eigenfunctions, and linear combinations thereof, which do not necessarily form a convenient basis to approximate reasonably general initial densities in the $L^1$ space of integrable functions). However several systems possess an associated functional basis consisting of a countable infinity of elements that is closed under the Perron-Frobenius operator, whose elements have the very special property that

$$U|i\rangle = [\sum_{k=0}^{i} W_{k\,i}|k\rangle] + W_{i+1\,i}|i+1\rangle$$

$$= \sum_{k=0}^{i} [W_{k\,i}|k\rangle] + (1 - \sum_{k=0}^{i} W_{k\,i})|i+1\rangle$$

Property A

where $W_{ki}$ is the associated transition matrix and

$$\sum_{k=0}^{i+1} W_{k\,i} = 1$$

An important property of this basis is that only certain transitions are allowed, in particular $W_{j\,i} = 0$ and $W_{i+1\,i} \neq 0$ $\forall$ $j > i+1$ and $i \geq 0$. Equivalently Property A can be viewed as a self-consistent method of constructing recursively higher order basis functions from lower orders _when they are known to exist_. Thus, one can define a construction operator $A|i\rangle = |i+1\rangle$, that is

$$A \,|i\rangle = \frac{\left\{ U|i\rangle - \sum\limits_{k=0}^{k=i} [W_{k\,i}|k\rangle] \right\}}{1 - \sum\limits_{k=0}^{k=i} W_{k\,i}} = |i{+}1\rangle \tag{4}$$

For example, if a particular normalized basis set is known to satisfy Property A, but only the zeroth order basis function is known explicitly, the rest of the basis can be calculated using (4). First the zeroth order basis function( $|0\rangle$ )is evolved with U, the result $U|0\rangle$ will consist of a coefficient times the zeroth order basis function $W_{00}|0\rangle$, and another term not proportional to $|0\rangle$ which equals $(1 - W_{00}) \, |1\rangle$, thus, $|1\rangle$ is obtained directly. The higher order basis functions are calculated continuing this process recursively. Moreover approximations can often be made where only a finite number of the elements are used to describe the evolving densities, the major source of error in the corresponding description being precisely this truncation.

One can prove that piecewise-linear maps, whose non-differentiable points **do not** fall in a finite number of iterations onto a periodic orbit/orbits, possess a basis of functions which essentially satisfy property A. The number of basis functions required for an exact description is countably infinite. In practice we can not deal with such a large basis, however, truncation wherein only a finite number are used is possible, in which case the evolution of initial densities which are linear combinations of a subset of these states is generated by a time independent matrix. Truncation will lead to an error in some of our predictions. In the following section we shall provide evidence that the truncation error is temporally well controlled and decreases rapidly as the order at which truncation is made is increased. A detailed analysis will appear in a forthcoming paper. The above functional basis is equivalent) via invertible linear transformations) to the set of characteristic functions associated with a partition of the phase space of the system constructed from the iterates of the non-differentiable points of the iterative map. If a finite number of iterates are used to define the partition then the partition is an approximate topological Markov partition, that is all but one cell will have the Markov property. However, the general method of construction of the former basis is more likely to be applicable to non-piecewise linear maps where the latter basis is not expected to be useful. Indeed, a heuristic proof together with a numerical analysis strongly indicates that property A is satisfied for the logistic map $f(x) = \lambda x(1-x)$ for several values of $\lambda$. The local expansivity almost everywhere, after a sufficient number of iterations, combined with the fact that at all but one point the curvature $f''(x)/(f'(x))^2$ is comparatively low, appears to play a vital role in the observed behavior of the dynamical system. We conjecture that one dimensional chaotic systems whose iterative functions are almost everywhere expansive and whose values of curvature are high only about a finite number of points should also obey property A. If true, this would permit a simple description of the evolution of initial densities and thus typical experimental point like initial conditions.

Let us now return to the piecewise linear maps of the last paragraph. We shall call those points where a given piecewise linear map has abrupt changes of slope angular points(ang) and denote the set thereof by {ang}. Property A and relation (4), strictly speaking, are satisfied, solely by piecewise linear maps which have only one angular point( e.g. symmetric and asymmetric tent maps). Property A and relation (4) become slightly more complicated for piecewise linear maps having more than one angular point, then one has families of basis functions $\{ \{|i_1\rangle\}, \{|i_2\rangle\}, \ldots, \{|i_r\rangle\} \}$ where each family $\{|i_1\rangle\}$, of linearly independent functions corresponds to a given angular point of which there are r in number. For the purposes of exposition we shall discuss the proof when there is only one angular point, generalization to several angular points is relatively straightforward. The basis can be constructed if the zeroth order basis function is known by recursive use of the construction operator, $A\ |i> = |i+1>$. From hereon let us denote the basis states $\{|i>\}$ by $\{\Phi_i(x)\}$. We choose here as the zeroth order basis function the normalized characteristic function which is nonzero in the region $\Omega$ where the iterative map is defined, and zero everywhere else,

$$\Phi_0(x) = \frac{\chi_\Omega(x)}{\Gamma_0} = \begin{cases} \dfrac{1}{\Gamma_0} & \text{if } x\ \varepsilon\ \Omega \\ 0 & \text{otherwise} \end{cases}$$

(5)

The basis functions then have the form

$$\{\Phi_i(x)\} = \left\{ \frac{1}{\Gamma_i}\Phi_0(f^{-1}{}_{a_1}\circ f^{-1}{}_{a_2}\circ...\circ f^{-1}{}_{a_i}(x)) \left|\frac{d}{dx}f^{-1}{}_{a_1}\circ f^{-1}{}_{a_2}\circ...\circ f^{-1}{}_{a_i}(x)\right| \right\}$$

(6)

where $\Gamma_i$ are normalization constants and i are positive nonzero integers. We define $\Phi_i(x) = 0$ for those x $\varepsilon$ $\Omega$ for which there does not exist a y $\varepsilon$ $\Omega$ such that $f_{a_i}\circ....\circ f_{a_2}\circ f_{a_1}(y) = x$, because for such points the above sequence of i consecutive inverse branch maps is ill defined and physically unrealizable. This definition can be applied to all terms in (2) and (3) which otherwise would be similarly ill defined. The sequence corresponds to the (symbolic) trajectory, monotonic region by monotonic region, followed by the angular point for i iterations. It is not hard to show that $\Phi_i(x)$ has a discontinuity at the $i^{th}$ iterate of the angular point $f^i(ang)$ and that another $\Phi_j(x)$ can have a discontinuity at $f^i(ang)$ only if $j \geq i$. Using the fact that the ranges of different inverse branches are mutually disjoint one can prove that the basis defined by (6) satisfies Property A. Now it is self-evident that if an initial density is a linear combination of the first $N_0$ basis functions constructed above and we know the transition matrix $W_{ji}$ for the first $N_1$ x $N_1$ block of elements with $N_1 > N_0$, we can describe the evolving density exactly for $N_1 - N_0$ iterations. However, if $N_1$ is large such a technique is impractical. We can have a truncated description of the evolving density for all time involving only the first $N_1$ basis functions, associated with this description there will of course be a truncation error. Indeed, one can easily prove that the truncation error decreases monotonically as the order of approximation($N_1$) is increased. Numerical evidence and analytical arguments indicate that for general piecewise linear maps, the truncation error for a given order of approximation, <u>temporally</u> at most oscillates within an envelope which is either shrinking or stationary for initial densities very far from their asymptotic states. It appears that upper bounds for the truncation error envelope can be readily established for all time by calculating the maximum truncation error for the first few iterations. The following figures show how the truncation error and average position for the tent map with slope 1.45, each oscillate within a corresponding envelope.

Once a basis has been found which incorporates the main contribution to evolving densities for a given system the spectral properties of the system can be investigated. These will indicate how long it takes the system to relax to it's asymptotic state(s), the competition between various eigenmodes the relative ergodicity /mixing of the system. We have recently obtained some interesting numerical and analytical results showing the spectral dependence of the tent map[9] on the absolute value of the slope, which indicate that, the number of eigenfunctions with eigenvalues equal to the roots of unity double successively as the absolute value of the slope is decreased. When the

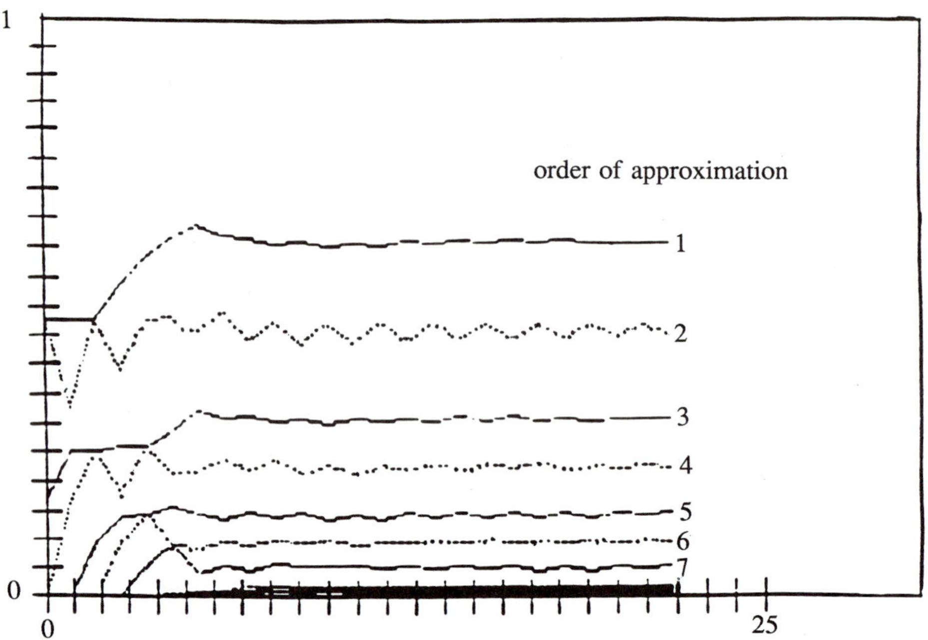

Fig. 1. Measure Error of Prediction Versus Time for Successive Orders of Approximation

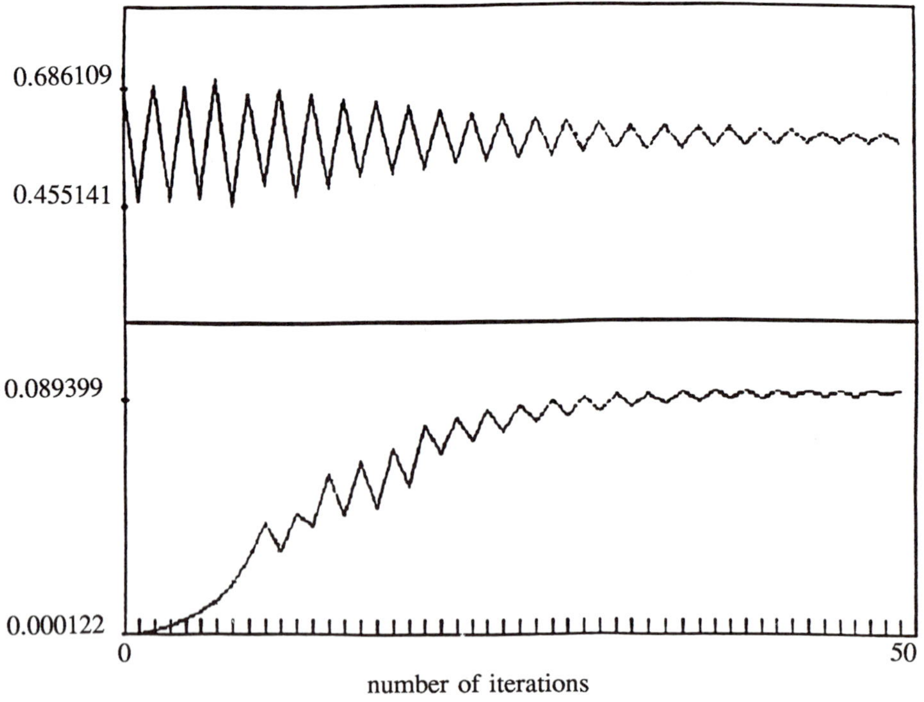

Fig. 2. Evolution of an Initial Point-Like Distribution, Average Position and Standard Deviation Versus Number of Iterations

number of such eigenfunctions equals $2^s$ then the region in which the dynamics asymptotically lies consists of $2^s$ disjoint pieces, s = 1,2,3... Moreover the critical values of the slope magnitude at which doubling takes place are $2^{1/2}$, $2^{1/4}$, $2^{1/8}$,.., $2^{1/r}$ where r = $2^s$. These results will be discussed more fully in a forthcoming paper.


Conclusion

We have presented analytical evidence that, for several chaotic systems, the evolution, iteration by iteration, of experimental point-like initial conditions can be described probabilistically in a computationally simple manner. Numerical results accord with the analytical evidence presented. We have argued that the evolution of the probability distribution initially representing a given initial condition and associated uncertainty, can be generated either exactly or up to a very good approximation by a time independent transition matrix W. This is achieved by projecting the probability evolution operator onto a finite basis of functions, particular to each dynamical system, which spans a functional space that is either exactly or up to a very good approximation invariant under action of the evolution operator. The initial distribution is projected then onto this set. The basis functions are particular in that they are constructed so that they share many of the symmetries possessed by the probability evolution operator of the given system. Several of the dominant eigenfunctions of the evolution operator can be calculated either exactly or approximately by finding the eigenvectors of W. Once the dominant eigenfunctions are known the spectral properties of the system can be investigated. Numerical and analytical investigations of the spectral dependence of tent maps on the absolute value of slope have indicated that the number of eigenfunctions with eigenvalues equal to the roots of unity double successively as the absolute value of the slope is decreased. We have proven in results to appear elsewhere that the first few critical values of the slope magnitude at which doubling takes place are $2^{1/2}$, $2^{1/4}$, $2^{1/8}$, $2^{1/16}$, $2^{1/32}$.... Numerical results and heuristic arguments suggest that it may be possible to generalize the results presented here to one dimensional chaotic systems whose iterative functions are almost everywhere expansive and whose values of curvature are high about only a finite number of points.


Acknowledgements

The author would like to thank I. Antoniou, P. Gaspard, J. Levitan and in particular G. Nicolis for interesting and informative discussions and critical reading of this text. The research of D.M. is supported by the Commission of the European Communities.


References

1.   A. Lasota and M. Mackey, "Probabilistic properties of deterministic systems", Cambridge University Press, New York.
2.   T. Li, Finite approximation for the Frobenius-Perron operator. A Solution to Ulam's conjecture; <u>Journal of Approximation theory</u>. **17**, 177-186 (1976).
3.   D. MacKernan and G. Nicolis, Probabilistic description of deterministic chaos: A local equilibrium approach, <u>in</u>: "Solitons and Chaos ", I. Antoniou and F. Lambert, eds., Research reports in physics, Springer-Verlag, Berlin, (1991).
4.   G. Nicolis, J. Piasecki and D. MacKernan, Toward a probabilistic description of deterministic chaos, <u>in</u>: "Festschrift for Péter Szépfaluski on the occasion of his sixtieth birthday," T Tél, ed, World Scientific, (in press).

5.  G. Nicolis and C. Nicolis,  Master-equation approach to deterministic chaos;  <u>Phys. Rev.</u> **A38**, 1 (1988).
6.  M. Courbage and G. Nicolis,  Markov evolution and H-theorem under finite coarse graining in conservative dynamical systems; <u>Europhys.Lett.</u> **11**, 1,  (1990).
7.  G. Nicolis, S. Martinez and E. Tirapegui,  Finite coarse-graining and Chapman-Kolmogorov equation in conservative dynamical systems; <u>Chaos and Solitons</u>, **1**,  1, 25-37,  (1991)
8.  S. Grossman and S. Thomae,  Invariant distributions and stationary correlation functions of one-dimensional discrete processes,  <u>Zeit. für Naturforschung a</u>, **32**,  1353-63,  (1977).
9.  R. Kluiving, H. W. Capel and R. A. Pasmanter,  Phase transition-like phenomena in a piecewise-linear map,  Instituut voor Theoretische Fysica,  Universiteit van Amsterdam,  (preprint).

RESONANT NORMAL FORMS

AND APPLICATIONS TO BEAM DYNAMICS

Graziano Servizi and Ezio Todesco

Department of Physics, University of Bologna.
Via Irnerio 46, 40126 Bologna Italy

## 1. INTRODUCTION

Recent developments in accelerator physics have raised new interest in the formalism of symplectic maps and the related perturbation theory, based on normal forms. The use of superconducting magnets in order to reach higher energies has caused the introduction of strong nonlinearities in the magnetic fields needed to focus the particles on the orbit. Therefore, modern accelerators are nonlinear machines which need new numerical and analytical tools to study problems like the maximization of the stability domain and the lifetime of the beam [1,2]. For hadron colliders since the synchrotron radiation can be neglected, the system is conservative and has many analogies with the well known problem of celestial mechanics concerning the stability of the orbits of the planets.

In this framework, the perturbative approach of normal forms for symplectic polynomial maps plays an important role. First introduced by Birkhoff [3], normal forms and related mathematical properties were extensively studied and applied to different problems, mainly in celestial mechanics [4]. During the last years the formalism of normal forms has been extended [5] to maps of $\mathbf{R}^{2d}$, and a Nekhoroshev-like theorem which gives an upper bound to Arnold diffusion has been proved [6]. Many applications to the analysis of the stability problem in accelerators have been made, including the estimate of the relevant nonlinear parameters and the optimization of the magnetic field of the corrector elements in order to recover a wider stability domain [1,7].

In this paper we review some of these results; in Section 2 we introduce the simplest nontrivial model (conservative Hénon map [8]) and the symplectic map which arises from the analysis of a magnetic lattice. Then we introduce the non resonant normal form, discussing the existence proof of the formal series. Some applications to accelerator physics [1,7] are outlined in Section 4. In Section 5 we show how to extend the formalism in order to take into account single resonances [9]. Some results on the divergence of the perturbative series are discussed in Section 6. Finally, we show how to compute from the interpolating hamiltonian of the resonant normal form the relevant parameters of a single resonance. Numerical checks of the analytical computations are given in Sections 4, 6 and 7.

## 2. THE MODEL: SYMPLECTIC MAPS

In this paper we deal with polynomial symplectic maps, which can be obtained from a Hamiltonian system by performing a Poincaré section. The reduction from the flow to the map is in general non trivial; nevertheless, in some cases it can be done exactly. The discrete model of symplectic maps has the big advantage of reducing the dimensionality of the problem; moreover a map is more easily implemented on a computer and therefore allows a deeper numerical analysis.

We first consider the hamiltonian:

$$H(p,x;t) = \frac{p^2}{2} + \frac{\omega^2 x^2}{2} + \sum_{n \in \mathbf{Z}} V(x)\delta(t-n), \tag{2.1}$$

which represents a kicked pendulum. We perform the Poincaré section at integer times, i.e. we set

$$x_n = x(t)|_{t=n_+} \qquad p_n = p(t)|_{t=n_+} \tag{2.2}$$

and we obtain the map

$$x_{n+1} = x_n \cos\omega + \left( p_n - \frac{\partial V}{\partial x}\Big|_{x=x_n} \right) \sin\omega$$

$$p_{n+1} = -x_n \sin\omega + \left( p_n - \frac{\partial V}{\partial x}\Big|_{x=x_n} \right) \cos\omega. \tag{2.3}$$

Setting $V(x) = x^3/3$ one obtains the conservative Hénon map [8]. We rewrite it using the standard notations for autonomous maps:

$$x' = x_{n+1}, \quad x = x_n, \quad p' = p_{n+1}, \quad p = p_n \tag{2.4}$$

and we introduce complex coordinates $z = x - ip$ in order to diagonalize the linear part of the map, which now reads:

$$z' = e^{i\omega}\left( z + \frac{i}{4}(z + z^*)^2 \right). \tag{2.5}$$

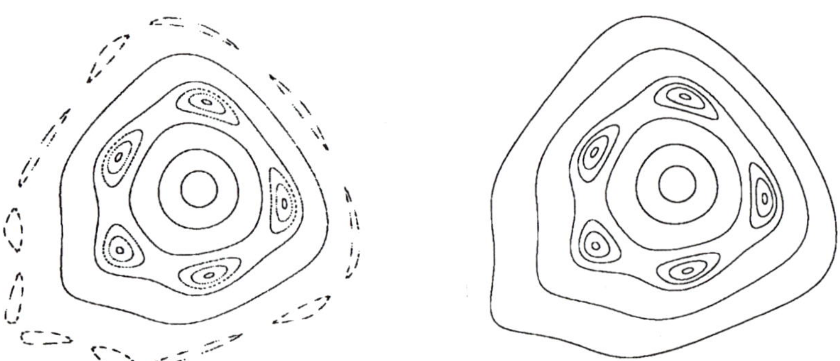

Fig. 1. Phase portrait of the Hénon map (left) and of its interpolating hamiltonian (right) at order 10 close to resonance 5.

This very simple model exhibits all the basic phenomena of nonlinear dynamics: KAM curves, islands, homoclinic intersections, chaotic regions (see Fig. 1, left). The polynomial character of the map implies, in general, the existence of a phase space zone where the higher order terms dominate: once a particle reaches this zone, it will rapidly escape to infinity. The boundary of the basin of attraction of the infinity is the so-called dynamic aperture (this definition is quite loose when the dimensionality of the phase space is greater than two, but a stricter definition goes beyond the aims of this paper).

A generalized polynomial map in two dimensions can be written

$$z' = F(z, z^*) = e^{i\omega}z + \sum_{n \geq 2} F_n(z, z^*) \tag{2.6}$$

where $F_n$ denotes homogeneous polynomials of order $n$ in $(z, z^*)$. As the map is the Poincaré section of an Hamiltonian flow, it has to satisfy the simplecticity conditions

$$[F, F^*] = 1, \tag{2.7}$$

where $[.,.]$ are the Poisson brackets with respect to the variables $(z, z^*)$.

A very good example of how a real system can be represented by a polynomial map is the magnetic lattice of an hadron accelerator [1,2]. In a circular accelerator a single particle has three degree of freedom: $s$, which is the coordinate along the fixed orbit, and $x$ and $y$ which are respectively the radial and the vertical displacement from the closed orbit. Under some assumptions one can decouple the motion in the $s$ direction, and the problem is reduced to the study of the following hamiltonian:

$$H(p_x, x, p_y, y; s) = \frac{p_x^2 + p_y^2}{2} + \frac{\omega_1^2 x^2 + \omega_2^2 y^2}{2} + V(p_x, x, p_y, y; s), \tag{2.8}$$

where $V$ is periodic in $s$, and is a function which depends on the nonlinear magnetic fields of the magnets. The standard approach of perturbative theory consists in expanding $V$ in a Fourier series in $s$. It has been pointed out that $V$ depends on $s$ in a discontinuous way: therefore the first orders of the Fourier series do not give a good approximation; in fact canonical perturbation theory for flows does not seem to provide a good mathematical framework to perform the analysis of this model [1]. Indeed, it is possible to build a transfer map for each element $M_n$ which propagates the position of the particle in the $(x, y)$ phase space from the beginning $s_n$ to the end $s_{n+1}$ of the element. These transfer maps are the natural generalization of transfer matrices [10] to the nonlinear case. Having computed all the maps relative to the elements of the lattice, one has to compose them in order to obtain the one turn map:

$$F = M_N \circ M_{N-1} \circ ... \circ M_2 \circ M_1. \tag{2.9}$$

We point out that in practice one can store only a finite number of polynomial coefficients: therefore $F$ is known only up to the order $N$. For this reason $F$ satisfies the simplecticity condition only up to the order $N - 1$. Here we have a four dimensional phase space, and the map has two components $F^{(1)}$ and $F^{(2)}$. Simplecticity conditions are the following:

$$[F^{(i)}, F^{(j)}] = 0 \qquad [F^{(i)}, F^{*(j)}] = \delta_{ij}. \tag{2.10}$$

## 3. NON RESONANT NORMAL FORMS

According to perturbation theory, one conjugates the model $F$ with a simpler model $U$, called normal form, which is integrable, symplectic, and which exhibits in a straightforward way the relevant nonlinear parameters of the dynamics. The transformation is carried out via a conjugating function $\Phi$ which must be invertible in an open neighbourhood of the origin. Both $\Phi$ and $U$ are written as Taylor series in the coordinates; the perturbation parameter is the distance from the origin. According to the type of normal form, one has different perturbation expansions; for the sake of simplicity, in this paper we will always consider the two dimensional case even if the extension to maps of $\mathbf{R}^{2d}$ is possible [5].

We first deal with the simpler case, i.e. the non resonant normal form, which is an amplitude dependent rotation:

$$\zeta' = U(\zeta, \zeta^*) = \exp\left(i\omega + i\Omega(\zeta\zeta^*)\right)\zeta. \tag{3.1}$$

The conjugation function $\Phi$ reads:

$$z = \Phi(\zeta, \zeta^*) = \zeta + \sum_{n \geq 2} \Phi_n(z, z^*); \qquad (3.2)$$

the unknown functions $U$ and $\Phi$ must satisfy the functional equation

$$F \circ \Phi = \Phi \circ U. \qquad (3.3)$$

The solution is achieved by a perturbative technique; having computed $\Phi$ and $U$ up to order $n - 1$, one projects (3.3) on the subspace of the homogeneous polynomials of order $n$

$$\Phi_n(e^{i\omega}\zeta, e^{-i\omega}\zeta^*) - \Phi_n(\zeta, \zeta^*) + U_n(\zeta, \zeta^*) = R_n(\zeta, \zeta^*), \qquad (3.4)$$

where $R_n$ depends on quantities which have been already computed. We introduce the operator $\Delta$

$$\Delta\Phi = \Phi \circ e^{i\omega} - e^{i\omega} \circ \Phi \qquad (\Rightarrow \Delta U = 0) \qquad (3.5)$$

and the projector $\Pi$ on the normal form subspace

$$\Pi U = U. \qquad (3.6)$$

Using these notations one can rewrite eq. (3.4):

$$\begin{aligned}
\Delta\Pi\Phi_n + U_n &= \Pi R_n \\
\Delta(1 - \Pi)\Phi_n &= (1 - \Pi)R_n.
\end{aligned} \qquad (3.7)$$

One can prove that the choice of $\Pi\Phi$ corresponds to the gauge freedom of the normal form transformation; therefore one can set it to zero, solving for $U_n$. The computation of $\Phi$ involves the inversion of the operator $\Delta$, which gives rise to the problem of the small divisors; let us consider the effect of $\Delta$ on a single monomial of $\Phi$:

$$\Phi_n = \sum_{k=0}^{n} \Phi_{nk}\zeta^k \zeta^{*n-k}. \qquad (3.8)$$

From the second equation of (3.7) one has:

$$\Delta\Phi_n = \sum_{k=0}^{n} (e^{i\omega(2k-n)} - e^{i\omega})\Phi_{nk}\zeta^k\zeta^{*n-k}. \qquad (3.9)$$

Therefore $\Delta$ can be inverted on the subspace generated by $1 - \Pi$ only if $\omega/2\pi$ does not satisfy any resonant condition up to order $n + 1$, i.e.

$$\frac{\omega}{2\pi} \neq \frac{p}{q} \qquad \forall p, q \in \mathbf{Z}, |q| \leq n + 1. \qquad (3.10)$$

Then, one has that if the linear frequency of $F$ does not satisfy any resonant condition up to order $N + 1$, one can formally conjugate $F$ with a non resonant normal form $U$ up to order $N$. In the case $\omega/2\pi$ irrational, this result states the existence of the formal series, i.e. the possibility of solving the functional equation (3.3) order by order without considering the convergence of the series. Indeed, the presence of the small divisor give rise to a factorial growth in $\Phi$ and $U$ which can be estimated using the method of majorant series. Simple topological considerations show that the series cannot converge: infact in the generic model $F$ the frequency of revolution around the origin varies with the amplitude, and according to the Poincaré Birkhoff theorem for every resonant value of the frequency one has a chain of islands [11]. On the other hand, the normal form exhibits only invariant curves which cannot be conjugated with islands. If the normal forms were to converge, there would exist an open neighbourhood of the origin without islands; but the islands of $F$ are dense on the origin, therefore $U$ cannot converge. Nevertheless the series truncated at a given order can provide a very good approximation of $F$, especially if $\omega$ is far from low order resonances.


## 4. APPLICATIONS TO ACCELERATOR PHYSICS

The application of perturbation theory for maps to accelerator physics allows the computation of the main non linear parameters of a magnetic lattice and the optimization of these quantities in order to recover a dynamic aperture as large as possible [1,7].

The standard numerical approach used in accelerator physics is called tracking [12]; it is based on the computation of the trajectory of a charged particle by applying all the maps relative to the single elements of the lattice. In this way one calculates the iterates of the exact one turn map for a given initial condition $z_0$:

$$z^{(l)} = F^{ol}(z_0, z_0^*). \tag{4.1}$$

The nonlinear parameters which are considered as good indicators of the nonlinearity are the tuneshift and the smear; the first one measures the shift in the frequency in its dependence on the amplitude:

$$\delta\nu(z_0) = \lim_{n\to+\infty} \frac{1}{n} \sum_{i=1}^{n} \alpha_i - \frac{\omega}{2\pi}, \tag{4.2}$$

where $\alpha_i$ is the angle in the complex $z$ plane between two successive iterates $z^{(i)}, z^{(i+1)}$. In principle, a high tuneshift endangers the stability of the motion as it can drive the frequency on resonant values. Indeed, one also has to consider the geometry of the orbits, which can be highly deformed even if the tuneshift is low. For this reason one defines the smear, which is the relative standard deviation of the amplitude of a particle along its orbit:

$$\sigma = \sqrt{\frac{<|z|^4> - <|z|>^4}{<|z|>^4}}, \tag{4.3}$$

where $< . >$ defines the time average:

$$< f(z) > = \lim_{n\to+\infty} \frac{1}{n} \sum_{i=1}^{n} f(z^{(i)}). \tag{4.4}$$

As the number of elements in a big accelerator is very high (of the order of $10^3$) and as one needs an high number of turns in order to have a good estimate of $\sigma$ and $\delta\nu$, the tracking approach is very expensive and slow, and it allows to scan only a few initial conditions in the phase space. The big advantage of normal forms is that they give an analytic expression of the nonlinear parameters which allows to make a complete analysis of the dependence on the initial conditions. In fact, from the expression (3.1) of the normal form it is easy to extract the tuneshift which reads as a power series in the nonlinear invariant $\rho = \zeta\zeta^*$:

$$\delta\nu = \Omega(\rho) = \sum_{i\geq 1} \Omega_i \rho^i. \tag{4.5}$$

The dependence of the tuneshift on the physical coordinates is obtained by the inverse conjugation function:

$$\rho = |\Phi^{-1}(z, z^*)|^2. \tag{4.6}$$

The smear is estimated by replacing the time average with a space average along the orbit, which according to the normal form theory can be parametrized as follows:

$$z(\theta) = \Phi(\rho e^{i\theta}, \rho e^{-i\theta}) \qquad \theta \in [0, 2\pi]. \tag{4.7}$$

The average of a dynamical variable is given by:

$$< f(z) > = \frac{1}{2\pi} \int_0^{2\pi} f(z(\theta)) d\theta; \tag{4.8}$$

this expression allows to compute the smear (4.3) in a perturbative way without the need of the knowledge of the iterates of the initial condition. These computations are clearly very quick if compared with the tracking methods. Estimates of the tuneshift and of the smear has been carried out for models of the Super Proton Synchrotron [13] and for the Large Hadron Collider [1,7]. The agreement with both the tracking simulations and, in the case of the SPS, experimental data is excellent.

Another powerful feature of the normal form method is that it allows an analytic optimization of the nonlinear parameters. An application has been carried out for the LHC case [7]; in this machine one has to insert some corrector magnets whose nonlinear fields should be set to the values which maximize the domain of linear motion. The problem of finding these values can be solved by the tracking method using a try and guess technique, which is quite slow and cannot guarantee that the achieved solution is the best one. On the contrary, the normal form approach allows to compute analytically the dependence of the coefficients of the nonlinear parameters on the magnetic fields of the corrector elements; therefore it is possible to compute the best values which optimize the lattice structure. Different layouts were considered and an exhaustive cross-check with the tracking method was carried out.


## 5. RESONANT NORMAL FORMS

In Section 2 we dealt with the concept of conjugation and we introduced the non resonant normal form as the simplest integrable map which can be formally conjugated to the model $F$. The limits of this normal form are evident: on the one hand, if $\omega/2\pi = p/q$, we have to stop the computation of the series at order $q - 2$, which can provide a bad approximation if $q$ is close to 1; on the other hand, the geometry of islands can be reproduced only by an averaging close trajectory: when the islands are thick this approximation is very bad.

We have seen that a non resonant normal form can be defined according to its group of symmetry, which is the group of continuous rotations:

$$\Delta U = U \circ e^{i\omega} - e^{i\omega} \circ U = 0 \qquad \forall \omega \in \mathbf{R}. \tag{5.1}$$

If we impose the commutation only with rotations of a discrete angle $\alpha/2\pi = p/q$, we have a resonant normal form which has the following polynomial structure [9]:

$$U(\zeta, \zeta^*) = e^{i\omega}\zeta \left(1 + \sum_{l,q}(\zeta\zeta^*)^s(u_{ls}^+\zeta^{lq} + u_{ls}^-\zeta^{*lq})\right). \tag{5.2}$$

If we consider the Fourier expansion obtained by setting

$$\zeta = \rho e^{i\theta} \qquad U(\rho, \theta) = \sum_{k \in \mathbf{Z}} e^{ik\theta} A_k(\rho), \tag{5.3}$$

then the nonresonant normal form has only the component $k = 1$, whilst the resonant one has all the components $k = lq + 1$, $l \in \mathbf{Z}$. The normal form allows to reproduce all the islands of period $q$ and its harmonics, giving a more faithful approximation of the model $F$. Clearly, the geometric reasons which cause the divergence of the series are not removed, as the normal form cannot reproduce islands of period different from $lq$, which indeed are dense around the origin. Nevertheless, the approximation given by the truncated normal forms is in most cases excellent and valid up to the dynamic aperture (see Fig. 1, right).

The construction of the formal series follows the same procedure of Section 2; one considers a linear frequency $\omega/2\pi = p/q + \epsilon$ ($\epsilon$ can be zero) and defines the operators

$$\Delta_\alpha \Phi = \Phi \circ e^{i\alpha} - e^{i\alpha} \circ \Phi \qquad (\Rightarrow \Delta_\alpha U = 0) \tag{5.4}$$

and the projector on the normal form space

$$\Pi_\alpha U = U. \tag{5.5}$$

Therefore one has

$$\Delta_\omega \Pi_\alpha \Phi_n + U_n = \Pi_\alpha R_n$$
$$\Delta_\omega (1 - \Pi_\alpha)\Phi_n = (1 - \Pi_\alpha)R_n, \tag{5.6}$$

where $\Delta_\omega$ now can be inverted on the subspace $(1 - \Pi_\alpha)$ at every order $n$ if $\epsilon = 0$ or $\epsilon$ irrational. The only difference with respect to the nonresonant case is that if we want to have a symplectic normal form the transformation $\Phi$ has to be symplectic ; therefore one has to add the condition

$$[\Phi, \Phi^*] = 1 \tag{5.7}$$

which, once projected at order $n$, gives

$$\left( \frac{\partial}{\partial \zeta}[\Pi_\alpha \Phi]_n + \frac{\partial}{\partial \zeta^*}[\Pi_\alpha \Phi^*]_n \right) = S_n, \tag{5.8}$$

where $S_n$ is a remainder which depends on the previous orders. This equation determines only a part of the coefficients of $\Pi_\alpha \Phi$; the remaining ones are free and correspond to the gauge invariance of the resonant normal form.

Nevertheless, we observe that $U$, truncated at order $N$, is symplectic only up to the same order; therefore it is necessary to build an interpolating hamiltonian to recover exact symplecticity. Such a hamiltonian has the level lines which correspond to the orbits of the map within an error which is of order $\zeta^N$. In order to compute the interpolating hamiltonian different methods has been proposed: we will not discuss them here, referring to [9]. The polynomial structure of the hamiltonian, which will be considered in Section 7, is:

$$H(\zeta, \zeta^*) = \epsilon \zeta \zeta^* + \sum_{l,s} (\zeta \zeta^*)^s (h_{ls}^+ \zeta^{lq} + h_{ls}^- \zeta^{*lq}). \tag{5.9}$$

## 6. DIVERGENCE OF THE RESONANT NORMAL FORM

In this paragraph we will review the results on the divergence of the perturbative series which arise from the resonant normal form [9]. Using the method of the Cauchy majorant series it is possible to give an upper bound to the growth of the coefficients of $\Phi$ and $U$. This upper bound clearly does not prove the divergence; nevertheless, numerical checks have shown that in this case the majorant series are not only an upper bound: actually, they give the right behaviour of the normal form series.

Having two formal series $\Phi$ and $\Psi$, we say that $\Psi$ majorates $\Phi$

$$\Phi \prec \Psi \tag{6.1}$$

if all the coefficients of $\Psi$ are greater than the modulus of the corresponding coefficients of $\Phi$. For a generic area preserving map we can always find a constant $c$ such that

$$F(\zeta, \zeta^*) \prec \frac{(\zeta + \zeta^*)^2}{1 - c(\zeta + \zeta^*)} \equiv m(\zeta + \zeta^*). \tag{6.2}$$

The basic idea is to build a formal series $h$ which satisfies a functional equation which keeps the features of the mechanism of divergence of $\Phi$ and $U$, and majorates them. We define $h$ according to

$$\|\Delta_\omega\| h(\xi) = m(\xi) + 2h(\xi)m'(\xi) + \kappa h(\xi)h'(\xi) \tag{6.3}$$

where the prime denotes the first derivative, $\kappa$ is a constant, $\|\Delta_\omega\|$ is the contribution of the divisors:

$$\|\Delta_\omega\| h(\xi) = \sum \eta_n h_n \xi^n \qquad \xi = \zeta + \zeta^*, \tag{6.4}$$

where $\eta_n$ is the smallest divisor of order $n$:

$$\eta_n = \min_j |e^{i\omega(j-1)} - 1|; \tag{6.5}$$

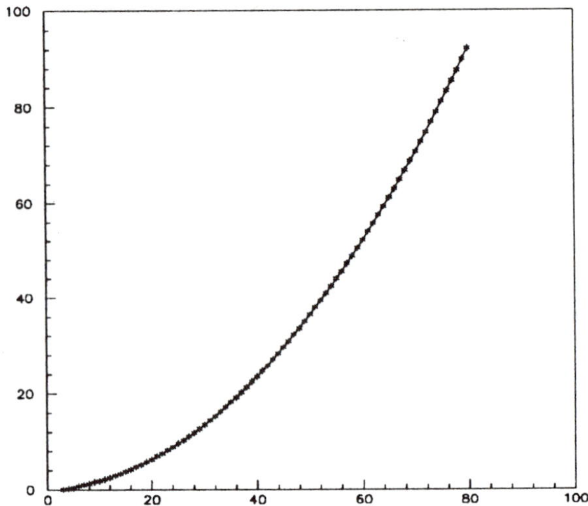

Fig. 2. Square of the apparent radius of convergence of the resonant normal form transformation for the Hénon map, resonance 3.

the minimum has been taken over all $j$'s smaller than $n$, excluding that ones which satisfy the resonant condition

$$e^{i(j-1)\alpha} = 1. \tag{6.6}$$

It is possible to prove that both $\Phi$ and $U$ are majorated by $h$. The functional equation (6.3) can be analysed, and leads to the following estimate:

$$h_n \leq \epsilon_n (n-1)! a^{n-2} \qquad n \geq 3 \quad h_2 = \frac{1}{\eta_2}, \tag{6.7}$$

where

$$\epsilon_n = \frac{1}{\eta_n} \max_{2 \leq k \leq n-1} \epsilon_k \epsilon_{n-k+1} \tag{6.8}$$

are the contributions of the small divisors which behave in the following way:

$$\begin{aligned} \epsilon_n &\approx (n!)^\alpha & \frac{\omega}{2\pi} \quad \text{diophantine} \\ \epsilon_n &\approx b^n & \frac{\omega}{2\pi} = \frac{p}{q}. \end{aligned} \tag{6.9}$$

Therefore, one reaches the result that also in the exactly resonant case, where all the small divisors are bounded from below, the series have a factorial divergence.

The numerical check was carried out by computing $\Phi$ up to order 80; we report in Fig. 2 the square of the apparent radius of convergence:

$$r_n^2 = \frac{\|\Phi_n\|}{\|\Phi_{n+2}\|} \tag{6.10}$$

as a function of $n$ for $\omega/2\pi = 1/3$. One observes a quadratic behaviour, in agreement with the analytical computation.

## 7. ESTIMATE OF THE NONLINEAR PARAMETERS OF RESONANCES

The analysis of the interpolating hamiltonian allows the computation of the parameters of a single resonance like the width of the islands, the distance to the fixed points, and the secondary frequencies. We rewrite the hamiltonian (5.9) in action-angle variables $\zeta = \sqrt{2r}e^{i\theta}$, keeping the lowest order resonant term:

$$H(r, \theta) = h(r) + r^{q/2}[B + A\cos(q\theta + \alpha_q)], \tag{7.1}$$

Table 1 - Scaling exponent of the area of the islands

| Resonance | $\log \Sigma / \log \epsilon$ | |
|---|---|---|
| | Numerical values | Theoretical values |
| 5 | $1.25 \pm 0.05$ | 1.25 |
| 7 | $1.72 \pm 0.08$ | 1.75 |

where

$$h(r) = \epsilon r + \Omega_2 r^2 + \ldots \tag{7.2}$$

and $B = 0$ if $q$ odd. We discuss the case $\Omega_2 \neq 0$ and $q \geq 5$: if $\epsilon\Omega_2 < 0$ there are $q$ elliptic and $q$ hyperbolic fixed points. If $\Omega_2 < 0$ and $A > 0$ then the elliptic fixed points are at $r = r_+$, $\theta = \theta_{2k}$, the hyperbolic fixed points at $r = r_-$, $\theta = \theta_{2k-1}$ for $k = 1, 2, \ldots, q$ where

$$\begin{cases} \theta_k = k\pi/q - \dfrac{\alpha_q}{q} \\ h'(r_\pm) \pm \dfrac{Aq}{2}r^{q/2-1} = 0. \end{cases} \tag{7.3}$$

Letting $r_0$ be defined by

$$h'(r_0) = 0, \qquad r_0 = -\frac{\epsilon}{2\Omega_2} + O(\epsilon^2), \tag{7.4}$$

we have

$$r_\pm = r_0 + O(\epsilon^{q/2-1}). \tag{7.5}$$

Therefore, one has the following scaling law for the distance to the fixed points:

$$|z| \propto \sqrt{r_0} \propto \sqrt{\epsilon}. \tag{7.6}$$

In order to compute the island parameters, we expand the Hamiltonian around $r_0$:

$$H(r, \theta) = h(r_0) + h''(r_0)\frac{(r - r_0)^2}{2} + Ar_0^{q/2}\cos q\theta. \tag{7.7}$$

This pendulum hamiltonian allows to estimate the area $\Sigma$ of the island, which is four times its width $\Delta$:

$$\Sigma = 4\Delta = 16\sqrt{\frac{A}{h''(r_0)}}r_0^{q/4}, \tag{7.8}$$

as well as the secondary frequency around the stable fixed points

$$\omega_{sec} = q\sqrt{Ar_0^{q/2}h''(r_0)}, \tag{7.9}$$

The scaling laws for these quantites are:

$$\begin{aligned} \Sigma &\propto \epsilon^{q/4} \\ \omega_{sec} &\propto q\epsilon^{q/4}. \end{aligned} \tag{7.10}$$

These laws were numerically checked; in Table 1 we give the scaling exponent for the area of the islands. The agreement with the perturbative calculation is excellent.

REFERENCES

1. A. Bazzani, G. Turchetti, P. Mazzanti and G. Servizi, Normal forms for hamiltonian maps and nonlinear effects in particle accelerators, Il Nuovo Cimento B102:51 (1988).
   A. Bazzani, G. Servizi, E. Todesco, G. Turchetti, Normal forms for symplectic maps and nonlinear theory of betatronic motion, CERN yellow report, in preparation (1991).
2. T. C. Bountis, C. R. Eminhizer and N. Budinsky, Resonance criteria for minimizing blow-up effects in colliding proton beams, Nucl. Instr. and Meth., 227:205 (1984).
   A. J. Dragt et al., Lie algebraic methods, Nucl. Instr. and Meth., A258:339 (1987).
3. G. D. Birkhoff, Acta Math. 43:1 (1920).
   G. D. Birkhoff, in: "Dynamical Systems," Ed. Am. Math. Soc. (1927).
4. C. L. Siegel and J. Moser, in: "Lectures in celestial mechanics," Springer Verlag (1971).
   F.G. Gustavson, On constructing formal integrals of a hamiltonian system near an equilibrium point, Astr. Journ. 71:670 (1966).
5. A. Bazzani, Cel. Mech. 42:107 (1987).
6. A. Bazzani, S. Marmi and G. Turchetti, Nekhoroshev estimates for isochronous nonresonant symplectic maps, Cel. Mech. 47:333 (1990).
   G. Turchetti, Nekhoroshev stability estimates for symplectic maps and physical applications, in "Number Theory in Physics," Ed. J. M. Luck and P. Moussa, Springer Verlag Berlin-Heidelberg (1990).
7. W. Scandale, F. Schmidt and E. Todesco, Multipolar correctors for LHC: an analytical approach using normal forms, Part. Acc. 35:53 (1991).
8. M. Hénon, Quarterly of Appl. Math. 27:291 (1969).
9. A. Bazzani, M. Giovannozzi, G. Servizi, E. Todesco, G. Turchetti, Resonant normal forms, interpolating hamiltonians and stability analysis of area preserving maps, preprint of the Department of Physics, University of Bologna, submitted to Physica D (1991).
10. E. Courant and H. Snyder, Annals of Physics 3:1 (1958).
11. J. Moser, in: "Stable and random motions in dynamical systems," Princeton Univ. Press (1973).
12. F. Schmidt, SIXTRACK: Single particle tracking code treating transverse motion with syncrotron oscillations in a symplectic manner, CERN SL/AP 90-11 (1990).
13. A. Bazzani and G. Turchetti, Normal form results for the 1988 dynamical aperture experiment, CERN-SPS/AMS 89-25 (1989).

# ANALYSIS OF SINGULARITIES OF THE STANDARD MAP

# CONJUGATION FUNCTION

Monica Malavasi

Department of Physics, University of Bologna
I.N.F.N. Sezione di Bologna
Via Irnerio 46, 40126 Bologna Italy

## INTRODUCTION

There are many different approaches to the determination of the breakdown of the invariant curves and the resulting stochastic transition for the area–preserving maps but essentially we can divide them in two classes. Given an invariant curve it is possible to detect its breakdown either analyzing the behaviour of the nearby resonance orbits, using the overlap criteria [1] or the Greene's method [2], or studying directly the properties of this curve [3,4]. The Kam theorem [5,6] tells us that the diophantine rotation number orbits of a slightly perturbed Hamiltonian system are analytically conjugated to the curves with the same winding number of the unperturbed system, until a certain value of the perturbation parameter is reached. The analytical function $\Phi$ that transforms the perturbed tori to the unperturbed ones is called conjugation function; it depends on the winding number and it is analytical in $\varepsilon$, strength of the perturbation, until $\varepsilon_c$, critical value. So a method to find the critical value at which the invariant curve with rotation number $\omega$ breakdown is to study the loss of analyticity of $\Phi$.

We perturbatively calculate the conjugation function for the golden mean torus of the Standard Map and numerically study the structure of its singularities in the complex $\varepsilon$ plane approximating $\Phi$ with Padé approximants. The $[N, M]$ Padé approximant of an analytic function $f(z)$ is a rational approximation with numerator and denominator of order respectively $N, M$, such that the Taylor expansion of $f$ and of the approximants coincide up to the order $N + M$. Under general considerations for $f$, this approximant exists and it is unique.

The reliability of the poles is checked considering the values of their residues and the stability in the presence of a small random noise that perturbs the coefficients of the series which define $\Phi$. This noise simulates the unavoidable numerical one and it is observed that only the poles with big residues are stable. This check is essential; in fact it was proven [7] that the unit circle is the natural boundary, with probability one, of the noise function. So if we wish to determine if there exists a natural boundary for $\Phi$ we have to distinguish between the "real" poles and those produced by the noise that are positioned on the same circle. The most stable poles are related to the contribution of the small denominators that affected the series.

The analysis of the Fourier spectra of $\Phi$ reveals the link between the resonances and the poles. The first pole on the positive real axis is related to $\varepsilon_c$. By means of the Padé approximants we detect it with an accuracy that is better than using Hadamard criteria but it is less than that given by more accurate methods. A deeper insight on the structure of the singularities of the perturbative series that define $\Phi$ will give us a better understanding of the mechanism of the breakdown and could explain the relation between the different methods used to determine it.

*Chaotic Dynamics: Theory and Practice*
Edited by T. Bountis, Plenum Press, New York, 1992

## PERTURBATIVE METHOD

The perturbative method allows us to investigate the behaviour of a single invariant curve of fixed winding number when $\varepsilon$ is varied.

The Moser's theorem [6] ensures that for every diophantine rotation number $\omega$ and $\varepsilon < \varepsilon_c$ there exists an analytical function

$$\Phi : (r, \theta) \longrightarrow (\omega, t)$$

that conjugates the Standard Map

$$M : \begin{cases} r_{n+1} = & r_n + \varepsilon \sin(\theta) \\ \theta_{n+1} = & \theta_n + r_{n+1} \end{cases} \tag{1}$$

with the rigid rotation

$$T_\omega : \begin{cases} \omega_{n+1} & = \omega \\ t_{n+1} & = t_n + \omega \end{cases} \tag{2}$$

where $M : R \times T \longrightarrow R \times T$ and $T_\omega : R \times T \longrightarrow R \times T$. The function $\Phi$ had to satisfy the functional equation

$$\dot{M} \circ \Phi = \Phi \circ T_\omega \tag{3}$$

If we express $\Phi$ as

$$\begin{cases} r = & \omega + v(t, \omega; \varepsilon) \\ \theta = & t + u(t, \omega; \varepsilon) \end{cases} \tag{4}$$

being $v, u$ periodical of period $2\pi$ in $t$, Eq. (3) reads

$$\begin{cases} u(t + \omega; \varepsilon) - 2u(t; \varepsilon) + u(t - \omega; \varepsilon) & = \varepsilon \sin(t + u(t; \varepsilon)) \\ v(t; \varepsilon) & = u(t; \varepsilon) - u(t - \omega; \varepsilon) \end{cases} \tag{5}$$

The function $u(t; \varepsilon)$ can be expanded as a double series in $\varepsilon$ and $t$

$$u(t; \varepsilon) = \sum_{n \geq 1} u_n(t)\varepsilon^n \qquad u_n(t) = \sum_{k=-n}^{n} u_{n,k} e^{ikt} \tag{6}$$

and substituting in the first equation of (5) we have

$$u_{n,k} = \frac{b_{n-1,k}}{2(\cos(k\omega) - 1)} \tag{7}$$

where $b_{n-1,k}$ depends on the $n-1, \ldots, 1$ orders. A convenient representation of $u(t; \varepsilon)$ is given by its Fourier expansion

$$u(t) = \sum_{k=-\infty}^{+\infty} U_k(\varepsilon)e^{ikt} \qquad U_k(\varepsilon) = \sum_{n=|k|}^{+\infty} u_{n,k}\varepsilon^n \tag{8}$$

We fix $\omega/2\pi = (\sqrt{5} - 1)/2$ and we calculate $u_{n,k}$ up to $n = 120$. $U_k(\varepsilon)$ is defined by a series in $\varepsilon^2$ so we can consider, for different $k$, various Padé approximants with $N + M \leq 60$ and we study the distribution of their poles in the complex plane.

The information on the loss of analyticity can be detected analyzing the decreasing of the Fourier Spectra, $U_k(\varepsilon)$. It decreases exponentially fast for $\varepsilon < \varepsilon_c$ and slower than linearly for bigger $\varepsilon$ [8]. The breakdown threshlod for the golden mean circle, determined with various techniques is $\varepsilon_c \simeq 0.971$. The best agreement with this value is obtained analyzing when the decaying ceases to be linear in $\varepsilon$. The value of the critical parameter obtained looking at the exponentially decrease and the position of the first real pole, for a perturbative expansion up to the order $n = 120$, have an error of 1%.

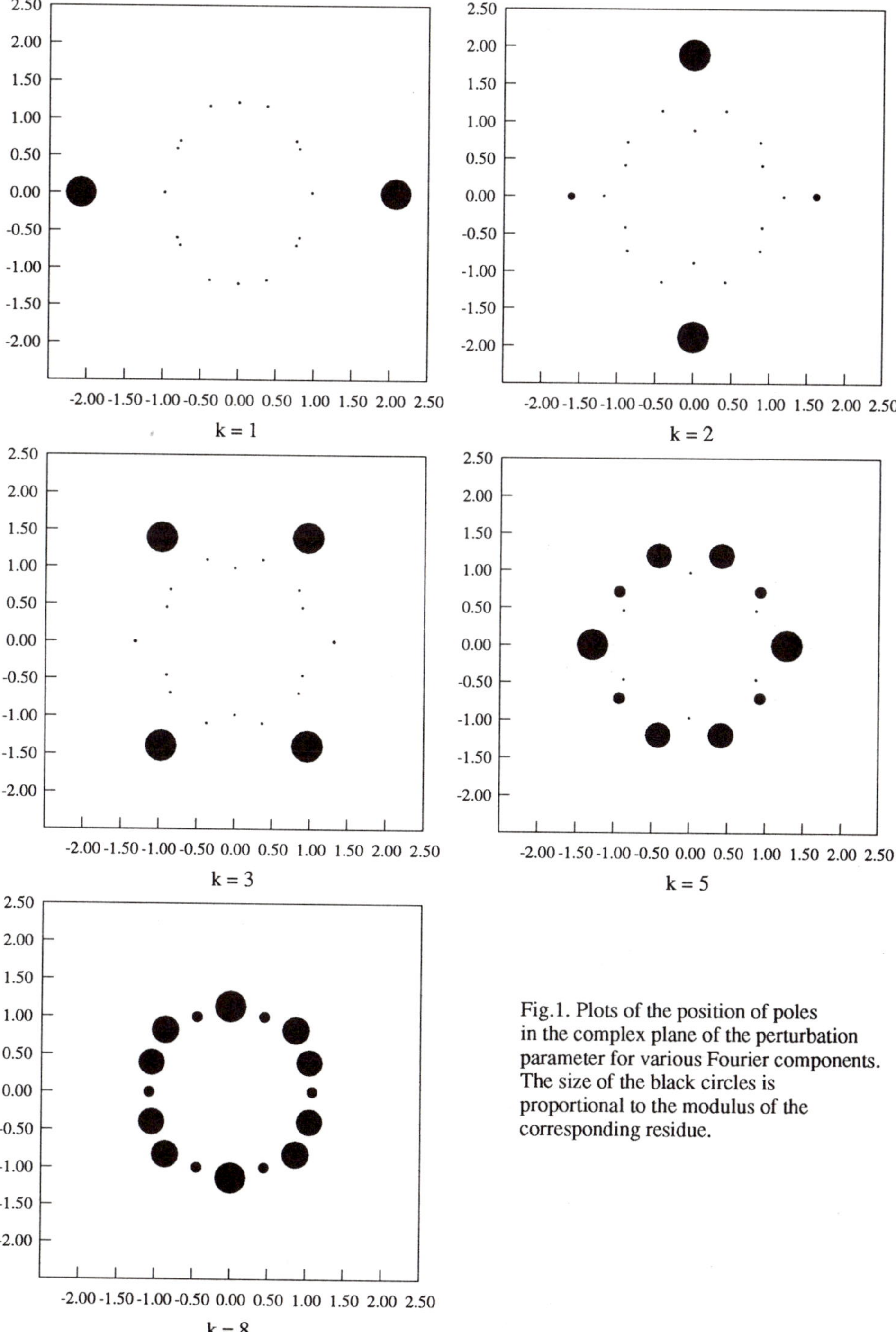



Fig.1. Plots of the position of poles in the complex plane of the perturbation parameter for various Fourier components. The size of the black circles is proportional to the modulus of the corresponding residue.

On the other side the sequence $|u_{n,k}|^{-1/n}$, calculated for a fixed $k$ exhibits a slow convergence from above to a value close to 1. The same analysis can be done for the other invariant circles in order to determine the critical parameter as a function of the winding number [9].

## RESULTS

The use of the Padé approximants in the analysis of the singularities of the conjugation function gives good results also in other problems such as the determination of the analytical domain for the Siegel problem [10]

In our case the Fourier spectra is the best tool to investigate the properties of summability of the perturbative series. A similar analysis done for the Standard Map [11] has shown that the singularities in $\varepsilon$ do not depend on the angle $t$. The clearest information that we can gather observing the structure of the poles is related to the small divisors contribution to $U_k(\varepsilon)$. The convergence mechanism of the series is related to the cancellation of different contribution of the small denominators. The relation between the poles and the resonances was found also in the polynomial maps where it was proven that the leading poles are located exactly at the fixed points of the corresponding leading resonance [12]. The most important contributions in the divisors appearing in the conjugation function are given by the rational approximations of $\omega/2\pi$. If we fix the order of the Padé or, in other words order of the perturbative expansion, we observe (see Fig. 1) that the number of the big poles, for different $k$, is always twice a Fibonacci number whose sequence is related to the continued fraction expansion of the golden mean.

The poles are represented by dots whose size are proportional to the value of the relative residues and the "bigger" poles are related to the dominant resonance. The poles situated inside the circle of radius $\varepsilon_c$ have residues zero. The value of the residues allows to distinguish between the "true" stable poles from the spurious ones produced by the numerical noise. If we fix $k$ and we plot Padé of different orders we can notice the appearance and the successive cancellation of the resonance. For example for $k = 3$, we can see (Fig. 2) that the resonance 2 is active in the Padé $[10, 10]$ and the resonance 3 appear only in the Padé $[25, 25]$. The delay in the rising of a determinate small divisor is in agreement with previously done estimates [4].

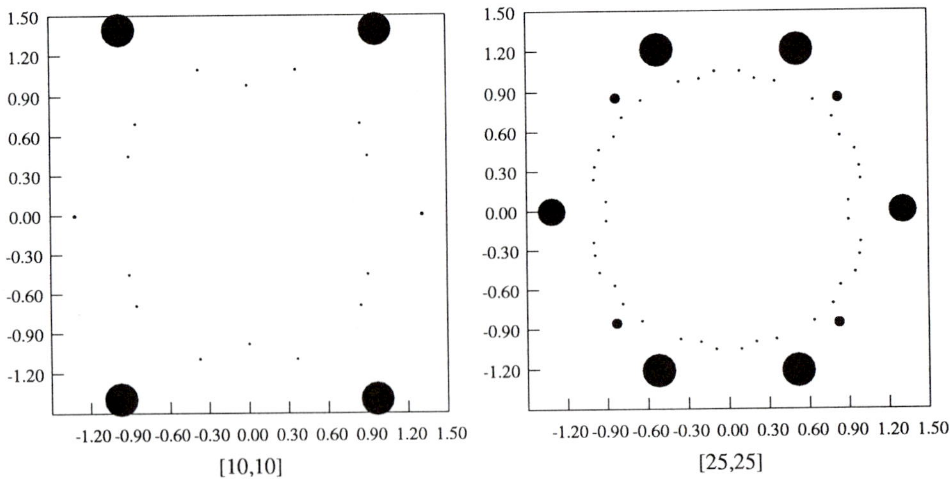

[10,10]  [25,25]

Fig 2. The [10,10] and [25,25] Padé for $k = 3$

We test the stability of the poles perturbing the series coefficients with a random noise

$$\tilde{u}_{n,k} = u_{n,k} + \gamma r_n \qquad 0 \leq r_n \leq 1 \qquad (9)$$

and studying the position of the new poles for different values of the amplitude $\gamma$ of the noise. All the poles are stable for $\gamma \leq 10^{-8}$. For $\gamma > 10^{-8}$ the poles with little residues, that lie nearby the unit circle, are quite unstable ehile the "big" ones, that are always located far away from $|\varepsilon| = 1$, are much more stable, Fig. 3.

Using the Padé approximants is very difficult to determine the existence of a natural boundary in $\varepsilon$ for $u$. In fact the noise function that affects our series has a natural boundary exactly on the circle that we wish to examine, as it was proven in [7]. In order to check the existence of a natural boundary we had to study the position of the poles for a higher order of perturbation calculating the coefficients and the poles with extended precision. In this way it will be possible to obtain a better estimate for the critical parameter. The numerical precision is in fact essential to produce reliable results at higher orders and to give a better estimate of the positions of the poles as it was observed in [10].

We will develop this analysis studying by means of the Padé approximants also the analytical boundary in $t$ and considerering other area–preserving maps.

**Acknowledgement**   The author wish to thank Prof. G. Turchetti for suggesting this investigation, Dr. A. Bazzani and G. Servizi for very useful discussions and Dr. R. Xie for the poles finder routine.

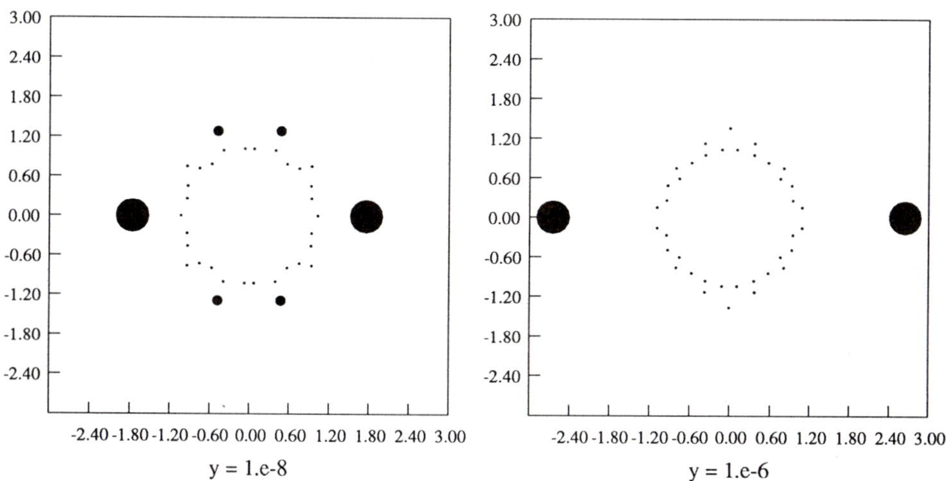



Fig.3 The Padè [20,20] for $k = 1$ perturbed by noise of amplitude $10^{-8}$ (left) and $10^{-6}$ (right)

**References**


1.  B. V. Chirikov, A universal instability of many–dimensional oscillator system, Phys Repts 52, 263:379 (1979).
2.  J. M. Greene, A method for determining a stochastic transition, J Math Phys 20, 1183:1201, (1979).
3.  S. J. Shenker, L. P. Kadanoff, Critical behaviour of a KAM surface Jou Stat Phys 27, 631:656, (1982).

4.  G. Servizi, G. Turchetti, Perturbative Expansions for area–preserving maps,
     Nuovo Cimento 95 B, 121:154, (1986).
5.  V. I. Arnóld, Proof of A.N. Kolmogorov theorem on the preservation of quasi–periodic
     motions under small perturbations of the Hamiltonian, Russ Math Surv
     18, 9:36, (1963).
6.  J. Moser, The analytic invariants of an area preserving mappings of an anulus,
     Nachr Akad Wiss Gott, 1:20, (1962).
7.  J.Gilewicz, B. Troung–Van, "Froissart doublets in the Padé approximation and noise",in:
     Constructive theory of functions'87, Sofia, 1988.
8.  G. Benettin,G. Turchetti, G. Zanetti, Critical behaviour of invariants curves in the Stan-
     dard Map: a perturbative approach, Phys Lett 105A, 436:438, (1984).
9.  M. Malavasi, in preparation.
10. R. Xie. in this volume.
11. Berretti, L. Chierchia, On the complex analytica structure of the golden invariant curve
     for the standard map, Nonlinearity 3, 39:44, (1990).
12. A.Bazzani G.Turchetti Singularities of normal forms and topology of orbits in area–pre-
     serving maps, to be pubblished,(1991)

# PADE APPROXIMANTS APPLIED TO THE ANALYSIS OF THE
# SINGULARITY STRUCTURE FOR CONJUGACY PROBLEM

Xie Ruifeng †

Department of Physics, University of Bologna
Via Irnerio, 46, Bologna

## INTRODUCTION

Padé approximants have been used to study the convergence problem in many areas. Though very little has been proved about their convergence and Gammel's counterexample shows that the poles of Padé approximants of an entire function may be distributed densely in the whole complex plane, the effect of Padé approximants for many practical problems is better than what one would expect from a theoretical viewpoint. Here we study numerically the singularity structure of the transform function of the conjugacy problems. In our program we have used the Multiple–Precision–Floating–Point Arithmetic which makes it possible to use a large number of digits. In fact we typically use the number of 160 digits in our program. This assures us that our numerical results are of high accuracy.

## SOME EXAMPLES

We have studied the following conjugacy problem

$$
\begin{array}{ccc}
z & \longrightarrow & z' = \lambda z + z^2 \\
\Phi \uparrow & & \uparrow \Psi \\
\xi & \longrightarrow & \xi' = \lambda \xi
\end{array}
$$

where $\lambda$ is complex number. Let

$$\Phi(z) = z + \sum_{n=2}^{\infty} a_n z^n$$

and

$$\Psi(z) = z + \sum_{n=2}^{\infty} b_n z^n$$

be the power series expansion of $\Phi$ and $\Psi$ respectively which can be determined by the conjugacy equations

$$\Phi(\lambda z + z^2) = \lambda \Phi(z)$$

---

† Supported by ICSC World Laboratory

*Chaotic Dynamics: Theory and Practice*
Edited by T. Bountis, Plenum Press, New York, 1992

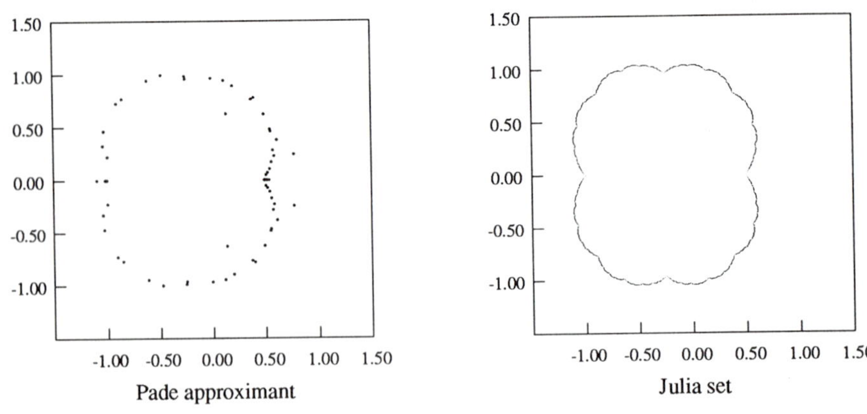

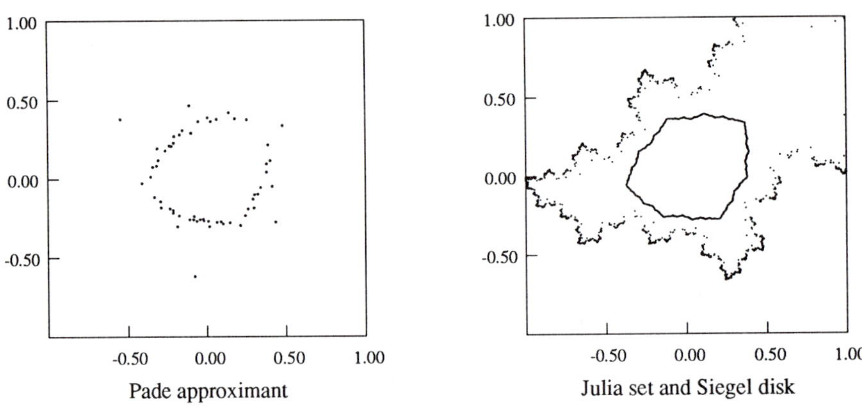

Fig.1 Padé approximant and Julia set, $\lambda = \frac{1}{2}$

Fig.2 Padé approximant and Julia set: $\lambda = e^{2\pi i \omega}$, $\omega = \frac{\sqrt{5}-1}{2}$

and

$$\Psi(\lambda\xi) = \lambda\Psi(\xi) + (\Phi(\xi))^2$$

The recursion formulas for the computation of $a_n$'s and $b_n$'s are

$$a_n = \frac{1}{\lambda - \lambda^n} \sum_{1 \le i \le \frac{n}{2}} \binom{n-i}{i} \lambda^{n-2i} a_{n-i}$$

and

$$b_n = \frac{1}{\lambda - \lambda^n} \sum_{k=1}^{n-1} b_k b_{n-k}.$$

One can estimate the growth of these coefficients by means of the major series method. But with this estimate we can get only the convergence disk of $\Phi$ and $\Psi$. It means that we know only that $\Phi$ and $\Psi$ are analytic in a disk. However, more information can be obtained by means of rational (Padé) approximants. In fact we may reconstruct the singularity structure of $\Phi$ and $\Psi$ with the poles of their Padé approximants.

On the left side of Fig.1, we give the picture of the poles of [60/60] Padé approximant for $\Phi$ with $\lambda = \frac{1}{2}$. One can compare it with the Julia set of the function $f(z) = \frac{z}{2} + z^2$ (see right side of Fig.1) which is claimed to be the natural boundary of the analyticity domain of $\Phi$.

For the case $|\lambda| = 1$, Padé approximants are also applicable. One can see from Fig. 2 where $\lambda = e^{2\pi i\omega}$ with $\omega = \frac{\sqrt{5}-1}{2}$ that the poles of $[60/60]$ Padé approximant of $\Phi$ approximate well the Siegel disk.

On the other hand, when we consider the Padé approximants of function $\Psi$, we find something new. Theoretically, we don't know how to find the singular points of $\Psi$. But the poles of Padé approximants show that all the singular points are located on the real axis, when $\lambda$ is real. We can numerically explain this with the numerical results of Table 1 which show that all the residues of the poles are positive (for $\lambda = 0.5$).

Table 1. [20/20] Padé approximants for real $\lambda$

| $\lambda = 0.5$ | | | | $\lambda = 0.1$ | | | |
| --- | --- | --- | --- | --- | --- | --- | --- |
| poles | | residua | | poles | | residua | |
| −1.143054 | 0 | 5.4454755E−02 | 0 | −5.290049 | 0 | 0.7686966 | 0 |
| −1.981334 | 0 | 5.3901281E−02 | 0 | 0.7318435 | 0 | −0.1055558 | 0 |
| −0.7472393 | 0 | 5.3275757E−02 | 0 | 0.2774316 | 0 | −3.0755084E−02 | 0 |
| −4.334184 | 0 | 5.0821919E−02 | 0 | −0.5620626 | 0 | 2.6421789E−02 | 0 |
| −0.5323230 | 0 | 4.7760639E−02 | 0 | 0.1615296 | 0 | −1.5311532E−02 | 0 |
| −16.51176 | 0 | 4.4591770E−02 | 0 | 0.1104397 | 0 | −9.1497377E−03 | 0 |
| −0.4014854 | 0 | 4.3577544E−02 | 0 | −0.3385800 | 0 | 6.0710097E−03 | 0 |
| −0.3171929 | 0 | 3.5295591E−02 | 0 | 8.2468212E−02 | 0 | −5.9723016E−03 | 0 |
| −0.2585214 | 0 | 2.8278532E−02 | 0 | 6.5227158E−02 | 0 | −4.0896712E−03 | 0 |
| −0.2169879 | 0 | 2.2907060E−02 | 0 | 5.3779613E−02 | 0 | −2.8736116E−03 | 0 |
| −0.1871293 | 0 | 1.6239740E−02 | 0 | 4.5789320E−02 | 0 | −2.0415476E−03 | 0 |
| −0.1637254 | 0 | 1.0721004E−02 | 0 | 4.0017586E−02 | 0 | −1.4493271E−03 | 0 |
| −0.1459885 | 0 | 7.4230228E−03 | 0 | −0.2724425 | 0 | 1.1449526E−03 | 0 |
| −0.1324075 | 0 | 5.1438850E−03 | 0 | 3.5752557E−02 | 0 | −1.0165605E−03 | 0 |
| −0.1219675 | 0 | 3.4959144E−03 | 0 | 3.2560315E−02 | 0 | −6.9539621E−04 | 0 |
| −0.1139730 | 0 | 2.2821943E−03 | 0 | 3.0164730E−02 | 0 | −4.5593191E−04 | 0 |
| −0.1079426 | 0 | 1.3937533E−03 | 0 | 2.8384887E−02 | 0 | −2.7892116E−04 | 0 |
| −0.1035419 | 0 | 7.5534038E−04 | 0 | 2.7100543E−02 | 0 | −1.5162259E−04 | 0 |
| −0.1005433 | 0 | 3.2876691E−04 | 0 | 2.6232269E−02 | 0 | −6.5811997E−05 | 0 |
| −9.8799169E−02 | 0 | 8.0605518E−05 | 0 | 2.5729794E−02 | 0 | −1.6219623E−05 | 0 |

Since $\Psi'(0) = 1$ by assumption, we may write

$$\Psi(z) = \sum_{n=1}^{\infty} \frac{\varepsilon_n z}{1 + \frac{z}{\alpha_n}}$$

where

$$\varepsilon_n \geq 0, \quad \sum_{n=1}^{\infty} \varepsilon_n = 1$$

and

$$\alpha_n \geq \alpha > 0.$$

From this one can easily prove that $\Psi$ is analytic on the whole complex plane excluding the half–line $[-\infty, -\alpha]$ on the negative real axis. Naturally we expect that the singular points of $\Psi$ lie on some ray lines if $\lambda = e^{2\pi i(\omega + ia)}$ where $\omega = \frac{p}{q}$ is rational.

This is verified by Fig.3 where $p = 1$, $q = 5$, $a = -0.1$. When $\omega$ is diophantine number, we get also some similar result: from Fig.3 one can see that there is some similarity between

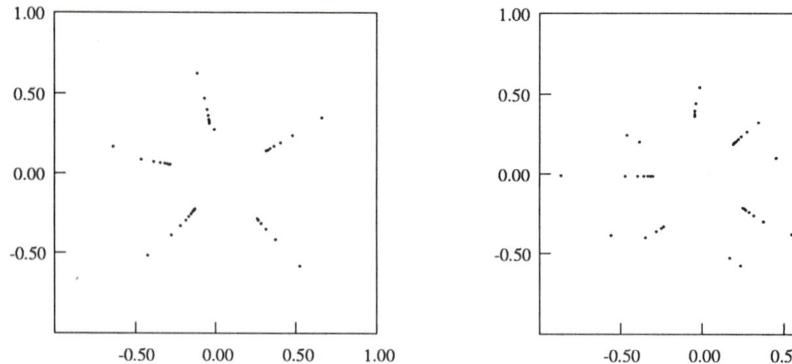

Fig.3 Padé approximant for $\Psi$ with $\lambda = e^{2\pi i\omega + a}$: $\omega = \frac{1}{5}$ on the left; $\omega = \frac{\sqrt{5}-1}{2}$ on the right.

the two pictures. The phenomenon that nonresanonce case is related to some resonance case is very interesting and deserve further study.

For the Siegel problem when $\lambda = e^{2\pi i\omega}$ with $\omega = \frac{\sqrt{5}-1}{2}$, one can determine the Siegel disk with the poles of Padé approximants, which is in good agreement with the theoretical result. For this we refer the reader to Fig.4 where $[60/60]$ approximant is used.

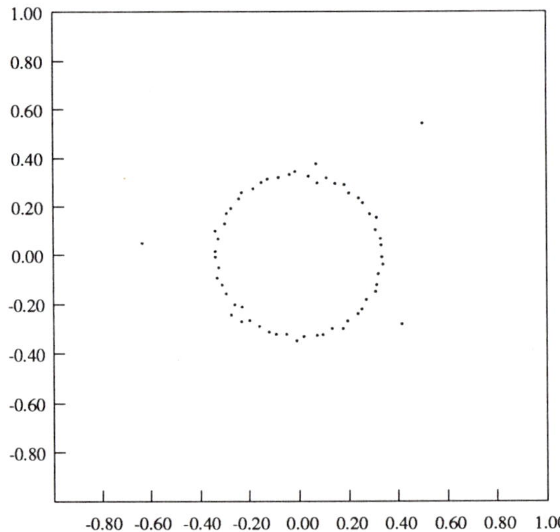

Fig.4 Padé approximant for $\Psi$ with $\lambda = e^{2\pi i\omega}$ and $\omega = \frac{\sqrt{5}-1}{2}$

We remind the reader that the numerical methods using usual double precision numbers for the study of Padé approximants may lead wrong results. For this we have studied the poles of Padé approximant of the function

$$f(z) = \sum_{n=0}^{\infty}(1 + 10^{-15}e^{2\pi i\varepsilon_n})z^n$$

where $\varepsilon_n \in [0, 1]$ are random variables. It is the pertubation of the function

$$g(z) = \frac{1}{1-z} = \sum_{n=0}^{\infty} z^n$$

398

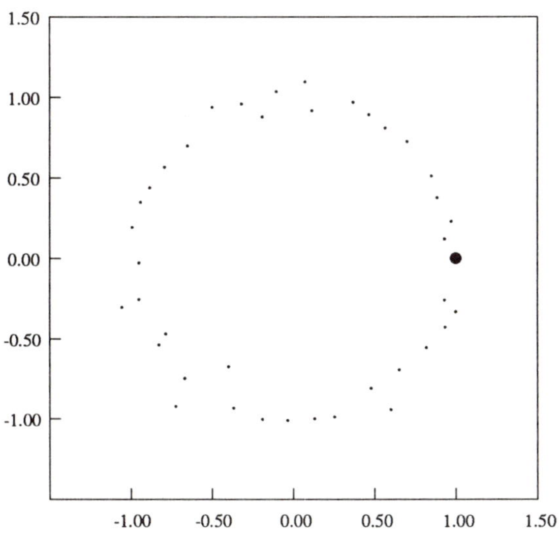

Fig.5. Poles of $[40/40]$ Padé approximant for $f(z)$

which has only one singular point on the whole complex plane. But the Padé approximant of $f(z)$ whose coefficients have a difference of order $10^{-15}$ from $g(z)$ have many poles distributing evenly on the unit circle(see Fig. 5).

Acknowledgements

I wish to thank G. Turchetti for his help in preparing this paper and G. Servizi for his help in the numerical works.

REFERENCES

1. J. L. Gammel, Continuation of functions beyond natural boundaries, Rocky Mountain J. Math., V4, N.2, 203:205 (1974)

2. J. L. Gammel and J. Nuttall, Convergence of Padé approximants to quasianalytic functions beyond natural boundries, J. Math. Ana. & Appl. 43, 694:696(1973)

3. J. Nuttall, The convergence of Padé approximants of meromorphic functions, J. Math. Ana. & Appl. 31, 147–153(1970)

4. Ch. Pommerenke, Padé approximants and convergence in capacity, J. Math. Ana. & Appl. 41, 775–780(1973)

# CONTRIBUTORS

ABARBANEL H.D.I          Institute for Nonlinear Science, R-002, University of California, San Diego, La Jolla, CA 92093 USA

ABLOWITZ M. J.           Department of Applied Mathematics, University of Colorado, Boulder, CO 80309 USA

ADAMOPOULOS A.         Medical Physics Lab., University of Thrace, Department of Medicine, GREECE

AIVALIOTIS A.          Iridos, 20 Halandri 15232, Athens, GREECE

ALVAREZ-PARRILLA A.    Mathematics Dept, University of California, Santa Cruz, CA 95064, USA

ANTONIADIS J.          Dept. of Physics, University of Chicago, IL 60637 USA

ARGOUL F.              CRPP/CNRS, Avenue Schweitzer F-33600 Pessac, FRANCE

ARNEODO A.            CRPP/CNRS, Avenue Schweitzer F-33600 Pessac, FRANCE

BADII R.               Paul Scherrer Institute, LUS CH-5232 Villigen PSI, SWITZERLAND

BAZZANI R.                    Department of Physics, Via Irnerio
                              46, 40100 Bologna, ITALY

BENETATOS G.                  Agias 80, Patras, GREECE

BERRY M.                      H.H. Wills Physics Laboratory,
                              Tyndall Avenue, Bristol BS8 1TL, UK

BEZERIANOS A.                 Dept. of Medical Physics, School of
                              Medicine, University of Patras,
                              26110 Patras, GREECE

BINTZIOU A.                   Solonos 22, Aegaleo, Athens, GREECE

BOASMAN P.                    University of Bristol, H.H. Wills
                              Physics Laboratory, Tyndall Avenue
                              Bristol, BS8 1TL, UK

BOUDOURIDES M.                Section of Physics and Appl. Mathem.,
                              Democritus University of Thrace,
                              67100 Xanthi, GREECE

BOURBOULAS TH.                Nik. Kapatou 39, 54629 Thessaloniki,
                              GREECE

BREEDEN J.                    Center for Complex Systems Research,
                              Un. of Illinois, Beckman Inst. 405
                              N.Mathews Ave, Urbana, IL 61801 USA

CASATI G.                     Dipartimento di Fisica, Universita
                              degli Studi di Milano, Via Celoria,
                              16 -I-20133 Milano, ITALY

CHRISTIANSEN F.               NBI Niels Bohr Institute, Blegdamsvej
                              17, DK 2200 Copenhagen, DENMARK

CHRISTIANSEN P.               Lab. of Appl. Mat. Phys. The Tech.,
                              University of Denmark, Building 303,
                              DK-2800 Lyngby, DENMARK

CHUA L. O.                      Dept. of Electrical Engineering,
                                University of California at Berkeley,
                                EECS Berkeley, CA 94720 USA

CONRADO CL.VIEGAS                The Niels Bohr Institute, (Ph.D
                                student), Blegdamsvej 17, DK2100
                                Copenhagen, DENMARK

CONTOPOULOS G.                  Department of Physics Lab. of
                                Astronomy, University Campus
                                Zografou, 15783 Athens, GREECE

CRAWFORD J. D.                  Department of Physics, University of
                                Pittsburgh, PA 15260 USA

DAHLQVIST P.                    Nordita, Blegdamsvej 17, DK-2100
                                Copenhagen, DENMARK

DEL RIO E.                      University of Madrid, (University
                                Complutense) Tres Peces, 34-Esc.3a,
                                2o B-Madrid 28012, SPAIN

DMITRIEV A.                     Institute of Radio Eng. Electronics,
                                Moscow, USSR

DOTTLING R.                     Institut fur Theoretische Physik, TU
                                Berlin, Hardenbergstr, 36 Sekr. PN 7-
                                1, D-1000, Berlin, GERMANY

ECKHARDT B.                     Universitat Marburg, FB Physik,
                                Renthof 6, 3550 Marburg, GERMANY

ECONOMIDOU O.                   Papageorgiou 5, 54635 Thessaloniki,
                                GREECE

EDGE A.                         School of Chemistry, University of
                                Leeds, Leeds LS2 9JT, UK

EISELE M.                       IFF, KFA D-W-5170 Julich, GERMANY

EVANGELOU S.N.                Physics Department, University of
                             Ioannina, Ioannina 45110, GREECE

FAIRHALL A.                  Dept. of Chemical Physics, Weizmann
                             Inst. of Science, Rehovot 76100,
                             ISRAEL

FARANTOS S.                  Department of Chemistry, University
                             of Crete, Iraklion 71110, GREECE

FEDORENKO V.                 Institute of Math., Ukrainian Academy
                             of Sciences, Repin str. 3, Kiev 252601,
                             USSR

FORDY A. P.                  School of Mathematics, University of
                             Leeds, Leeds LS2 9JT, UK

FROUSAKIS CH.                Department of Chemical Engineering,
                             Princeton University, Princeton, N.J.
                             08544 USA

FRANTZESKAKIS D.             Dept. of Physics Section of Applied
                             Physics, University of Athens, Athens
                             Panepistimiopolis 15783 Zografos,
                             Athens, GREECE

FÜLOP A.                     Department of Solid State Physics,
                             Eötvös University, Budapest, HUNGARY,
                             H-1088 40

GALLAVOTTI G.                Physics Department, University of
                             Roma P. Moro 2, 00185 Roma, ITALY

GEISEL T.                    Inst. fur Theoretische Physik Johann
                             Wolfgang Goethe-Univesitat Robert-
                             Mayer-Str. 8-10, P.O.Box 111932 6000
                             Frankfurt am Main 11, GERMANY

GEROYANNIS V.                Astronomy Lab. Department of Physics,
                             University of Patras, 261 10 Patras,
                             GREECE

GERSHENFELD N.                    Department of Physics Harvard
                                 University, 15 Oxford Street
                                 Cambridge, MA 02138 USA

GERSTENBERG M.                    LAMF, Technical University of
                                 Copenhagen, Building 303, DK-2800
                                 Copenhagen, DENMARK

GHIKAS D.                         Department of Physics, University of
                                 Patras, Patras GREECE

GIANNONI M-J.                     Division de Physique Theorique, Univ.
                                 Paris Sud IPN-Bat. 100-91406 Orsay
                                 Cedex, FRANCE

GLASS L.                          Department of Physiology, McGill
                                 University 3655, Drummond Street,
                                 Montreal, Quebec H3G 1Y6, CANADA

GORIELY A.                        Service de Physique Statistique ULB
                                 CP231, Bvd du Triomphe, 1050
                                 Bruxelles, BELGIQUE

GOZDZIEWSKI K.                    Institute of Astronomy, Nicolaus
                                 Copernicus University, Torun, POLAND

GRAMMENOS TH.                     Athens University, Physics
                                 Department, Nuclear & Particle
                                 Physics, Athens 15771, GREECE

GREBOGI C.                        Lab. for Plasma Physics Research
                                 University of Maryland, College Park,
                                 MD 20742 USA

GROUSOUSAKOU E.                   Spetson 36 Nea Sepolia, 12132 Athens
                                 GREECE

GUTKIN E.                         Mathematics Department of  University
                                 of Southern California, Los Angeles
                                 CA 90007, USA

GYORGYI G. Institute for Theoretical Physics, Eotvos University, Budapest, Puskin U 5-7, H-1088 HUNGARY

HADJISPYROS SP. School of Mathematics Queen Mary and Westfield College, Mile End, Road E1 4NS, London, UK

HANSEN K. Niels Bohr Inst., Blegdamsvej 17 DK-2100, Copenhagen, DENMARK

HATZIMIHAIL A. National Health System P.O. Box 56, 66100 Drama, GREECE

HENRICH E. University of California at los Angeles, 3320 Highland Ave. Hermosa Beach, CA 90254 USA

HIZANIDIS K. Dept. of Electrical and Computer Engineering National Technical University of Athens, GREECE

HOVEIJN I. Mathematisch Instituute Budapestlaan 6 Postbus 80.010 3508 TA, Utrecht, NEDERLAND

HYDE J. Department of Applied Mathematical Studies @ Physiology, The University of Leeds, Leeds LS2 9JT. UK

IACHELLA STEF. University of California, Berkeley 3084 Lester Road, Martinez CA. 94553 USA

IBISON P. University of Leeds, School of Chemistry, Leeds, LS2 9JT UK

ICHTIAROGLOU S. Department of Physics, University of Thes/niki 54006 Thessaloniki, GREECE

IOANNIDOU H.                    Dept. of Mathematics, University of
                                Patras, Patras, GREECE


IVANOV A.                       Math. Inst., Univ. Munchen
                                Theresienstr. 39, D-8000
                                Munchen 2, GERMANY


JOHNSON B.                      University of Leeds, School of
                                Chemistry, LEEDS, LS2 9JT UK


JUNG C.                         Fachbereich Physik, Universitat
                                Bremen d-2800, Bremen, GERMANY


KADANOFF L. P.                  James Franck Institute 5640 S. Ellis
                                Avenue, University of Chicago IL
                                60637 USA


KALAMATIANOS ST.                Nikolaou Plastira 5 Triandria,
                                Thessaloniki, GREECE


KALVOURIDES T.                  NTU of Athens,  Heroes of
                                Polytechniou Ave.,5, Athens, 15373,
                                GREECE


KAPITANIAK T.                   Inst. Appl. Mech, Technical,
                                University of Lodz Stefanowskiego
                                1/15, 90-924 Lodz, POLAND


KARANIKAS A.                    Physics Department, Univerity of
                                Athens, Nuclear and Particle Phys.
                                Sect., Athens 15771, GREECE


KAPOPOULOS A.                   Dania 7, Agios Nikolaos, Filopapou,
                                Athens, GREECE


KARANTONIS A.                   Lab. of Physical Chemistry,
                                Department of Chemistry Aristotle,
                                University of Thessaloniki, GREECE

KATSANOS D.                  Physics Dept, University of Ioannina,
                             45110 Ioannina, GREECE

KATSIKAS T.                  Univ. of Patras, Dept. of
                             Mathematics, Patras 261 10, GREECE

KEVREKIDES I.                Department of Chemical Engineering
                             Princeton University  Princeton, N.J.
                             08544  USA

KING G. P.                   Mathematics Institute, University of
                             Warwick, Coventry CV47AL U.K.

KIOUSTELIDIS J.              Dept. of Mathematics, National
                             Technical University of Athens
                             Philotimou 18, Athens 11363, GREECE

KLUIVING R.                  Institute for Theoretical  Physics
                             Valckenierstr. 65, 1018 XE Amsterdam,
                             THE NETHERLANDS

KOMINEAS S.                  Makedonias 3, 24100 Kalamata, GREECE

KONSTANTINEA S.              Dania 7, Agios Nicolaos, Filopapou
                             Athens, GREECE

KONSTANDINIDIS SAV.          Korinthou 186, Patras, GREECE

KONSTANTOUDIS V.             Theoretical and Physical Chemistry
                             Institute, National Hellenic Research
                             Foundation, 48 Vas Constantinou Ave.
                             11635 Athens, GREECE

KOUGIAS CH.                  Fachbereich Physik, Universitat
                             Oldenburg, Postfach 2503, D-2900
                             Oldenburg, GERMANY

KOURAKIS Y.                  Lab OSC, UFR Sciences et Techniques
                             Univ. de Bourgogne, 6 Blvd Gabriel
                             21100 Dijon, FRANCE

KOVACS Z.                    Institute for Theoretical Physics,
                             Eotvos University 1088 Budapest
                             Puskin UTCA 5-7, HUNGARY

KRAUSE J.                    University of California, Santa Cruz
                             Department of Mathematics, UCSC,
                             Santa Cruz, CA 95064 USA

KRUSKAL M.                   Department of Mathematics, Rutgers
                             University, New Brunswick, NJ 08903
                             USA

KUPKA I.                     Department of Mathematics, University
                             of Toronto, Toronto M5S1A1, CANADA

LAKSHMANAN M.                Department of Physics, Bharathidasan
                             University, Tiruchirapalli, 620024
                             Tamilnadu, INDIA

LAMB J.                      Institute for Theoretical Physics,
                             University of Amsterdam,
                             Valckenierstraat 65, 1018 XE
                             Amsterdam, NETHERLANDS

LAMBROPOULOU S.              Mathematics Institute, University of
                             Warwick, Coventry CV4 7AL U.K.

LIEBERMAN M.                 Department EECS, University of
                             California at Berkeley, Berkeley CA
                             94720 USA

LINSAY P.                    MIT Plasma Fusion Center NW17 241,
                             175 Albany St, Cambridge MA 02139,
                             USA

LIU J.                       School of Chemistry, University of
                             Leeds, LS2 9JT, UK

MACIEJEWSKI A.               Institute of Astronomy, Nicolaus
                             Copernicus University, Torun, POLAND

MACKERNAN D.                Universite Libre de Bruxelles Faculte
                            des Sciences, Campus Plaine CP321
                            1050 Bruxelles, BELGIUM

MALAVASI M.                 Dipartimento di Fisica Universita di
                            Bolognia, Via Irnerio 46, 40126
                            Bolognia, ITALIA

MANTICA G.                  Service de Physique Theorique C.E.N.
                            Saclay F-91191 Gif-sur-Yvette Cedex,
                            **FRANCE**

MARDER M.                   Department of Physics, University of
                            Texas at Austin, Austin, Texas 78512
                            USA

MASTORAKI M.                Laboratory of Astronomy Dept of
                            Physics, Panepistimiopolis Zografos,
                            Athens, GREECE

MAVRAGANIS A.               Nat. Techn., University Athens, 5
                            Heroes of Polytechniou Ave. Athens,
                            157 73, GREECE

MAVROMATIS A. S.            30 Prox. Koromila str. 54622
                            Thessaloniki, GREECE

MELETLIDOU E.               Room 12, Luke House, 7-9 Canton
                            Street, Poplar E14, London UK

MEYER T. P.                 CCSR, Beckman Inst. 405 N. Mathews
                            Ave., Champaign IL 61820, USA

NICOLAIDES C.               Theoretical and Physical Chemistry
                            Institute, National Hellenic Research
                            Foundation,
                            48 Vas. Constantinou Ave.,116 35
                            Athens, GREECE

NICOLIS J.                  Department of Electrical Engineering,
                            University of Patras, Patras, GREECE

OTT E.                          University of Maryland Lab. for
                                Plasma Research, College Park, MD 20742
                                USA

PAGITSAS M.                     Lab. of Physical Chemistry,
                                Department of Chemistry, University
                                of Thessaloniki, GREECE

PAPADAKI E.                     Messogion 90-92 K.T 115 27, Athens,
                                GREECE

PATSIS P.                       Krisila 15-17 116 35, Athens, GREECE

PAVLOPOULOS TH.                 Laboratory of Applied Mathematical
                                Physics, Technical University of
                                Denmark,  DK-2800 Lyngby, DENMARK

PAVLOS G.                       Polytechnic   School, Section of Space
                                and Telecommunications,
                                Democritos University of Thrace,
                                Xanthi, GREECE

PAVLOU ST.                      Department of Chemical Engineering,
                                University of Patras, 261 10 Patras,
                                GREECE

PAVONE A.                       Niels Bohr Institute Blegdamsvej 17,
                                DK-2100 Copenhagen, DENMARK

PANAGOPOULOS P.                 Dept. of Mathematics, University of
                                Patras, Patras 261 10, GREECE

PAPAIOANNOU G.                  122 Argostoliou St., Agios Dimitrios
                                17342 Athens, GREECE

PAPAGEORGIOU V.                 Dept. of Mathematics, Clarkson
                                University, Potsdam, N.Y. 13676 USA

PEDERSEN H.C.                   LAMF, Technical University of
                                Denmark, DK-2200 Lyngby, DENMARK

PERCIVAL I. C.            School of Mathematical Sciences,
                         Queen Mary College, Mile End Road,
                         London E1 4NS, UK

PETEK P.                 Department of Mathematics, University
                         of Ljubljana, Jadranska 19, 61000
                         Ljubljana, YUGOSLAVIA

PICKERING A.             School of Mathematics, University of
                         Leeds, Leeds, LS2 9JT, UK

PINOTSIS A.              Dept. of Astronomy, University of
                         Athens, Panepistimiopolis 15783
                         Zografos, Athens, GREECE

POLYMILIS C.             Dept. of Astronomy, University of
                         Athens, Panepistimiopolis 15783
                         Zografos, Athens, GREECE

PROSMITI R.              Post-Graduate Student, Department of
                         Chemistry of Crete, P.O. Box 1470
                         71110 Iraklio, GREECE

PSILLAKIS Z.             Department of Physics, University of
                         Patras, 261 10 Patras, GREECE

QUISPEL G. R. W          Depart. of Mathematics, La Trobe
                         University, Bundoora, Victoria 3083,
                         AUSTRALIA

RABINOVICH N.            Inst. of Appl. Physics,  46 Uljanov
                         str., N. Novgorod (Gorky) 603600,
                         USSR

RAJASEKAR S.             Department of Physics, University of
                         Bharathidasan, Tiruchirapalli 620 024,
                         INDIA

RAPTIS S.                Dept. of Physics, University of
                         Patras, Patras 261 10, GREECE

RAUH A.                     Fachbereich Physik, University of
                            Oldenburg, Oldenburg 2900, GERMANY


ROBERTS J.                  Institute for Theoretical Physics,
                            University of Amsterdam
                            Valckenierstraat 65, 1018 XE
                            Amsterdam, NETHERLANDS


RODRIGUEZ-LOSANO A.         Isla Toja, 138 (Urb. Montelomas)
                            Boadilla, Madrid 28.660, SPAIN


ROMANENKO E.                Institute of Math., Ukrainian Acad. of
                            Sciences, USSR, Repin str.3, Kiev
                            252601, USSR


ROMEIRAS F.                 Centro de Electodinamica Instituto
                            Superior Tecnico 1096, Lisboa Codex,
                            PORTUGAL


ROSS D.                     University of California, Santa Cruz
                            Mathematics Department, Santa Cruz,
                            CA 95064 USA


ROTOLI G.                   Department of Physics, University of
                            Salerno, I84081 Baronissi (Salerno),
                            ITALY


SAMARDZIJA N.               E.I Dupont de Nemours & Company, Inc.
                            Experimental Station, E357,
                            Wilmignton, Delaware 19898 USA


SARAFOGLOU G.               Dept. of Astronomy University of
                            Athens Panepistimiopolis 15783
                            Zografos, Athens, GREECE


SAZOU D.                    Lab. of Physical Chemistry,
                            Department of Chemistry, University
                            of Thessaloniki,  Thessaloniki, 540
                            06, GREECE

SEGUR H.                      University of Colorado at Boulder,
                              Campus Box 526, Boulder, Colorado
                              80309-0526 USA

SEMPIO P.                     Physics Department Via Marzolo 8,
                              Universita di Padova, Padova, ITALY

SHARKOVSKY A.                 Institute of Math. Ukrainian Acad. of
                              Sciences USSR, Repin Str 3, Kiev
                              252601, USSR

SIAFIARIKAS P.                Department of Mathematics, University
                              of Patras, Patras 261 10, GREECE

SIMONI F.                     Krathion Egialias 25006, Akrata,
                              GREECE

SKALTSAS D.                   Aratu 28-30, 26221 Patras, GREECE

SKOKOS CH.                    Dept. of Astronomy, University of
                              Athens, Panepistimioupolis, 15783
                              Zografos, Athens, GREECE

SOERENSEN M.                  Lab. of Appl. Math. Phys., The
                              Techn.  University of Denmark, Bldg.
                              303, DK-2800 Lyngy, DENMARK

SOURLAS D.                    H. Trikoupi 26A, 26222, Patras,
                              GREECE

STEFANOU G.                   Dept. of Civil  Engineering
                              University of Patras, Patras, 261 10,
                              GREECE

TODESCO E.                    University of Bologna, Department of
                              Physics, Via Mascarella 77/5 40126
                              Bologna, ITALY

TOMASCHITZ R.                 Dipart. di Matematica Pura et Appl.
                              via Belzoni 7, Universita di Padova
                              35131, Padova, ITALY

TURCHETTI G.                  Dipartimento di Fisica 46 Via Irnerio
                              Universita di Bologna 40126 Bologna,
                              ITALY

VAN DER HEIJDEN G.            Mathematics Institute, University of
                              Utrecht, Budapestlaan 6, P.O Box 80
                              010, 3508 TA Utrecht, THE NETHERLANDS

VAN DER WEELE J.              Theoretical Physics, University of
                              Twente, P.O Box 217, 7500 AE Enschede,
                              THE NETHERLANDS

VASSILIADIS D.                University of Maryland, Department of
                              Physics, College Park, MD 20742, USA

VAVOUGIOS D.                  Physics Dept, University of Patras,
                              Patras 261 10, GREECE

VOGLIS N.                     University of Athens, Department of
                              Physics, Section of Astronomy
                              Panepistimiopolis, Zografos, Athens,
                              15783, GREECE

VOUROS G.                     Lykourgou 50 12351, Athens, GREECE

VOYATZIS G.                   Theoretical Mechanics,  Department of
                              Physics, University of Thessaloniki,
                              Thessaloniki, 54006 GREECE

VLAHOS E.                     University of Patras, Department of
                              Physics, Patras 261 10, GREECE

XIE R.                        World Lab. Project and I.N.F.N.
                              Dipartimento di Fisica, Via Irnerio
                              46, Bologna, ITALY

YAO D.                        FB Physik, Philipps-Universitat
                              Marburg Renthof 6, D-3550 Marburg,
                              GERMANY

YACHNAKIS E.      University of Crete, School of Health
Sciences, Department of Medicine,
Heraclion, Crete 71409, GREECE

YIANNAKOPOULOS A.   Physics Department, University of
Warwick, Coventry CV4 7AL, U.K.

ZACHILAS L.     Department of Chemistry, University
of Crete, P.O Box 1470, 71110
Iraklion, GREECE

ZAHARIOU H.     P. Faliro 9818346 Pandrosou 26, GREECE

ZDETSIS A.      Physics Dept, University of
Patras, Patras 261 10, GREECE

ZOURIDAKIS G.    Electrical Engineering Department,
University of Houston, Houston TX
77204-4793 USA

# INDEX